Advances in Fashion and Design Research II

Joana Cunha · Ana Cristina Broega ·
Helder Carvalho · Bernardo Providência
Editors

Advances in Fashion and Design Research II

Proceedings of the 6th International Fashion and Design Congress, CIMODE 2023, October 4–6, 2023, Mexico City, Mexico

Springer

Editors
Joana Cunha
School of Engineering, 2C2T-Centre for Textile Science and Technology
University of Minho
Guimarães, Portugal

Ana Cristina Broega
School of Engineering, 2C2T-Centre for Textile Science and Technology
University of Minho
Guimarães, Portugal

Helder Carvalho
School of Engineering, 2C2T-Centre for Textile Science and Technology
University of Minho
Guimarães, Portugal

Bernardo Providência
School of Architecture, Art and Design, Lab2PT-Landscape, Heritage and Territory Laboratory
University of Minho
Guimarães, Portugal

ISBN 978-3-031-43936-0 ISBN 978-3-031-43937-7 (eBook)
https://doi.org/10.1007/978-3-031-43937-7

This Springer imprint is published by the registered company Springer Nature Switzerland AG
The registered company address is: Gewerbestrasse 11, 6330 Cham, Switzerland

Paper in this product is recyclable.

Preface

This book presents the research papers selected for the 6th edition of CIMODE—International Fashion and Design Congress—under the category of full papers in English language.

In this 6th edition of CIMODE, 66 works were selected for oral presentation, of which 48 were selected for this book.

A total of 88 full papers and 2 posters were submitted to this edition of CIMODE and were subject to a double-blind review selection process. This process was only possible due to the invaluable support of the members of the CIMODE Scientific Committee that selected the final 66 research papers of the utmost research in the field of Fashion and Design.

The present edition has an extraordinary character, as it takes place only one year after the last edition, instead of two. This exception is due to the will of the organizing committee to honour the commitment to carry out the edition in Mexico with the partners of IBERO—Universidad Ibero de la Ciudad de México, originally scheduled for 2020, but that due to the emergence of the SARS-CoV-2 pandemic crisis was cancelled.

The theme chosen for this 6th edition was "HILAR LA DIVERSIDAD", inspired by the premise that design aspires to be a humanistic and universal discipline easily accessible to all people, equally respecting their differences and integrity. However, this is not always carried out in practice. The complexity of what it means to be human implies going beyond the standards and the dominant culture without alluding to the exclusion, marginalization, and/or discrimination of people who deviate from established aesthetic canons, such as people of plus size, seniors, disabled, non-binary, or a variety of ethnicities. "HILAR LA DIVERSIDAD" invites us to see design from a universal and inclusive perspective that embraces minorities to promote plurality.

This book, following the congress format, is organized around seven major themes: Fashion and Communication; Fashion, Identities, and Cultures; Fashion and Product Design; Marketing and Consumption; Teaching and Education; Sustainability in Fashion and Design; Emotional Design and Fashion.

CIMODE is a platform for research exchange in Fashion and Design, promoting the discussion of ideas between researchers, academics, designers, and other professionals in the fields of Fashion and Design. Through an interdisciplinary dialogue and intercultural perspective, CIMODE aims to generate and present new scenarios on the present and future of Fashion and Design.

Being an inter- and trans-disciplinary conference, CIMODE explores Fashion and Design at the social, cultural, psychological, and communication perspective, seeking to bring together different approaches and perceptions on the practice, education, and culture of Design and Fashion.

It is our understanding that the themes to be discussed during the conference, both in plenary and in the parallel sessions, will provide a valuable basis for HILAR LA

DIVERSIDAD of Fashion and Design and will allow to establish closer cooperation links among researchers and academia, opening space for new reflections.

Joana Cunha
Ana Cristina Broega
Helder Carvalho
Bernardo Providência

Organization

Executive Commitee

Ana Cristina Broega	University of Minho, 2C2T, Portugal
Cynthia Gómez Ramírez	Universidad Ibero de la Ciudad de México, México
Joana Cunha	University of Minho, 2C2T, Portugal
Maria Eugenia Rojas Morales	Universidad Ibero de la Ciudad de México, México

Organizing Committee

Alfonso Luna Soto	Universidad Ibero de la Ciudad de México, México
Ana Cristina Broega	University of Minho, 2C2T, Portugal
Bernardo Providência	University of Minho, Lab2PT, Portugal
Cynthia Gómez Ramírez	Universidad Ibero de la Ciudad de México, México
Helder Carvalho	University of Minho, 2C2T, Portugal
Ileana Jalil Kentros	Universidad Ibero de la Ciudad de México, México
Joana Cunha	University of Minho, 2C2T, Portugal
Maria Eugenia Rojas Morales	Universidad Ibero de la Ciudad de México, México
Noemi Vidal Tapia	Universidad Ibero de la Ciudad de México, México

General Secretary

Helder Carvalho	University of Minho, 2C2T, Portugal

Scientific Commitee

Fashion and Communication

Acã Lopes	UFA-Brazil
Alessandra Vaccari	Università IUAV di Venezia, Italy
Ana Claudia Mei Alves de Oliveira	PUC-SP, Brazil
Benilde Reis	ULusófona, Portugal
Carlos Figueiredo	FAUL, Portugal
Catarina Vasques Rito	ULusófona, Portugal
Caterina Cristina Fiorentino	UCLV, Italy
Cyntia Tavares	UFC, Brazil
Fausto Viana	ECA-USP, Brazil
Flávio Glória Caminada Sabrá	IFRJ, Brazil
Inês Vasconcelos Amaral	DET-UMinho, Portugal
João Barata	ULusófona, Portugal
Kathia Castilho	ABEPEM, Brazil
Mara Rubia Sant'Anna	UDESC-SC, Brazil
Moisés Lemos Martins	DCCOM-UMinho, Portugal
Nelson Pinheiro	IADE-UL, Portugal
Renata Pitombo Cidreira	UFBA, Brazil
Rossana Gaddi	POLIMI, Italy
Silvana Mota Ribeiro	DCCOM-UMinho, Portugal
Solange Riva Mezabarba	UFF, Brazil
Taisa Vieira Sena	PUC-PR, Brazil
Zara Pinto Coelho	DCCOM-UMinho, Portugal

Fashion, Identities, and Cultures

Albertino Gonçalves	DCCOM-UMinho, Portugal
Aline Monçores	UESC, Brazil
Bárbara Pino	FAAD-UDP-Santiago, Chile
Claudia Cano	IBERO, Mexico
Cristiane Mesquita	UAM, Brazil
Diana Lucía Gómes-Chacón	UCM, Spain
Elisabeth Murilho da Silva	UFJF, Brazil
Eduarda Guimarães	SENAC, Brazil
Francisca Mendes	UFC, Brazil
Helena Pires	DCCOM-UMinho, Portugal
Herbert González	UComplutense Madrid, Spain

Laura Luceño	UPM, Spain
Laura Zambrini	FADU-UBA, Argentina
Maria Claudia Bonadio	SENAC/SP, Brazil
Maria Eduarda Araujo Guimarães	SENAC/SP, Brazil
Ornella Cirillo	UniCampania, Italy
Pedro Bessa	DCA-UA, Portugal
Rita Morais de Andrade	FAV/UFG, Brazil
Rita Ribeiro	ICS-UMinho, Portugal
Rossana Gadi	UNICH, Italy
Tula Fyskatoris	CPS/PUC-SP, Brazil
Verónica Noemi Vidal Tapia	IBERO, Mexico
William Afonso Cantú	ESECS-IPL, Portugal

Fashion and Product Design

Aguinaldo dos Santos	UFPR, Brazil
Alba Cappellieri	POLIMI, Italy
Ana Margarida Fernandes	ESART-IPCB, Portugal
Andrea Saltzman	FADU-UBA, Argentina
Araguacy Filgueiras	UFC, Brazil
Arturo Dell'Acqua Bellavitis	Italy
Daniel Raposo Martins	ESART-IPCB, Portugal
Deborah Cristo	UFRJ, Brazil
Demétrio Ferreira Matos	IPCA, Portugal
Emanuele Magnus	Brazil
Gianni Montagna	FAUL, Portugal
Guillermo García-Badell	UPM, Spain
Hélder Carvalho	DET-UMinho, Portugal
Iara Braga	UFP, Brazil
Isabel Cristina Gouveia	UBI, Portugal
Isabel Maria Dias Cabral	2C2T-UMinho, Portugal
Maria Alice Vasconcelos Rocha	UFRPE, Brazil
Maria Antonietta Sbordone	UniCampania, Italy
Maria Celeste Sanches	ABEPEM, Brazil
Marizilda dos Santos Menezes	UNESP, Brazil
Mônica Moura	UNESP, Brazil
Patricia de Mello Souza	UL, Brazil
Patrizia Ranzo	UNICAMPANIA, Italy
Paula Miguel	FADU-UBA, Argentina
Paula Trigueiros	EAAD-UMinho, Portugal
Rafaela Norogrando	UBI/IPV, Portugal
Renato Bispo	IPL, Portugal

Rita Salvado	UBI, Portugal
Silgia Costa	EACH – USP, Brazil
Teresa Franqueira	DCA-UA, Portugal

Marketing and Consumption

André Robic	IBModa, Brazil
António Dinis Marques	DET-UMinho, Portugal
Carlos Alberto Miranda Duarte	IADE/EU, Portugal
José Ferro Camacho	IADE, Portugal
Luciane Robic	IBModa, Brazil
Maria Carolina Garcia	UAM, Brazil
Maria da Graça Guedes	DET-UMinho, Portugal
Mário de Araujo	UMinho, Portugal
Mercedes Rodríguez	UPM, Spain
Olga Pépece	UEM, Brazil
Patrícia Sant'Anna	Tendere, Brazil
Sylvia Demetresco	ISA Vevey, Switzerland
Valter Cardim	CEAUL, Portugal

Teaching and Education

Alexandra Cruchinho	ULusófona, Portugal
Ana Moreira da Silva	CIAUD-UL, Portugal
Bárbara Pino	FAAD-UDP-Santiago (Chile)
Cátia Rijo	ESELX, Portugal
Chiara Scarpitt	UNICAMPANIA, Italy
Flavio Gloria Caminada Sabrá	IFRJ, Brazil
Giovanni Maria Conti	POLIMI, Italy
Madalena Pereira	DCTT-UBI, Portugal
Mara Rubia Sant'Anna	UDESC, Brazil
Márcia Merlo	UAM/SP, Brazil
Maria de Fátima da S. C. G. de Mattos	CUML/RP, Brazil
Marly Menezes	FASM/SP, Brazil
Rui Miguel	DCTT-UBI, Portugal

Sustainability in Fashion and Design

Ana Cristina Broega	DET-UMinho, Portugal
Anne Anicet Ruthschilling	Uniritter, Brazil
Bruna Lummertz Lima	IFSC, Brazil

Carla Morais	FAUL, Portugal
Cristina Figueiredo	UTAL, Portugal
Desamparados Pardo Cuenca	ESDValencia, Spain
Evelise Anicet Ruthschilling	URGS, Brazil
Isabel Cantista	U.Lusíada – Porto, Portugal
Maria José Marques Abreu	DET-UMinho, Portugal
Roberto Liberti	UNICAMPANIA, Italy
Sirlene Costa	USP, Brazil
Suzana Barreto Martins	UEL, Brazil

Emotional Design and Fashion

Bernardo Providência	EAAD-UMinho, Portugal
Fernando Moreira da Silva	FAUL, Portugal
Germannya Araújo Silva	UFPE, Brazil
Joana Cunha	DET-UMinho, Portugal
Leandro Miletto Tonetto	Unisinos, Brazil
Livia Laura Matté	UTFPR, Brazil
Luís Carlos Paschoarelli	UNESP, Brazil
Luisa Mendes	2C2T-UMinho, Portugal
Maria Cecilia Loschiavo dos Santos	USP, Brazil
Mónica de Matos Romãozinho	LABCOM-UBI, Portugal
Syomara Duarte	UFC, Brazil
Wellington de Medeiros	UFCG, Brazil

Contents

Fashion and Product Design

Marketing and Consumption

Teaching and Education

Fashion and Communication

The Problems of a Visually Impaired User in the Process of Buying Fashion Products

João Teixeira(✉), Joana Cunha, and Inês do Amaral

Department of Textile Engineering, University of Minho, Guimarães, Portugal
pg42810@alunos.uminho.pt, {jcunha,inesamaral}@det.uminho.pt

Abstract. In today's society, sight is the most relevant sense, as we rely heavily on it to gather information from the external environment (Davidoff 2001). However, given the high rate of ageing of the population, we often speak of users with visual impairment (Baker et al. 2002). The world of fashion is critical for the visually impaired user, namely in his socialization process, however, since we live in a contemporary world that manifests itself in a predominantly visual way, there are some communicational barriers, which make the ability to guide the visually impaired user to the act of buying fashion products be impaired (Schneider et al. 2017). Thus, starting from this basic principle that the world of fashion is extremely important for the visually impaired user, it should be democratized and inclusive. All communicational changes to be developed must always be accompanied by accessibility and value. Therefore, this article seeks to disseminate and present what are the main problems faced by the user in the process of buying fashion products, to potentiate future research regarding the development of design strategies to overcome the problems presented.

Keywords: Visual Impairment · Fashion · Buying Process · Communication Design · Accessibility

1 Introduction

Today, we live in a society where, according to a report launched by the World Health Organization in 2019, there are about 75 million blind users and about 2.2 billion users with disabilities associated with visual perception (Anon 2019). In any case, despite the large number of users with visual impairment, we are still far from its peak, given the gradual increase in the world's ageing population (Baker et al. 2002).

As a society, we inhabit a world that expresses itself in a predominantly visual way, where there are very specific cultural constructions (Oliveira 2007). This way, when a visually impaired user tries to connect with the world of fashion in his daily life, he faces some communicational barriers, namely regarding color perception and the understanding of any message (Schneider et al. 2017; Costa and Prette 2012).

In contemporary times, a significant conceptual difference is interpreted between the interactions established with the world of fashion by users with and without disabilities, with users with visual impairment being the most impaired concerning the culturally constructed communication system in the world of fashion.

Therefore, this communicational barrier between the two leads to the visually impaired user not being able to complete even the simplest tasks in the process of buying a fashion product (Kaufman-Scarborough 2001).

Having said that, inserted in this problematic scope between the fashion world and the visually impaired user, this article seeks to disseminate and present the problems faced by the user in the process of buying fashion products, with a view to potentialising future research, namely concerning the development of possible design strategies, to overcome some of the problems disseminated here.

2 Get Know the Visually Impaired Consumer

2.1 The "Incapacity"

According to Martín, Bueno and Pedro (2003), the visual impairment is defined by the total or partial absence of the visual sense.

In the vast sensory range vision is the most important remaining sense, as it interconnects the body with the world, thus creating constant information and immediate verification (Santin and Simmons 1977). In this way, it becomes inevitable to point out that visual impairment is irremediably correlated with a reduction in the quality of life of a human being (Allen 1989).

On reflection, it is important to consider how dependent we are on the visual sense, namely in contacting and obtaining information from the external environment (Davidoff 2001).

2.2 The "Sensory Vision"

The "sensory vision" allows us to understand that a visually impaired user depends on his remaining senses to help identify certain variables, which in general are designed for the visual sense (Cattaneo and Vecchi 2011).

As a whole, the remaining senses that make up "sensory vision" are tools of the highest level of detail for the visually impaired user, they also can take users to different states of mind (Snowden et al. 2012; Solomon et al. 2002).

However, it should be noted that in "sensory vision" the greatest dependence of the visually impaired user falls namely on the tactile sense, and there is a high probability of the user "seeing" through touch because in general, the tactile sense elaborates a very detailed appreciation forming images of what is touched (Kastrup 1997).

From the beginning of their disability, the visually impaired user is trained daily to use the tactile sense, both in the recognition process and in the learning process. Within the panorama of the fashion world, the tactile sense is a critical tool for the visually impaired user, as its characteristics allow the user to differentiate and evaluate different properties of products (Batista 2005).

Given what has been previously mentioned, the tactile sense also can modify how the visually impaired user relates to the product, however, it is pertinent to recognize that the tactile sense does not directly replace the visual sense (Kastrup 1997).

In the wide range of the remaining senses that are incorporated into "sensory vision", the auditory sense is also very important for the visually impaired user, as it transmits information that changes both in space and time (Hirsh 1988).

The auditory sense is based on high threshold values for discriminating both frequency and intensity (Capelle et al. 1998). Thus, the auditory sense is capable of dealing with complex and rapidly changing sound patterns, thus providing information about the speed, location and interaction of objects in real time (Hirsh 1988).

The studies of Stevens and Weaver (2006), and of Wan et al. (2010), prove that visual impairment confers advantages on the auditory sense related to attention, memory and language. These advantages can provide a greater degree of independence to the visually impaired user in carrying out his daily (Wan et al. 2010).

In the fashion world, the visually impaired user can acquire, through the auditory sense, references that compensate for the lack of vision, thus improving his degree of autonomy in analyzing the properties of products.

The sense of taste, within "sensory vision", is recognized as working in conjunction with the other remaining senses, particularly the olfactory sense (Auvray and Spence 2008). However, in the fashion world, it is verified that both the sense of taste and the sense of smell are not directly related to the visually impaired user, though, they can be worked on in a way that positively contributes to the strengthening of a brand's image.

Finally, it is necessary to understand that the construction of "sensory vision" on the part of users with visual incapacity is related to the capacity of attention before the stimuli and not to the capacity for development of the sense itself (Hatwell 2003). According to Hatwell (2003), the loss of sight does not immediately result in the potentiation of the other senses, but by learning to redirect attention to them, the visually impaired user can take advantage of the senses that until then were not part of his cognitive domain.

2.3 His Relationship with the Fashion World

In contemporary times, there is a lack of knowledge in terms of the relationship between the visually impaired user and the world of fashion.

However, this lack of knowledge does not become a barrier for the visually impaired user, as he tries to achieve autonomy by overcoming any impact resulting from visual limitation, be it total or partial, as well as from socially conceived stigmas related to visual impairment.

The visually impaired user feels the need to be autonomous, which encourages personal independence and freedom, thus providing a concrete capacity to make choices and carry out actions, based on self-reflection. Still, within this panorama, communication assumes a relevant role because if there is minimally inclusive communication, the visually impaired user will have an easier time developing his degree of autonomy.

In the world of fashion, when the visually impaired wearer is making direct contact, his thinking will be that his disability is irrelevant, he cannot limit himself from doing anything he wants because if he is deprived of wearing the clothes, he would like he might lose part of his personality (Alcantara et al. 2017; Kaufman-Scarborough 2001). According to Pizzimenti and Roberts (2005), no matter how much the weakness of a visually impaired user is considered, he develops an inexhaustible list of adaptive strategies providing him with a greater degree of autonomy.

The communicational power of fashion is used by the visually impaired user as an ignition to his socialization process because inevitably there is a care in the way he presents himself to society. After all, the way he presents himself is also the way he communicates to others.

As a foundation to what was previously mentioned, in the world of fashion, the visually impaired user also searches for information related to trends and colors. This research arises from the importance for the visually impaired user to feel good about what he wears because for him buying something aesthetically appealing promotes well-being.

In short, the communicational power of fashion assumes a relevant daily role for the visually impaired user because through a good presentation, it can promote socialization, working in reciprocity, that is if the user is well with his presentation, he feels that people around him are also well.

2.4 His Consumer Profile

In society, consumption is a strategy regularly used by the most different social groups involving rights, identities and lifestyles (Campbell and Barbosa 2006). Apart from theoretical questions, in accordance with Cardoso (2007), there is a recognition of the importance of consumption in the routine of the visually impaired user, ending up having a direct influence on his personal life and the construction of his identity.

In the practice of consumption, the vulnerability of a visually impaired user is felt when he feels a state of powerlessness arising from an imbalance in the interactions carried out with the fashion market (Baker et al. 2005). This vulnerability is increased by the enormous difficulty that the visually impaired user has in shopping independently, which consequently generates exclusion, isolation and stigma, directly affecting self-image and the perception of inclusion (Elms and Tinson 2012).

However, despite the various problems that visually impaired user still faces in the process of buying fashion products, their consumer profile is characterized by their willpower. As much as society may consider that a visually impaired user hides his weakness by avoiding challenging tasks, what happens is exactly the opposite, that is, the visually impaired user reinvents himself by conquering his autonomy.

In another aspect that characterizes the consumer profile of a visually impaired user, the lines of desire and pleasure play a very important role because a significant part of the time, the purchasing process of a visually impaired user is born from an aesthetic motivation (Baker 2006).

2.5 His Purchasing Process

The practice of consumption, as noted earlier, is a strategy regularly used by many different social groups, however, it can be a very challenging task for those with disabilities, including the visually impaired user (McCracken 1990).

In the purchasing process, the visually impaired consumer feels a state of powerlessness arising from the blocking of his independence by the lack of understanding of the most diverse communications (Kaufman-Scarborough 2001).

As can be seen in Table 1, the purchasing process of a visually impaired consumer is defined in seven different stages, five of which are common to all consumers (Ramatla and Mastamet-Mason 2013).

Table 1. Description of the Buying Processes by Ramatla and Mastamet-Mason 2013

NORMAL CONSUMER	VISUALLY IMPAIRED CONSUMER
PROBLEM RECOGNITION	PROBLEM RECOGNITION
INFORMATION SEARCH	INFORMATION SEARCH
EVALUATION OF ALTERNATIVES	POINT OF SALE BRIEFING
BUYING DECISION	PRE-PURCHASE EVALUATION
POST-PURCHASE EVALUATION	EVALUATION OF ALTERNATIVES
	BUYING DECISION
	POST-PURCHASE EVALUATION

The first stage of the buying process is recognizing the problem (Ramatla and Mastamet-Mason 2013). At the time of recognition, the problem can assume a double meaning, because it can be interpreted as a pleasurable problem of aesthetic motivation, and it can also be interpreted as a need, where the consumer with visual incapacity to recognize this problem in a fashion item will rely on his tactile sense, to feel for example a worn piece (Ramatla and Mastamet-Mason 2013).

Regarding the search for information on the product during the purchasing process, given the lack of inclusiveness of some information, such as color, type of print or type of pattern of fashion items, it is inevitable that the visually impaired consumer resorts to third parties, whether family members or friends (Ramatla and Mastamet-Mason 2013). Given this, the visually impaired consumer is denied a chance to obtain information about the product independently (Ramatla and Mastamet-Mason 2013).

In the third stage of the purchasing process, there is always the need for the visually impaired consumer to carry out a briefing on the point of sale, that is, the consumer should analyze the point of sale from his "sensory vision", looking for sensory clues which make him feel comfortable carrying out his purchasing processes, such as the comfortability of the floor, the music, the smell and the texture of the displays (Ramatla and Mastamet-Mason 2013). However, if the point of sale does not stimulate the "sensory vision" of the visually impaired consumer, he will rarely make a purchase (Ramatla and Mastamet-Mason 2013).

The pre-purchase evaluation, the fourth stage of the purchasing process, is in line with what was previously mentioned, that is, in the pre-purchase evaluation, the visually impaired consumer evaluates all the existing alternatives according to the pleasantness indexes to his tactile sense (Ramatla and Mastamet-Mason 2013).

After a pre-selection comes the detailed evaluation of all alternatives, in this evaluation the visually impaired consumer is forced to rely on third parties, given the lack of inclusiveness of some information on fashion items, such as size, price, composition and even maintenance care (Ramatla and Mastamet-Mason 2013).

The purchase decision is perhaps one of the most important steps in the purchasing process unfortunately for the visually impaired consumer, variables such as the display of the item at the point of sale, the shop environment and the shop assistants can create problems due to lack of vision (Ramatla and Mastamet-Mason 2013). The visually impaired consumer must be helped to interpret certain information, such as size, price, composition and color of fashion items, because these are important factors that can directly influence the purchasing decision (Ramatla and Mastamet-Mason 2013). It is also important to note that these problems will be disseminated and presented in detail below.

Finally, we have the post-purchase evaluation, in this last stage, the visually impaired consumer must evaluate all the parameters mentioned above and conclude whether or not their buying process was satisfactory (Ramatla and Mastamet-Mason 2013).

3 Qualitative Methodology

3.1 The Interviews

The research carried out in this article follows a qualitative methodology, thus allowing for a more complete and in-depth perception of the topic under study. It should also be noted that the methodology used involved a survey conducted through exploratory interviews.

The qualitative research method, based on exploratory interviewing, allows for the combination of life story interviews with interviews informed by assumptions drawn from phenomenology (Bertaux 1981). Consequently, exploratory interviews grant an understanding of the meaning that interviewees make of their experience on a particular subject (Seidman 1991).

That said, fourteen exploratory interviews with visually impaired users were conducted in this research. Of the fourteen interviewees, 30% belong to the male gender and 70% to the female gender. Overall, 58% of the interviewed users acquired a visual impairment throughout their lives, on the other hand, 42% of the users were already born with a visual impairment. It is also possible to ascertain that 58% of the users interviewed suffer from total visual impairment, while 42% suffer from partial visual impairment.

The structuring of an exploratory interview is a very important procedure, Seidman (1991) argues that, an exploratory interview should always be divided into three different parts, with a different objective for each of the parts.

Therefore, the first part of an exploratory interview is concerned with the interviewee's life story, that is, here it is important that the interviewee tells as much as possible about himself in the light of the topic of the moment (Seidman 1991).

In the second part of the exploratory interview, the details of the experience are presented, where the interviewee should focus especially on the concrete details of the experience by reconstructing these details (Seidman 1991).

In the third and last part of the exploratory interview, the interviewee presents a reflection on the meaning of his experience, in this case, his process of buying fashion products (Seidman 1991).

Finally, the main purpose of using this qualitative methodology is to understand the experience lived by users with visual impairment in the process of buying fashion products in order to conclude what the real problems are that they face in this process.

3.2 The Content Analysis

With the maximum purpose of analyzing all the content coming from the fourteen interviews carried out, in this research, we opted for a method of analysis called categorical analysis. Bardin (2011) points out that, in the conjunctural range of content analysis techniques, categorical analysis, in practice, is the most widely used in research.

After defining the analysis method to be used, we proceeded to the process of creating categories according to the formula advocated by Bardin (2011). Within the scope of the construction of the categories, it is also relevant to point out that the whole construction process was carried out in articulation with the data collected in the fourteen exploratory interviews carried out, that is, a progressive categorization was carried out as a floating reading of the data was undertaken.

In the fullness of the research, given the density of the data collected, it was felt necessary to create nine categories in the categorical segmentation. Given this, as it is possible to verify in Table 2, the nine categories that are essential to the organization of information are presented in a summarized way, four of them being essential in the identification of the main problems faced by the visually impaired user in the process of buying fashion products.

Table 2. Representation of Categories for Content Analysis

CATEGORIES
"FEEL THE FASHION"
"IMPORTANCE OF FASHION"
"THE DAILY LIFE OF VISUALLY IMPAIRED IN FASHION"
"THE REASON FOR PURCHASING"
"THE PERCEPTION OF THE BRANDS' POSITION"
"THE PURCHASING PROCESS"
"PROBLEMS IN THE PURCHASING PROCESS"
"PURCHASING EXPERIENCES"
"IMPROVEMENT POINTS"

After defining the categories, we proceeded to fragment the interviewees' statements in the fourteen interviews in order to categorise the most diverse fragments.

Bardin (2011) argues that, in content analysis, the last phase is born from the relationships between the most diverse categorical themes. Therefore, in order to discover the themes most frequently addressed by the interviewees in this research, for each fragment of speech extracted from the interview, it was necessary to create keywords. Together, the keywords obtained gave rise to word clouds, where the main objective is to ascertain the themes most addressed by the interviewees.

At the core, the word clouds developed became very important, as they not only contributed to the realization of the most diverse thematic relationships but also collaborated in the definition of some critical points that became quite important in the course of the whole research.

4 Results

After the achievement of the analysis of all the content from the fourteen exploratory interviews carried out, we proceed to the presentation of the other results obtained, namely concerning the problems faced by the visually impaired user in the process of buying fashion products.

4.1 The Main Problems in the Purchasing Process

There is a lack of autonomy, on the part of the visually impaired user, in the process of buying fashion products (Hopkins 2000). Therefore, it can be seen that this lack of autonomy is fueled by some problems mentioned by the user, namely concerning the labelling of the articles, the identification of the color of the articles, the online purchasing process, the payment process and mobility in the physical store.

Complementing, there is an urgent need to develop design strategies that emphasize accessibility, to combat these critical problems.

The Labelling

How does a visually impaired user independently know the size of a fashion item or even a simple price?

The answer is simple, you don't know the difficulty in reading the written information on a label is undoubtedly the main obstacle faced by visually impaired users during the process of buying fashion products.

This problem, related to the reading of labels, makes the visually impaired user dependent during the purchasing process, as he needs a second person to verbalize the information.

However, even in the face of this problem, the visually impaired user, in a sense of self-education in the conquest of autonomy, tries to create by himself all the possible conditions to know information about the product.

With the technological advance experienced nowadays, some applications such as OrCam Read and Wizo have appeared which make it possible to audio-read written information, however these applications, apart from their high cost, work better in a digital scenario and not in a physical one, thus continuing to persist with the problem.

Faced with this problem, the visually impaired user can identify some possible points of improvement to be developed to provide them with a greater degree of autonomy during the process of buying fashion products.

Therefore, one point for improvement could be the use of Braille language in the development of labels to enable the user to read some written information, such as size and price.

According to research conducted by Okazaki et al. (2012), Qr Codes are increasingly integrated into products all over the world, so Qr Codes can be a design strategy target of a study, becoming a possible point of improvement to provide a higher degree of autonomy to the user with visual impairment in the process of buying fashion products.

The Colour Identification

In their daily lives, visually impaired users encounter some barriers in the process of identifying the color of their clothes. Although there is a concern for the user to search for color combinations autonomously, the user cannot identify the color of a garment.

In contemporary times, we witness a remarkable technological advance with the development of technological applications that help the visually impaired user to identify colour. However, after conducting qualitative research, it is possible to conclude that technological applications for colour identification are hardly feasible and expensive.

As a point of improvement, faced with the problem of colour identification, visually impaired user tries to get around the problem by developing their identification strategies. At this particular point, the interior labels of fashion garments take on a critical role, as they help the user to develop their identification strategies, namely through cutting and writing.

At its juncture, the colour identification problem leaves three gaps to be addressed. Firstly, despite the existence of technological applications to aid the identification of colour, they still do not present full viability in the representation of its function.

Secondly, as a consequence of the lack of inclusiveness, there is an enormous dependence of the visually impaired user on colour identification, namely in the course of their fashion product purchasing process.

Finally, even if there is someone to verbalise the colour of a fashion product to the visually impaired user, it has to start from their self-taught sense, the creation of a strategy, so that they will never forget that a certain garment has a certain colour.

The Online Purchasing Process

Concerning the online purchasing process, one of the questions we wanted to reflect on from the interviews conducted was: Is the degree of autonomy of a visually impaired user greater in the online shopping process?

The answer is no, however, it should be noted that the purchasing process does not always have to take place at the physical store.

In the online purchasing process, the visually impaired user is confronted with descriptions that are not very detailed and not at all representative, thus questioning to what extent the user can even imagine the product.

Apart from this descriptive problematic, it is significant to highlight that the visually impaired user does not use the online store only to buy most of the time the user navigates through websites that allow it because not all websites provide a good accessibility in navigation, working as a pre-process purchasing strategy.

In a future panorama, as a point of improvement, it is necessary to work on the development of design strategies, where the focus has to be directed towards accessibility in the navigation of fashion brands' websites.

The Payment Process

The payment process belongs to the last step of the purchase process of a visually impaired user (Karg et al. 2022).

However, despite being the last stage of the purchasing process, in the payment process, the visually impaired user faces some problems, in particular in payments using the ATM, where they have great difficulty in identifying the numbers on the terminal.

Previously, in the "sensory vision", the importance of the auditory sense in the relationship of the user with visual impairment with the fashion world was already referenced, so the auditory sense can be used to assist an improvement point to the payment process within a design strategy.

The Mobility

At the level of mobility, the following was reflected upon after the interviews were conducted: To what extent can the visually impaired user be autonomous in their mobility within a physical store?

At present, visually impaired users are classified as a small niche of consumers, thus, there is not a high degree of concern on the part of brands regarding the organization of physical stores. That said, this possible carelessness triggers some mobility problems for the visually impaired user.

Therefore, during the purchasing process, the visually impaired user is faced with the inexistence of an inclusive identification for example, the user does not know where the fitting rooms are located, which means he is always dependent on third parties.

Sometimes it may go unnoticed by the eye, but another problem faced by the visually impaired user is identifying obstacles, such as mirrors, irons and decorative elements.

Given these problems, some points for improvement have been identified. The first of these is related to the development of inclusive signage strategies, thus identifying the most diverse obstacles within the physical store.

As another potential point for improvement, we may have the development of identification plates in Braille for several different areas of the physical store to respond to the dependence of the visually impaired user on their location.

To conclude the potential points for improvement, there is just one nuance regarding the increase of the circulation space within the physical store. Nowadays, with the covid-19 pandemic, physical stores have a larger circulation space as a consequence of the pandemic restrictions, however, it will be essential to maintain this larger circulation space, giving greater comfort to the visually impaired user in his mobility inside the physical store.

5 Conclusions

The existing disability makes the visually impaired user unable to understand visual elements, which inevitably makes it difficult to understand any message. Therefore, nowadays, in the world of fashion, the user faces some difficulties when buying (Oliveira et al. 2017).

There is a vulnerability felt by the visually impaired user in interacting with the fashion world (Baker et al. 2005). However, being a visually impaired user doesn't mean you can't be fashionable.

After carrying out the whole methodological process through qualitative research, it is possible to conclude that there is a communicational barrier between the visually impaired user and the fashion world, which consequently is reflected in the following problems faced by the user: the process of reading labels, identifying the color of articles, navigating and understanding descriptions at the online store, the payment process and mobility within the physical store.

With an emerging purpose, the response to these problems starts from the principle of inclusion. According to Sassaki (1997), inclusion is a system by which society adapts to be able to include, in general, people with special needs. Inclusion advocates a new stance adopted by society, where there is a change in thinking, respecting all differences equally, thus making society as a whole (Filho 2000).

It is extremely necessary to associate inclusion with the world of fashion, inevitably, fashion is something we identify with daily (Pereira 2016).

Therefore, all this research becomes effectively relevant because, faced with the urgent need to look at these problems, this article has the purpose of potentiating future research, regarding the development of design strategies in order to respond effectively to all the problems raised.

Finally, it is also important to highlight that the research presented in this article also contributes to fashion brands having more knowledge and information about the visually impaired user, thus being able to develop possible design strategies, to confer a greater degree of autonomy to the user, during the process of purchasing fashion products.

References

Alcantara, A.C., Silva, A., Soares, A.: A Moda e os Portadores de Acondroplasia: Um Estudo Comparativo Através da Modelagem de Calça. Campinas, Galoá, s.n (2017)

Allen, M.: The meaning of visual impairment to visually impaired adults. Lead. Glob. Nurs. Res. **14**(8), 640–646 (1989)

Anon. World Health Organization (2019). https://www.who.int/publications/i/item/9789241516570. Acedido em 3 Jan 2023

Auvray, M., Spence, C.: The multisensory perception of flavor. Conscious. Cogn. **17**(3), 1016–1031 (2008). https://doi.org/10.1016/j.concog.2007.06.005

Baker, M., Gentry, J., Rittenburg, T.: Building understanding of the domain of consumer vulnerability. J. Macromark. **25**(2), 128–139 (2005)

Baker, S.M.: Consumer normalcy: understanding the value of shopping through narratives of consumers with visual impairments. J. Retail. Dezembro **82**(1), 37–50 (2006)

Baker, S.M., Stephens, D.L., Hill, R.P.: How can retailers enhance accessibility: giving consumers with visual impairments a voice in the marketplace. J. Retail. Consum. Serv. **9**(4), 227–239 (2002)

Bardin, L.: Análise de Conteúdo. 1ª ed. São Paulo: Edições 70 (2011)

Batista, C.G.: Formação de conceitos em crianças cegas: questões teóricas e implicações educacionais. Psicologia: Teoria e Pesquisa **21**(1), 07–15 (2005). https://doi.org/10.1590/S0102-37722005000100003

Bertaux, D.: Biography and Society: The Life History Approach in The Social Sciences. Sage Publications, Thousand Oaks (1981)

Campbell, C., Barbosa, L.: Cultura, Consumo e Identidade, 1st edn. FGV, Rio de Janeiro (2006)

Capelle, C., Trullemans, C., Arno, P., Veraart, C.: A real-time experimental prototype for enhancement of vision rehabilitation using auditory substitution. IEEE Trans. Biomed. Eng. **45**(10), 1279–1293 (1998). https://doi.org/10.1109/10.720206

Cardoso, A.C.M.: Tempos de Trabalho, Tempos de não Trabalho: Vivências Cotidianas de Trabalhadores, São Paulo: s.n (2007)

Cattaneo, Z., Vecchi, T.: Blind Vision: The Neuroscience of Visual Impairment (2011)

Lopes, C.S., da Costa, A., Prette, D.: Estudo Comparativo de Observação de Habilidades Sociais de Gêmeas Com e Sem Deficiência Visual. Revista Educação Especial **26**(47), 75–88 (2012). https://doi.org/10.5902/1984686X3552

Davidoff, L.: Introdução à Psicologia. 3º ed. Makron Books, São Paulo (2001)

Elms, J., Tinson, J.: Consumer vulnerability and the transformative potential of internet shopping: an exploratory case study. J. Mark. Manag. **28**(11–12), 1354–1376 (2012). https://doi.org/10.1080/0267257X.2012.691526

Filho, J.F.B.: Inclusão: Projeto para o Acesso à Saúde Mental. WVA, Rio de Janeiro (2000)

Hatwell, Y.: Psychologie Cognitive de la Cécité Précoce. Dunod, Paris (2003)

Hirsh, I.: Auditory perception and speech. In: Handbook of Experimental Psychology. Wiley, New York (1988)

Bradley, R., et al.: A study of the influence of visual impairment on the purchase of clothing. Br. J. Vis. Impairment **18**(2), 79–81 (2000). https://doi.org/10.1177/026461960001800209

Karg, A., et al.: Strategic Sport Marketing, 5ª Routledge, London (2022)

Kastrup, V.: A Invenção de si e do Mundo: Uma Introdução do Tempo e do Coletivo no Estudo da Cognição (1997)

Kaufman-Scarborough, C.: Accessible advertising for visually-disabled persons: the case of color-deficient consumers. J. Consum. Mark. **18**(4), 303–318 (2001). https://doi.org/10.1108/07363760110392985

Martín, M.B., Bueno, S.T.: Deficiência Visual: Aspectos Psicoevolutivos e Educativos. Santos, São Paulo (2003)

McCracken, G.: Culture and Consumption: New Approaches to the Symbolic Character of Consumer Goods and Activities. Indiana University Press, Indiana (1990)

Okazaki, S., Li, H., Hirose, M.: Benchmarking the use of QR code in mobile promotion: three studies in Japan. J. Advertising Res. Janeiro **52**(1), 102–117 (2012)

de Oliveira, J.V.G.: Arte e Visão - A Questão Da Cegueira (2007). https://www.deficienciavisual.pt/r-ArteVisualidade-JVGanzOliveira.htm. Acedido em 20 Jan 2023

Oliveira, T., Nickel, E., Cinelli, M.: Tactile color systems: comparative study of advantages and limitations for blind people. Santa Catarina, Blucher Design Proceedings (2017)

Pereira, A.: Moda Inclusiva: A Necessidade da Moda Inclusiva no Mundo de Hoje, São Paulo: s.n (2016)

Pizzimenti, J., Roberts, E.: The low vision rehabilitation service. Internet J. Allied Health Sci. Pract. **3**(3) (2005)

Ramatla, A., Mastamet-Mason, A.: The decision-making processes of visually impaired consumers in an apparel retail environment. In: Tshwane University of Technology, South Africa, pp. 220–229 (2013)

Santin, S., Simmons, J.N.: Problems in the construction of reality in congenitally blind children. J. Vis. Impairment Blindness 425–429 (1977)

Sassaki, R.: Em: Inclusão: Construindo uma Sociedade para Todos, p. 3. WVA, Rio de Janeiro (1997)

Schneider, J., Gomes, M., Ramirez, A., Santos, C.: Etiquetas Têxteis em Braille: Uma Tecnologia Assistiva a Serviço da Interação dos Deficientes Visuais com a Moda e o Vestuário. Estudos em Design **25**(1), 65–85 (2017)

Seidman, I.: Interviewing as Qualitative Research: A Guide for Researchers in Education and the Social Sciences, 3rd edn. Teachers College Press, New York (1991)

Snowden, R., Thmpson, P., Troscianko, T.: Basic Vision: An Introduction to Visual Perception. Oxford University Press, Oxford (2012)

Solomon, M., Bamossy, G., Askegaard, S., Hogg, M.: Consumer Behaviour - A European Perspective, 5° ed. Pearson Education Limited (2002)

Weaver, K.E., Stevens, A.A.: Auditory gap detection in the early blind. Hearing Res. **211**(1–2), 1–6 (2006). https://doi.org/10.1016/j.heares.2005.08.002

Wan, C.Y., et al.: From music making to speaking: engaging the mirror neuron system in autism. Brain Res. Bull. **82**(3–4), 161–168 (2010). https://doi.org/10.1016/j.brainresbull.2010.04.010

A Perspective on the Future of the Fashion and Textile Sector in Portugal with the Use of AI, AR and VR Technological Tools

Catarina Filipa Moura Carvalho(✉) and Maria José Abreu

2C2T – Center of Textile Science and Technology, Universidade do Minho, Guimarães, Portugal
ccarvalho2901@gmail.com

Abstract. As a result of AI, AR, and VR technologies, consumers can have a multisensory, entertainment-rich experience, which will play a crucial role in the future of fashion, given the current situation - after the COVID-19 pandemic and the economic and climate crisis. For this sector of activity, these technologies will prove to be a profitable investment, as they will provide valuable insight into consumer modifications and fragilities occurring worldwide in different countries - resulting in numerous advantages in product production, communication, marketing, storage, and distribution. Accordingly, a qualitative and quantitative research project was undertaken to examine Portuguese companies and consumers in this sector. These findings indicate that there is little implementation and development of these technologies in Portuguese companies and little knowledge about them among Portuguese consumers. These tools are still largely underutilized in sales and communication channels. Lastly, the advantages and disadvantages of these technological tools are discussed both during the buying process and as they relate to companies and retailers.

Keywords: Design · Fashion · Consumer · Artificial Intelligence · Augmented Reality · Virtual Reality

1 Introduction

Considering the global pandemic and current economic difficulties, technology can be a great opportunity for the fashion industry - resulting in the development of more productive, competitive, and profitable processes. With the advent of virtual technologies, the fashion industry is becoming more flexible and accessible, as well as enhancing a new wave of creativity across the entire supply chain. As a result, companies must rethink their marketing and communication strategies and develop these technologies, such as Artificial Intelligence (AI), Augmented Reality (AR), and Virtual Reality (VR) to address consumer uncertainty, increase sales, and create new digital experiences. In an increasingly digital world, human life is becoming more intertwined, making it essential that the Fashion industry adapts to the latest technological advancements to increase the efficiency of manufacturing, production, marketing, and communication.

To conclude, the topic is relevant to the future strategies of the companies. An initial review of the national landscape revealed a dearth of information on the topic. Therefore, the present study aims to analyze whether technological tools could play a role in the future in the fashion and textile national industry.

2 State of Art

Following the growth of communication, design, and technology, the fashion sector has involuntarily become a mirror of today [5]. It is the fashion industry's responsibility to adapt to new consumer behaviors, from habits to lifestyles, on a continuous basis, thus incorporating society's and the world's current interests.

Currently, after the pandemic, the economies of countries remain fragile and consumers still worry about financial uncertainty, which is a result of geopolitical conflicts that damage the economies of countries, increasing inflation, a lack of raw materials, and an energy crisis [1]. For innovation to develop, it is necessary to implement new technologies that assist companies in understanding the drastic changes in consumer behavior that are being experienced in the textile industry, such as the accelerated adoption of e-commerce, the division of spending between online and offline shopping, and the increased consumption of sustainable brands [7].

In 2021, companies in the fashion sector invested between 1.6% and 1.8% of revenues in technology, while for 2030 this figure is expected to rise to between 3.0% and 3.5%. The increase in innovation results in a competitive advantage for the customer - in creative, sensory, and entertainment activities, as well as in product development and distribution [1]. This industry will be able to reap the benefits of Artificial Intelligence, Augmented Reality, and Virtual Reality because these technologies will aid in understanding consumer modifications and weaknesses occurring worldwide [8].

AI helps companies improve their business models, from forecasting trends to managing inventory to enhancing collaboration between different channels of business [2]. During the forecast period from 2019 to 2027, the global AI market for fashion is expected to grow at a CAGR of 36.9%, according to Statista 2023.

With AR, customers can experience products virtually, increasing post-purchase satisfaction, reducing exchanges and returns, and increasing brand loyalty [9]. By 2020, 32% of consumers used AR while shopping. Technology such as this offers a unique opportunity to engage customers and increase their satisfaction levels [4].

Lastly, virtual reality technology provides a multi-sensory experience that is defined as real-time induction graphics with multidimensional structures, enabling the creation of virtual experiences based on simulated environments [10]. IDC 2020 reports that retail will invest the most in the implementation of this technology until 2024, with $7.3 billion.

3 Methodologies

For this paper, a case study was selected as the methodology, to understand the consumer behavior towards the concerned technologies: consumer surveys were conducted to assess consumer attitudes toward these technologies. A case study is considered a scientific research method in which two or more objectives are selected for in-depth analysis [6]. Six brands will be analyzed in the case study, beginning with Wonther brand and Nutch Sustainable Lifestyle which employ AI tools. Next, we will examine the production company Trimalhas and Teresa Cabreira which make use of VR technology as well as the brands Zippy and *As Portuguesas* that have used AR in physical and online shops. By studying how they behave in the companies' sales and communication channels, and by examining how this immersive technology was applied in the company's plan, we can make direct comparisons between collected data.

The multiple case study will analyze two brands in this case rather than one in order to examine the phenomenon of technologies within the context of fashion businesses and compare the results of the two. On the other hand, online questionnaires are generally associated with quantitative research plans. For instance, using an online questionnaire when you want to survey a large number of people in order to characterize a group should be conducted using probable samples [3]. This study focuses on the consumer's perception of how technology is being used in the fashion industry.

4 Results

4.1 Case Studies

Through the analysis of the case studies, it was possible to determine how these technological tools were incorporated into the marketing strategies of the Portuguese brands, in what contexts they were used, and for what purposes. The following tables provide a summary of the information obtained from each company and the technology used by them.

The impact of AI on two Portuguese brands was examined in a study (Table 1). Wonther is a Portuguese company that produces unisex jewellery that is ethical and sustainable. The brand was founded by Olga Kassian at the end of 2019. Alternatively, Nutch was founded in 2021 by Margarida and Francisca who produce sustainable activewear for women.

For both Portuguese brands, chatbots are an important tool for selling their products online: customers can obtain information about sizes, products, materials, and orders from them. As a result, the customer will be able to complete the purchase process more quickly and efficiently. Chatbots assist these two national brands in clarifying their doubts to their consumers, and both brands are primarily oriented towards the digital market, without having physical stores. Overall, chatbots facilitate the buying process and enhance the relationship between the brand and the client by clarifying doubts in a more personalized manner. It is located on the lower right side of the digital platforms, making it easy for the customer to access (Fig. 1 and 2).

Table 1. IA - Comparative Analysis of Wonther and Nutch Companies

Technology	AI	
Companies	Wonther	Nutch
Objective	To create a tool that could talk to the customer. Facilitating the buying process and helping the interaction between the brand and the customer	The Nutch is an entirely digital brand, the chatbot turns out to be a fundamental element in the online shop, as it helps the customer and is available 24 h a day
Advantages for the company	Reduction of service costs It is immediate and more contextual It creates a more personalised purchasing process	Personalised communication
Advantages for the consumer	Immediate availability Accessibility to information Effectiveness in response Convenience Curiosity for the digital experience	Speed of response Access to all the information required for a better purchasing decision

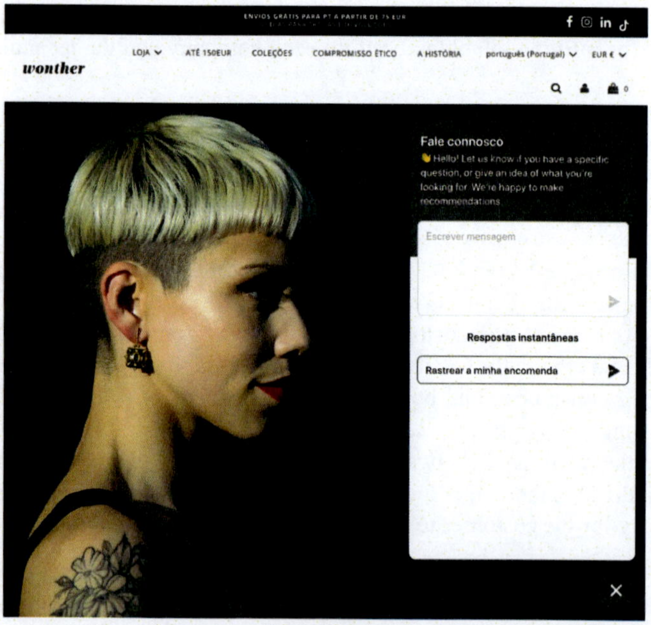

Fig. 1. Chatbot of the Wonther brand website. Source - https://wonther.com/

Fig. 2. Chatbot of the Nutch brand website. Source - https://www.nutch-brand.com/

For the study of AR in Portugal, Zippy and *As Portuguesas* were selected (Table 2). Zippy is a Portuguese children's brand that was founded in 1996 in Porto. This brand is a member of the Zeitreel group and sells clothing, interiors, footwear, and accessories for children. In 2014, Pedro Abrantes created As Portuguesas, a portuguese brand that produces sustainable footwear for women and men.

Table 2. Comparative Analysis of the companies Zippy and *As Portuguesas*

Technology	AR	
Companies	ZIPPY	*As Portuguesas*
Objective	Replacing a traditional fitting room - viewing products virtually in real-time. Captivating the consumer inside the physical shop	Helping customers to visualize the models during the purchase in the online shop Bringing the brand closer to consumers

(continued)

Table 2. (*continued*)

Technology	AR	
Companies	ZIPPY	*As Portuguesas*
Advantages for the company	Greater movement in the shop Facilitates the employee's work Enables stock reduction Reduction in returns Positioning Customer loyalty	Innovation in the purchasing process Greater interaction with the consumer Reinforces the brand's presence in the digital world Positioning Consumer loyalty
Advantages for the consumer	Personalised service Creates immersive and complete experiences 3D visualization of the entire collection Comfortable product visualization	Speed in the process of viewing the product Comfortable viewing of the product on the body Reduces displacement New digital experiences Ease in the purchase decision process

AR is a valuable tool for these two brands, particularly in sales and communication of products during a pandemic event. In Zippy's case, the brand created an AR application with 5G internet inside Norte Shopping, where customers were able to visualize the products through AR and try them out (Fig. 3). *As Portuguesas,* on the other hand, has developed an app for the smartphone that enables customers to try on any model of the brand from the comfort of their own homes and place orders online (Fig. 4).

Based on the brands under study, it can be concluded that the implementation of these technological tools provides consumers with the opportunity for a personalized and immersive experience, becoming a competitive advantage against national companies, standardizing digital communication, simplifying the process for both the customer and the company, as fewer stocks will be held in the physical store and fewer returns will be made. From product visualisation to customer service, it facilitates the entire purchasing process.

Fig. 3. AR at Zippy's physical shop, Norte Shopping. Source - https://www.ntech.news/loja-zippy-usa-5g-para-potenciar-retalho-inteligente/

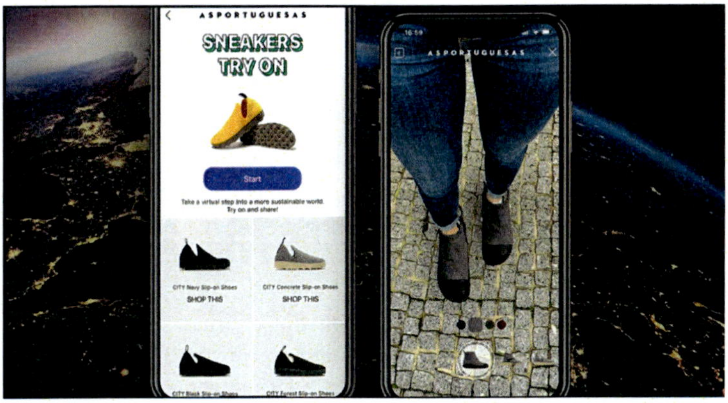

Fig. 4. AR application for smartphone of the brand *As Portuguesas*. Source - Official App

The study on VR technology was conducted on Trimalhas and Teresa Cabreira (Table 3) Trimalhas - Knit Inspiration S.A. from Guimarães is a new and fast-growing company founded in 2002 but with more than 40 years of know how. Teresa Cabreira is a company that emerged in 2002 and has two physical ready-to-wear multi-brand shops in Póvoa de Varzim.

Table 3. VR - Comparative Analysis of the companies Trimalhas and Teresa Cabreira

Technology	VR	
Companies	Trimalhas	*Teresa Cabreira*
Objective	Replacing a traditional fitting room - viewing products virtually in real-time. Captivating the consumer inside the physical shop	Helping customers to visualise the models during the purchase in the online shop Bringing the brand closer to consumers
Advantages for the company	Greater movement in the shop Facilitates the employee's work Enables stock reduction Reduction in returns Positioning Customer loyalty	Innovation in the purchasing process Greater interaction with the consumer Reinforces the brand's presence in the digital world Positioning Consumer loyalty
Advantages for the consumer	Reduces displacement Presents the collections in an interactive way Creation of digital experience Allows a generalised visualisation of the products Comfortable viewing Greater ease and security in the purchasing process	Reduces displacement Creation of digital experience Comfort in the purchasing process Personalised service

Virtual experiences serve the purpose of creating a virtual space where products can be displayed in a fully digital manner, enabling customers to visualize the products. In the case of Trimalhas the company created a virtual showroom where one can see the materials of the collections, from photographs and videos properly subtitled, thus being allowed to see the colours, textures and thickness of textile materials (Fig. 5). In the case of Teresa Cabreira, while the customer can only see a generalized view of the products, it presents a chatbot which allows the customer to have a more personalized experience while browsing (Fig. 6). By implementing these tools into these Portuguese companies' strategies, they can reap the benefits, taking into account that consumers often use digital channels, reducing the displacement of the same, and finally considering the consequences that the pandemic has brought to the market.

As a result of the Covid-19 pandemic, companies had a difficult time maintaining active sales and communication channels. Creating new digital experiences is essential to replace international fairs and physical points of sale, providing consumers with new ways to view products while maintaining coherence and coordination among digital channels. To conclude, the benefits of implementing these technologies are the possibility of creating digital experiences for consumers, thereby providing a competitive advantage over national competitors. In addition to facilitating the entire purchasing process, from the visualization of the product to customer service, convenience, and internationalization, it will be possible to standardize digital communication, assuring that the company has a more strategic positioning.

Fig. 5. Trimalhas 2021 Virtual Showroom. Source - https://mpembed.com/show/

Fig. 6. Teresa Cabreira 3D Website. Source - https://www.teresacabreira.pt/

4.2 Consumer Questionnaire

Based on Raosoft, Inc's 2004 Sample Size Calculator software, 377 participants answered the questionnaire, which is a significant sample. We conducted this survey over a period of two months, starting in June and going through July 2022, taking into consideration a population of 20000 people with a margin of error of 5% and a confidence level of 95%. In this survey, the majority of respondents are between the ages of 20 and 30, are enrolled in higher education, and reside primarily in the northern region of Portugal, in Porto and Braga.

Regarding the use of technological tools, specifically the use of Augmented Reality (AR) in immersive experiences while shopping in fashion brands, it was determined that about 80.1% (n = 302), or more than half of the sample, had never experienced AR in fashion brands, whether in real-life shops or online. Thus, a large percentage of Portuguese consumers have not experienced the benefits that this technology can offer in the purchasing process and in visualizing products. Moreover, 19.9% (n = 75) have already used this technology to visualize how the product would fit them. Following this technology, only with respondents who had used AR, the question was asked, if AR affected the decision to purchase fashion products. In the sample, 75.7% (n = 56) of respondents responded positively, while 24.3% (n = 18) said it did not influence their purchasing decisions. All the answers of the questionnaire regarding AR is in the Table 4.

Table 4. AR - questions about this technology to the Portuguese consumer

Augmented Reality (AR) - Questions	Nº Respondents	YES	NO
Have you ever used Augmented Reality in your online shop?	377	19.9%	80.1%
What role has Augmented Reality (AR) played in your decision to purchase fashion products?	75	75.7%	24.3%
Would you like to experience MORE products from Augmented Reality (AR) technology?	377	86.5%	13.5%

Based on the answers in this table, there is no doubt that consumers are interested in experimenting with AR products. The following table 5 describes an important factor that companies in this industry should consider, which is the consumer's curiosity about experiencing new immersive experiences that will enable them to read and visualize fashion products in a different manner.

In the first question regarding Virtual Reality technology (Table 5), the respondent is asked whether or not they have ever watched a virtual fashion show. In this sample, it is evident that most consumers are unfamiliar with such digital events. Eighty-three percent (n = 313) of respondents do not know and seventeen percent (n = 63) do. Even though -most fashion shows during the pandemic were broadcast via digital media, most Portuguese consumers did not watch them.

Table 5. VR - questions about this technology to the Portuguese consumer

Virtual Reality (VR) - Questions	Nº Respondents	YES	NO
Have you ever seen a Virtual Fashion Show?	377	17%	83%
Have you ever visited a Virtual Showroom of Fashion/Textile Brands?	377	15.4%	84.6%
Have you ever bought digital clothing just to use in an avatar or a digital character?	377	5.1%	94.9%

In the digital showroom section, we first questioned the respondent if he had ever seen a virtual showroom of Fashion/Textile brands, which we found to be the majority, representing 84.6% (n = 319) of respondents never saw and 15.4% (n = 58) states had already experienced this type of event. By analyzing in detail, the responses of those respondents who said they had already experienced a virtual showroom, it can be concluded that the majority of these respondents are Portuguese, with ages between 20 and 30, and working in this field at a professional level, including Product Designers, Fashion Designers, Online Traffic Managers, Computer Engineers, Designers, Projectionists, Motion Designers, as well as students and researchers. Accordingly, it is evident that most respondents who recognize this type of technology are directed towards areas where these tools are used; on the other hand, the general public is not aware of these virtual experiences. To expose consumers who are not familiar with this area to these digital events, it will be important to communicate these events and advertise them.

In addition, the respondent was also asked if he/she/it has acquired any digital clothing, either as an avatar or as an AR filter. Among the samples obtained in this case, 94.9% (n = 357) had never purchased a product of this type, whereas 5.1% (n = 19) had previously purchased a digital fashion product.

The use of Artificial Intelligence, specifically chatbots, in the fashion brand's online shops can be observed in Table 3, where approximately 71.4% (n = 269) of the sample does not use a chatbot to inquire about their doubts when browsing the online shops; 28.6% (n = 108) of the sample uses a chatbot to access the fashion brands' websites.

Lastly, to conclude the theme of chatbots in online stores, 95.4% (n = 103) agree that the use of chatbots is usually effective in answering questions they have during the purchase process, while 4.6% disagree websites (Table 6). According to the study conducted by Deriva (2021), most consumers (87.2%) report neutral or positive experiences with chatbots, while only a small percentage (12.8%) report negative experience. An additional 45.9% of consumers said they used chatbots for immediate responses in the same study.

Table 6. AI - questions about this technology to the Portuguese consumer

Artificial Intelligence (AI) - Questions	Nº Respondents	YES	NO
Is it common for you to use chatbots on the websites of fashion brands when browsing fashion online shops?	377	28.6%	71.4%
Are chatbots usually effective at answering questions at the time of purchase?	163	95.4%	4.6%

As a conclusion, do you believe that the technological tools discussed in this survey are necessary for the current fashion industry?

As this question presents a direct approach, the purpose is to obtain a sample that is defined in exact values, identifying whether the consumer considers it appropriate to use these technologies in the Fashion/Textiles industry. As a result, 80.9% of the respondents (n = 305) consider the technologies addressed in this survey to be necessary for the current fashion market, whereas 19.1% (n = 72) do not. These results indicate that the majority of respondents think that the use of these tools is indispensable for the future of fashion, thus making the consumer one of the most important pieces of the future for fashion. When the customer experiences this type of digital interaction with the brands, as well as through the online and physical channels, it can be realized that they will be curious and willing to do so.

5 Conclusions

Even though the study area is relatively new and there are a limited number of studies, this study aims to contribute to the development of knowledge about the use of these technological tools in the fashion industry and its impact on sales channels and product communication. The purpose of the study was to identify and analyze technologies such as artificial intelligence, augmented reality, and virtual reality and how they can be implemented in this field. Companies must often invest significantly to implement these technologies, making them a very complex strategy. It is usually necessary for a company to invest significant human, financial, and technological resources in the development of these technologies; these technologies can, however, have several benefits for the organization's customers as well as for its long-term success. Since these tools were used in the sales and communication channels, these six companies were able to provide a more personalised approach to the customer in a more digital way, considering the pandemic situation. Even so, these technologies are still in their infancy when it comes to the use of sales and communication channels in the industry, especially at the national level.

On the other hand, from the consumer's perspective, the use of these technological tools is still very low, because there is still little widespread use of AR, VR, and AI in direct contact with customers via sales and communication channels. As can be seen, consumers consider the use of these technologies to be important, since they provide extraordinary convenience, facilitate the purchase process, and enable businesses to gain and retain new customers. The national consumer, however, continues to use traditional methods of purchasing and only a small percentage has turned to online shopping, but due to the pandemic, there has been a noticeable change in consumer behavior.

Finally, by analyzing these two methods of data collection, it is possible to verify that national brands invest minimal resources in adopting these technologies, which results in consumers having a limited immersive experience.

Acknowledgment. The authors would like to acknowledge the funding of the project UID/CTM/00264/2021 of 2C2T - Universidade do Minho, Centro de Ciência e Tecnologia Têxtil, provided by the National Funds through the FCT/MCTES.

References

1. Amed, I., et al.: State of Fashion Tech 2022, p. 5 (2022). Accessed 15 Mar 2023
2. Cao, L.: Artificial intelligence in retail: applications and value creation logics. Int. J. Retail Distrib. Manag. 13 (2021). Accessed 15 Mar 2023
3. Coutinho, C.P.: Metodologias de Investigação em Ciências Sociais e Humanas: Teoria e Prática. 2a ed. Coimbra: Almedina (2013)
4. Future of AR (2022). https://www.wikitude.com/sdk-de-realidade-aumentada/. Accessed a 15 março 2023
5. Gauri, D.K., et al.: Evolution of retail formats: past, present, and future. J. Retail. 19 (2021). Accessed 15 Mar 2023
6. Meirinho, M., Osório, A.: O estudo de caso como estratégia em educação. Eduser (2010)
7. Ngoh, C., Groening, C.: The effect of COVID-19 on consumers' channel shopping behaviors: a segmentation study. J. Retail. Consum. Serv. 16 (2022). Accessed 15 Mar 2023
8. Shankar, V., et al.: How technology is changing retail. J. Retail. 15 (2021). Accessed 15 Mar 2023
9. Watson, A., Alexander, B., Salavati, L.: The impact of experiential augmented reality applications on fashion purchase intention. Int. J. Retail Distrib. Manag. 19 (2020). Accessed 6 Mar 2023
10. Xue, L., Parker, C.J., Hart, C.: How to design fashion retail's virtual reality platforms. Int. J. Retail Distrib. Manag. 20 (2020). Accessed 6 Mar 2023

#1st_Composition# - The Artistic Performance as an Integral Part of a Fashion Show

Sara Vaz[1(✉)] and Alexandra Cruchinho[2]

[1] CICANT, Lusófona University, Lisbon, Portugal
sara.vaz@ulusofona.pt
[2] Lusófona University, Lisbon, Portugal

Abstract. The presented project explores the interconnection between fashion, performance art, and new technologies, whose ultimate goal is to develop a new presentation model of a fashion show. The ephemerality of a fashion show becomes a crucial aspect to consider since the live experience is often what makes a show remarkable and memorable. However, the inclusion of technology, in light, sound, video, and other media, can also help to transmit a stronger, emotional message, with more interaction between the audience and the presentation that is taking place, making the moment more impactful and unforgettable.

The interdisciplinary approach is important to help reach answers that meet the project's main concept. By being connected with the role of the spectator during a fashion show, the experience, and emotions are felt during the show. Also presenting a creative methodological line (Fashion Thinking) to create new ideas and approaches for the project proposed.

Keywords: Fashion · Media Art · Multimedia · Artistic Performance · Fashion Show

1 Introduction

This research project aims to connect two areas of interest and curiosity of the Ph.D. student at a creative level, which will involve the development of a fashion show characterized by a performative act, presented in a physical space, and conducted by a panoply of audiovisual objects. These objects will also contribute to a digital presentation, thus using the image, video, and sound characteristics to present and highlight the fashion phenomenon, linked to modernity, digitalization, and the methods used by it digitally.

Firstly, the area of academic research and the space of experimentation that will complement the parade will be highlighted, which goes against the area of Media Art, thus providing a total involvement between performer and camera.

The valorization of the performative act during the fashion show thus becomes the focus of the concept development. That revolves around the unveiling a moment characterized by various emotions transmitted through the interpreter to the present audience. This project was born, which was designed to meet the need to provide sensations/emotions to the public through performance. That will somehow replace the conventional parade to give way to an interpretation of a narrative performed initially by the garments presented and then by all its involvement with the space and the cameras. In this universe, the creative experience is one of the most relevant elements to consider when developing an idea of the final goal that brings together what we consider the best of both worlds: the analog/physical and the digital.

From here, we start to the analysis of the role of performance as an artistic element during the fashion show. Leading this study to the definition of the problem related to the impact that this element may have before a spectator who goes to the space with the idea that he is going to watch a fashion show and ends up being involved in a performative act characterized by various audiovisual elements.

This leads to the question: could the integration of a performative act in a fashion show presentation be the result of a new model presenting a fashion show?

The main lines of study will be presented throughout the article starting by developing research on themes such as fashion shows, performances, multimedia, and spectator. The answers to these questions will imply distinct methods and different approaches, which use interdisciplinarity to drive the connection of fashion with technology and Media Art.

2 Theoretical Framework

2.1 Fashion Show

Framing fashion definition, fashion refers to styles and customs in a particular time and place regarding clothing, footwear, accessories, make-up, hairstyles, and body proportions. It is a form of expression that reflects the time's cultural, social, and economic context. Fashion constantly evolves, with new trends emerging each season in response to cultural influences, technology, and innovative designs. According to Lipovetsky (2010), fashion is not for all civilizations. However, it is multifaceted and can be understood in different ways and perspectives. It can be associated with ideas and forms of expression and self-expression. It can elicit complex emotional responses from groups, and sectors of society and even shape the behaviors, values, and attitudes of groups or societies (Hopkins 2021).

The fashion show has evolved and is an important part of popular culture and the fashion industry.

Fashion shows are momentous events, long-awaited, highly anticipated, and rarely repeated. They usually last less than ten minutes but impact those who attend and view them. They are designed to create a desire for the designer or brand, and have become an important part of popular culture, serving multiple purposes such as entertainment, and communication, and are often a stage for social and political criticism. The fashion show becomes a multi-sensory and immersive experience collectively consumed as part of our contemporary cultural zeitgeist. The fashion show is also considered an artistic medium, while it is also a more commercial presentation functioning as a venue for communication between companies and consumers. Please note that the first paragraph of a section or subsection is not indented.

In Staging Fashion - The Fashion Show and Its Spaces (Ferrero-Regis 2020), the authors describe the fashion show as the intersection of clothing, body, and space, commanding different staging practices in different spaces, represented through various media over the years of fashion history. The organization of a fashion show (Fig. 1) requires exploring how its manifestations have been represented through commercial, cultural, and social trends throughout history. As such the runway show can be seen as a discursive space, where rules, ruptures, bodies, forms, objects, clothing, and practices come together to create a narrative.

Fig. 1. Nehera | Spring Summer 2018 Full Fashion Show | Exclusive. Font: https://www.youtube.com/watch?v=6gVW5fRlWVM

2.2 Performance

Performance is an artistic or cultural presentation that can transmit political, social, cultural, and artistic messages and establish an emotional and physical connection between the performer and the audience. Performance is a living and dynamic art form that is different in every performance, with the interaction between performer and audience making each performance unrepeatable and unique.

According to Diana Taylor (2015, p. 43), art historians such as RoseLee Goldberg, consider the antecedents of performance with the work of the Futurists, Dadaists, and Surrealists who focus on the creative process rather than the final artistic product. For (Taylor 2015), the complexity of performance and its multiple possibilities is highlighted in performance. The performance may meet many things, which sometimes become paradoxical. For the author, performance can be visible or invisible. It can be ephemeral and long-lasting, it can lead to the normalization of behaviors, but also shocks and test the role of the spectator in a frontal and direct way. It can also be radically unstable, depending on the framework, the occasion, and even who it is for.

In short, according to the author, it is a form of artistic expression that can be interpreted in different ways and have different impacts on the audience.

2.3 Multimedia

Multimedia is a concept that will be the combination of various media, such as text, audio, video, images, and animations, that will be used together to create an interactive media experience.

According to Mayer (2014, p. 2), multimedia may take us to varied paths and meanings. It can praise a hands-on experience if we refer to watching a podcast on the mobile phone or even playing a game on the tablet; it can also be thought of in sitting in a room where images are displayed on one or more screens, along with sounds coming out of speakers. Other possibilities could include a video on a TV screen while listening to the corresponding words, music, and sounds or even watching a presentation together with audio.

In Multimedia Foundations (Costello 2016), we take away some relevant concepts for the theoretical and practical development of the project:

Graphic Components

Graphics support compelling and impactful visual elements of the experience, encompassing a wide range of components such as digital photos, illustrations, clipart, and any other still-image that can be displayed on a monitor or digital screen. Software such as Illustrator and Photoshop are considered to create and edit graphics. Photoshop is used to edit digital photos and bitmap graphics, where pixels are the building blocks of a digital image. Illustrator is used to edit vector graphics, which will be defined by pathways formed by points, lines, curves, and shapes.

Video

In Multimedia Foundations (Costello 2016), the author tells us that before the era of digital production, the terms film and video referred only to specific analog methods for recording moving images. The video was an electronic platform that relied on magnetic tape, optical disk, and television broadcasting to disseminate programs. The use of video as an expressive medium and as an artistic medium in the fashion area thus establishes its language. Praising characteristics that meet the capture and transmission of moving images, enabling editing and manipulation of images and sounds, and using technical resources with visual and sound effects.

Referencing Pierre Cardin in 1970, one of the pioneers within the fashion area to explore the video format runway show, presented us with a fashion show going against the pop theme.

In 2020 Alessandro Michele for Gucci (Fig. 2), under the direction of Gus Van Sant presented a miniseries in seven parts, thus exposing the collection "Ouverture of Something That Never Ended", taking as a starting point two timely questions for the pandemic moment: "What are the new horizons that arise when fashion leaves its comfort zone?", "What life do clothes gain, when they stop being paraded on the catwalk?" (Phelps 2020).

Fig. 2. GUCCI- At Home | Ouverture Of Something That Never Ended| Font: https://www.youtube.com/watch?v=zKqbG6TLYnc

Photography
According to author Costello (2016), photography is a process of image fixation through the action of light. In traditional chemical processing, photographic images are created by exposing a light-sensitive emission in a controlled environment.

Currently, photography can be used in a personal and artistic way, facilitating the creation of unique and impactful images for amateur and professional photographers. Through lighting techniques, framing, and composition, the photographer can create his narratives and atmospheres, following his artistic language.

Fashion photography is a form of art and visual communication that can be used to express ideas and concepts and promote products. It is a very creative form of photography, aiming to create compelling and impactful images with technical and artistic skills.

In Fashion's Double (Adam Geczy 2015), fashion photography is portrayed and modeled from components of the outside world to create a visual hypothesis of what that world could become. According to the author unlike art, which is meant to serve some higher purpose that goes beyond the material conditions of what it represents. The fashion image's perseverance in being special is inscribed in the demands and expressiveness of the message itself.

Spectator
In Fashion, Performance, and Performativity - The Complex Spaces of Fashion (Andrea Kollnitz 2021) the authors reflect on fashion's relationship with various media. Including literature, voices, and images, discussing the performativity of fashion and how it can destabilize the distinction between the descriptive and the performative. Austin's importance in defining an "unhappy" performative act is highlighted, a moment of failure of the performative, "when something goes wrong and the act... is therefore at least to some extent a failure".

Fashion performance spectators might lead us to a specific fashion audience interested in attending a fashion event involving performative elements, like a fashion show or even an interactive fashion installation.

3 The Physical and Presential Object

The presented draft will lead the spectator to a narrative told from the clothing present in the space itself, by the involvement of sound and the presence of cameras and television sets to capture movements characterized by the moment that being staged.

It is presented as a performance created to support an idea defended by the doctoral student, which meets the concept that revolves around the definition of a Fashion Show. For Gill Stark (2018), the fashion show is characterized by the creative vision of what it is to create a show. Still, it also meets the practical production, always accustoming us to meet the main objective of a show: to stimulate the desire for the designer's pieces or brand, extolling the idea of pleasure and promise of gratification when we buy products.

The concept becomes essential for developing the main idea - the artistic performance as a language in a fashion show, using multimedia tools.

Physical and in-person objects refer to any object or item that exists in the physical world and can be seen, touched, or interacted with in person. This could include anything from a book to a work of art, a building, or furniture. Physical objects can be observed with the naked eye or instruments and can be manipulated, moved, or modified in some way. They are distinct from digital or virtual objects that exist only in the digital world and cannot be directly interacted with physically.

Hibu (Fig. 3) presented her Autumn/Winter 23/24 collection at the 60th ModaLisboa, 1000 mg, focusing on 1990's grunge and the balance between streetwear and delicacy, keeping the minimal aesthetics. The designer even risked an installation of static bicycles where some models pedaled to the rhythm of the music that accompanied the show. Imprinting rhythm to the presentation, managing to capture the audience's attention not only for the garments. Also, what was happening in the room where the static bicycles were?

In short, physical and present objects are essential aspects of our material culture and play a significant role in our daily lives, from the clothes we wear to the works of art we admire.

Fig. 3. HIBU I Fall Winter 2023\2024 I Full Show. Font Show I Exclusive. Font: https://www.youtube.com/watch?v=0e0HreOEFJg

4 The Digital Object

Digital fashion refers to designing and creating of garments and accessories for the digital world. This can include creating virtual garments for video games, avatars on social networks, or even virtual reality fashion platforms. Virtual fashion is a point of intersection between fashion and new media because it combines fashion elements with technology. Virtual fashion allows people to create clothes and accessories in virtual environments, share their creations, and interact with others over the internet. This can lead to greater creative freedom, access to new ideas and forms of expression.

In Feldner-Busztin (2022) Chanel, Dior, and Louis Vuitton continued their efforts to return to normal life and not go fully digital. In the end, these were the most successful of the season, according to data monitoring company Launchmetrics, which measures success through Media Impact Value (MIV). MIV aims to represent the financial impact, generated by media placements, and mentions made by influencers in the fashion, luxury, and beauty space. This allows companies to track the return on investment of their marketing strategies. The success of in-person shows is not surprising, as the fashion glitterati still seems to crave a physical show. Still, these shows challenged the challenge of inventing something new within the digital methodology that was not just temporary. Being physically present at a fashion show can be an exciting and unforgettable experience, with all the energy and adrenaline the environment provides. However, the emergence of digital fashion has also brought benefits to this sector regarding the possibility of making the event more accessible and sustainable. However, the emergence of digital fashion has also brought benefits to this sector regarding of the possibility of making the event more accessible and sustainable.

A hybrid of both digital and in-person fashion could be an ideal solution for the fashion industry today, as it allows more people to access designers' creations while maintaining the physical experience for those who prefer to see them in person.

This article highlights the On|Off project that explores the coexistence of physical artifacts and their digital representations to foster creative possibilities. Three designers, Taskin, Oscar Keene, and Balaclava, work with new technologies and take their vision to new levels with the support of On|Off. The organization will host a "real-time virtual fashion show" at LFW, to explore new technologies for a live audience at the Swiss Church in Covent Garden. Viewers can interact with the physical garments as they are streamed live to the metaverse, along with their digital avatars. The initiative represents a new way to experience fashion, where reality and fantasy meet.

5 #1st_Composition# Exhibition

Bruno Munari was an Italian designer, artist, and writer who developed a creative methodology to stimulate creativity and innovation. Munari's methodology emphasizes the importance of play, experimentation, and thought in the creative process. It was therefore decided to present the project (Fig. 4) through a scheme concerning Bruno Munari's methodology, taking into account aspects that were part of the starting point: Observation as the key to stimulating creativity believing that by observing what surrounds us we can obtain inspiration for new ideas and creative solutions; Experimentation of the creative process, with new materials, techniques, and approaches to find innovative solutions; and prototyping, managing to develop quick and simple prototypes, testing them and trying to identify opportunities for improvement (Fig. 5).

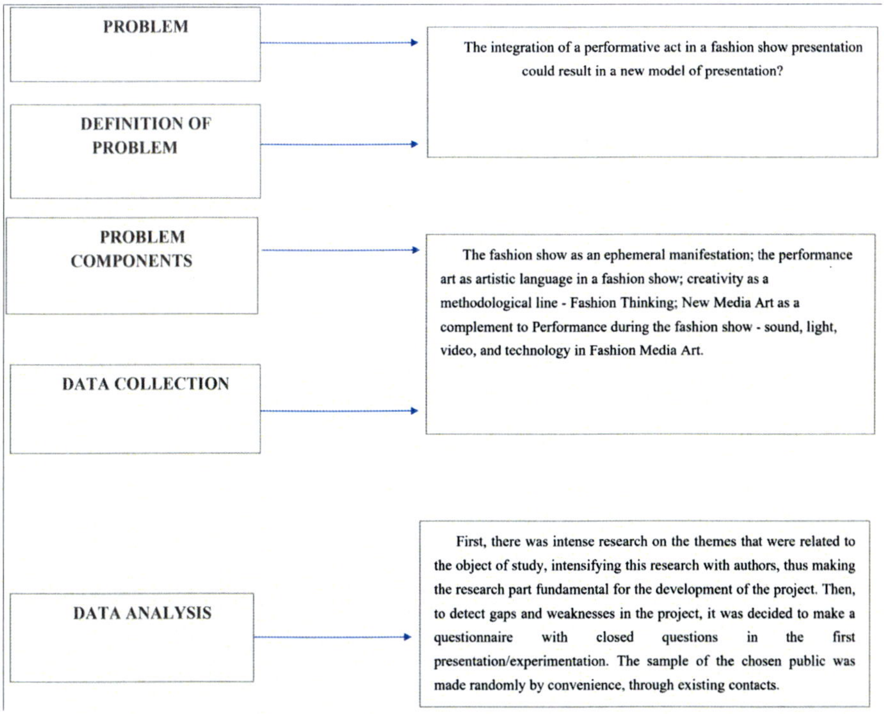

Fig. 4. Design methodology adaptation by the author Bruno Munari_1. Font:Author

CREATIVITY →

The development of the project was based on the methodological line of creativity (Fashion Thinking), always trying to find innovative and original solutions to respond to creative challenges. It was born from brainstorming, generating a large amount of ideas. In the idealization of the performance #1st_composition#inserted in a fashion show context, it is inevitable not to discuss and deepen the theme of the relationship between the body, the pieces themselves, and the human relationship. The interdisciplinary approach is fundamental for this analysis, going towards the analysis of people's perceptions and emotions, the cultural and historical practices of each participant, as well as technological and environmental influences.

MATERIALS & TECHNOLOGY →

Equipment to be used: Polaroid / Instax; Computer screens/ computers; Television (Large screen); Spotlights; Speakers; Mobile phones to connect live for the public to watch; Distance sensors for speakers to increase and decrease the sound intensity; Software: Adobe Audition | Audacity; Clo 3d (fashion) animation; Camo Studio; Arduino – Platform | Arduino IDE 2.0.3; Puredata

EXPERIMENTATION →

Fig. 5. Design methodology adaptation by the author Bruno Munari_2. Font:Author

6 Conclusion

In conclusion, the presented project aims to explore the interconnection between fashion, performance art, and new technologies to develop a new presentation model for a fashion show, focusing on the valorization of the performative act during the fashion show as a way of providing emotions and sensations to the public, going beyond the conventional format of the show, combining elements that make a connection between them, the analogical/physical and digital, and exploring the impact of performance as an artistic element in the fashion show.

The interdisciplinary approach is essential to achieve answers that meet the project's main concept, focused on the role of the spectator during a fashion show and the emotions experienced during it. The project also presents a creative methodological line, Fashion Thinking, to create new ideas and approaches for the project presented. Ultimately, this project has the potential to contribute to the development of a more impactful and unforgettable fashion show experience that combines performance, technology, and fashion.

References

Kollnitz, A., Pecorari, M.: Fashion, Performance, and Performativity - The Complex Spaces of Fashion, 1st edn. Bloomsbury Visual Arts (2021)

Costello, V.: Multimedia Foundations - Core Concepts for Digital Design. Routledge (2016)

Feldner-Busztin, E.: The interline. Obtido de Where Digital and Physical Collide: The New Era of Digital Fashion Shows (2022). https://www.theinterline.com/2022/06/30/where-digital-and-physical-collide-the-new-era-of-digital-fashion-shows/

Hopkins, J.: Fashion Design: The Complete Guide (2021). https://www.perlego.com/book/2740472/fashion-design-the-complete-guide-pdf?queryID=36e01282f757b9e30b6642a469f9d2b3&index=prod_BOOKS&gridPosition=38. Accessed 5 Feb 2023

Lipovetsky, G.: O Império do Efémero - A moda e o seu destino nas sociedades modernas. 2° ed. Alfragide: Dom Quixote (2010)

Mayer, R.E.: The Cambridge Handbook of Multimedia Learning. Cambridge University Press, Cambridge (2014)

Phelps, N.: Vogue (2020). https://vogue.globo.com/moda/noticia/2020/11/alessandro-michele-e-gus-van-sant-falam-sobre-minisserie-da-gucci.html. Accessed 5 Mar 2023

Taylor, D.: Performance. Duke University Press Books (2015)

Ferrero-Regis, T., Lindquist, M.: Staging Fashion - The Fashion Show and Its Spaces (2020). https://www.perlego.com/book/2035512/staging-fashion-the-fashion-show-and-its-spaces-pdf?queryID=1a027d79559627b58cc3126ef7dfff0a&index=prod_BOOKS&gridPosition=3. Accessed 5 Feb 2023

Fashion, Identities and Cultures

Fashion as and Aid to the Body Image of Women with Breast Cancer

Jéssica Tuelen Krein(✉) and Gabriela Poltronieri Lenzi

UNIFEBE, Brusque, SC, Brazil
{jessica.krein,gabriela.lenzi}@unifebe.edu.br

Abstract. Women with breast cancer have to endure consequences on their body image mainly due to treatments, which, in turn, point out the disease, making it social. Thus, this research suggests that fashion can be an aid to this group's body image and, because of that, had as its main objective to analyze how clothing and fashion elements can serve as an aid to women with cancer. This study is applied research, qualitative in regard to the problem and exploratory in relation to the objectives. Bibliographical and documentary research were carried out, in addition to data collection through semi-structured interviews with women with breast cancer and with specialists working in this environment. Using the data obtained through this study, it was possible to observe the difficulties experienced by the patients and how the disease and treatments affect identity and interfere with self-esteem. It was also identified how fashion is present in this context, serving as an aid to the sick women, thus reducing the physical, emotional and social consequences and allowing identity, self-esteem and body image rebuilding.

Keywords: Fashion · Body Image · Identity · Self-esteem · Female Breast Cancer

1 Introduction

According to the Brazilian National Cancer Institute [1], breast cancer particularly affects women. Not only women are affected by it, but they are the most affected group. It is noteworthy that this disease does not only affect the patients' body, but involves mental, emotional, psychological and social issues [2].

This way, when facing the disease, the woman with cancer needs to reframe many things, including body representation, the rebuilding of her identity and her social roles [3]. Thus, it is believed that fashion, by means of clothing, has the role of rebuilding the female image and identity, in order to serve as a tool for the individual to be able to maintain her private and social space throughout cancer and its treatments. In this sense, clothing goes beyond creating distinction and becomes a process of creating identity and maintaining self-esteem, from which one can develop an image for oneself and another for the others, that is, it is a way for one to resignify oneself before oneself and before the society in which one participates [4].

Therefore, the research aimed at gathering information in order to answer the following question: how can fashion serve as an aid to women undergoing treatment for breast cancer? Thus, the main objective was to analyze how clothing and fashion elements can serve as an aid to women with breast cancer. As specific objectives, we aimed at: a) understanding the impact of side effects of cancer treatments on the body image of the woman with cancer; b) identifying how fashion and clothing can influence the identity and self-esteem of women undergoing chemotherapy and c) verifying which fashion and clothing elements mainly help women undergoing cancer treatment.

This research is justified in view of the number of women affected by this type of cancer [1] and by the side effects of the treatment that directly interfere with their identity, self-esteem and body image. It is believed that by recognizing the problems faced by women with breast cancer it is possible to understand how fashion, by means of clothing, can help them cope with the disease and thus serve as a tool to provide well-being for the female patients.

We can point out the lack of studies and research in the field of fashion for this specific group, for which we believe not only related content is needed, but also products and services designed and intended for these women. Therefore, the research aimed at stimulating and promoting dialogue about fashion and women with cancer, in order to raise interest in the subject in fashion design professionals and researchers.

2 Methodological Procedures

2.1 Research Methodology Classification

This research is applied, a qualitative approach in regard to the problem and exploratory in relation to the objectives.

Applied research seeks to solve problems and needs that are found in society [5]. Following this method, it takes a qualitative approach, since it is characterized by the detailed analysis of a certain phenomenon and finds the source for data collection in natural environments [6]. For Gil [5], exploratory research has the "objective of providing an approximate overview of a given fact."

As technical procedures these were adopted: bibliographical research, documentary research and data collection through semi-structured interviews.

For Marconi and Lakatos [7], bibliographical research is considered a safe and reliable source of data, which makes use of materials, such as books, for the research of a given subject. Thus, its objective is to bring the researcher closer to all types of existing content on the subject studied. On the other hand, documentary research is characterized as the study based on primary, unedited sources, which is restricted to documents, whether written or not [7].

Semi-structured interviews were also conducted with women with cancer and specialists working in this environment. The interview is defined by Martins and Theóphilo [8] as a technique that aims at understanding, from the interviewees' perspective, the meanings attributed to certain circumstances and topics.

2.2 Field Contextualization and Research Subjects

Data collection was carried out at the Women's Cancer Network (Rede Feminina de Combate ao Cancer) in the town of Brusque, Santa Catarina State, Brazil. It is a non-profit institution which works for the prevention of breast and cervical cancer, in addition to giving support to patients with cancer. The entity has been based in Brusque since 1989 and is supported by donations and promotional events [9].

This philanthropic entity not only made available the place for carrying out the semi-structured interviews, but also offered help with inviting patients able to meet the research requirements. This way, semi-structured interviews were carried out with women with cancer and also with specialists working for this group's well-being.

To achieve this goal, the interviews included question blocks with open answers, allowing the interviewees to narrate their personal thoughts and feelings without being limited or induced by pre-established answers. The interviews took place both in person and online. The face-to-face interviews were carried out at the headquarters of the Women's Cancer Network in Brusque, whereas the online interviews made use of an Internet application as a communication channel.

In total, ten women with breast cancer between 45 and 67 years old participated in the research, dealing with three question blocks on average. These blocks were organized into different sections: a. personal data; b. self-esteem, identity and body image and c. fashion and clothing elements present throughout the treatment.

In addition to this group of women, two volunteers from the entity were also interviewed so that data could be collected from specialists who have empirical knowledge about the self-esteem of women with breast cancer. The volunteers interviewed were: the coordinator of the support group for women with cancer (who has been working at the entity for 16 years) and the institution's psychologist (who started her activities at the Brusque Women's Cancer Network in March 2022). They were asked to answer three question blocks with these sections: a. personal data; b. perceptions about the self-esteem and identity of women with cancer and c. experience with fashion and clothing elements used throughout cancer treatments.

All participants agreed to the Informed Consent Form (ICF) and therefore agreed to participate in the research. However, their identities were kept confidential. The interviews were recorded and later transcribed for the purpose of analysis.

3 Analysis of the Results

In order to carry out the analysis of the data collected, the semi-structured interviews were divided into two categories, according to the participants' profile: women with cancer and specialists. Such a division was based on the assumption that, depending on the place where things are said, the participants' discourse could have specific characteristics. This way, the analysis of the results shown below were separated as follows: *3.1 Fashion and Clothing Elements as an Aid to the Body Image of Women with Breast Cancer* and *3.2 Fashion and Clothing throughout Treatment: a Specialist's View on the Body Identity of Women with Breast Cancer*. The first subsection consists of data collected from sick women, while the second subsection brings information from specialists.

During data analysis, not only words were taken into account, but also emotions, intonation, silences, emphases in the subject's speech, whose aspects were used to go beyond the collected information.

3.1 Fashion and Clothing Elements as an Aid to the Body Image of Women with Breast Cancer

In order to identify how women with cancer felt about being diagnosed with the disease, they were asked about the first feeling that the diagnosis caused. The answers obtained involve sadness, denial, desolation and doubts. However, they claim that, despite the difficulty experienced, they understood that it was necessary to move on, fighting the disease.

Regarding self-esteem and identity, they report that both are strongly affected, undergoing changes as a result of cancer. Image is affected mainly during the treatment of the disease, a period in which the most drastic changes occur in the body, such as: changes in the skin, hair loss and breast removal. These can therefore be seen as symbolic elements which directly affect the patient's image. As a result, women who face chemotherapy-induced alopecia and/or undergo mastectomy feel their identity shaken and end up having damage to their self-esteem. In this stage, they mention that they find it difficult to look at themselves in the mirror, usually finding strangeness in the image they see, as reported by Interviewee 3: "*You look and you don't believe it, right? Because I've had long hair all my life. Now, all of a sudden you find yourself with no hair, no eyebrows, which is horrible, even your eyelashes fell out. [...] we feel bad, even when people say 'wow, you look pretty, you look cool like that,' but you really don't see that.*"

However, they understand that this stage is part of the healing process and see it as a necessary evil. They argue that the most important thing is, in fact, to get rid of the disease.

The physical changes in their bodies caused by the treatments also affect the social relations of women with cancer. Some say that they receive strange looks and that they are able to notice people's pity, thus starting to feel weak as a result. In this sense, they reinforce that this is not the image they want to convey, feelings like this only worsen the state they are in. They report that they want to continue to be seen as ordinary people, who are just facing a disease, as told by Interviewee 1: "*I always said: 'you don't need to treat me differently because I am undergoing cancer treatment. [...] I'm still normal, I'm normal, I'm just undergoing treatment.' But people see you very differently.*" This condition, when the person is seen as sick before being seen as an individual, stigmatizes the person socially and is initially due to the physical aspects [10] that manifest the illness in the person's body.

By analyzing the group's words, one understands the importance of social support when coping with cancer, as reported by Andrade, Panza and Vargens [11]. It is also evident how social factors influence this process, as argued by Interviewee 3 in relation to this topic: "*If you do not have support, something to raise you up, to help yourself is difficult, alone it is difficult. No! Alone I believe that the person can't, no matter how much the person has, I have a lot of faith, [...] because it is not easy.*"

As it is considered a social phenomenon, fashion, by means of clothing, has the capacity to establish the person's identity in his/her environment [12]. During the interviews with patients and volunteers, the search for aesthetic and fashion elements is clearly an aid to self-esteem and identity in the personal and social context. As Interviewee 10 declares: "*...I didn't want those looks from people thinking, thinking I was going to die. I thought that was the worst thing, right? I thought like this: 'If I put on makeup, if I try* *to get as close to who I am, right?, people won't even notice that I'm sick' [...] And that was what I wanted. I wanted people to treat me as an equal.*"

Regarding fashion and clothing, when asked if clothes and accessories helped with self-esteem and identity throughout the disease treatment, all responses obtained were positive. When challenged, Interviewee 2 replies: "*I think they can help a lot, because it seems that the person's self-esteem improves, right? [...] And some makeup or a scarf, something like that, right? It always makes you look better and it seems that the person is happier that way, with more self-esteem, right? [...] Both in terms of the disease and, right?, self-esteem, right?*".

The participant adds that fashion and clothing elements have the ability to increase self-esteem, bringing emotional benefits and thus favoring the healing process.

Interviewee 10 explained that when she put on makeup, scarves, wigs, earrings, that is, when she adorned herself, her well-being increased and this was directly reflected in the exams. According to her, fashion helps experience the disease and treatment in a more serene way. She adds that self-esteem is fundamental for women at any time in life, but during cancer treatment it becomes even more essential. It can therefore be seen that fashion and clothing elements are essential in this process, as they help women cope with the disease, serving as an aid in the search for female self-esteem.

In addition to clothing, patients mention the experiences they had with makeup, which helped them throughout cancer. They report that eyelashes and eyebrows are also susceptible to hair loss. In this context, they recommend using false eyelashes and eyebrow pencils. They also state that the use of blush helps reduce the paleness of the face resulting from the treatment, adding color to the skin. Elements like these involve indication of stigma, allowing women with cancer to be protagonists of their identity.

In general, participants stated that the use of turbans, scarves and wigs are essential to ease the lack of hair. Interviewee 3 reports that the scarf served to cover the bald head, but was also essential to warm it, protecting it from the cold she felt. Thus, for example, scarves go beyond aesthetics, also showing functionality. This reinforces that the use of adornments has importance during the treatment process, due to their ability to maintain the patients' self-esteem, in addition to promoting a positive boost to coping with the adversities of the disease.

In relation to clothing, they comment that they opted to wear more colorful clothes after the treatment, in order to minimize the paleness caused by the disease and the treatments. Some interviewees talk about breast removal and how it impacts the body image conveyed by clothes: "*...I put on a T-shirt and it looks a little crooked*" (Interviewee 8). In this context, there are clothing elements capable of minimizing the impact of the lack of breasts after mastectomy. The removable external breast prosthesis, for example, is mentioned by numerous interviewees. Women claim that the use of these prostheses helps reduce the visual aspect caused by breast removal surgery. They affirm that, through fashion and clothing, there are several ways to mitigate the impacts of treatment in the social sphere: "*...there are girls who turn beautiful, put on makeup, put on that bra that has that thing (prosthesis), right? I do think they look really pretty. We don't even notice that they have cancer, right? Because there are many of these fashion models out there on the catwalk with these turbans, these things, right? So it's not even noticed.*" (Interviewee 9).

Through the reports, it is possible to see that women undergoing chemotherapy feel the need to make use of fashion and aesthetics in order to mitigate the visual impacts resulting from the disease, as suggested by Reis's research [13]. They do not change one's personal style; however, they adapt to the patient's new needs. Because of this, patients show preference for vibrant colors, as they believe that they convey happiness, for example.

After being diagnosed with cancer, some interviewees state that their identity underwent internal changes. They mention that they started to feel more willing to socialize, as Interviewee 5 comments: "*I don't want to stay there alone, quiet, because I have cancer. No! I want to fight, I want to do everything I used to do before cancer and much more. And I do a lot more now than before. Because we learn how to appreciate things.*" They say they have become stronger owing to coping with the disease, feeling more alive and giving greater value to moments and to their bodies the way they are.

3.2 Fashion and Clothing Throughout Treatment: a Specialist's View on the Body Identity of Women with Breast Cancer

The entity's volunteers, who work with patients with cancer on a daily basis, were also asked questions about how women react to the disease. When asked about how patients dealt with the diagnosis, both answered that the uncertainty caused by the disease and the fear of what is ahead makes this a difficult time. They emphasize that women experience a period of emotional instability, as explained by Volunteer 2: "*…It's a mix of doubt. It's a mix of fear, right? Not knowing if it will work or not. If you're going to die, if you're not. So, it goes up and down a lot. This is a winding journey. There are days when they are happy and confident, and the following week they feel sick, they think they are going to die.*"

According to Silva [2], this reaction occurs due to the emotional and psychological impacts caused by the disease, creating fear of death. Since it is an aggressive disease, patients suffer not only its physical impacts [14], but also emotional ones.

They believe that patients realize that they were affected by the disease when the side effects of the treatment appear. That is, it is at the stage of hair loss and other consequences of chemotherapy, radiotherapy and mastectomy that women realize that they have cancer, and thus their self-esteem and identity begin to be harmed. When they look in the mirror and see themselves without elements that until then characterized their appearance, such as hair, eyebrows, eyelashes, breasts, among others, this is the moment when they realize they are sick. This is also the moment when society starts to label them as sick, as these missing physical symbols betray the disease. In this context, the volunteers reported that hair loss is the moment of greatest impact for a woman with cancer. Prates [15] argues that one of the worst cancer treatment effects is chemotherapy-induced alopecia, as it directly affects the patient's image. Since it is external and is a sign of the disease, the lack of hair can result in insecurity, creating social problems and emotional disorders.

Since they worked directly with the group of women with cancer in the search for fashion and clothing elements, with the goal of reducing the stigmas of the disease, the volunteers believe that it is fundamental that the patients have access to clothes, accessories, makeup or any artifice that may be used in order to rebuild the body image. Thus, they mention elements such as wigs, scarves, turbans, removable prostheses, appropriate bras and other accessories that help improve the patients' image and reduce the visual impact. The experts interviewed also consider that these alternatives help cope with the disease, contributing to the healing process, providing well-being and self-esteem for women. Baitelo, Reis and Gradim [16] state that being able to face the disease in an optimistic way can be beneficial in the cancer treatment process. They add that the use of alternative fashion elements such as scarves, wigs and other accessories serve as an aid to patients, helping them in the search for integration and social belonging.

The entity where the interviews were carried out offers several treatment options for women with cancer. Treatments aimed at the patients' health and well-being stand out among them. They believe in the power of fashion and clothing. Therefore, they have a considerable collection of scarves and wigs available for loan to those undergoing chemotherapy, in addition to offering bras with removable prostheses for women who have undergone breast removal surgery. They also offer lectures and psychological assistance, addressing issues related to their self-esteem and identity.

4 Final Comments

This research had the main objective of analyzing how clothing and fashion elements can serve as an aid to women with cancer. Thus, we aimed at understanding how the side effects of cancer treatments affect the body image of women with cancer. In this study, it was possible to verify that women who undergo treatments for breast cancer (chemotherapy, radiotherapy, mastectomy, etc.) suffer the side effects of these therapies, which are directly reflected in physical aspects of identity.

Breast removal and hair loss are mentioned as the main elements that harm body image, causing changes that affect their emotions. The loss of eyelashes and eyebrows, the skin color, especially on the face, marked by paleness, are also characteristics resulting from the treatment. It can be observed that such transformations in their bodies go beyond exclusively personal issues and become public, since these physical traits betray the disease. Women recognize that they are sick when they notice the effects of treatments manifested in their bodies. These body modifications make the disease evident, thus becoming a public declaration of cancer.

Therefore, we aimed at identifying how fashion and clothing could influence the self-esteem and identity of women undergoing chemotherapy. Fashion is involved in this context as an essential aid to the process of identity rebuilding, seeking help for the feeling of personal and social belonging with alternative elements. Through fashion, which encompasses clothing, accessories and makeup, women with cancer find ways to show themselves as individuals with their own desires and traits, softening their self-image as sick people. These resources grant them a social representation as human beings, instead of being categorized as patients. Through the patients' reports, this can be seen as a fundamental process for them to feel safe throughout treatment.

The feeling of safety made available through these elements directly affects their self-esteem, reinforcing their perception of social belonging and also of personal recognition. Adornment objects let them look in the mirror and find themselves there. It does not mean that they see themselves the same way they saw themselves before the disease, but help them feel good about the new self that is born. This way, for many of them, the contribution of self-esteem to the healing process is evident, allowing them to reinvent themselves in the new reality they are experiencing.

When analyzing which fashion and clothing elements most help women undergoing cancer treatment, elements to cover the bald head, caused by chemotherapy, are emphasized, such as: scarves, wigs, turbans and caps. On the other hand, for women who underwent mastectomy, the removable external prosthesis is commonly found, as it reduces the visual impact caused by the surgery. The use of makeup is also common to minimize the paleness caused by the treatment and to draw again lost eyebrows and eyelashes as well. In addition to this, there is a preference for the use of accessories, such as earrings, for example, and clothes with vibrant colors. They link them to joy and energy, being a symbol of well-being for them.

Based on this study, we can conclude that the intended objectives were achieved. It is therefore possible to verify that clothing and fashion can act as allies for women with cancer, helping them cope with the disease.

There is still a need for further research, with a specific focus on dealing with functional and symbolic issues of clothing and fashion elements used with patients. It is also evident the difficulty of finding current studies on the subject. Due to this, a multidisciplinary and interdisciplinary approach was chosen in order to provide support for the theoretical foundation. Finally, research in the area of fashion design is recommended, to study elements of clothing with the specific aim of raising interest in the creation of products and/or services for this public, since they need adornments with unique features to cope with the disease.

Acknowledgments. Thanks to the Brusque Women's Cancer Network, in particular to the volunteers who carried out the interviews presented above and those who participated in them.

References

1. Instituto Nacional de Câncer. Incidência (2022). https://www.gov.br/inca/pt-br/assuntos/gestor-e-profissional-de-saude/controle-do-cancer-de-mama/dados-e-numeros/incidencia. Accessed 19 Oct 2022
2. Silva, L.C.: Câncer de mama e sofrimento psicológico: aspectos relacionados ao feminino. Psicol. Estud. **13**(2), 231–237 (2008). https://doi.org/10.1590/S1413-73722008000200005
3. Palmeira, L.V.S.: Corpo feminino e (re) significações da beleza: um estudo sobre mulheres com câncer de mama em um grupo de apoio – Fortaleza/CE (2015). https://scholar.google.com.br/scholar?hl=pt-BR&as_sdt=0%2C5&q=Corpo+feminino+e+%28re%29+significa%C3%A7%C3%B5es+da+beleza%3A+um+estudo+sobre+mulheres+com+c%C3%A2ncer+de+mama+em+um+grupo+de+apoio&btnG=. Accessed 12 July 2022
4. Sant'Anna, M.R.: Teoria de moda: sociedade, imagem e consumo. Estação das Letras, Barueri (2007)
5. Gil, A.C.: Métodos e técnicas de pesquisa social. Atlas, São Paulo (2019)
6. Zanella, L.C.H.: Metodologia de pesquisa, 2nd edn. Departamento de Ciências da Administração, UFSC, Florianópolis (2013)
7. Marconi, M.A., Lakatos, E.M.: Técnicas de pesquisa, 9th edn. Atlas, São Paulo. Available via Biblioteca de e-books (2021). https://integrada.minhabiblioteca.com.br/reader/books/9788597026610/epubcfi/6/10[%3Bvnd.vst.idref%3Dhtml4]!/4/38/2. Accessed 1 Oct 2022

8. Martins, G.A., Theóphilo, C.R.: Metodologia da investigação científica para ciências sociais aplicadas. Atlas, São Paulo (2007)
9. Rede Feminina de Combate ao Câncer de Brusque (2022) Quem somos. http://www.redefemininabrusque.com.br/site/. Accessed 2 Nov 2022
10. Goffman, E.: Estigma: notas sobre a manipulação da identidade deteriorada. LTC, Rio de Janeiro (1988)
11. Andrade, G.N., Panza, A.R., Vargens, O.M.C.: As redes de apoio no enfrentamento do câncer de mama: uma abordagem compreensiva. Ciênc. Cuid. e Saúde **10**(1), 82–88 (2011). https://doi.org/10.4025/cienccuidsaude.v10i1.10609
12. Crane, D.: A moda e seu papel social: classe, gênero e identidade das roupas. Editora Senac, São Paulo (2006)
13. Reis, A.P.A.: Alopecia: cotidiano da mulher com câncer de mama em tratamento quimioterápico. Master's thesis, Universidade Federal de Alfenas (2015). http://bdtd.unifal-mg.edu.br:8080/bitstream/tede/328/5/Dissertacao%20de%20Ana%20Paula%20Alonso%20Reis.pdf. Accessed 4 Nov 2022
14. Furtado, H.M.S., et al.: Repercussões do diagnóstico de câncer de mama feminino para diferentes faixas etárias. Ciênc. Saúde **10**(1), 1–8 (2016). https://doi.org/10.15448/1983-652x.2016.1.21813
15. Prates, A.C.L.: Influência da imagem corporal na autoestima de mulheres em tratamento do câncer de mama. Master's thesis, Universidade Federal de Goiás (2014). https://repositorio.bc.ufg.br/tede/handle/tede/3958. Accessed 4 Nov 2022
16. Baitelo, T.C., Reis, A.P.A., Gradim, C.V.C.: A atuação da enfermagem na alopecia da mulher com câncer de mama: revisão integrativa. Rev enferm UFPE on line **9**(11), 9898–9905 (2015). https://periodicos.ufpe.br/revistas/revistaenfermagem/article/download/10783/11933. Accessed 24 Sept 2022

Fashion, Women's Everyday Life and Consumption During the Covid-19 Pandemic

Solange Riva Mezabarba[1](✉) and Thais L. Pinto Vieira[2]

[1] Faculdade de Design de Moda Senai Cetiqt, Av. Luis Carlos Prestes 230, Rio de Janeiro 22775-055, Brazil
roupaesociedade@gmail.com

[2] Universidade Federal do Espírito Santo, Av. Fernando Ferrari, 514, Goiabeiras, Vitória, ES 29075-910, Brazil

Abstract. Humanity has faced many epidemics and some pandemics, all with some common elements among them. What made the last pandemic different from the others – COVID-19 in the years 2020 and 2021 – was the fact that we had internet spread across most of the affected territories. For a while, we changed our lifestyle and retreated to the domestic space. Some work modalities, previously exercised in person, migrated to the domestic space. If the house and the street are symbolic spaces, and women, by conquering the domains of the street, have become consumers of fashion and clothing, what change in dressing practices in everyday life can be observed in times of online interactions? We will see how interactions (although public), but mediated by videoconferencing platforms, generated a new way of thinking and consuming fashion and clothing, especially among Brazilian women.

Keywords: Clothing · online interactions · house and street · women

1 Introduction

The house and the street are symbolic spaces where we act in different ways. The house represents comfort, the safe environment, while the street, opposed to the house, is the domain of work, of struggle [1]. If the weekend has a family meaning, and therefore, it is structured from the space of the house; the middle of the week – from Monday to Friday – must be dedicated to work. At the time of the pandemic, the cleavage between the spheres of the home and the street deepened. Regarding the days of the week, however, we could observe an amalgamation, as if the days were repeated without the distinction between workdays and rest days. It is in this sense that we perceive how the interactions, predominantly in the domestic space during the quarantine, impacted on the pleasure and consumption practices of clothes and accessories. Thus, we reaffirm how much interactions in the public space (especially among women) were decisive for the increase in the consumption of goods aimed at self-presentation. But what happens when the office enters our home, and we submit ourselves to public scrutiny via videoconferencing?

In light of the work of Mica Nava [2], it is possible to see how public space occupation (especially by bourgeois women in the 19th century) impacted not only the volume of clothes purchased, but also the consolidation of the consumer society [3]. The detachment of work from the domestic space displaced 19th century men from the domains of the home, leaving the primacy of the home to female authority. This is how a binary thinking was consolidated that occupied the imagination of modern societies in which the street and paid work were established as a "male space", while the home (and unpaid domestic activities) is established as the woman's space. Nava deconstructs this imaginary, describing how bourgeois women, in Paris in the 19th century, occupied the public space, especially for leisure, socialization and consumption purposes. This phenomenon observed by the author can be related to the idea of presentation of self in the public space [4], highlighting, once again, the dichotomy between the way we deal with appearance at home and on the streets. Nava [2] recovers bourgeois women on the streets of Paris in the 19th century and concludes that this is a moment in which the presentation of self in the public space becomes the object of greater care with the appearance on the part of these women.

The following analysis deals with the consumption of fashion and clothing, specifically during the years in which we experienced the COVID-19 pandemic, and evidence the extent to which the everyday interactions that migrated from the streets to the domestic space impacted the way a group of women began to elaborate their choices and fruition of fashion-clothing. We are referring, notably, to women who, during the pandemic period, stopped performing their work duties in their offices and transferred their activities to the domestic space, interacting with their peers through videoconferencing platforms on the internet. It is important to emphasize that the world has already gone through several pandemics and that, analyzing the advent of COVID-19, we can list points of contact between them through historical reports [5–7]. However, the 2019–2022 epidemic brought a big difference in relation to the others: the internet. According to research published in September 2019 by the ITU [8], that year, about 51% of the world's population had access to the internet in some way. Since then, the relationship of a large part of this population with the internet has intensified, bringing new ways of interacting in everyday life and establishing new consumption habits. According to a report published by *Meio & Mensagem* magazine in December 2022, signed by Caio Fulgêncio, who disseminates the study "Consumer behavior post-COVID 2022" prepared by the Marco agency, about 98% of Brazilians increased their *online* purchases after the years lived in confinement due to the COVID-19 pandemic[1]. Fashion and clothing products occupy the 7th position among the goods that had an increase in this acquisition modality. Barriers that, before 2019, prevented or created some insecurity in *online* shopping practices, such as in relation to payments and eventual exchanges [9] were being knocked down as individuals found themselves in the situation of seeking alternatives to on-site interaction in stores and markets.

Given these assumptions, this work aims to highlight some fashion-clothing consumption practices observed during the pandemic. To support this analysis, we resorted to different surveys. Among the studies will be seen: monitoring three families residing in

[1] For more details see: https://www.meioemensagem.com.br/marketing/e-commerce-brasileiros-aumentaram-compras-online-na-pandemia. Accessed on: 27/04/2023.

Rio de Janeiro during the years 2020 and 2021, from an ethnographic perspective, interacting with the help of video conferencing platforms such as Zoom, Teams and Google Meet, as well as with audio and recorded conversations by Whatsapp; an autoethnographic work; an research-action carried out with students; and an image survey on the use of masks in public space[2].

We organized this text as follows: first, we will make a quick account of the time in question and the specificities experienced by a large part of the world's population, emphasizing the implications brought by the pandemic, and in particular, for women who worked in high qualification positions, both in which refers to their daily interactions, but also the new consumption practices engendered in this specific period; next, we will analyze the findings of the research carried out, problematizing the role of the internet as a mediator in the interactions that we call "hybrid" – in other words, they are carried out from the domestic space, whilst achieve visibility almost equivalent to the on-site routine in their office and work environments. In this sense it is important to rely on [10] concept of the public sphere, or what is public, that is, the space where what is exposed to those present is the object of their judgment. Finally, we will weave some final considerations about the experience lived by all of us.

2 COVID-19: The First Pandemic of the 21st Century

The first known epidemic occurred in Athens between 430 and 427 B.C. and became known as the "Great Epidemic of Athens" [5]. After that, we can list some equally or more devastating: the Justinianic Plague, between 527 A.D. and 565 A.D.; the Black Death in Europe in 1347; the Spanish Flu in 1918. More recently, in the 1980s, the world knew about the HIV virus. All these epidemics have common characteristics and have impacted collective behavior. What makes the COVID-19 epidemic, in terms of social life, different from other epidemics is the fact that the internet is widespread and is already part of the daily lives of most of the affected individuals. Technology journalist Kelly [11] recalls that when personal computers were designed, there was little interest in producing these items. The collective (and corporate) interest appears when the internet offers new possibilities for these devices, generating an increase in the development of other devices until we reach the portability of cell phones.

As a measure to mitigate the spread of COVID-19, some governments have decreed population lockdown, a practice that suggests "confinement". In Brazil, although there was no consensus, several categories of workers benefited from the modality called "home office", that is to say, the house also became a part of the office where these people work. For Sennett [16], the fact that some classes of workers may be confined (and others may not) generates an important social divide – that is, there are those who can be confined, and those who cannot, and need to expose themselves to the virus. On the other hand, the home-street dichotomy gains dramatic contours with the pandemic,

[2] Ethnography presupposes frequent and/or immersive contacts with individuals from a certain social group, leaving the concept of "sample" to quantitative studies, which is statistical. For more information on ethnography see: FONSECA [12]; CLIFFORD and MARCUS [13]; SUDERLAND and DENNY[14]; BECKER [15] (the latter addressing research in social sciences).

highlighting the street as the place of the impure, of danger, and the house as an aseptic, safe space, to resume the discussion by Douglas [17] in her work Purity and Danger. In this sense, the period experienced during the pandemic is marked by new forms of interaction, especially those mediated by videoconferencing platforms accessed from the home environment.

Freyre [18], when mentioning the observations of the English traveler Maria Graham in the 19th century in Brazil, recalls how the observer odd the fact that Brazilian women from the so-called "good society", in the domestic space, dressed carelessly. When they are encouraged to visit the public space, however, this changes.

The massive presence of women in the public space and in the labor market, especially from the end of the 19th century [19], generated new ways of dressing, notably the incorporation of items considered "masculine" in women's wardrobes. Then the 1970s marked a time when Brazilian women began to occupy a higher percentage of vacancies in universities than men [20] and, consequently, they conquered space in working life in jobs that demand a high level of education. Women's clothing for executives or for skilled workers has been the subject of dispute for a long time. For Crane [19], incorporating elements considered "masculine" in "feminine" work clothes reveals a silent manifesto in favor of marking the position of women in spaces of power. On the other hand, Hollander [21] ponders that the men's suit has become an appropriate garment for men in their role as workers in the corporate world, the respectful, diplomatic garment of the man of state. Boltansky [22] recalls how even the very definition of "executives" is related to the way in which those who make up this working class present themselves, based on men's clothing. Therefore, while among male workers, this self-presentation in the corporate world was consolidated in the image of the suit, among women, there are still doubts. In any case, more recently, it is possible to observe that working women in command positions or in highly qualified positions follow codes and postures often given by their companies, or even by a pedagogical media that recommends how these women should dress for consecrate themselves in their work positions [23]. All this in a world where work interactions were predominantly on-site. And what happens when these interactions take place in the domestic space, but expose these women to public scrutiny in the moment they access platforms and turn on their cameras to act in meetings and other work encounters? It is in this sense that we see these interactions as "hybrid", when although protected by domestic life, we talk about public exposure from interactions mediated by the internet – in particular, those involving images.

The home office scheme has blurred the boundaries of schedules, entering periods that would previously have been dedicated to leisure or rest. This evidenced an experience of excessive work – even more so for the women who started to accumulate tasks, especially those with children. We noticed, however, that in each case, each family experienced the quarantine differently. The distinct ways of dealing with the pandemic and the quarantine signaled different behaviors in terms of buying and using fashion-clothing.

The material culture of the pandemic involved alcohol gel, pocket bottles for carrying alcohol gel, hypochlorite solutions for cleaning provisional purchase items, gloves and, most importantly, masks. These were the subject of controversy, but also of creative use by the people observed.

In the next section, we will address the research findings that supports this article.

3 Self-presentation in "Hybrid" Interactions During the Pandemic

When asked to list the clothes and accessories used in the on-site work mode and in the online modality, the interlocutors revealed some practices and choices of everyday dressing that strongly marked the differences between on-site interactions in the public space, and the interactions that we now call "hybrids" established from the domestic space, but with visibility to the work environment. Thus, this section is divided according to insights from the fieldwork: first, we will talk about the two-year research with three families in Rio de Janeiro; then we will post our own experiences of dressing for work online during the pandemic period; then, we will talk about a research-action with fashion college students; and finally, we will discuss the use of masks in public spaces, an item that became a common accessory and on which, at that time, the possibility of its permanent use was being discussed.

3.1 "Now I Only Live in Pajamas": Three Families from Rio de Janeiro During the Pandemic

Housing structures can be listed as one of the factors that facilitated the proliferation of the virus that would cause the first recorded epidemic, in Athens between 430 and 427 B.C. [5]. Today we know that the way people move around in large urban centers [16] and the precarious living conditions of communities, as in the case of Rio de Janeiro (with basic sanitation problems and reduced space housing for entire families) persist and exacerbate the risks. The families accompanied by us, therefore, can be considered privileged, since, belonging to the middle urban segments, they live in salubrious residences and had the possibility of living in social isolation during a good part of the years 2020 and 2021. Below we list the characteristics in Table 1, of the families that were observed[3].

Lucia reveals the frustration she felt when she finished renovating her office and had to go into quarantine. Asthmatic, the husband, since the beginning of the confinement measures (second half of March/2020), has been locked up in his house, an apartment of about seventy square meters. In order to "change the environment", she decided to go to the work place in the recently renovated office, even though she cared for her patients online. The relationship with the house changed, and her purchases, according to her, mostly through e-commerce, began to include more frequently bed linen, rugs and curtains. Still according to her, "it tires the eyes" to see the "same house every day". In this sense, she reports that, despite maintaining her "tidying up ritual" [3] as always, since she drives to her office in the same neighborhood where she lives, Lucia lowered her purchase of clothes and accessories, especially for using more resources to make the domestic space more welcoming.

Claudia, for the first time, bought her family's Christmas presents via e-commerce. But, according to her, there is still a certain distrust, which is why she chose to shop at stores she already knows, such as Hering, C&A and Lupo. This family describes how

[3] For ethical reasons, the names have been changed.

Table 1. Characteristics of the observed families

Couples	Age	Neighborhood	School level	Employment
Lucia and Augusto, no children at home. Augusto has a son from his first marriage who moves between one house and another	52/55	Tijuca	College degree (both)	Psychologist and journalist
Andrea and Mario Sergio, with a 10-year-old son at home	48/49	Copacabana	Master (both)	Economist and Engineer
Claudia and Lucio, with 4 children: a teenage girl, Martina, aged 18, and two boys of higher age at home. The eldest son moved between his parents and his girlfriend's house	55/57	Laranjeiras	College degree (her) and Masters (him)	Journalist and filmmaker

their social life was hectic before the pandemic, as Claudia's brother is a musician and performed frequently. The pandemic led everyone to seclusion and, with that, to what she called "living in our pajamas". An article by Vogue Magazine (O Globo) [24] published on July 19, 2021 informs that, according to a Google search (Beauty & Fashion Trends), some clothing items had increased sales. Among them, pajamas and sweatshirts. Contact with Claudia's family, however, signaled that pajama, for the family, were a concept. "Wearing pajamas" is an expression that, for Claudia, designates "staying at home", but not necessarily wearing pajamas. With the pandemic, however, pajamas have become an option to wear in the home environment and sufficiently composed for "hybrid" interactions. Being a researcher in the field of journalism for a large TV station, Claudia had her work interactions entirely moved to the domestic environment. In her words, "a T-shirt is enough, even if it has already been worn", something very different from the codes she needed to abide by when working on-site. Her 18-year-old daughter, Martina, took advantage of the confinement to get a haircut that she didn't feel comfortable doing before. The girl practically shaved her locks and colored the very short strands with blue.

Andrea, having gone to the city of Buzios[4], took only light summer clothes in her suitcase, considered suitable for the informality of a beach town. That was in March 2020, still with a very high temperature in the state of Rio de Janeiro. As the quarantine extended, Andrea had to go back to her house in Copacabana to take more clothes – this time, it included hoodies and coats. Like Claudia, Andréa stated that, for her interactions

[4] Buzios is a coastal city in the state of Rio de Janeiro, located about 3 h by car from the capital, it is a typical vacation and summer spot.

mediated by Zoom, "a shirt, which didn't even have to be that good", was enough, that is, of good quality.

Insights from women from the three families who were followed during the pandemic suggest that, to some extent, purchases have retracted, as use was directly affected. The exception is 18-year-old Martina, Claudia's daughter. Martina declared that she did not stop buying new clothes – starting to buy them in virtual stores, and via social networks. According to her, during the pandemic period, many brands carried out sales. To seize the moment, she bought even more than she would have otherwise. In other way, her mother, who started buying clothes via e-commerce during the pandemic, decided not to take a risk with little-known brands. To buy Christmas gifts, she ventured into online stores, but opted for big and familiar brands like C&A and Hering[5]. According to her, since they were brands she was familiar with, the risk of receiving clothes that did not please her was considered low. Still, this behavior represented the breaking down of a barrier that previously prevented her from buying clothes online: the lack of confidence in delivery and doubts about quality and size. As for Lucia, although she continued to dress as usual, she began to spend more money on items for her home, because, although she continued working in her office, she maintained isolation at home, avoiding, for example, meetings in the sphere of leisure.

3.2 The Material Culture of Online Classes

In an effort to operate the observation of our own behavior, in an autoethnographic perspective, we developed the tables below (Figs. 1 and 2) to reflect on what it meant, in terms of choices for the presentation of oneself, the migration from on-site work to the modality home office. Ellis, Adams and Bochner [25] define autoethnographic work as a possibility of research and writing that seeks to systematically describe and analyze the personal experiences (of the researcher) in a way that contributes to the understanding of the cultural experience within a specific society, at a specific time.

At first, what catches our attention in the images is the amount of clothing and accessories, more numerous for on-site work compared to the online modality. Shoes and handbags were practically suppressed during the period.

Another interesting aspect is the rescue of clothes with damages – holes, worn or frayed fabric, formation of "small fabric balls" – for use in front of the canvases. In other words, clothes that were no longer suitable for the public space, and which had already been "downgraded" to be used at home space, earn survival by being designated for online interactions.

Autoethnography provokes reflections on apparently natural attitudes that turn out to be specific to a context or moment. One of the researchers (Fig. 2) always made use of comfortable clothes, with malleable fabrics, a priority in her clothing choices, not forgetting aesthetics in self-presentation, but leaving it in the background. She realized that when the work ambient was brought to home, she could make even more unconcerned self-presentation choices, as part of the body would be hidden. Could work barefoot, as usually at home, and even taught one of her online classes wearing a swimsuit, as the time between the end of the class and the swimming lesson would be short. Aesthetic

[5] Brazilian T-shirt brand.

Fig. 1. Clothing inventory - comparing on-site and online work for author's 1

Fig. 2. Clothing inventory - comparing on-site and online work for author's 2

care was concentrated on accessories and on improving the image composition in the wall behind her at the work room. Perceived that she had been more concerned with creating a scenario for her classes, when, on the occasion, she framed posters related to the course she teaches, which composed the wall during the videoconferences.

3.3 Research-Action: Consumer Research Coordinating Student's Groups

With the aim of developing insights into research on the choices and uses of fashion clothing in a time of pandemic, three groups of students at Senai Cetiqt Fashion Design

School were encouraged to undertake research-action on different topics on women who were mothers and worked in occupations where they could work remotely[6]. The research-action has been an activity widely used in the academic context because it is a resource for reflection on research itself [26].

One of the group's observations pointed to the multiple uses that clothes began to acquire among mothers working at home. Nightwear began to lend itself to remote work and gym clothes became everyday clothing in the home space. Another important observation by the group was related to the use of makeup – if it used to be constant in the daily lives of the women observed, during the pandemic it became a last resort. Meetings and conferences in which they played a leading role became the only motivation for wearing make-up, after all, the face is more in evidence. The women approached by the group also described replacing closed shoes, tailored trousers, blazers and dress shirts with everyday use of sweatshirts, knitwear, tricot, trapeze-cut dresses, flip-flops or mules.

As a conclusion on the observations, the group realized that there was a significant change from what they designated workwear to loungewear among these women, even during work interactions (which were mediated by the internet).

3.4 If You're Going to Wear a Mask, Let it be "Me"!

In the first year of the pandemic, until mid-2021, the sanitary product industries were still unable to produce enough surgical masks for use by the population in addition to hospitals. Thus, people used their own resources to make fabric masks for everyday use when they needed to go out. Despite being confined, many people needed to go out occasionally for minor purchases, or even to go to the hospital for an appointment, not to mention those who could never stop going out. But some came to believe that the use of masks would be mandatory forever.

In this way, masks appeared in sizes, colors, shapes and prints as varied as the phenotype of passers-by in large cities. This phenomenon was observed, recorded and analyzed by us[7] between March 2020 and March 2021 [27], and included interviews with mask users.

At that time, we observed few shades of neutral colors on the streets. There was a profusion of themes printed on the fabrics that showed a need to overcome the facial homogeneity generated by a piece of fabric covering large part of personal features.

Among more than three hundred images of masks collected, we mapped some large groups, which were named according to the explicit content, as can be seen in Fig. 3: the **Ideological** ones, stamping political positioning or dissemination of some belief or cause

6 The group of students who developed this work was: Anna Catharina Silva, Kelly de Oliveira, Arilson Gomes Júnior and Stefanie Simili, whom we thank for their collaboration. They accompanied 15 women who were mothers for a semester and who, with the pandemic, started working online.

7 Research carried out by Thais Vieira, Ana Paula Carvalho, Maristela Pessoa and Gisela Monteiro in the cities of Rio de Janeiro-RJ and Vitória-ES. The research, in its entirety, addresses the use of masks, by men and women, but, for the purpose of this work, we selected only the female insights.

that mobilizes the user; the **Institutional** ones, with fashion logotypes emphasizing their brands, or companies that print their brands by adding the accessory to their employees' uniforms; **Playful** ones, with images or funny phrases; **Mimetics**, which simulate faces generating an optical illusion effect, with the intention of simulating personal face traces, or moking facial features; the **Religious**, with illustrations of saints, gods, phrases or symbolic images of a religious nature; and those for **Football Fans**, containing emblems, colors or names notoriously related to traditional football teams.

Fig. 3. Collection of masks between March 2020 and 2021

In the interviews, we noticed the concern with the combination between the mask and their clothes, even taking care of the adequacy according to the occasion, or with the place where they would circulate. These factors, among the interlocutors, were considered to define the color, shape and even the material of the chosen mask, since each one admitted to having their own collection of the object.

The face mask began to be used as a fashion accessory, going far beyond its function of protecting against the disease. People chose the style according to the clothing. As Silvano [28] well recalls, "generally, it can be said that the affirmation of personal identities universally passes through the face" (2021: 181).

4 Final Considerations

During the critical years of the pandemic, 2020 and 2021, everyday interactions in public spaces, notably in the street domains, were compromised. Masks have become mandatory accessories. The domestic space, on the other hand, gained new forms of usufruct, namely those related to online work. If the house is configured as a space of intimacy, it was necessary for us to learn to control our behavior in front of cameras, which could put us in embarrassing situations. This, obviously, was reflected in the way women started to deal with their clothes in their work interactions mediated by the internet. In this study, supported by investigations of a qualitative nature, we were able to verify different ways of dealing with the moment, and with the consumption of fashion-clothing. Returning to the work of Mica Nava [2], which reveals the female occupation of public spaces in the 19th century, we could see that at that moment, consumption related to entertainment, but above all, to self-presentation, gained expression among bourgeois women in Paris. This premise leads us to believe that the dichotomy of public space and private space significantly impacts the way we choose to dress or even how we enjoy our clothes and accessories. At a time when the home also becomes a workspace, accessed online and exposing our most intimate way of life – the decoration of our home, the customs of each family, the clothes we wear – we begin to choose ways of dressing compatible with the scrutiny to which we will be exposed in work situations. Thus, some strategies were listed. On the streets, however, the thematic diversity of the masks signals a willingness to single out the face to some extent.

During the pandemic experience, there was a recurrent talk of a "new normal", giving the impression that some customs would consolidate in everyday life. Now, in 2023, we see that some customs have indeed consolidated – such as shopping for clothes via e-commerce, for example. Others, such as the use of masks in public spaces, have not. Regarding online interactions, we can say that part of the workers who were able to adhere to this modality returned to on-site mode, interacting in the public space as they always had. One of our informants, who worked completely online during the confinement period, never fully returned to her on-site work. She continued to work at home two days a week, which has been reflected in her way of dressing, and consequently, in her post-pandemic clothing fashion consumption practices. At home, work doesn't always demand online interactions. In this case, shorts and T-shirts become work clothes. And why not?

References

1. DaMatta, R.: A casa e a rua: espaço, cidadania, mulher e morte no Brasil. Rocco, Rio de Janeiro (1997)
2. Nava, M.: Modernity's disavowal: women, the city and the department store. In: Falk, P., Campbell, C. (eds.) Sage, London (1997)
3. MacCracken, G.: Cultura e consumo: novas abordagens ao caráter simbólico dos bens e das atividades de consumo. Mauad, Rio de Janeiro (2003)
4. Goffman, E.: Comportamento em lugares públicos: notas sobre a organização social dos ajuntamentos. Vozes, Petrópolis-RJ (2010)
5. Ujvari, S.C.: A história e suas epidemias: a convivência do Homem com os microorganismos. SENAC, São Paulo (2003)
6. Goulart, A.: Revisitando a gripe espanhola: a gripe pandêmica de 1918 no Rio de Janeiro. Revista História, Ciências, Saúde – Manguinhos, Rio de Janeiro **12**(1), 101–142 (2005)
7. Spinney, L.: Pale Rider: The Spanish Flu of 1918 and How it Changed the World. Hachette Books, New York (2017)
8. State of Broadband Report. Geneva: International Telecommunication Union and United Nations Educational, Scientific and Cultural Organization, 2019. Licence: CC BY-NC-SA 3.0 IGO (2019). https://broadbandcommission.org/Documents/StateofBroadband19.pdf. Accessed 04 Apr 2023

9. Cavalcanti, S.S.B., Ferreira, L.B.: Barreiras ao uso do comércio eletrônico. Universitas, Gestão e TI **3**(2), 21–29 (2013)
10. Habermas, J.: Mudança estrutural da esfera pública: investigações quanto a uma categoria da sociedade burguesa. Tempo Brasileiro, Rio de Janeiro (2023)
11. Kelly, K.: Inevitável: as 12 forças tecnológicas que mudarão o nosso mundo. HSM, São Paulo (2017)
12. Fonseca, C.: Quando cada caso não é um caso. Revista Brasileira de Educação **10**, 58–78 (1999)
13. Clifford, J., Marcus, G.: Writing culture. The poetics and politics of ethnography. University of California Press, California (1986)
14. Sunderland, P., Denny, R.M.: Doing anthropology in consumer research. Left Coast Press, California (2007)
15. Becker, H.: Métodos de pesquisa em ciências sociais. Hucitec, São Paulo (1994)
16. Sennett, R.: Les villes dans la pandémie. In: Analyse, Opinion, Critique (2020). https://aoc.media/analyse/2020/04/16/les-villes-dans-la-pandemie. Accessed 20 May 2020
17. Douglas, M.: Pureza e perigo: ensaio sobre a noção de poluição e tabu. Edições 70, Lisboa (1991)
18. Freyre, G.: Sobrados e Mucambos: decadência do patriarcado e desenvolvimento do urbano. Global, São Paulo (2003)
19. Crane, D.: A moda e seu papel social. Classe, gênero e identidade das roupas. Senac, São Paulo (2006)
20. Bruschini, C., Puppin, A.B.: Trabalho de mulheres executivas no Brasil no final do século XX. Cadernos de Pesquisa **34**(121). Fundação Carlos Chagas, São Paulo (2004)
21. Hollander, A.: O sexo e as roupas: a evolução do traje moderno. Rocco, Rio de Janeiro (1996)
22. Boltansky, L.: Les Cadres: la formation d'um groupe social. Les éditions de minuit, Paris (1982)
23. Mezabarba, S.: Vestuário e cidades: Ethos, consume e apresentação de si no Rio de Janeiro e em São Paulo. Thesis (Antropology PhD) – PPGA – Programa de Pós-Graduação em Antropologia) Universidade Federal Fluminense, Niterói-RJ (2012)
24. 91,4% de aumento: confira qual produto de moda disparou em buscas durante a pandemia (2021). Revista Vogue, published in O Globo newspaper in 19 July 2021. https://vogue.globo.com/Vogue-Negocios/noticia/2021/07/914-de-aumento-confira-qual-produto-de-moda-disparou-em-buscas-durante-pandemia.html. Accessed 29 Apr 2023
25. Ellis, C., Adams, T.E., Bochner, A.P.: Autoethnography: an overview. Hist. Soc. Res. **36**(4), 273–290 (2011)
26. Tripp, D.: Pesquisa-ação: uma introdução metodológica. Educ. Pesqui. **31**(3), 443–466 (2005). https://doi.org/10.1590/S1517-97022005000300009
27. Vieira, T.L.P., et al.: As máscaras na vida cotidiana: como estão sendo inseridas nos projetos estéticos pessoais. In: Encontro Nacional de Estudos da Imagem V Encontro Internacional de Estudos da Imagem, 2021, Online. Caderno de Resumos Eneimagem. Universidade Estadual de Londrina, Londrina, pp. 148–149 (2022). https://eneimagem2021.com/caderno-de-resumos.html. Accessed 17 June 2023
28. Silvano, F.: Antropologia da Moda. Lisboa, Documenta (2021)

The Importance of Adornments in Insular Traditional Costumes: The Case of Jewellery

Sylvie Castro[1(✉)], Joana Cunha[2], and Leonor Sampaio da Silva[1]

[1] Department of Languages, Literatures and Cultures, University of the Azores, Ponta Delgada, Portugal
sylvie.alves.castro@gmail.com, maria.ls.silva@uac.pt
[2] University of Minho, Guimarães, Portugal
jcunha@det.uminho.pt

Abstract. The paper proposes an analysis of the symbolic importance of adornments in insular traditional costumes. Garments, being contextualised on a specific time and space, constitute a relevant cultural element for the definition of individual and collective identity. Our study draws on this premise for a reflection on the traditional female costumes of the regions of the Azores, Madeira and Corsica. We intend to consider these costumes in relation to the adornments (or the lack of them) and to create new visual objects aligned with the identitarian legacy of these archipelagos, as well as with their environmental and endogenous resources. For this purpose, we start by exploring the nexus between culture, traditional costume, and contemporary jewellery. The jewels were conceived with the aim of revitalising the traditional costumes as well as to disseminate and promote the cultural legacy to which they belong. The adornments will connect the garments, the peoples and the regions here considered, besides expanding the knowledge of their identity abroad, which will increase their reputation and relevance.

Keywords: Identity · insularity · traditional · culture · adornments · jewellery

1 Introduction

> … I think that there is only one way to science – or to philosophy, for that matter: to meet a problem, to see its beauty and fall in love with it; to get married to it and to live with it happily, till death do ye part – unless you should meet another and even more fascinating problem or unless, indeed, you should obtain a solution. But even if you do obtain a solution, you may then discover, to your delight, the existence of a whole family of enchanting, though perhaps difficult, problem children, for whose welfare you may work, with a purpose, to the end of your days (Popper 1956, p. 42).

This quote by Popper reflects the motivation for the present study, whose research was conducted as part of an ongoing process of creating female adornments. In the present case, our aim is to especially design jewels for the traditional costumes of the archipelagos previously mentioned. In one instance (Azores), there is no ornament present in the traditional costume; the other tow instances (Madeira and Corsica) manifest an irregular use of adornments. Therefore, we propose the inclusion of two pieces of jewellery, for Corsica and Madeira, namely: one for the necklaces of Corsica's regional costume – pendent with a cross in the centre and ornamented with corals –; and the other for the female cap (or *carapuça*) on Madeira's costume – a pendent with spheres of polished stone at the bottom of the traditional triangle. There never was in the Portuguese insular costumes the tradition of adding ornamental jewellery, a circumstance that presents a striking contrast with the case here taken as the Portuguese reference of adornment that provides emphasis to collective identity – Minho. As John Carl Flügel (1966) elucidates, even in the most primitive civilizations we find peoples without clothes but not without adornments.

The focus of this paper is, therefore, the addition of jewellery to the insular Portuguese traditional costumes, given the absence of ornaments in them. To pursue it, we devised two priorities: to follow contemporary aesthetics and to respect the identitarian legacy of the regions. These are two intertwined concerns, since there are many cases of gradual, occasional and miscellaneous aggregation of ornaments that present-day tolerance will eventually incorporate, even if they don't respect the origins from which costumes stemmed. Thus, an important motivation for the present study is to unveil such pernicious changes and to validate the inclusion of the jewellery hereby proposed, based on the intent to throw off the influence of non-identitarian elements and to supply instead artworks that, respecting the history of the costumes, may embellish them without mischaracterizing.

2 Regional Adornments: From Identity to the Piece of Jewellery

This study will present the traditional costumes both in general and in their specific particularities. We'll favour a comprehensive approach as opposed to an explicative one (Cunha 2006, p. 45), and we'll prioritise an interpretative methodology as well as the relation between fields of knowledge, as we share the belief that the analysis of culture "is not an experimental science in search of law" but, rather, "an interpretive one in search of meaning" (Geertz *apud* Cunha *idem, ibidem*). We therefore endorse Kilani's view (Cunha *idem,* p. 46) about the author's place within the production of meaning in the institution: it is inevitable to assume "authorship" as generating meaning (*idem, ibidem*).

The regional costumes will be contrasted with the one from Minho – a case of great complexity and richness of adornments; we take it as the basis for the present study as it provides a scientifically relevant contrast with the selected insular costumes, being also representative of a non-insular and economically robust region. Corsica, Madeira and Azores, on the other hand, are three insular regions, two of the Portuguese and one French. Given their diversity, the analysis gains a larger foundation for comparing traditions and fostering cross-cultural knowledge.

Our approach to the regional costume prioritises awareness of local identity elements as well as a keen consciousness of how important it is to preserve local traditions. Garments are a means of cultural expression com high symbolic relevance. Through clothing, a region expresses affection, religiosity, festivity rituals and daily practices. Its importance for the preservation of rural and regional memory goes beyond the scope of the place it belongs to, since it is often exhibited in urban contexts, such as during activities of folklore groups in which past and present merge. On such occasions, we become aware of its origins, materials (wool, cotton, silk, straw, wood, leather), as well as of the methods employed in working them: handmade techniques combined with industrial machinery that allows the perpetuation of know-how, taste and knowledge across generations.

In the course of time, several other materials have been added to the national traditional ones, due to international exchanges:

> Thanks to the abundance of gold and gems originating from a vast overseas territory which extended from the East to Brazil, across Africa, Portugal benefitted from a situation particularly suited to the development of goldsmithery and silversmithing from the 16th to the end of the 18th centuries (AAVV 2020).

Gold has been a recurrent presence in Minho's costume up to the present days. The weight and glitter of the gold adornments are fully displayed in parades such as the *Cortejo da Mordomia* or the most famous national Ethnographic Parade dedicated to Our Lady of Agony, in Viana do Castelo. We can say that the most expressive and invaluable adornment of the woman wearing the Vianese custom is "the gold of the filigree earrings, of the necklaces, of the pendants, and of the rings glimmering in shiny amber, like flakes of sun" (Pimentel 1906, p. 19). Another author, D. António Costa, travelling across Minho in 1874, wrote that the chest of the *minhota* "is a starry sky". Listing the vast array of jewels, he stresses:

> Thick gold necklaces of all shapes, hearts of gold of a size that exceeds a span, flowers that evoke the great Spanish badges of Carlos III, filigree earrings that reach the shoulders, huge crucifixes, huge figures of Virgins of the Conception, complete pieces of calvaries artworks, containing, in addition to the three large crosses of Jesus and the thieves, the group of the Marys and the tragedy scene (Costa 1874, pp. 258–259).

As we compare the regional insular costumes with that of Minho, we notice a striking difference regarding the presence of adornments. It is, however, necessary to mention that despite Madeira regional costume is devoid of ornament, that of the '*saloias*' – inspired by it – is richly adorned.

There is no precise date marking the first appearance of the '*saloias*', but it is noticeable that when the Liberal Revolution occurred (1820), there was a preference for garments characterised by simplicity, a trend that continued during the 19th century (Vieira 2006). During this period "women's clothing consisted of shirt, striped skirt, corset, cape, small cap and flat boots" (Vieira, s.p.). However, Rebelo (2019) refers the existence of two types of costume associated to the cultural traditions of the island and of the persons known as '*saloias*' – young girls between the ages of 9 and 12, who "join Easter activities known by «Espírito Santo»" (Rebelo 2019 p. 211). These young girls (Fig. 1) show on their head a cap decorated with gold chains and, similarly to Minho's costume, several borrowed necklaces on the neck. According to the sociologist Moisés Espírito Santo, the loans belong to a religious tradition of the ancient Hebrews, who.

> believed that the jewels could capture the essence of "God's presence" in a sacred place. Wearing jewels resulted in bearing virtues, as jewellery had among them the same function as amulets have nowadays among us. For this reason, everybody volunteered to lend their jewellery, because the more pieces of jewellery they lent, the richer in magic virtues they'd get (Espírito Santo 1988, p. 98).

The loans to the "*saloias*" symbolize, therefore, the presence of the divine and blur, even if transitorily, the frontier between the "rich" and the "poor", as they enable the latter to partake, if not in fact but in appearance, of the wealth enjoyed by the former. This is an example of how material culture relates to identity – in a region culturally impacted by religion, the costume reflects devotional aspects:

> [...] there is a practice related to Religion, which establishes an internal and external movement in the insular space leading to devotion concerned with the 'salvation of the soul' and which determines actions that show devotion, like pilgrimages [...] (Vieira 2018, p. 5).

Fig. 1. The "*Saloias*" of the "*Espírito Santo*" Easter activities - Ribeira Brava[1]

In this context, the people creates a space of exception, through a religious festivity (Cunha 2006, p. 57) which allows to some of its members – the "*saloias*" – to symbolically enter the realm of the divine and of the abundance.

1 https://cultura.madeira.gov.pt/visitas-virtuais1/148-visitas-pascais-as-saloias-do-esp%C3%ADrito-santo.html.

To exhibit wealth provokes a contrast regarding Madeira's traditional costume, and alters the general impression of seriousness, as the incorporation of gold is visible in the ornamented caps used by women wearing the regional costumes (Fig. 2). The richly ornamented element in the "*saloias*" appearence is mainly present in the *carapuça*, the complement to the costume to which we created a jewel.

Fig. 2. Folklore group of Madeira Island. The ornamented cap is visible in one of the women wearing the traditional costume. Photo by: António Martins[2]

While representative of a regional and local culture, the costume reflects the dynamic character of the identity, which "is in constant transformation" (Varela 2021, p. 31). Thus, it is normal for regional costume to adapt to new identity elements that time brings to the community. Madeiran regional costume is no exception to this rule. In it, polychrome and monochrome blended over time, "between the explosion of joy and the sense of interiority" (Teixeira 2009, p. 376).

The design of the jewel that we present here will allow us to imagine other jewels to integrate the costumes of Madeira, in the cap, and that of Corsica, in the coral necklaces. We intend that the jewels incorporate the costumes as identity elements and aggregators of meanings, enriching them without de-characterizing them. It should be noted that the case of Corsica is distinguished by the fact that its costume is already ornamented, as we will present later.

The inclusion of ornaments will be based on an asymmetrical, inclusive and aggregator of identity meaning perspective. The first piece to be designed had the Azorean costume in mind. A collar was designed (Fig. 3), suitable for the anatomy of the female shirt, which did not overlap neither with the frills nor with the embroidery. The piece, presented in the article "The Azorean Traditional Costume as a Sign of Regional Identity

[2] https://www.rotasturisticas.com/fotos_37414_funchal_portugal_trajes_tipicos_da_ilha_da_madeira.html.

and Culture: From Clothing to Jewellery" is composed of nine elements, which represent each of the islands of the archipelago, with the island of São Miguel highlighted in the centre. We propose to disaggregate this central element and adapt it to the symbolic elements of the history of Corsica, converting it into a pendant to place on the traditional coral necklaces.

Fig. 3. Drawing of woman dressing the rich costume from the island of *São Miguel*, Azores, with the proposed jewel. Right: vector design of the jewel. Source: the author.

The chosen typology – the necklace – dialogues with the Portuguese tradition and with the reference case – the Minho costume. In Santa Marta de Portuzelo, Viana do Castelo, the "*Mordomas*" – formerly girls who would soon get married and today the main characters of the Ethnographic Parade – "are characterized by the use of imponent plastrons" (Mota 2015, p. 179). The position of the ornamentation is known as "*Santa Marta*":

> «very neat», being the local plastrons the most elaborate in terms of symmetry, quantity of pieces and careful distribution of them, aiming at an attentive and particular aesthetic effect. Among the forms they adopt, we highlight the use of a reasonable number of bead necklaces arranged transversally on the golden mass made up of countless threads and ornaments, an ornamentation system that we characterize as transversal (Mota 2015, p. 179).

According to the same author, the "decoration is the heir to the importance that beaded necklaces had in the late 19th century and first half of the 20th century" (ibidem). At this stage of history, costumes were characterized by an abundance of gold (Fig. 4), with special emphasis on the laps of *lavradeiras* [peasant woman], even if they are always placed next to the neck, [...], revealing this fact an evolution and renewal in the use of this chain so characteristic of Alto Minho. The density of the golden mesh constitutes the dominant characteristic of the plastron of this locality and all types of popular gold are associated to form a kind of gold plate. Compact and long, the *Portuzelo* plastrons progressively decreased in length, following the general trend shown in the region's pageants, which rejects cords below the waist (ibidem).

Fig. 4. "Mordoma" from *Santa Marta de Portuzelo* with patron build with 207 pieces. Photo by José Barroso. 2000[3]

Like the costume we assume as a matrix, the Corsican costume is also quite ornate. Even today, precious stones are, on the island of Corsica, an important addition to the costume, in particular the red coral stone, considered the treasure of the island. This stone is already present in the adornments of the costume: the necklaces, the cross and the earrings. The jewels are made from real red coral, natural from the Mediterranean Sea. We attend to the detail in the centre of the pendant, a branch of coral adorning the cross. Once again, a robust relationship is noted between the costume and an identity strongly linked to religious sentiment: "...in the 20th century, the mourning dress was considered the national costume of Corsican women" (Giannesini 2012, p. 48), which may not exactly correspond to the historical truth, but may have been incorporated into the insular identity from a literary work – *Colomba* (1840), by Prosper Mérimée –, which interpreted the female regional costume as being a mourning dress. The grounded

[3] https://ciencia.ucp.pt/ws/portalfiles/portal/31105575/21.pdf.

research of ethnologist Rennie Pecqueux-Barboni (2008), which focuses on the study of traditional and typical costume models, does not subscribe to the idea that mourning was central to the regional costume. The models, according to the author, quite colourful, refute this thesis.

With regard to the six coral beads present in the pendant created by us (Fig. 5), they symbolize the capital, Ajaccio, and the second most populous city on the island, Bastia, which is the capital of Upper Corsica. The other four important localities are Porto-Vecchio, Borgo, Corte and Calvi.

As in the Azorean case, here we also considered insular identity characteristics related to the geography of the place. If, in the case of the Azores, the colours of the islands were respected, here red has the dual function of reinforcing tradition and expressing the weight of corals.

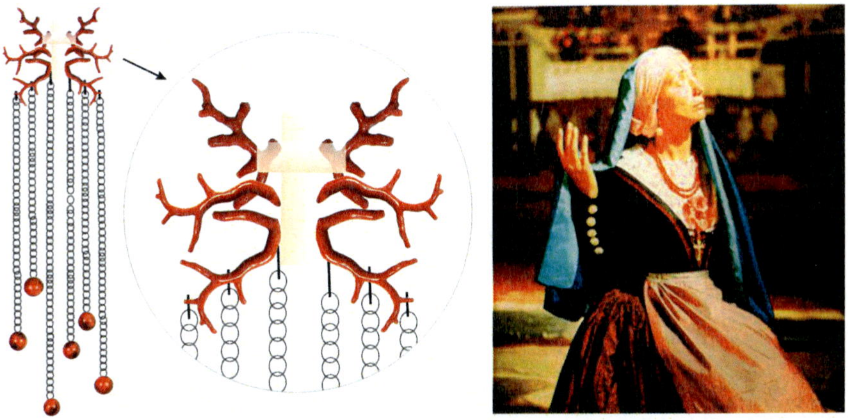

Fig. 5. On the left: design of the vector jewel with emphasis on the coral branch detail with the cross in the centre, in pink coral. Source: the author. On the right: Photograph from the Conservatory of the *Cap Corse*[4] costume.

On the island of Madeira, there are no jewels in gold or precious stones associated with the costume.

> One of the reference ornaments in the regional costume is the Carapuça– a conical cap used in the 18th and 19th centuries, influenced by the medieval hood and Portuguese caps. From a hood that covered the entire head, it evolved into an elegant form, almost an adornment. Barrow, who was in Madeira in 1790, says that the women "wore a helmet on their heads" (Barrow 1790), no more than the antecedent of the *carapuça*. This is confirmed by the indispensable iconography: "*Antiga carapuça*" [Ancient *Carapuça*], W. Combe, 1821; "Dress of the Country People of Madeira", N.C. Pita, 1802 (Cardoso 2009 s/p).

[4] https://www.tripadvisor.fr/Attraction_Review-g2291766-d8599621-Reviews-Conservatoire_du_Costume_du_Cap_Corse-Canari_Brando_Haute_Corse_Corsica.html.

The Madeira archipelago is made up of four island groups (Madeira Island, Porto Santo Island, Desertas Islands and Selvagens Islands), replicated in the pendants proposed as an addition to the *carapuça* (Fig. 6). We hope, with this, to evoke formal and identity affinities: the pendant, based on the "*Strelitzia Reginae*" flower, in the costume from Madeira (Funchal), dialogues with the pendant from Corsica, placed on the coral necklaces, while maintaining formal similarities with the Azorean necklace.

Fig. 6. On the right: vector drawing of the jewel with emphasis on the detail of the *Strelitzia Reginae*, the island's iconic flower. On the left: blue Madeiran fabric *carapuça* with the proposed jewel. Source: the author.

All proposals seek to match contemporary aesthetics. If it is true that we cannot get to contemporary jewellery without going through traditional jewellery, not just for the sake of a timeline, but because our past is projected into the present, it is also undeniable that contemporaneity has its own characteristics. According to Skinner (2013), contemporary jewellery represents the jeweller's temporal desire to "be of his time". Our proposals are especially sensitive to two priorities of our time: giving visibility to peripheral identities and underlining the role of culture for the sustainability of a way of life that is intended to be dynamic and respectful of its legacy. To this end, we adopted the creation of pieces of jewellery as a basis for contributing to the cultural dynamism of these island regions, in the conviction of the multiple functions of these objects – a way of understanding the position of a community in time, an expression of the individual and collective self, and a way of telling stories:

> (...) avant-garde jewel, which positions itself as radically ahead of dominant ideas; modernist or modern jewellery, which aspires to reflect the spirit of the time in which it was made; studio jewellery that emphasizes the artist's studio over the craft workshop; new jewel, which assumes an ironic posture with the past; and finally contemporary jewellery, a term that represents a perfect balance between innovation, personal language and recognition by established circuits of galleries, museums and collectors (Gaspar 2007, p. 12).

> Jewellery often functions as a symbol of the self-symbol, as a signifier of aspects of identity, as a means of transporting us to other times, places and people, and as a receptacle for our feelings towards this associated other (Koulidou 2018, p. 22).
>
> The narratives that accompany the piece add value to it. They are often incorporated into the form and can trigger memories that are meaningful to the wearer and/or the maker (Koulidou 2018, p. 28).

At the intersection of these three functions, we sought to establish a critical, conscious and reflected relationship between each piece and each of the three communities that, given their geography, are peripheral, if not even ultraperipheral, in the European context, showing the identity, the relationship between past and present and contributing to the narrative continuity of island experiences and memories, in which regional, national and international influences intersect.

3 Final Considerations

In this article, we explore the insular context, specifically traditional costumes, and the potential applications of contemporary jewellery, considering clothing and adornments in their significant dimensions as material culture.

Based on the costumes of folklore groups, we present two examples of jewellery that depart from the initial form of the necklace created for the costume of the Azores (São Miguel), and that contains the memories, meanings and places of various islands. We seek to add value to how craft practices and contemporary aesthetics can create poetic and emotionally rich interactions.

Through the creation of this collection of jewels, we also intend to contribute to bringing the regions closer together, through ornaments whose conception, inspired by local ethnography and endogenous materials – basalt and coral –, establishes bridges between different geographies, favouring the cultural interaction.

On the other hand, we also assume the field of research as authorial, which is why we have advanced to the creation of two jewels to be included in traditional costumes from Madeira and Corsica, starting from the central element of the jewel already created to be included in the Azorean costume. We assume culture and identity in a perspective of creative renewal and potentially founding new alliances between islanders.

We have created space for awareness of regional, national and international links. The proposed jewels not only reflect the cultural vitality and dynamism of the identity but are also expected to collaborate towards strengthening identity ties and accentuating archipelagic knowledge, promoting renewed interactions, at national and international level.

Acknowledgments. This work is financed by Project M3.1.a/F/024/2022 of FRCT – Fundação Regional de Ciência e Tecnologia and CHAM Açores.

References

AAVV, Apontamentos Culturais: Joias e Objetos de adornos. Museu Quinta das Cruzes (2020). https://mqc.madeira.gov.pt/noticias/apontamentos-culturais-joias-e-objetos-de-adorno. Accessed 8 Mar 2023

Cardoso, C.: Trajes de Portugal (2009). http://trajesdeportugal.blogspot.com/2009/02/carapuca-madeira.html?fbclid=IwAR3tTJf0sCC4npvuYkfrVam7LcpfJvDbZmJCPvIFuc-21XPHNmbkMVzLEGE. Accessed 8 Mar 2023

Costa, D.A.: No Minho, Lisboa: Imprensa Nacional (1874)

Cunha, L.: Memória Social em Campo Maior. Usos e percursos da fronteira. Lisboa, Portugal: Dom Quixote (2006)

Espírito Santo, M.: Origens orientais da religião popular portuguesa. Lisboa: Assírio e Alvim (1988). https://cultura.madeira.gov.pt/visitas-virtuais1/148-visitas-pascais-as-saloias-do-esp%C3%ADrito-santo.html. Accessed 24 Apr 2023

Flügel, J.C.: A psicologia das roupas. São Paulo: Mestre Jou (1966)

Gaspar, M.: Contemporary jewelry in post-historical times. In: BERSEGIO, MC Time tales: time's perceptions in research jewellery. Barcelona: Grupo Duplex (2007)

Giannesini, G.: Mythes et Symboles de Corse. Patrimoine culturel. Editions Jean-Paul Gisserot (2012)

Koulidou, N.: Why should jewellers care about the digital? J. Jewellery Res. 01 (2018)

Mérimée, P.: Colomba. Publié avec le concours de l'Éducation nationale, Rectorat de l'Académie de Corse dans le cadre du contrat de projets État/CTC 2014–2020 (2010). http://www.educorsica.fr/phocadownload/pdf/culomba.pdf. Accessed 3 Apr 2023

Mota, R.: O ouro sai às ruas: A presença de ornatos áureos nas festividades populares do sec. XX e alvores do sec. XXI. Porto: CITAR (2015). https://ciencia.ucp.pt/ws/portalfiles/portal/31105575/21.pdf. Accessed 23 Jan 2023

Pimentel, A.: Romarias portuguezas: I. Nossa Senhora da Agonia em Viana do Castelo, Lisboa, Antiga Casa Bertrand – José Bastos (1906)

Popper, K.R.: Acerca da inexistência do método científico, em Prefácio da edição de 1956 do livro "O Realismo e o Objetivo da Ciência", Publicações Dom Quixote (tradução), Lisboa 1987 (1956)

Pecqueux-Barboni, R.: Costumes de Corse, Pannu è panni, Éditions Albiana (2008)

Rebelo, H.: Contributo para a definição de "saloia" no âmbito do património linguístico: função das "saloias" na festividade do espírito santo no Arquipélago da Madeira in CHAVES, Duarte Nuno (coord.), Memória e identidade insular: religiosidade, festividades e turismo nos Arquipélagos da Madeira e Açores, pp. 239–248 (2019)

Skinner, D.: Contemporary jewelry in perspective. Asheville: Lark Crafts (2013)

Teixeira, M.B.: O Traje Regional Português e o Folclore, vol. VII (2009). https://www.om.acm.gov.pt/documents/58428/182327/1_PI_Cap7.pdf/803cd114-b7e8-448a-8dea-aeacb67866fe. Accessed 9 Feb 2023

Varela, R.: O Folclore Regional Madeirense: Peculiares Memórias. Grupo de Folclore da Casa do Povo da Camacha. Instituto Politécnico de Viana do Castelo (2021). http://repositorio.ipvc.pt/bitstream/20.500.11960/2569/1/Rolando_Varela.pdf. Accessed 2 Feb 2023

Vieira, A.: Bordado da Madeira. Centro de Estudos de História do Atlântico (2006). https://www.researchgate.net/publication/310952197_Bordado_da_Madeira. Accessed 10 Apr 2023

Vieira, A.: Porto Santo e Madeira. Território. Poderes e Instituições. 1418–1974. Dados para uma Visão Interpretativa. Projeto "Memória Autonomia", Cadernos de Divulgação do CEHA, pp. 1–91 (2018)

Body, Fashion, and Artificial Intelligence: Reflections about Authorship and Style in Creation

Renata Pitombo Cidreira[1(✉)], Beatriz Ferreira Pires[2], and Etevaldo Cruz[3]

[1] Universidade Federal do Recôncavo da Bahia, Salvador, Brazil
[2] Escola de Artes, Universidade de São Paulo, Ciências e Humanidades, São Paulo, Brazil
beatrizferreirapires@usp.br
[3] Universidade Federal da Bahia, Salvador, Brazil

Abstract. From a comprehensive perspective and in the context of celebrating artificial intelligence, this article seeks to reflect on the relationship between authorship and style in intertwining among body, fashion, and technology itself. Artificial intelligence has helped the production of the most varied images, including images of virtual models. In general, virtual models wear clothes created by designers so far and make it much easier and less expensive to produce a catalog or even a complete advertising campaign for a new collection. This whole process and all these dematerialization culminate with the arrival of ChatGPT in November 2022. ChatGPT is a virtual assistant that operates with artificial intelligence, developed by OpenAI. In the theoretical-conceptual arc, authors such as Sennett, Pareyson, Merleau-Ponty, Rocha, among others, will be mobilized.

Keywords: Body · Artificial intelligence · Fashion

1 Introduction

In Artificial Intelligence, Interaction, performance and illusion (2021), Marcio Alves da Rocha offers us an intriguing interdisciplinary journey on the subject of AI (artificial intelligence) and its relationship with human intelligence, based on the notion of Human-Human Interaction. Computer (IHC) and the limitations and problems that such interaction poses. According to Rocha, recent AI studies offer "important theoretical support for understanding the human mind and human and machine intelligence" (Rocha 2021, p. 16, translated by the authors).

Its disciplinary condition allows us to approach it differently from other fields of analysis of the theme, as it evokes some complex and interdisciplinary texture with other areas of knowledge, such as philosophy, thus it allows us to approach the HCI theme in a more humane way and earthly. In this sense, the approach starts from the integral assumption, involving the body, the mind, and the world, considering the interaction between humans and computer in a mundane way.

The notion operationalized by the author to trace the hermeneutic path on the subject is that of the "embodied mind", whose understanding offers a broader analytical key to the interactions of humans with machines and computers when perceiving the relationship between mind, body, and of the world as an interconnected event.

AI must be understood more broadly involving art history, mythology, and folklore studies that revolve around artificial machines, which offers us a significant comprehensive framework of AI and its historical and epistemological contingencies. The history of AI brings significant contributions to our anthropological and phenomenological dimensions by highlighting "our human passion and our impulse to create artificial life" (Rocha 2021, p. 73, translated by the authors).

Anchored in this perspective, we must understand that intelligence is not just a human characteristic, but it can be "conceptualized as a relative property, which is situated and is sensitive to the context [...] not belonging to agents, human or non-human, but it emerges from the organic interaction between them (that is, between human beings, machines, objects, and artifacts)" (Rocha 2021, p. 74, translated by the authors). The idea of intelligence as an extra-human property, computable and capable of manipulating symbols is a limited conceptualization of intelligence, as it neglects the new developments of cognition and embodied intelligence according to Rocha.

Regarding the historical character, the initial notion of AI was influenced by the conception that our mental model is similar to an information processor, where mind, body, and world worked independently, which means a disembodied mind of the world and physical and contingent experiences in our daily lives.

Rocha highlights that the computer scientist, logician, and cryptologist, Alan Turing (1912–1954) strongly influenced such ideas with the concepts he developed, especially regarding the understanding of computing as an innovative form of intelligence, whose destiny would be to surpass human intellectual capacity.

The classical model of artificial intelligence was based on a Cartesian epistemological framework of appropriate symbolic representation. This model sought to symbolically represent and formalize common sense to make it computable, says Rocha (2021).

However, this classic model found its limit in the very human contingency of worldly experiences, given its complexity and dynamism. Idiosyncratic characteristics of our ability to offer sense and meanings to our culturalized productions. Unlike us, the automaton cannot act outside its materiality, as it does not dynamize outside its own nature, since it depends on an external causal agent.

The ambition to "materialize human-centered intelligence in a machine created the false notion that intelligence could be replicated quickly, and took the discipline into the Dark Ages of Artificial Intelligence" (Rocha 2021, p. 74, translated by the authors). The author reminds us that machines possessing superior intelligence similar to ours is yet another imaginary attribute superior to our technological capacity to carry it out.

In other words, it is our everyday life that we experience in a full and non-representable way, our systems of beliefs and values, our expressive institutions that, in the end, lead to what is unique and unrepeatable in us, that is, our style, our form that emerges from strength as an expressive capacity of a circumstantial, historical and space-time perception crossed by sociability and communicability, whose dynamics are woven in the emergence of everyday life.

In, "Technology as an Ally or Enemy of human creativity: the discussion around the ChatGPT" (2023), Teresa Henrique shows us another perspective on the relationship between human creativity and technology. Although the author considers the positive dimension of technology as a means for the development and improvement of human creativity, the author calls our attention to another aspect, pointing out how technological dependence can reduce our critical and creative capacity. In addition, technological dependency throws us into an atmosphere where product efficiency is out of step with innovative solutions.

Therefore, if on the one hand technology can open promising paths for the improvement of our creative processes, on the other hand, the dependence or understanding of technology as an end in itself can affect what our shaping brand is, that is, the our creative ability. Therefore, the author reminds us that "human creativity is still the source of innovation and originality, and it is necessary to make the most of the technological tools available" (Henrique 2023, p. 07, translated by the authors).

In the context of celebrating artificial intelligence, this article seeks to reflect on the relationship between authorship and style in the intertwining of the body, fashion and technology itself. In the theoretical-conceptual arc, authors such as Sennett, Pareyson, Merleau-Ponty, Alves, among others, will be mobilized.

1.1 Artificial Beauty

Recently, a version of the painting Girl with a Pearl Earring - by the Dutch painter Johannes Vermeer -, from 1665, was the subject of intense controversy. The fact that the current painting was produced with the aid of artificial intelligence (AI) based on the boldness of contemporary artist Julian Van Dieken was what effectively gave rise to the debate.

It all started with a competition promoted by the Mauritshuis Museum (Mauritius), in The Hague, Netherlands, which houses the original work. The famous painting was loaned to the Rijksmuseum, located in Amsterdam, until June 4th. During this period, it would be replaced by a re-reading. This initiative alone is enough reason for a heated debate. But what exactly provoked intense discussion was the fact that the chosen frame was produced with the interference of artificial intelligence.

The original canvas (Fig. 1a) portrays the enigmatic face of a young woman, in a play of light and shadows, producing an effect of great realism and has a pearl earring as the focal point, and is considered one of the masterpieces of humanity. We can say that it presents us with a poetics of delicacy, which enchants us and, from the canvas, we share an aesthetic experience. (Cidreira; Vieira 2019). The painting and the girl portrayed deserve attention for some essential elements, the moistened lips, the pearl earring and the fabric of the turban in shades of blue and yellow that cover the hair. The face instigates us! (Cidreira; Vieira 2019).

As supposed, the rereading (Fig. 1b) modifies many aspects, but it maintains proximity to Vermeer's famous painting in a certain way. But especially some of those details that make the canvas something special are nullified; a certain porosity is erased in an apology for the polish and smoothness of the digital image, to paraphrase Byung-Chul Han (2020) The dampness of the lips disappears, the dimmed and mysterious eyes give way to two vivid circles in blue tones, the contrast of the turban's colors is eliminated and the pearl is simply replaced by an artifact that even duplicates. Effectively, it is not about questioning the legitimacy of the new image, but debating how vivid and unique a digital

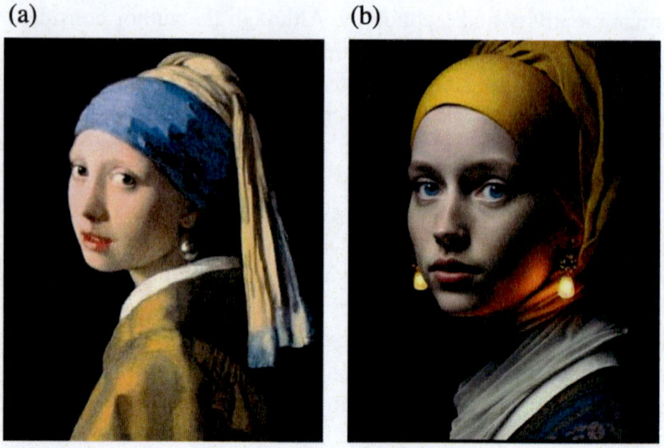

Fig. 1. (a) Girl with pearl earring (1665), by Vermeer. (b) Girl with pearl earring (2023), by Julian van Dieken. Fom: https://vogue.globo.com/cultura/arte/noticia/2023/03/museu-na-europa-e-criticado-apos-trocar-quadro-famoso-por-arte-feita-por-inteligencia-artificial.ghtml

image can be and to what extent it affects us, without the interference of human activity. As Byung-Chul Han (2020, p. 28, translated by the authors) attests, we experience an 'aesthetics' of complacency that "confirms the subject in his autonomy and self-conceit, instead of impacting him".

Another recent development has sparked a similar discussion in the fashion world. Artificial intelligence has helped to produce the most varied images, including images of virtual models (Fig. 2). In general, so far, virtual models wear clothes created by designers and make it much easier and less expensive to produce a catalog or even a complete advertising campaign for a new collection.

As we witness the debate about the creation of virtual models made by AI in the field of fashion, on the other hand we must pay attention to another broader aspect that concerns the creation of collections through AI that facilitate the development of visual prototypes that support the work of creators. This process is powered by algorithms. An example of this is the Brazilian brand The AirTificial which brings in its collections an authorial futuristic fashion profile. Another brand that used AI in its products was Newfacet, which created an abstract fashion campaign, where the models and pieces of the campaign did not really exist, but derived from existing products in the brand's collection[1]. These examples bring a perspective on how AI can aid the creative process of fashion designers[2].

This alternative has generated both favorable and unfavorable reviews. Some argue that artificial intelligence allows the great benefit of a more restricted cost for a small

[1] Available at: Futuro da moda: marcas usam inteligência artificial nas criações | Metrópoles.(metropoles.com). Access in June 30th 2023.

[2] Available at: Roupas projetadas por inteligência artificial podem ser futuro da moda, diz.startup de Hong Kong | Tecnologia | Época NEGÓCIOS (globo.com). Acess in June 30th.2023.

producer, for example, in addition to experimenting with multiple possibilities of circumscribed images in different contexts, without needing a production team to carry out the photo shooting. On the other hand, there are those who criticize exactly the fact that the professionals involved in the creation of the photographic essay are cancelled, in which many jobs end up being eliminated, which generates the precariousness of the sector's professionalization; in addition to the criticism of the fact that, as the virtual models are generated by algorithms, they can produce images that are very similar to each other. The physical characteristics produced by an algorithm can be very limited and even stereotyped, leading to an erasure of representativeness in the fashion industry.

Fig. 2. Picture – Artificial Intelligence (Karen Merilyn). From: https://stealthelook.com.br/inteligencia-artificial-na-moda-o-que-nos-espera-no-futuro/

Aiming to contribute to the debate, we can include an interesting reflection. If we return to the Pareysonian perspective (1989) that there is no law of art, we can infer that every form of poetry is valid for experiencing an aesthetic. In this sense, even a 'creative' production performed by Artificial Intelligence could generate a potent aesthetic experience. Anyway, Pareyson (1989) himself gives us some clues to think about the creative process. For the author, art is a mix of construction, knowledge and expression. And if so, does artificial intelligence absorb all these dimensions? Apparently, the embodied teckné would not be contemplated in AI, much less the expressive dimension, in which a dialogical relationship between force and form takes effect. We could only recognize that there is an archive of knowledge acquired and sedimented by man himself.

Again, it is perhaps worth reiterating that it is not a matter of questioning the effectiveness of these virtual images or even whether they are capable of promoting an aesthetic experience, as we have already mentioned. But it is about trying to understand the extent to which they manage to restore the sublimity of beauty, capable of reaching us intensely in our affection, in our feeling, as Byung-Chul Han (2020) would say; whether they are capable, in the last resort, of snatching us away.

Such captivating capacity transforms us and broadens our horizons, perhaps due to the recognition that it is through human activity, that is, a human training effort that brings together technical dexterity, knowledge acquisition, and the expressive manifestation of emotions, that perform the beautiful. This triad is the necessary construct for the development of a creative and authorial formative process, which even generates its style. In art and fashion, creation, style, and authorship have always been valued elements, which legitimize and validate artists and designers. It is, in fact, still the human presence with its riches and weaknesses that promotes admiration and indifference, beauty and pain.

1.2 Body - Organic, Inorganic

We, humans, are space-time beings. There is no physical possibility of, materially, being in places where our bodies are not and of not perceiving the effect of time, both in our bodies and in the difference between the precise moment in which an event occurs and its preceding and subsequent periods. Linked to this equation, we organize our lives, cultures, societies, policies, economies, etc. In the material sphere, everything happens according to these axes.

Thinking about the body-space relationship, we know that in their physicality each one is linked to the place where they are. From this premise, we can say that the individual is engendered and shaped by the environments in which he transits, lives, frequents, among other factors. Different spaces lead to different movements, gestures, postures, perceptions, reflections, feelings, sensations, and imaginations. People who live in snow-covered areas, such as those located around the Arctic Circle, perceive the world and interact with it in a different way than people who live in large Brazilian cities, for example.

For the vast majority of the contemporary urban Western population, space is composed of an infinity of rectangles that overlap, intertwine, and intersect almost non-stop. Buildings intended for the most varied purposes, such as residences, schools, culture and leisure centers, shops, industries, etc. are projected through the articulation of this geometric form. Buildings, bedrooms, classrooms, offices, transport, access roads, books, cell phones, notebooks, keyboards, screens…mostly, all have a rectangular shape. The use of the same format for making different things, in addition to helping dull sensations/perceptions and facilitating body control, meets the market's logic of generating fewer expenses and more profits.

The individual develops its gestures, its way of perceiving, thinking, feeling, and acting modulated by this quadrilateral formed by right angles, which in no way resemble the existing angles in nature.

Unlike objects built and produced from the precise development of previously elaborated projects, the human body is born and developed, presenting an innumerable range of variables and aesthetic possibilities.

Adapting diverse bodies to momentarily stipulated beauty standards imposes significant wear and tear on the individual whose desire follows towards that. As in general the objective is never reached, since the standards are not perennial, in these cases, the body can become an insoluble problem that steals the attention and interest of the individual in relation to things and forms such as the world around them gets organized. Disinterested and focused on his own body, the individual exempts itself from knowing, apprehending and acting on several of factors that make up the world. In addition, by moving in this direction, the individual generates significant profits for market sectors that operate in the body aesthetics segment.

In a society that stipulates body standards and values those who are similar to them, the dissimilar occupies segregated social zones. For Sennett, our society does not respect "the dignity of human bodies and their diversity" (Sennett 1997, p. 15, translated by the authors).

Ideal images of the human body lead to mutual repression and insensitivity, especially among those with different and nonstandard bodies. In a society or political order that generally praises "the body", there is a risk of denying the needs of bodies that do not fit the paradigm. (Sennett 1997, p. 22, translated by the authors).

Aspiring to join those or what the social consensus determines and enshrines as an ideal model, a significant part of those who have deviant bodies want to adapt their bodies to existing standards and approach privileged social zones.

We know that the relationship between body and city has changed a lot over the centuries. In the current period, the biggest change has occurred with the arrival of smartphones. In the immediately preceding period, devices aimed exclusively at communication, cell phones along with computers, had already played the role of drastically modifying everyday life.

With smartphones, the idea that the individual starts to extrapolate his bodily limits and perform functions in an extracorporeal way is realized. More than communication between two or more people be done via video call, at any time, from anywhere that has a network signal, but several other functions can be performed using this device such as searching and obtaining information about any subject, being geographically located, being reminded of appointments, drinking water, bedtime, being woken up and having recorded the number of steps taken per day.

Simultaneously to these functions, we follow the dematerialization of books, bodily changes in terms of gestures, the way of writing, walking, and positioning the head, shoulders, and spine, for example.

Through the screens of these devices, the individual can have visual contact with himself and record his image as many times as he wants. We've never seen each other so much.

Part of the body is only seen by others so the face steals the look once dedicated to the face of its interlocutor. Immersed in his own image and in the brightness of the screen, the individual seeks his best angle, the best light, the best scenery, the best pose. The term instagrammable appears. Now the lives of some users of these applications begin to allow them to replace clothing with costumes, everyday sounds with a personalized soundtrack - made possible by headphones -, personal objects with props and private places with scenography made public. For some of these, satisfaction does not come with the act of doing, performing, creating, moving to unknown places, but rather the conduct of showing, sharing, reaching a large number of views (views) and receiving many compliments (likes).) for the posts.

This whole process and all these dematerializations culminate with the arrival of ChatGPT, which is a virtual assistant that operates with artificial intelligence, developed by OpenAI in November 2022. With it, the acts of reasoning, imagining, creating, interpreting written and/or imagetic texts, acquiring knowledge, developing critical capacity, etc., are transferred from the organic body to virtual programs and applications.

The temporal moment we are in, in which attitude is replaced by passivity, the concrete by the virtual, and the body by the avatar, is the perfect moment for the process of immateriality to reach the body and fulfill its goal of replacing the human with the humanoid.

Such a substitution would resolve Bataille's observation that the flesh is in us that "excess that opposes the law of decency." (Bataille 1987, p. 87, translated by the authors).

Theoretically, beings devoid of flesh, in addition to not presenting questions and problems related to organic convulsions, would be totally and completely controllable. Opinions different from those exposed above were elaborated by several writers, screenwriters, and film directors, who pointed out to us in their science fiction works, various inventions and situations that exist and occur today. Among them, we can mention: Stanley Kubrick with the film *2001- A Space Odyssey* (1968) and Ridley Scott with the film *Blade Runner* (1982). In both, artificial intelligence gained autonomy and rebelled against the human.

1.3 AI, Body Experience and Style

When dealing with the space-time relationship of the body immersed in the world, we refer to the considerations of Maurice Merleau-Ponty, in which the philosopher marks our experience of visible and seer in confusion, "inherence of the one who sees what he sees, of the one who touches the that it touches, from the sentient to the sense - a self that is taken, therefore, among things, that has a face and a back, a past and a future" (Merleau-Ponty 2014, p. 20, translated by the authors). In other words, our experience is the expression of the convergence between the sediments that delineate the physiognomy of their way of perceiving the world and the contemporaneity of the power to institute a new horizon of senses, where the body engages in the creative movement, opening itself if in possibilities.

This is how Merleau-Ponty begins the chapter on the spatiality of one's own body in the Phenomenology of Perception (2019):

> The outline of my body is a frontier that ordinary space relations do not cross. This is because its parts originally relate to each other: they are not unfolded alongside each other, but involved in each other (Merleau-Ponty 1999, p. 145, translated by the authors).

My relationship with space happens in the involvement of my body with the parts that surround it, where my outline is the border. This undivided condition is structured from what the author conceptualizes as a body schema. The notion of body schema is complex and ambiguous. In principle, it was understood as a kinetic a priori that allows us to adjust our bodily changes to each preconceived idea, corresponding to the stimuli apprehended in childhood in a kind of visual translation of movements, whose images are at the center of the preconceived process. Articulation.

Thus, it is possible to think, as Merleau-Ponty suggests, about the subject who, among familiar artifacts, has them in a "naturally" structured experience in the spatial system of his own body; since his powers are mobilized by the perception of artifacts, through "intentional threads" that links it to given objects, including the dynamic modalities of readaptation and reinvention of relational modes that cannot be performed by AI.

Therefore, the complex statement that "to experience the agreement between what we aim at and what is given, between intention and effectuation - and the body is our anchorage in a world" becomes more evident (Merleau-Ponty 1999, p. 200, translated by the authors). In other words, the relationship between body and space is a rooted relationship where they are familiarly confused.

Nevertheless, if the integral experience of the body with the world and with others marks the aspect of how we understand and express our worldly presence, we are confronted with the issue of style. The unrepeatable and unique way of forming that evokes not only our ability to establish new paradigms, but the style is the bearer of our experiences and marks, histories and repertoires that mark the way we perceive the world, including our artistic and creative expression.

The style corresponds to the interchangeability between the incorporeal in which we are embedded and the more personal trait that we mark in ourselves, revealing our condition as beings loaded with those who preceded us. From this aspect, it is possible to correlate the dimension of intelligence and style, since the first, as Rocha states quoting Piaget on the notion of intelligence and mental life: "is assimilation insofar as it incorporates all the data of experience into its structure. There can also be no doubt that mental life is also accommodation to the environment" (Piaget apud Rocha 2021, p. 86, translated by the authors).

Style emerges when individual life abandons its inherence, ceasing to enjoy itself, and becomes a universal means of understanding and making understood, of seeing and giving to be seen, as Merleau-Ponty says. Now, it is the communal diffusion of the artist that, when expressing himself, highlights the way of perceiving, resulting from his experience "in which an operative and latent sense found for itself the emblems that should free it and make it manageable by the artist and at the same time accessible to others" (Merleau-Ponty 2001, p. 79, translated by the authors).

If style is the unveiling of expression, it cannot be understood in the manner, in the tics or processes that are part of the artist's "personality", or even more so in the "artist's own predilections". In a broader sense, style corresponds to a mode of formulation that meets the request of the artist's perception as an artist. With this, we can resume Malraux's expression highlighted by Merleau-Ponty when he states that "perception already stylizes".

Style, therefore, corresponds to the interweaving of the elements of the world and how the artist and the fashion designer organize what they seek to make expressible. Thus, the convergences of the visible vectors are already outlined in the artist's perception. Now, but does that mean that the style apprehends the fullness of things? Not at all. What he has access to are certain concavities, fractured zones, fissures, figures and backgrounds, and daydreams. Therefore, what is allowed to be expressed by the work, or by the language.

In this way, style is for the fashion designer and each painter, as Merleau-Ponty states, "the system of equivalence that he constitutes for this work of manifestation, the universal index of "coherent deformation", the convergent point of its perception-expression. Style is in the clashes of our institution in the world, as the artist is immersed in the world of things, his work does not emerge from a laboratory, but in the very texture of life, experiences, and clashes that are their own.

2 Final Considerations

After this comprehensive journey, we can see that human passion marks our enchantment for the representation of our image, the contours of an external reproduction in our inner manifestation, umbilically linked to us by the extension of belonging, whose strength derived and still derives from our creative imagination, materialized by technology.

As Rocha shows in his research, the production of this artificial exteriority has a history as old as humanity. From the Golem, a creature similar to an animated human, a formless and imperfect mass of Jewish folklore, to Greek mythologies with Hephaestus, Pygmalion and Galatea, to children's stories like Pinocchio, replication seems to move our creative pathos, revealing something suggestive about the place of the body in life experience in and with the world in its space-time relationships.

In any case, we risk saying that the insistent transfer of the organic body to virtual programs and applications cannot do without the body's protagonism in and for the realization of everyday and aesthetic experience, whether in the broader scope of life, in the universe of art and even in the sphere of production and reception of fashion artifacts. We believe that it will still and always be the human presence, with its porous and frictional corporeality that will manifest its style in its actions and expressions.

References

Bataille, G.: O Erotismo. L&PM, São Paulo (1987)

Cidreira, R., Reis, G.O.: prazer do belo na tela e na aparência In Cidreira, R. (Org.) (2019). O Belo Contemporâneo: Corpo, Moda e Arte. J. Andrade, Aracaju

Han, B.C.: A Salvação do Belo, Vozes, Petrópolis (2020)

Henriques, T.: Tecnologia como aliada ou inimiga da criatividade humana: a discussão em torno do ChatGPT. The Trends Hub, (3) (2023). https://parc.ipp.pt/index.php/trendshub/article/view/5096. Accessed 01 julho 2023

Pareysin, L.: Os problemas da estética, 2ª ed., Martins Fontes, São Paulo (1989)

Sennett, R.: Carne e Pedra, Record, São Paulo (1997)

Merleau-Ponty, M.: Fenomenologia da percepção.,Martins Fontes, São Paulo (2014)

Merleau-Ponty, M.: Fenomenologia da percepção. Martins Fontes, São Paulo (1999)

Merleau-Ponty, M.: A linguagem indireta e as vozes do silêncio. In O olho e o espírito. Cosac Naify, São Paulo (2013)

Rocha, A.M. Inteligência Artificial, Interação, Performance & Ilusão: Uma abordagem transdisciplinar sobre a mente e a interação com o mundo, máquinas e objetos. Goiânia-Goiás: Tipo de recurso (Kindle) (n.d). https://www.amazon.com.br//Intelig%C3%AAncia-Artificial-Intera%C3% A7%C3%A3o-Performance-Ilus%%C3%A3o-ebook/dp/B094RFWVCX. Acesso 29 Abril 2023

The Beauty: From Philosophical Thought to Fashion

Ana Claudia Alcantara[1(✉)] and Mónica Romãozinho[2]

[1] Universidade da Beira Interior, Rua Marquês de Ávila e Bolama, 6201-001 Covilhã, Portugal
aclaudia.alcantara@gmail.com
[2] Labcom-UBI, Universidade da Beira Interior, Rua Marquês de Ávila e Bolama, 6201-001 Covilhã, Portugal

Abstract. This article discusses the philosophy that portrays beauty and how each philosopher theorized beauty in their narratives, namely Plato, Aristotle and Kant. The essay aims to report how beauty was initially described and what were the aspects that transformed these thoughts into a science that shaped aesthetics as far as we know it. The research moves on to the philosophy of the 20th century with authors such as Gilles Lipovetsky and Luc Ferry who were instrumental in understanding the progressive affirmation of Beauty through difference and the expression of the self. The analysis is of an investigative and exploratory nature accentuated in the process of analysis of relevant literature around the proposed theme.

Keyword: Fashion · Aesthetics · 20th century · Beauty · Philosophy · Art

1 Introduction

We still interpret beauty today as something that transcends mere physical appearance and includes values such as harmony and balance. With these definitions, one can understand how Beauty contributes to humanity. We have as strong examples, the beauty in the form of works of art, music, and literature, which inspire feelings of admiration, empathy, and critical reflection. These emotions can lead people to connect with the experiences and perspectives of other cultures and societies. Eco, [1] reflects the relationship with the word Beautiful, which also uses similar words such as: "graceful", "charming", or even "sublime", "wonderful" and "superb" and that these expressions are adjectives that we use frequently to suggest something we like. In this sense, it is suggestive that what is Beautiful is equal to what we consider Good, and at different historical times this close connection between Beauty and Good was formed and how this has shaped the aesthetic history of objects to the present day.

In this context, fashion can be considered a strong means of expressing the beauty of each time and society. This is due to the simple fact that fashion is considered a representation of what is beautiful, historically in the form of objects and aesthetic construction, whether through clothing, behavior or body. But what is considered Beautiful and not? This theme has well-articulated aspects in the areas of philosophy, art, anthropology, and semiotics that seek to explain, through their theories and studies, how this imaginary is constructed. Chalumeau, [2] argues that Greek philosophy, during its anthropological period, sought to note the reasons why human activities have a faithful relationship with values related to beauty. The author points out that since the beginning, the idea of beauty has been intertwined with the production and transformation of nature. With that, Bayer, [3] reports that the sea was a very familiar element for the Greeks, where they lived, bathed and practiced their trade, thus, the most beautiful line for their culture was the wavy line, which corresponded to the natural movement from the philosophical point of view. This representation of lines and the sea is very present in Greek works, through paintings and sculptures. In this way, it becomes visible how Beauty is a personification of what is important for each culture, society and time.

Thus, the present study aims to understand the initial theories about Beauty and how this shaped philosophical thoughts for the 20th century. Thus, it is necessary to go through the initial philosophies such as Platonic, Aristotelian and Kantian to know a little of the reflections that were initially formed to explain beauty. And as in modernity, specifically the twentieth century, which provided a redefinition of aesthetic paradigms. As pointed out by Sant'Anna, [4] the aesthetic experience is the emergence of rational depth and, as such, pleasure that is realized by the absence of the represented that, abstracted in its concreteness, provides the apex of its contemplation. This demonstrates the relevance that Beauty has in the creation of artistic and aesthetic movements from antiquity to contemporaneity.

2 Beauty According to Platonic to Kantian Philosophy

Beauty has been a central approach in philosophy since ancient Greece. There are different philosophical theories that address what beauty is and how we can understand it. In the present research, the Platonic, Aristotelian and Kantian philosophies will be initially approached to contextualize the first ideas about the Beauty that has reports.

According to the Platonic theory of beauty, proposed by the philosopher Plato (427–347 BC), describes that beauty is a universal and transcendental idea that exists independently of the things we consider beautiful. For Plato, perfect beauty is not found in the physical world, but is a pure form that is understood by reason. The philosopher believed that beauty was not just a matter of appearance, but a fundamental quality of things. He argued that beauty was an objective quality that could be found in everything from physical objects to abstract ideas and concepts [5].

Plato argued that beauty was one of the most important forms of knowledge, and that the contemplation of beauty led to an understanding of truth and reality. He believed that beauty was a path to enlightenment and to understanding perfect and eternal ideas. Plato also emphasized the importance of harmony and proportion in beauty. He argued that beauty was found in the proper harmony and proportion of the parts of an object or idea, and that this harmony was perceived by the beholder as a sense of aesthetic pleasure [6].

For the philosopher Plato, the Beautiful was deeply linked to the Good, this Greek ideal of perfection is the sharing of the Beautiful with the Good. The world of ideas, of forms, only contains Beauty, in so far as it is perfect. The true beauty considered by the philosopher is that which has its own existence in the supersensible world and is, therefore, a superior, absolute, divine beauty, the only true beauty and which the sensible world is full of through its reflections and participations [7]. With this, Plato sought to relate utility to the idea of beauty. Affirming the existence of "Beauty in itself", a base, present in the "world of ideas", responsible for everything that is Beautiful [8].

Entering the Aristotelian theory of beauty, proposed by Aristotle (384–322 BC), the philosopher argues that beauty is an objective quality of things. According to Aristotle, beauty is not just a matter of appearance, but also of functionality and purpose. Aristotle argued that beauty is a subjective experience, but that there are objective standards by which we can judge beauty. He believed that beauty tends to be judged in relation to the purpose of the object in question. Bayer, [9] points out that, for Aristotle, the Idea has no existence in itself, it is abstracted by us. What is important is reality. To know it, it is necessary to be able to reduce it to its causes, and in this way, a research dedicated to this pure science was elaborated. The author points out that Aristotle represented causal research in four elements, namely: "the material cause (that of which the object is made); the motive or efficient cause (what gave rise to that object); the formal cause (what gave the object its form); the final or teleological cause (that in view of which, what an object aims at); here, aesthetics could be introduced. The true nature is known only when the final cause is known." Aristotle also emphasized the importance of proportion in beauty. He argued that correct proportions must be used in all parts of the object to create overall harmony [10].

The studies idealized by Immanuel Kant (1724–1804), German philosopher, who argued that beauty is a subjective experience that involves the perception of a harmony between imagination and understanding. The philosopher believed that beauty is a universal experience that transcends cultural and individual differences. Kant believed that beauty is not an objective quality of things, but rather a subjective response to them [11]. Kant, argued that beauty was not found in things themselves, but in the way people perceive them. He emphasized the importance of the aesthetic experience as a way of appreciating beauty. Kant also emphasized the importance of the disinterested contemplation of beauty. He argued that beauty should be appreciated without any practical or utilitarian interest, and that the contemplation of beauty should be a disinterested and purely aesthetic experience [12]. But if we consider that beauty is a feeling that emerges in front of works, landscapes and people, from a Kantian perspective, we agree that beauty is not in the film, in the painting… since Beauty is not an intrinsic property of an object, landscape or person. Beauty, as Kant warns us, is precisely the feeling that invades us when we are faced with something and someone that gives us pleasure [13].

Kant, will lay the foundations of a conception of taste that will go beyond this opposition between rationalism and materialism to found the essence of the theories of genius taken up by romanticism. Beauty is not true, as the classics think, nor pleasant, as empiricists want [14]. For the philosopher Kant, beauty presented two kinds of aesthetic judgment, the Beautiful and the sublime. Beauty as a "disinterested, serene and pure sensation" and the sublime as an "aesthetic feeling mixed with pleasant sensations and terror" [15]. Thus, the Kantian theory of beauty argues that beauty is a subjective experience that involves the perception of a harmony between imagination and understanding. Beauty is not an objective quality of things, but a subjective response to them.

3 The Reframing of Beauty in the 20th Century in Philosophy and Fashion

The theory of beauty as expression, proposed by philosophers such as Friedrich Nietzsche (1844–1900) and Benedetto Croce (1866–1952), argues that beauty is an expression of human life and emotions. Thus, beauty is not just a characteristic of objects, but is created by the observer in response to his experience. The authors, Praxedes and Reegen, [7] report that taking art as an example, "beauty within aesthetics includes the bitterness and harshness of Goya's dark phase, the Gothic gargoyles, the tragic and the comic, as also Beauty, considering that aesthetics does not only investigate Beauty in the usual sense, but everything that influences us aesthetically, which leads us to believe that Beauty and the sublime can only be found in the spirit of those who contemplate it". Valverde [16] points out that the approach to the aesthetic experience in terms of unity and correspondence is what makes it possible to reconcile the two divergent meanings that the word "aesthetics" has assumed since it came to be used, in the 18th century, as a theory or science of Beauty from Baumgarten (1714–1762) and as an analysis of the conditions of the possibility of sensitive apprehension in the sense of Kant's "transcendental aesthetics". This approximation allows us to recognize the congeniality between creation and fruition, thus reintroducing the problem of taste in the flow of experience.

In the 20th century, the strength of the artistic vanguards that erupted in various geographies is unavoidable, the utopias move and intersect them, their contradictions, the unanimous proclamation of a new function of the "artist" in society and of a new concept of Beauty that it will move away from a mere formal innovation, propelling a new artistic order with pursuits in fashion. In Le Sens du Beau, Luc Ferry resorts to Wassily Kandinsky (1866–1944), because he considers that the theme of avant-garde ideology is embodied by his literary work *On the Spirituality of Art* (1912). The avant-garde artist is the man who consciously assumes the break with all tradition, the man who rises to the highest apex of a triangle, divided into unequal sections, with the smallest and most acute at the top, which metaphorically represents spiritual Life. The base of this same triangle symbolizes the popular mass, trapped in conformism, which does not understand the creative impetus of that man. «This triangles is not stratic: where the apex was "today," the second segment is going to be "tomorrow," that is to say, that which today can be understood only by the apex, and which to the rest of the triangle seems an incomprehensible gibberish, tomorrow forms the true and sensitive life of the second segment.» [17].

The role of contemplative observer is reneged on to make way for the collective task of transforming the world. The individuality of the vanguards only achieves meaning when it is driven by a collective and "militant" will, which justifies the almost simultaneous outbreak of artistic movements from Cubism onwards. For the author, the avant-garde metaphor actually finds the military accents that inspired it (the expression avant-garde appears for the 1st time in the military vocabulary) and, quoting Kandinsky again, states that those superior men, "(…) who ever so often, overstep the border of safety and die, while yet conquering another stronghold of science, in the manner of self-sacrificing soldiers forgetting all caution in their desperate fight, to win the stubborn fortress wall:" [18].

This struggle process leads to a Vanguard as an expression of the Self, what Kandinsky designates as "pure expression of inner life", an individuality, which according to Ferry, brings together elitism and historicism for its originality or difference that takes us to Friedrich Nietzsche (1844–1900). We are talking about a historicism that results from the struggle against the Socratic dialectic, whose objective was according to Ferry, to refute the opponent's positions with a view to arriving at a superior truth, that of the Idea (of the intelligible) [19][1]. Nietzsche "exchanges" this dialogic attempt at convergence for the radical multiplicity of life, for the multiplicity of interpretations. Truth ceased to mean Identity, Transparency, Harmony, to become this pure difference that is the multiplicity of vital forces and, from this perspective, in "Nietzschean" art, beauty is none other than the hierarchical expression of this multiplicity [20].

The assertion of difference, of overcoming the past and the present, makes the avant-garde succeed one another quickly and, in this sense, we witness the birth of a tradition, representative of a historicism that, at first glance, we would not associate with innovation from the vanguards: to the tradition of negation, of the new or of the Nietzschean difference [21].

Therefore, the logic of this paradoxical individualism is born from the break with classical norms, the logic of classicism leads, on the contrary, to the imposition of rules. The contradictory relationship between an ideal of beauty, which corresponds to the machine proclaimed by the International Style or Constructivism, and the attitude of permanent negation on the part of these same movements, characterizes this new classicism that seeks a truth (no longer absolute) and the expression of a chaotic (no longer harmonious) reality. The art work will not be autonomous and universal if it does not keep it´s deep connections to reality: it will be a reality in itself, more intense and truer than real objects, because it expresses, like the classics, the essence that it represents in such a way that the elements, that compose it do not refer to arbitrariness, nor to caprice, neither to imagination, nor to good decorative taste [22]. Ferry concludes that if, for Nietzsche, art is true, is it not because it is adequate to reality, also much more adequate than this illusion that we are used to designating, at least after Plato, under the name of truth? [23]. Nietzsche can be considered "classical", not in the Cartesian sense, but in the sense in which, for him, art continues to establish a direct connection with truth [24]. Therefore, parallel to individualism, we can speak of a "hyper-classicism", since more than ever, art has the function of translating a reality more real than that of the ancient classics, a reality that is no longer rational, harmonious, Euclidean (as in classicism) but illogical and chaotic [25].

Global attitudes of the vanguards, such as ceasing to imitate nature, come precisely from the need to break with tradition, to express a multiplicity of interpretations in the face of the world, at the same time a subjectivity inherent in the spiritual path that is

[1] Nietzsche's radicalism is only imposed because, after Plato, philosophy never ceased to devalue the sensible world in relation to the intelligible world. The instrument of this discrimination was called "dialectic". In a certain sense, the Nietzschean moment entertains itself with the negation of the Platonic-Hegelian dialectic of art, a relationship analogous to that which the Kantian moment establishes with classical rationalism. In both cases, it is about conquering or regaining the autonomy of sensibility. Luc Ferry - Le sens du Beau. Paris: Le Livre de Poche, 2002, p. 145.

at the base of individualism in the face of an uncertain world. The first of these currents corresponds, according to Ferry, to Abstract Art, propagated by Robert Delaunay, Kandinsky or Piet Mondrian (1872–1944) but it is impossible for us not to contemplate the suprematist Kasimir Malevitch, who rejects any attempt to reproduce the real in the name of an art without an object or Sonia Delaunay (1885–1979) in the field of Fashion. Philosophical or mystical doctrines, opposed to dominant materialism, namely Theosophy, would boost the pictorial explorations of Kandinsky, Mondrian and Malevitch. The geometric gestures that structure Mondrian's paintings, in particular the orthogonal lines, take on a profound symbolism: the celestial perpendicular, the horizontal line of the terrestrial base, the vertical associated with the masculine principle, the horizontal with the feminine principle and the cross as intersection, mythical concept of life and death [26]. Cubism, in particular, when breaking with naturalism, contradictorily seeks a classic ideal of an objective or realistic art – the new realism -, linked to the principles of figuration, even when it contradicts the laws of Euclidean perspective through the fourth dimension. Ferry highlights this rupture with the Euclidean perspective, which translates into the exploration of this new dimension, which determines the assertion of the individualism of the vanguards [27].

The extreme fragmentation of form operated by Pablo Picasso and Georges Braque would definitively mark this dimensional rupture with the existing modes of representation. However, both would not remove reality from their works, unlike Marcel Duchamp (1887–1968) and Léger, who would seek to explore a new realism that was increasingly distant from figuration. In the case of Fernand Léger, this bond with reality was betrayed by his fascination with modern life, with an ideal of dynamism, with the metallic and cylindrical surfaces of machines. For the Modernism that marked design and architecture in the 20th century, Beauty is largely influenced by the appreciation of the universe of machines, functionality and rationality.

Marcel Duchamp would not fully integrate himself into either the Cubist or the Futurist groups and in *Nu descendant un escalier* (1912), he would explore a deformation that he would call elementary parallelism, a formal decomposition into linear sheets that follow parallel while deforming the object, while they form movement, an abstract movement embodied in the painting by the viewer's eye. Movement constituted, in this way, the answer to the exploration of a fourth dimension only accessible by the imagination, and in this sense Duchamp sought to express the idea of an invisible fourth dimension, in the same way that a three-dimensional object causes a projected shadow in two dimensions, Duchamp would consider that the fourth dimension could project an object in three dimensions, that is, that every three-dimensional object would be the projection of something unknown in four dimensions [28].

However, isn't the essence of classicism – "expressing the real" – in progress in the avant-garde through these indirect routes? Didn't the Cubists, with their New Realism, intend to be more realistic than the classics, that is, more classic than they themselves were? This is a classicism of difference that translates into the present, expressing reality in its most subjective aspects and no longer a classicism of identity associated with harmony. In Cartesian aesthetics, truth was defined as rationality, so defended by Friedrich Hegel (1770–1831), while for Nietzsche it was defined by difference. The "classicism" of the vanguards is taken to its ultimate consequences in the name of a new realism, which can be chaotic and different, the "Dionysian" pointed out by Nietzsche, which symbolizes chaos or fragmentation, but which accepts new rules.

Regarding beauty, Lipovetsky, [29] points out that it plays a fundamental role in modern culture, being considered a central value in today's society. According to Lipovetsky, beauty is no longer just an attribute of works of art or nature, but a value that extends to all areas of life, including fashion, design, advertising and body aesthetics. He argues that the pursuit of beauty has become a social and psychological need in contemporary society, where image and appearance are of great importance.

For Lipovetsky, the sequence of catastrophes witnessed in the 20th century led to a questioning of reason as an instrument of responsible and bureaucratic control, and our relationship with the times and specifically with the future, is already marked by this criticism, even if they persist remnants of the optimistic past, especially in the techno-scientific field [30]. On the other hand, with the diffusion of the logic of fashion throughout the entire social body, the entry into the post-modern era takes place, a moment in which we witness the expansion of the sphere of individual autonomy, the multiplication of individual differences, the detranscendentalization of social regulatory principles and the dissolution of the unity of ways of life and opinions [31]. However, Lipovetsky, [32] also points out that the notion of beauty is something subjective and changeable over time, varying according to the values and beliefs of each era and society. He points out that, in modern culture, beauty is increasingly associated with youth, thinness and body symmetry, which can generate aesthetic pressure and standardization of appearance.

Fashion plays a critical role today on beauty standards, Sant'Anna, [33] addresses the essence that fashion conveys very well in recent decades. Which refers to trends, that no one follows to stop being who they think they are, but to adapt what they project as their "I" to their appearance. The author refers that this effort to be another is actually to be your even better self. "Therein resides the core of the modern poetics of appearance, the power of individualization, and massification of Fashion and the seduction of images on us, modern subjects". Thus, we can understand that what is considered beautiful in 20th century Fashion is related to an individual choice to perfect their "I" that is inserted in a great collective. Thus forming the standards of beauty in fashion. Almeida [34] argues that Lipovetsky does not reflect fashion as an isolated phenomenon, but from a scientific rationality. Thus, it can be understood that the creation of Fashion aesthetics has its consequences in different contexts and areas.

4 Final Considerations

In the observation about Beauty initially in philosophy, its construction in the understanding of man with his natural environment was visible. These reflections through the philosophers Plato and Aristotle are notorious, where there is an association of the Beautiful with the Good. However, reflective lines were shaped as new thinkers shaped the characterization of Beauty. Over the following centuries, the conception of beauty continued to evolve and transform, passing through various artistic movements, such as Baroque, Rococo, Neoclassicism, Romanticism and Modernism. In each of these movements, beauty was understood differently and often challenged and reinvented.

The conception of beauty is even broader and more diverse, with space for a multitude of styles, forms and artistic expressions, including Fashion. In this way, beauty is no longer limited to rigid and universal standards, but rather recognized as something subjective and relative, which depends on the perspective of each individual and their culture and social context.

The general theory of beauty that was formulated in ancient times stated that beauty consists in the proportions of the parts, to be more precise, in the proportions and ordering of the parts and in their interrelationships. According to Tatarkiewicz, [35] this theory persisted for centuries, stating in its most limited version, which is the quantitative one, that beauty is only found in those objects whose relations between the parts maintain a ratio analogous to that observed in proportions. Praxedes and Reegen, [7] report that beauty is not a property of the object, but an elaboration of the contemplating spirit, governed by the feeling of pleasure or displeasure, judgments of taste coexist in which something that may seem ugly to some is considered beautiful to others. Kant showed us how beauty is in our reaction to something that generates feelings and emotions. The philosopher suggested that beauty is not directly in the object, but in what it generates to its observer. In short, the history of the conception of Beauty demonstrates how this idea has always been influenced by historical, natural and cultural factors, evolving and transforming itself over the centuries and today being understood as something more plural and diverse. Especially with regard to Fashion, due to its connection with other areas of study and its amplification in modern life. Therefore, Fashion is not exclusively linked to an object or clothing, but to behaviors and social, artistic and cultural manifestations.

References

1. Eco, U.: História Da Beleza. Translated by A. Maia da Rocha, p. 08. Círculo De Leitores, Lisboa (2002)
2. Chalumeau, J.L., Taipas, P.: As teorias da arte: filosofia, crítica e história da arte de Platão aos nossos dias, p. 25. Instituto Piaget, Lisboa (1997)
3. Bayer, R.: História Da Estética. Translated by J. Saramago, p. 26. Editorial Estampa, Lisboa (1995)
4. Sant'Anna, M.R.: Teoria de Moda: sociedade, imagem e consume, 2nd edn., p. 20. Estação das Letras e Cores, Barueri (2009)
5. Oliveira, J.V.G.: Estética em Patão? Phoinix **11**(1), 90–101 (2005)
6. Bayer, R.: História Da Estética. Translated by J. Saramago, p. 38. Editorial Estampa, Lisboa (1995)
7. Praxedes, H.N., ter Reegen, J.G.J.: A beleza do urubu. Kairós **13**, 91–95 (2016)
8. Chalumeau, J.L., Taipas, P.: As teorias da arte: filosofia, crítica e história da arte de Platão aos nossos dias, pp. 26–27. Instituto Piaget, Lisboa (1997)
9. Bayer, R.: História Da Estética. Translated by J. Saramago, p. 47. Editorial Estampa, Lisboa (1995)
10. Freitas, S.P.R.: da "grande teoria da beleza": harmonia como ordem, proporção, número e medida objetiva. Revista Música Hodie **12**(1), 04 (2012)
11. Bayer, R.: História Da Estética. Translated by J. Saramago, p. 196. Editorial Estampa, Lisboa (1995)

12. Cecim, A.M.: Baumgarten, Kant e a teoria do belo: conhecimento das belas coisas ou belo pensamento? Paralaxe **2**(1), 2–19 (2014)
13. Cidreira, R.P.: O Belo Contemporâneo: Corpo, Moda e Arte, p. 26. J. Andrade, Aracaju (2019)
14. Ferry, L.: Le sens du Beau, p. 39. Le Livre de Poche, Paris (2002)
15. Cidreira, R.P.: O Belo Contemporâneo: Corpo, Moda e Arte, p. 18. J. Andrade, Aracaju (2019)
16. Valverde, M.: A dimensão estética da experiência. Textos de Cultura e Comunicação **37**(8), 14 (1997)
17. Kandinsky, W.: Do Espiritual da Arte, p. 38. Publicações Dom Quixote, Lisboa (2002)
18. Kandinsky, W.: Concerning the Spiritual in Art, p. 38. Dover Publications Inc. (2000)
19. Ferry, L.: Le sens du Beau, p. 145. Le Livre de Poche, Paris (2002)
20. Ferry, L.: Le sens du Beau, pp. 145–147. Le Livre de Poche, Paris (2002)
21. Ferry, L.: Le sens du Beau, p. 208. Le Livre de Poche, Paris (2002)
22. Ferry, L.: Le sens du Beau, pp. 267–268. Le Livre de Poche, Paris (2002)
23. Ferry, L.: Le sens du Beau, p. 180. Le Livre de Poche, Paris (2002)
24. Ferry, L.: Le sens du Beau, p. 140. Le Livre de Poche, Paris (2002)
25. Ferry, L.: Le sens du Beau, p. 149. Le Livre de Poche, Paris (2002)
26. Moszynska, A.: Abstract Art, 2nd edn., p. 50. Thames & Hudson, London (2004)
27. Ferry, L.: Le sens du Beau, p. 255. Le Livre de Poche, Paris (2002)
28. Cabanne, P., Duchamp, M.: Engenheiro do Tempo perdido, p. 60. Assírio & Alvim, Lisboa (2002)
29. Lipovetsky, G., Serroy, J.: O capitalismo estético na era da globalização, p. 27. Edições, 70, 490, Lisboa (2014)
30. Charles, S., Lipovetsky, G.: Los tiempos hipermodernos, p. 14. Editorial Anagrama, Barcelona (2006)
31. Charles, S., Lipovetsky, G.: Los tiempos hipermodernos, p. 20. Editorial Anagrama, Barcelona (2006)
32. Lipovetsky, G., Serroy, J.: O capitalismo estético na era da globalização, p. 54. Edições, 70, 490, Lisboa (2014)
33. Sant'Anna, M.R.: Teoria de Moda: sociedade, imagem e consume, 2 edn., p. 44. Estação das Letras e Cores, Barueri (2009)
34. D'Almeida, T.: Moda em diálogos: entrevistas com pensadores, p. 25. Memória Visual, Rio de Janeiro (2012)
35. Tatarkiewicz, W.: Historia de Seis Ideas: arte, belleza, forma, creatividad, mímesis, experiencia estética, p. 157. Editorial Tecnos, Madrid (2002)

Trend Studies and the Operational Concept of Cultural Trend as Change(s): A Semi-systematic Review

William Afonso Cantú(✉) and Nelson Pinheiro Gomes

CEAUL/ULICES University of Lisbon Centre for English Studies, Lisbon, Portugal
williamcantu@edu.ulisboa.pt

Abstract. Future Studies and Trends Studies operate various concepts and terminologies rooted in a prospective basis of action. While Trend Studies look from the past to the present in search of patterns, Future Studies focus and aim at scenarios and possibilities for narratives to come. The convergence of areas regarding the use of 'trend' terminology has been explored in different ways in the literature. This requires a clarification of this concept, which is addressed in this text with the support of a semi-systematic review, to build an operational definition of trend. This article seeks to explore, in the first phase, the approach to the concept of trend. In the second part, we develop a systematic literature review to define the concept and present an operational definition in the last part. The contributions of this research are related to the definition of the term "trend" and the categorization of authors who operates it.

Keywords: Sociocultural Trend · Trend Studies · Future · Trends

1 Introduction

Trend Studies have seen an increase in scientific production. This has prompted a certain attention, autonomy, and awareness of a more individual identity. As a consequence, Trend Studies begin their process of legitimation and self-definition as a specific approach, associated by many with culture and the study of culture (Gomes et al., 2021; Gomes et al., 2018; Powers, 2018, 2019, 2020; Rech & Nascimento, 2017), in a relationship already established with disciplines such as marketing (Higham, 2009) or design (Lopes, 2019; Rech, 2016). Studying the development of a sociocultural trend is seeing cultural development and how to do 'cultural research' (Powers, 2018, p. 3). In these developments, both a relationship with the future and the benefits of projecting futures becomes clearer, as well as a disciplinary difference between Trend Studies, as a specific approach, and Future Studies. Focused on the present and the past, Trend Studies seek to understand the evolution of sociocultural patterns, while Future Studies intends to cast possible and emerging future scenarios (Bengston, 2019; Dragt, 2017; Powers, 2019). However, they do not fail to see in their analyses of the past and present the patterns, repetitions and stabilities, or instabilities, in mindsets that allow us to build solutions based on more emerging or stable/long term trends.

Given the importance of understanding this approach to the analysis of sociocultural patterns, the definition of trend, which assumes an important role in the projection of futures, must be clear. We developed our research anchored in a semi-systematic (or narrative) literature review to understand how the term 'trend' is seen by several leading authors who have published works on the subject. Snyder says that this typology of literature review contributes to understanding the evolution of certain topics over time (Snyder, 2019, p. 335). It is important to underline the role of this review, given the interdisciplinarity that guides the studies of trends and futures, which requires a greater openness in investigation and in the search for works and references. In this sense, it contributes to the clarification of the term and a better conceptualization of it in the approaches.

2 Trend as a Direction of Change

The development of Trend Studies and its relationship with the study of the future is inevitably associated with the main object of study of the former: trends. Sociocultural trends are the materiality that link mindsets, patterns, and cultural objects. They make it possible to activate knowledge in the design of solutions and to understand the evolution of society in general. But it is also in trends, in their history and actuality that we find the seeds and narratives/storytelling of the future (Lopes 2019). In fact, the most consensual characteristic in the definition of trend is "change" (Dragt, 2017; Gomes et al., 2021; Gomes et al., 2018; Higham, 2009; Raymond, 2020; Vejlgaard, 2008). The change with the past that translates into a present that can give indications about the future, even if in the short term. The desire to anticipate outcomes to foresee problems or opportunities has led to curiosity and stimulated the creation of scenarios so that, strategically, individuals could prepare their daily lives and projects and build the future. According to Kongsholm and Frederiksen, "understanding trends can be a crucial competitive parameter. If you can see and understand the trends that are on their way, and you can respond to them, then you can adapt your products, services, design, and communication and ensure a higher probability of success" (Frederiksen & Kongsholm, 2018, p. 19).

Trend Studies do not define future scenarios, but they provide guidance and contribute to an analysis on the threshold of change, facilitating strategic creation and understanding of the future. In Trend Studies, we work within the scope of a critical and present analysis of the most diverse perspectives and objects capable of representing what Williams (1961) and Eliot (1948) assume as 'a way of life' and Hall (1997) as 'signifying practices'. Perspectives increasingly oriented towards the tension between ways and styles of life. The sociocultural patterns associated with trends are an instrument to understand how they represent the collective imagination and thus transfigure into specific practices that carry meanings for specific lifestyles.

To continue this discussion, it is important to delve deeper into the nature of the concept of 'trend' within different approaches. Vejlgaard refers to the imprecision of the definition of this term and criticizes its glamorized use, which results from new creations within design (Vejlgaard, 2008). Kongsholm & Frederiksen (2018) also reflect on the imprecision of this concept, which was initially used to indicate the changes present in statistical curves and in the economy. The same authors also reflect on the ephemerality of trends and the great impact they have on people's lifestyles, both consciously and unconsciously (Kongsholm and Frederiksen, 2018). Likewise, Devon Powers (2018) explains that in the 1990s the use of this terminology fell into popular discourse and trends consolidated themselves as the 'lingua franca' of cultural changes. The popularization of 'trend forecasters' helped to underpin strategic cultural analysis to generate solutions for markets (Powers, 2018). In Dragt (2017) we see a plurality of characteristics for the definition of trend that are important to highlight: (1) the direction of change (which refers to understanding the trend dissemination and dissemination groups); (2) values and needs (which represent the mentalities and attitudes of the population); (3) the forces (that inhabit the sociocultural and symbolic system where the trend is analysed); (4) the manifestations of the trend (which are related to social changes and are transversal to people, culture and segments) and (5) to specific groups (to understand how the trend is disseminated in society; who is responsible for adopting behaviours earlier and even who is responsible for identifying and sharing new manifestations with communities) (Dragt, 2017, pp. 36–39).

As Cramer et al. refers, if we 'ask ten random people what a trend is and we will get ten different definitions' (Cramer et al., 2016). The term develops around the idea of change and adapts to the specific language of different areas (Dragt, 2017; Higham, 2009; Kongsholm & Frederiksen, 2018; Powers, 2019; Vejlgaard, 2008). However, the oscillations between the definition of change indicate that the issues of alterations, directions, variations, or oscillations (broadly) seem to be associated with the trend definition itself. Vejlgaard tells us that "to a trend sociologist, a trend is not something that has happened, but rather a prediction of something that is going to happen in a certain way—specifically, something that will be accepted by the average person" (Vejlgaard, 2008, p. 7), that is, the speculation of evolution is also latent in the sociocultural directions and changes that emanate from society. This is present in some of the main authors of the approach such as Raymond (2020, pp. 14–18); Kjaer, (2014); Mason et al. (2015, pp. 52–55); Dragt (2017, pp. 38–41); (Kongsholm & Frederiksen, (2018, pp. 25–37); Powers (2019, pp. 6, 9–12).

3 Perspectives on Defining What is a Trend

In this sense, it is important to confront the different perspectives on the concepts and understand how they impact the nature of Trend Studies and their relationship with the future. Campos analysed different authorial perspectives on the categories of trends, arriving at a discussion on the categories of macro and micro trends (Campos, 2020). Our study addresses the trend-setting characteristics themselves. Table 1, based on our semi-systematic review, seeks to summarize the various definitions of trend worked over the last two decades by different authors:

Consequently, we can observe that in addition to the issue of change already mentioned, the understanding of what a trend is involves understanding what goes on in people's minds (Waters, 2000) and that social phenomena are a trigger to understand the change that is at the origin of the research of trends (since a moment of rupture in the routine attracts attention), as we see in Dragt (2017); Erner (2010); Kjaer (2014) and in Gomes et al. (2018). In this way, although we are dealing with sociocultural trends and

Table 1. Trend definition by different authors who have published books on the subject. Developed by the authors.

Waters (2000)	"Trends are indicators that point to what's going on in the hearts and minds of consumers." (p. 1)
Caldas (2013)	"(...) the word and the concept are trivialized to the point of being empty of meaning. [...] we agree, without much reflection, with the idea that almost everything in the world is in motion and tends towards some other position [...] In short, the concept of 'trend' that has become widespread in contemporary society was built based on the ideas of movement, change, representation of the future, evolution, and quantitative criteria" (pp. 7–8)
Gloor e Cooper (2007)	"We see a promising trend: companies tapping into the swirling network of ideas around the globe that may not even correspond directly to the business the executives think they're in." (p. 3) "For example, a music lover and hobby musician on the jury had a tendency to overrate educational music productions. The other jurors noticed this trend" (p. 42)
Erner (2010)	"The same word [trend] serves to designate the movements [...] of society and subterranean phenomena, perhaps condemned to remain invisible. Even more: the notion of trend can designate both futile phenomena [...] as well as much more serious issues. [...] this term designates commercial phenomena, it can also designate objects that ignore any economic logic." (p. 13)
Vejlgaard (2008)	"To a trend sociologist, a trend is not something that has happened, but rather a prediction of something that is going to happen in a certain way—specifically, something that will be accepted by the average person." (p.7); "a trend sociologist will talk about a trend moving from the trendsetters to mainstream. In this case, trend refers to a process of Change." (p. 8)
McCracken (2008)	"Circuit City, recently deceased, failed to take advantage of the 'Geek Squad' trend that helped lift competitor Best Buy to greatness." (p. 12); "Property developers were surprised by the loft condo trend that hit American cities in the 1980s" (p. 12); "Sometimes status is the wind beneath the wings of a trend. For example, the single-malt scotch trend of the 1990s was driven in part by status consideration" (p. 03); Consider the artisanal trend that changed the way Americans thought about chocolate, beer, and bread (p. 8); "in Chapter 4, I offered a picture of the hippie trend" (p. 9)

(continued)

Table 1. (*continued*)

Higham (2009)	"A trend is defined as 'A line of general direction or movement… a prevailing tendency or inclination… The general movement over time of a statistically detectable change… [or] a current style or preference" (pp. 14–15); "a trend is: 'a change that occurs among consumers'" (p. 15); a trend is best defined as: 'a long-term change in consumer attitudes and behaviours that offers marketing opportunities' (p. 16)
Mártil (2009)	"Trend: it is the prelude to fashion. We speak of a trend when something new begins to be adopted by a critical mass of consumers capable of generating in the group the feeling that this novelty should be adopted." (p. 31)
Raymond (2010)	"A trend can be emotional, intellectual, and even spiritual. At its most basic, a trend can be defined as the direction in which something (and that something can be anything) tends to move and which has a consequential impact on the culture, society, or business sector through which it moves. The word 'trend' is an old one, and its origins can be traced back to Middle English and High German, where it meant 'to turn' or 'to spin' or 'to revolve'." (p. 14)
Mackinney-Valentin (2010)	"(…) the term trend is not without problems. As mentioned earlier in this chapter, there are trends in most everything – also terminology. It might even be argued that the term trend was a result of a trend from the point of view that the fashion media, designers, and consumers grew tired of using fashion to describe changing styles and adopted trend as a new and fresh term." (p. 36); "the Retro Trend is defined on this premise" (p. 116)
Maioli et al. (2013)	"The concept of trend is very complex, but we can say that they are evolving phenomena that indicate new elements in the sociocultural system and that, in the real world, imply a path: the world, the house, the body, fashion, the object". (p. 22–23)

(*continued*)

Table 1. *(continued)*

Rehn & Lindkvist (2013)	"Trends are synonymous with booms, the new product, the hip thing. Trends make us think about progress and rising curves" (p. 10); "*Microtrends* are the little things that happen all around us all the time, the tiny shifts that occur in everything from the clothes we wear and the snacks we eat to the way we work, play and love." (p. 10); "*Macrotrends* are, depending on whom you ask, either aggregated microtrends or more sweeping changes that are affecting society— things we can already see" (p. 11)
Kjaer (2014)	"(…) the process of managing trends involves observing specific changes or advances, as well as considering the general direction in which society is moving" (pp. 1–2)
Mason et al. (2015)	"a consumer trend is a new manifestation among people in behaviour, attitude, or expectation of a fundamental, human need, want or desire." (p. 46)
Cramer et al. (2016)	"A trend is about a change, which means it is aimed at innovation, the "new."; A trend can both be quantitative and qualitative in nature […]; The focus on the qualitative dimension, changes and the "new" also means that basing a trend on historical data is not sacrosanct. We should not think that a trend is predictable in advance. After all, historical data do not always continue into the future. […]; A trend has to "have a future," which relates to the expectation that a trend will last for a while. […]; a trend has a certain degree of uncertainty, because it is uncertain in advance whether a trend will in fact continue into the more distant future. […]; The subjectivity of a trend is also expressed in the concept of counter-trend […]; trend can be linked to an "event," i.e., a seemingly separate occurrence that can be a symptom of a trend, and therefore a "weak signal"." (pp. 43–43)
Dragt (2017)	"A trend is a direction of change in values and needs which is driven by forces and already manifests itself in various ways within certain groups in society." (p. 14)

(continued)

Table 1. (*continued*)

Holland & Jones (2017)	"Trends have been around for many years - some academics date the beginning of trends to the fifteen centuries. Since then, they have been informed by different individuals and groups, and shaped by external factors such as technology and politics […] For centuries, the progression in styles of dress was driven by changes in the ruling classes. […] Today, trends are mostly heavily influenced by fashion professionals and by consumers' own lifestyle […] we are more exposed than ever to changes in design, lifestyle and power - and to a more diverse range of influences, meaning that trends can now change quickly." (p. 9)
Gomes et al. (2018)	"(…) trends themselves act as forces of change, causing changes among themselves and in associated mentalities" (p. 56); "the trend is considered as a direction of changes and values represented by mentalities, which will manifest itself in society in different ways, from objects, patterns of behaviour, among others." (pp. 60–61)
Kongsholm and Frederiksen (2018)	"There are vast numbers of suggestions to what the term means, but perhaps also an indicator that it is not something that there is a precise definition for, and therefore many struggles with the concept." (p. 44); "Trend continuously emerge, disappear and have an impact on everything from the way we live and work to what we eat and drink and the clothes we wear, as well as what we spend spare time on. Trends affect the values and that characterise our attitudes and actions - a direct (conscious) or more indirect (subconscious) impact, depending on whether we recognise a trend's existence or not" (p. 44)
Powers (2019)	"(…) trend is a label for change. The nature and extent of the change are variable, as is the way in which those changes are documented or seen. But a trend does not exist unless something has transformed, will transform, or is in the process of transforming." (p. 9); "Trends summarize the imminent now and illustrate the distant future; they magnify tiny percolations and distil immense chaos." (p. 12)

within the framework of collective mindsets, the conception of trend changes based on the sector and the context in which it is applied (Vejlgaard, 2008). As such, it is important to contextualize the discourses where they are applied (Erner, 2010; Gloor & Cooper, 2007; Kongsholm & Frederiksen, 2018) to understand the paths and changes that are currently taking place in society (Powers, 2019). In this way, in a proposal of cultural analysis, understanding trends is not directly related to future scenarios, but to the visions of the present that allow us to draw strategies or foresee possible positions based on a diachronic analysis of culture (Gomes et al., 2018; Vejlgaard, 2008). However, it is also important to emphasize that some authors confuse the definition of trends with that of the object (the result of trends), that is, their manifestation, as is the case of Vejlgaard (2008), Erner (2010); McCracken (2008) or Mackinney-Valentin (2010). This must be considered when the operationalization of concepts is necessary, as the analysis of texts (cultural objects) will allow us to understand the meanings and, consequently, the readings of the patterns behind these objects, thus identifying the DNA of the trend. Thus, we can say that a trend is "a process of change that (sometimes) comes about because of product development that (sometimes) results in new products", (Vejlgaard, 2008, p. 8) and which is represented by a social pattern that spreads, becoming widespread (Powers, 2018) and revealing social values and mindsets (Dragt, 2017).

4 Final Considerations

In this text we reviewed several definitions of the concept of trend by several relevant authors through a semi-systematic review of the literature. The various contributions that have formed the field that defines the methodological proposal of Trend Studies deserve to be disseminated and studied. This work sought to contribute to a clarification of a concept widely used in several areas, such as Future Studies.

Through this literature review we contributed to the consolidation of the trend definition through the various compiled works. It was possible to perceive that a trend is a representation of what goes on in people's minds, a specific pattern of mindsets in a temporal context. It is also the system behind social phenomena, dynamics and flows that must be mapped continuously. A trend stands for a network of meanings that propagate through the collective conscious and unconscious creating common mindsets. A transmission system via communication and contact with the visible aspects of behaviors/rituals, representations, discourses, and artefacts. A system that highlights the preferences of each moment, with a long-term nature, filled with changes (smaller and larger), and prospects of future development. A trend can show movements with specific directions, that may start in the fringes and the more invisible contexts of lifestyle groups, which will become impactful and generate a process of change with new elements, affecting a large number of individuals and their values/world views, underlining what is eminent.

At the same time, we understand it as a malleable concept from sector to sector, which emphasizes the need of contextualizing the concept when using it. In the future, this research can look for more scientific works on the subject since it is a topic of extreme relevance in the specific literature and that has been expanded over time. We see a future approach to this definition through a systematic review of the literature as beneficial, highlighting the difficulties that will arise, given the plurality of areas and existent materials focused on this topic in the market.

References

Bengston, D.N.: Futures research methods and applications in natural resources. Soc. Nat. Resour. **32**, 1099–1113 (2019). https://doi.org/10.1080/08941920.2018.1547852

Caldas, D.: Observatório de Sinais: Teoria e Prática da Pesquisa de Tendências. E-ODES, São Paulo (2013)

Campos, A.: Taxonomia dos tipos de tendências: uma proposta concisa. DAT J. **5**, 313–328 (2020)

Cramer, T., van der Duin, P., Heselmans, C.: Trend analysis. In: van der Duin, P. (ed.) Foresight in Organizations, pp. 40–58. Routledge, New York (2016)

Dragt, E.: How to Research Trends: Move Beyond Trend Watching to Kick Start Innovation, 1st edn. BIS Publisher, Amsterdam (2017)

Dragt, E.: How to Research Trends Workbook. BIS Publisher, Amsterdam (2018)

Eliot, T.: Notes Towards the Definition of Culture. Faber and Faber, London (1948)

Erner, G.: Sociología de las tendencias. Editorial Gustavo Gilli, Barcelona (2010)

Gloor, P., Cooper, S.: Coolhunting: Chasing Down the Next Big Thing. AMACOM, New York (2007)

Gomes, N., Cohen, S., Cantú, W., Lopes, C.: Roteiros e modelos para a identificação de tendências socioculturais e a sua aplicação estratégica em produtos e serviços. Modapalavra e-periódico **14**, 228–272 (2021). https://doi.org/10.5965/1982615x14322021228

Gomes, N., Cohen, S., Flores, A.: Estudos de Tendências: contributo para uma abordagem de análise e gestão da cultura. ModaPalavra **11**, 049–112 (2018). https://doi.org/10.5965/1982615x11222018049

Hall, S.: Representation: Cultural Representations and Signifying Practices. SAGE, London (1997)

Higham, W.: The Next Big Thing: Spotting and Forecasting Consumer Trend for Profit. Kogan Page, London (2009)

Holland, G., Jones, R.: Fashion Trend Forecasting. Laurence King Publishing (2017)

Kjaer, A.L.: The Trend Management Toolkit. Palgrave Macmillan (2014). https://doi.org/10.1057/9781137370099

Kongsholm, L., Frederiksen, C.: Trend Sociology v.2.0. Pej Gruppen, Herning (2018)

Lopes, M.V.: The discourse of fashion change: trend forecasting in the fashion industry. Fashion Style Popular Cult. **6**, 333–349 (2019). https://doi.org/10.1386/fspc.6.3.333_1

Mackinney-Valentin, M.: On the nature of trends: a study of trend mechanisms in contemporary fashion. Ph.D. dissertation. The Danish Design School, Copenhagen (2010)

Mártil, V.: Coolhunting: El arte y la ciencia de descifrar tendencias. Ediciones Urano, Barcelona (2009)

Mason, H., Mattin, D., Luthy, M., Dumitrescu, D.: Trend Driven Innovation. Wiley, Hoboken (2015)

McCracken, G.: Chief Culture Officer. Basic Books, New York (2008)

Powers, D.: Towards a futurist cultural studies. Int. J. Cult. Stud. **23**, 451–457 (2020). https://doi.org/10.1177/1367877920913569

Powers, D.: On Trend: The Business of Forecasting the Future, Chicago (2019)

Powers, D.: Thinking in trends: the rise of trend forecasting in the United States. J. Histor. Res. Mark. **10**, 2–20 (2018). https://doi.org/10.1108/JHRM-09-2016-0021

Raymond, M.: The Trend Forecaster's Handbook, 2nd edn. Laurence King, London (2020)

Rech, S.: Trends to innovation: an investigative method in trends studies. In: 14° Colóquio de Moda. Associação Brasileira de Estudos e Pesquisas em Moda, pp. 2–16 (2016)

Rech, S.R., Nascimento, J.: Triângulação Cultural nos Estudos de Tendências. In: 13° Colóquio de Moda, pp. 1–8. UNESP, Bauru (2017)

Rehn, A., Lindkvist, M.: Trendspotting - The Basics. Booktango, Bloomington (2013)

Snyder, H.: Literature review as a research methodology: an overview and guidelines. J. Bus. Res. **104**, 333–339 (2019). https://doi.org/10.1016/j.jbusres.2019.07.039
Vejlgaard, H.: Anatomy of a Trend. McGraw-Hill, New York (2008). https://doi.org/10.1036/0071488707
Waters, R.: The Trendmaster's Guide. Portfolio, New York (2000)
Williams, R.: The Long Revolution. Penguin Books, London (1961)

Axé Ilê Obá: Garments of Candomblé

Aymê Okasaki(✉)

University of São Paulo, São Paulo, Brazil
ayme.okasaki@usp.br

Abstract. This article seeks to reflect on the clothes of the Queto Candomblé house, Axé Ilê Obá. The objective is to investigate the garments present since the foundation of the terreiro, throughout the succession of its three religious' leaders: Father Caio of Xangô (1950–1985); Mother Sylvia of Oxalá (1986–2014); and Mother Paula of Iansã (2015-) and identify whether there are aesthetic changes in the clothing worn in the house during the tenure of each religious leader. For a comparative study of the attire in the three moments of the terreiro, photographs from the periods of Father Caio and Mother Sylvia were collected through researchers written by members of the terreiro (especially Mother Sylvia's book, *O perfil do Aché Ile Obá*, 1980), as well as the photographic collection of Axé Ilê Obá, which was consulted. To analyze the current attire, in addition to field research conducted since 2017, interviews were also conducted with clothing manufacturers that produce garments for the terreiro, images were requested from the official photographers of the house, and an interview was conducted with Mother Paula. The importance of analyzing the changes in these moments is to verify the visual impact that the religious leaders in Candomblé promote in the attire of everyone, according to their personal taste, the divinity of this religious leader, and their personal relationships with the clothing producers. Thus, we understand how these leaders of a Candomblé act as dress code guide for the Candomblé house, impacting this market of Candomblé garments, the axós.

Keywords: Axó · Candomblé · Axé Ilê Obá

1 Axós: Candomblé Attire

The term "axó" refers to clothing or fabric in the Yoruba language. In the context of Candomblé, it specifically denotes the garments worn by the members or initiates of this religion. Candomblé is an Afro-Brazilian religion that worships African deities. Candomblé terreiros/houses are divided into different nations, each with distinct cosmogonies. For example, the Queto-Nagô nation worships the orixás, the deities of Yoruba culture (an ethnic-linguistic group from the regions of Nigeria, Benin, and Togo), while the Congo-Angola nation venerates the inquices, deities from the Bantu macro-group. The Jeje nation, on the other hand, worships the voduns, deities from the ancient Kingdom of Dahomey, among other nations. These nations also exhibit variations in their clothes. However, it is important to highlight how different African cultures converge and transform within Afro-Brazilian religions, fostering interaction and reciprocal modifications between the nations in Candomblé. Additionally, these religions receive influences from the practices of Brazil's indigenous peoples, Catholicism, and other religions. In the Queto nation (the focus of this research), the worship is dedicated to the orixás, anthropomorphic deities from Yoruba culture. Candomblé has a hierarchical structure with well-defined initiation and passage rituals that determine an individual's position within the Candomblé terreiro. Terms such as "terreiro," "abaçá," "house of axé," and "roça" are used to refer to the places where Candomblé ceremonies take place. The anthropologist and professor Vagner Gonçalves (1995) identifies various positions within Candomblé, ranging from the initial stages to leadership roles, as presented in Table 1:

Table 1. Positions in Candomblé. Adapted from: Silva (1995) [11].

Abiã (from Yoruba, meaning the chosen one)	Uninitiated. Wears only white clothes in reverence to the orixá of creation, Oxalá
Iaô (from Yoruba, meaning spouse)	Initiated, up to the 7th year of initiation. After this initiation rite, the individual becomes *Ẹl'ẹgùn*, entering a trance state and embodying the energy of the deity. In the skirts of women, one can find ribbons sewn at the hem, with each ribbon representing a year of fulfilled obligations (up to 7 satin ribbons)
Ebomi (from Yoruba, meaning *older sibling*)	Initiated with more than 7 years of initiation. Also, an *Ẹl'ẹgùn*
Equede (from Yoruba, meaning *second*)	Women who do not enter into trance. Assist the mother of the saint during rituals and public festivities, taking care of dressing the orixás. Equedes and ogãs are "raised" to their positions without undergoing the initiation rite like iaôs
Ogã (from Yoruba, meaning *to promote* or *to exalt*, and from Fongbe, meaning *chief*)	Initiated individuals who do not enter into trance and may have different functions
Babalorixá (from Yoruba, meaning *father*) or Ialorixá (from Yoruba, meaning *mother*)	Ebomis who have initiated other followers and usually have their own terreiro (father or mother of the saint, respectively)

These different positions are reflected in the attire worn. As a religion that values tradition and seniority, the longer one has been in Candomblé, the higher their status. Following the hierarchy of positions within Candomblé, individuals with higher seniority are allowed to wear more elaborate garments, adorned with additional ritual necklaces, and so on. Additionally, there is a second division of attire that corresponds to different moments within the terreiro. There are everyday garments worn during routine activities, attire for public festivities (which are more elaborate and denote hierarchical ranking, following the colors associated with the orixá being honored on that day), and attire specific to the orixás themselves. The orixá's garments are designed with the colors and emblems associated with each deity and are accompanied by the tools and symbols that represent the orixá's specific myths. This attire is worn exclusively during trance rituals, when the deity is embodied and performs dances that symbolically recall their respective myths.

2 Candomblé Terreiro Axé Ilê Obá

The Candomblé terreiro Axé Ilê Obá (a term in Yoruba meaning "the strength of the house of the king") belonging to the Queto nation. Currently, this house has a great importance in São Paulo. It stands out not only for its expansive physical dimensions, which are notably large for the context of the State of São Paulo (with a total area of 4000 m^2), but also for the considerable number of members it has, as well as being the first terreiro to be recognized as a historical heritage in São Paulo. In 1990, it was officially listed by the Council for the Defense of Historical, Archaeological, Artistic, and Touristic Heritage of the State of São Paulo - CONDEPHAAT. As a result, this space attracts both: religious members of Candomblé and researchers of Afro-Brazilian culture in São Paulo. This is evident through numerous visual records and research studies conducted on the house and its members, providing a means to analyze its historical processes. In this particular study, we will focus on the attire of the Abaçá (the temple of Xangô) Axé Ilê Obá, examining its evolution since its foundation and spanning the succession of its three religious' leaders. The aim is to understand the significant changes in the clothing worn and how these continuities and variations in attire can serve to document and narrate the history of the house [2].

Like many terreiros in the city of São Paulo, this house began as an umbanda terreiro, which was the predominant Afro-Brazilian religion in the early second half of the 20th century, in São Paulo. On July 22, 1950, the Beneficent Spiritist Congregation Father Jerônimo Umbanda terreiro was founded by Caio Egydio de Souza Aranha, known as Father Caio of Xangô, along with a group of women, in the Brás neighborhood, where it remained until the mid-1950s. Saint Jerônimo is syncretized with the orixá Xangô in São Paulo, referring to Caio Egydio's orixá. One of the many differences between Umbanda and Candomblé (particularly Queto Candomblé) is the worship of entities such as caboclos (see Fig. 1 - nomenclature of the Brazilian colonial period for the person resulting from the miscegenation between the Brazilian indigenous and the black), pretos-velhos (refers to elderly black people who lived during the Brazilian slave period), exus (not the orixá Exu, but the entity, who provides consultations to people), pombas-gira (considered a female Exu), and others. Father Caio's Umbanda terreiro maintained these practices from its inception in 1950 until the present day, having transformed into a Queto Candomblé terreiro [2].

In the year 1960, the terreiro underwent a change of location, moving to Macuri Street in the Jabaquara neighborhood, and underwent a transition, becoming the Axé Ilê Obá Candomblé house. The relocation was motivated by the desire to establish themselves in a larger space, with closer proximity to nature and farther away from the city center and police presence (at that time, there was still significant police repression against Afro-Brazilian religious terreiros). In 1974, Father Caio initiated yet another relocation to 77 Azor Silva Street in Vila Facchini, Jabaquara, and the spacious facilities were officially inaugurated in 1977 [6].

Fig. 1. Headdress with natural feathers for the "caboclo Rompe Mato" spirit of Father Caio (still influenced by Umbanda, the origin of this and many houses in São Paulo) (Photograph courtesy of Axé Ilê Obá Archive, 2023)

The *religious crossroads*[1] that took place in Axé Ilê Obá was evident in the attire worn, even during the Umbanda terreiro days. Umbanda caboclos and forest deities often don plumage and feathers in their headdresses, as well as fabrics adorned with leaf and flower patterns, a tradition that has been maintained since the time of Father Caio, through the era of Mother Sylvia, and up to the present day. It is interesting to note that some of the older members of the terreiro still preserve at least the characteristics or even the actual garments from the time they entered the house (such as the attire worn by Jacira of Oxum, see Fig. 2, initiated by Father Caio, who wear the clothing of Oxum from Father Caio's era). Consequently, the older members of the house possess some distinct attire compared to the current members, with a greater use of chita fabric, for instance, in the caboclos' costumes.

2.1 Caio Egydio de Souza Aranha (November 25, 1925–February 15, 1985)

Father Caio of Xangô was initiated into Candomblé in 1941 in Salvador by Aunt Massi, from *White House of Old Sugar Mill* terreiro in Bahia. Father Caio was a son of Xangô and

[1] In the culture of Candomblé, the term "crossroad" is associated with the orixá Exu, who governs the paths and intersections that connect the spiritual and earthly planes. The concept of "crossroad" is expanded upon by various authors (such as Leda Maria Martins, Luiz Rufino, among others) to signify the blends and unions that occur within the context of Candomblé. Thus, I employ the term "crossroad" as a substitute for concepts such as syncretism, mixture, and association, as I consider that these associations present their own complexities and issues, but ultimately result in new contexts and pathways.

Fig. 2. Attire of Oxum worn by Jacira, the oldest initiated woman who is still present and active in Axê Ilê Obá. There is a little change in the fabrics: in the painting by artist Agnes Farias dos Santos, made between 1980's-1990's, displayed on the walls of the terreiro, Oxum wears a skirt made of yellow/gold satin (or fabric with a satin-like sheen) and blue lace; in the central image, the skirt and bow are made of lamé fabric, with more shine; and in the more recent image on the right, the bow is made of transparent fabric with embroidery, and the skirt once again features lace or embroidery on tulle, in blue (Photograph courtesy of Axé Ilê Obá Archive, 2023)

had Oxum as his juntó orixá, his second orixá (Oxum being a female orixá associated with gold, which is reflected in the attire of Father Caio, as seen in Fig. 3, with brocade fabric woven with metallic threads, Lurex). At the time, the initiation of men was uncommon in the terreiros of the original Bahian houses, which are the first recorded Candomblé terreiros [2].

In 1949, Father Caio receives the initiatory seniority of 7 years, represented physically by a string of beads, from the renowned Menininha of Gantois. Thus, he becomes a babalorixá, the leading priest of a Queto Candomblé terreiro [2].

Father Caio was a popular singer of samba and operettas at *Spell Nightclub*, located on Avenue São João in São Paulo during the 1950s–1960s, which allowed him to meet many artistic and important personalities, such as the humorist Dercy Gonçalves [2]. These contacts also helped increase his clientele for the practice of divination with cowrie shells and other spiritual consultations and services. This is an important factor to demonstrate that the terreiro has always been closely connected to prominent individuals in São Paulo, whether in the artistic or political sphere, as we will see in the phase of Sylvia Egydio. This aided in the continuity of the house and its significant growth.

When Father Caio passed away in 1985, without leaving any biological children, the house faced many legal issues. Initially, a dispute arose over who would inherit the physical space of the terreiro, followed by who would assume the leadership of the house. Two years before Caio's death, he had initiated his niece, Sylvia of Oxalá. After numerous visits to other terreiros and consultations with cowrie shells, Sylvia of Oxalá assumed the position of *ialorixá*, woman leader of a Candomblé terreiro. Additionally, there was a legal dispute over the ownership of the land, which Sylvia was able to resolve by acting and requesting the recognition of the house as a historical heritage site [2].

During this period, the inspiration of the clothing worn in Bahian terreiros on the attire of Axé Ilê Obá is evident in photographs from that time. Men, especially the Ogãs

Fig. 3. Father Caio of Xangô in his red velvet chair at Axé Ilê Obá, wearing metallic brocade attire (Egydio, 1980, p. 29) [3]

(a male role that does not incorporate the orixás and may have various functions such as playing the drums of Candomblé, organizing the room during public rites, making offerings, etc.), wore white suits during public festivities (see Fig. 4). This also relates to Bahian terreiros, as many of the original houses did not initially initiate men, leading them to wear clothing not necessarily tailored for candomblé but rather "Sunday best attire" (the best outfits that, especially in the 19th century, were reserved for use in Catholic Sunday masses). In an interview with ialorixá Mother Paula of Iansã, one possibility for the use of suits could be to avoid appearing as practitioners of Afro-Brazilian religions, due to prohibition and strong religious prejudice during that period[2].

The women, on the other hand, appear in attire commercially known as "baiana", featuring a triangular silhouette with greater volume in the skirts due to starched petticoats. The traditional female baiana costume typically includes petticoats; a long pleated skirt with a waistband and drawstring (usually patterned or made of colorful fabrics in the colors of the orixás of the public celebration, or white for non-initiates); a underskirt called "calçolão," petticoats, and "quebra-goma" to give volume to the skirt; a blouse

[2] Egydio, Paula Regina Maria. **Interview**. [05 jul. 2023]. São Paulo. Interview given to the author.

Fig. 4. Ogãs of Axé Ilê Obá (Sylvia, p. 34, 1980) [3]

(for ebomis and higher-ranking positions) or lace or embroidered camisole; a "pano-da-costa" (a fabric worn across the back); a headscarf (made of the same fabric as the "pano-da-costa"); sandals or mules, in addition to beads or ilequês [12].

Clothing and fabrics serve as markers and differentiators of hierarchy, not only during the time of Father Caio but also in the subsequent phases of Sylvia of Oxalá and Paula of Iansã, with the use of "richelieu" embroidery by high-ranking positions. This openwork embroidery can feature various cut-out figures (leaves, flowers, arabesques, or even letters and insignias of orixás - as mentioned by Bianca Almeida in an interview[3] regarding a blouse worn by the former ialorixá of Axé Ilê Obá, Sylvia of Oxalá, adorned with dove figures, which has been inherited by the current ialorixá of the terreiro, Paula of Iansã – see Fig. 5) outlined.

In the French court of Louis XIII, richelieu embroidery was primarily used for Catholic priestly garments (a fabric associated with Cardinal Richelieu Armand-Jean du Plessis, the prime minister of Louis XIII's court) and the nobility. In the context of candomblé, the fabric also adheres to a hierarchy and specific moments of use, being predominantly reserved for festive attire, divine garments, and high-ranking positions (such as ialorixás, equedes, and ebomis). Within terreiros, richelieu embroidery is executed on sturdy cotton fabric [5]. It serves as a distinctive technique denoting hierarchy.

The more complex the design, the more expensive the fabric will be, especially if it is hand-embroidered or machine-stitched. The fabrics that signify hierarchy within terreiros need to be visually appealing and of high quality, although they do not necessarily have to be artisanal or excessively costly. Richelieu embroidery is predominantly found in the garments of elders or heirloom attire, as well as in decorative details for festive ceremonies.

In general, these hierarchical fabrics tend to be more expensive. It is worth noting that in 1749, in Brazil, the use of fine cotton fabrics was prohibited for black individuals according to the Extravagant Law. Furthermore, in 1785, Queen Maria I banned the

[3] Almeida, Bianca Batista. **Interview**. [06 sep. 2019]. São Paulo. Interview given to the author.

Fig. 5. Attire of the ialorixá Paula of Iansã, in richelieu embroidery, featuring dove motifs - left; and attire of Paula of Iansã with seashell motifs, both worn during Oxalá celebrations (Photograph courtesy of Eduardo Cancissú, 2019).

manufacturing of fabrics in Brazil, allowing only the production of coarse fabrics that were used by enslaved people. Therefore, there is a revival and utilization of some fine fabrics in candomblé that were previously forbidden to the entire black population [1].

During the time of Father Caio, the attire of the orixás combined a rich blend of African, European, Catholic, and Umbanda influences, often adorned with a significant amount of shimmer and shine. Fabrics with Lurex threads, satin materials, and visually similar tools to those found in the candomblé terreiros of Bahia were commonly used. These tools were crafted from materials such as tinplate, brass, copper, and nickel-plated zinc [2].

The ialorixá who would take the place of Father Caio, Mother Sylvia of Oxalá, emphasized that the most important aspect at that moment was for the house to be colorful and that they sought to maintain traditions (referring primarily to the practices carried out in the terreiros of Bahia):

> Without losing its simplicity, it brings luxury and beauty to its accommodations. Everything is meticulously crafted and cared for, presenting, in a colorful visual display, the true image of African deities. [...] By adapting to the reality of an urban society, opening its doors for progress to enter with its technologies and benefits, it sought to do so without altering its customs and traditions (Egydio, 1980, pp. 9-10) [3].

2.2 Sylvia Egydio (July 15, 1938 – August 8, 2014)

In 1986, Sylvia Egydio assumed the position of ialorixá at Axé Ilê Obá. This also marked a strong aesthetic division in the attire of the terreiro, as the ruling orixá of the ialorixá became a fun-fun orixá, that is, associated with the color white. It is noticeable that there was a lightening of the tones used in the clothing and an increased use of pastel colors (as pointed out in an interview with Mother Paula of Iansã[2]).

Sylvia Egydio pursued studies in Business Administration and International Trade. It is evident how her academic trajectory in administration contributed to the management of the terreiro and to her political connections, which helped in maintaining the structure of the terreiro, especially during the period of preservation at CONDEPHAAT. To prevent the loss of the land, Sylvia initiated a preservation process with the Council for the Defense of Historical, Artistic, Architectural, and Touristic Heritage of São Paulo - CONDEPHAAT, declaring it as the first terreiro to be listed as heritage in São Paulo in 1990 [6].

In addition to her political relationships, in which her activist role was instrumental, Sylvia Egydio's religious connections with the mother houses of candomblé in Bahia (White House, Gantois, Silver Pestle, etc.) also contributed to another important terreiro, reaffirming its status as a queto nation house. In this confirmation as a Queto nation terreiro, the pursuit of titles in Nigeria becomes an element of differentiation among the houses in São Paulo. In 1995, Mother Sylvia traveled to the city of Oxobô in Nigeria, where she received the title of Ianifá, a woman knowledgeable about the oracular divination system of Ifá [2]. These titles are marked by specific clothing and attire, such as the yellow pano-da-costa (the color of the orixá Oxum, deity of the city of Oxobô), the adê headdress, the staff of authority, and the bead necklace that Mother Sylvia wears in Fig. 6 upon her return to Brazil. The yellow fabric of her pano-da-costa is an axó oke, an artisanal fabric woven on a narrow loom by Yoruba men. Her professional and academic experience in international relations and commerce aids in her travels to the African continent.

This is not an isolated movement of Axé Ilê Obá. Already in the 1980s, the terreiros in São Paulo began to seek greater autonomy, reaffirming themselves as Yoruba Candomblé and seeking their references not from the mother houses in Bahia or from Umbanda, but by establishing direct connections with Nigeria and Benin. Many candomblé members underwent initiations and received titles during the various trips made to Nigeria and Benin. With African titles, the terreiros gained renown, attracting clients, and expanding their establishments. It is a process of (re)africanization that shapes the identity of candomblés in São Paulo [9]. In these recurrent trips to Africa, it is not only titles that are transported but also fabrics (such as industrially printed wax fabrics and axó oke fabrics), clothing, and information about material culture. The terreiros in São Paulo accumulate information - without eliminating it - combining the aesthetics of Umbanda and Bahian Candomblés with what is found in Africa during this period. And the clothing is the exponent that demonstrates these changes, phases, and adaptations of the Candomblés to the urban contexts of the city of São Paulo.

Many candomblé terreiros began to identify themselves as (re)africanized. In the (re)africanized terreiros, both male and female attire undergoes a different composition: the grand bou bou, agbada, or riga tunics; the buba blouse; the shokoto pants; the *filá* or

Fig. 6. Sylvia of Oxalá with pano-da-costa made of axó oke fabric, and wearing insignias she received in Nigeria (right) representing her title of Ianifá (Peres, p. 17, 1997; Negrão, p. 61, 2018) [6, 7]

iketè hat and *kufi*; the *isiagu* blouse; the *dàńṣíkí* blouse; the *jalabiya* tunic with Islamic influence; the axó oke fabric outfits; the *iro* wrapped skirt; the *gele* headwrap; the *pele* cloth on the shoulder; the *kaftan*; the *abeti-aja* hat, among other garments [10]. These clothes bear influence from Islamic attire used in the African continent (embroidery and elongated tunic styles, traditional in Islamic African garments).

In an interview with Mother Paula of Iansã, currently only two equedes wear a *kaftan*, with her preference being the so-called "complete Bahian" style[2]. The Axé Ilê Obá terreiro does not self-identify as a (re)africanized terreiro, meaning that they do not seek to eliminate elements considered syncretic (especially Catholic ritual elements) in pursuit of a cult with exclusively African rites and elements.

However, the overall movement of (re)africanization does influence the terreiro, particularly in terms of fabrics from Africa and the adoption of specific male clothing ensembles.

Starting from 1975, researcher Barbara Plankensteiner (2013, p. 18–25) notes an increased use of lurex threads, adding more shine to the fabrics used in Nigeria [8]. We can draw a parallel with what was happening in Candomblé terreiros, with the use of shiny fabrics, partly influenced by African garments as well. During Mother Sylvia's time, there was extensive use of satin due to its affordability and ability to provide shine to the attire. Additionally, there were variations of brighter fabrics such as lamé, as well as the use of sequins.

The presence of shiny garments is a characteristic that varies according to the region, candomblé nation, and hierarchy within the terreiro itself. It is permitted for higher-ranking positions, while fabrics with less shine are typically used for abiã, for instance.

The choice of fabric also depends on the deity present (for example, Oxum, as the orixá of gold, uses fabrics with more shine, while orixás from the straw family, voduns, such as Omolu, do not typically utilize shiny materials).

In the Axé Ilê Obá terreiro, shiny fabrics adorned with sequins often served as substitutes for metals, as was the case with Mother Regina's Xangô attire (see Fig. 7).

Fig. 7. Attire of Xangô in fabric, worn by ebomi Regina, between the 1980s and 1990s (Photograph courtesy of Axé Ilê Obá Archive, 2019)

During Mother Sylvia's time, the insignias, and tools of the orixás that traditionally utilized metal still incorporated tinplate, brass, and copper, as practiced in the Bahian tradition since the time of Father Caio. However, in the 1980s, there was an introduction of shimmer through the application of sequins on the garments. This material became more affordable due to the availability of new synthetic raw materials, and the sequin manufacturers themselves recognized the consumer demand within the terreiros, as demonstrated in an interview with Joaquim Silveira, a director of Francotex (a company employing 450 workers and producing approximately 150 tons per year of sequins, beads, brocades, and ribbons), published in *O Estado de São Paulo* newspaper (Faleiros, 2005, p. B6): "We sell a lot for festivals dedicated to the orixás." [4].

2.3 Paula Regina Egydio (1989-)

After Sylvia's death in 2014, her eldest daughter, Mother Paula of Iansã, assumes the position and has been serving as the ialorixá (the leading priestess of a Queto Candomblé terreiro) from 2015 to the present day.

Following in the footsteps of Mother Sylvia, Mother Paula is also an academic who has devoted herself to Candomblé (holding a degree in Physiotherapy and an MBA in Business Management). However, in addition to the presence of the house's ialorixá, her younger brother, Baba Péricles of Oxaguiã, was an influential figure in the attire of the terreiro too, particularly because he was engaged to the main clothing supplier of the house, Bia of Oxum from the Patuá atelier. This atelier, which served the terreiro until the year 2022, brought forth intriguing characteristics, such as the slim African ensembles for men (as we can see in the outfit on the right in Fig. 8). These are shorter, more fitted robes made of wax print fabric, with influences from tailoring, due to the fact that Bia of Oxum's mother, who is part of Patuá Clothing, already works in the party wear/tailoring fashion segment[1]. However, with Péricles of Oxaguiã's departure from the terreiro, the clothing ceased to be produced by the Patuá atelier. The new recommended clothing workshop by the house became Okan Rere atelier, run by Luana of Iansã and Isabela of Iemanjá. As a result, new techniques started to gain prominence, such as the barafunda embroidery, which the new atelier introduced to the ebomis' and orixas' clothing.

Fig. 8. Xangôs and ialorixá Paula of Iansã in 2019 (Photo courtesy of Eduardo Cancissu)

The ateliers of young women in the terreiro have brought specific fabrics and textile techniques, such as the revival of barafunda lace, an artisanal technique of removing threads from fabric to create patterns. This fabric already appeared in photographs of black women in the 19th century, in images of candomblé in Bahia in the first half of the 20th century, and currently, it is a technique that has been reintroduced in candomblé terreiros. One example is the daughters Luana Rampazi and Isabela Rezende, who learned barafunda stitching techniques online and now produce custom-made pieces or present them to elder members of the house[4].

The young women in the terreiro have also introduced another type of fabric purchased from the African continent, which is the wax print fabric. These industrially printed fabrics, produced in the Netherlands, England, and even China, became widely used in West Africa from the 20th century onward and are marketed in Brazil as African fabrics. Two other fabrics that are often imported from the African continent to the ateliers are African guipure lace and axó oke striped fabrics. The atelier Odó Iná, led by ebomi Oiá Somikan (Georgia Prado[5]), has sought to aesthetically strengthen a movement of (re)africanization of candomblé by incorporating these fabrics into their creations.

This atelier of Georgia Prado, which serves the terreiro Axé Ilê Obá, has a smaller production scale and conducts research on an ancestral black aesthetic in candomblé clothing. In an interview with Georgia Prado, it was also possible to perceive differences from other terreiros in São Paulo, which work with more vibrant clothing, greater use of trims, many gold-plated orixá tools, and more accessories and even makeup.

> Less is more, especially in candomblé. I believe that what should draw attention is perhaps my skirt, not whether I'm wearing eyeliner or mascara [...]. Because there, I am not myself, there I am a representative of Oiá. I am a representative of Oxóssi. [...] Because it's not about me, it's about the energy that dwells within me, and it is that energy that should stand out, not my beauty (see footnote 4).

This statement also highlights an important point, which is who wears the clothes. Although they are worn by the candomblecista, the clothes of the orixás belong to the deities (in studies such as that of Patrícia Ricardo Souza in 2007, the author recounts an orixá who did not accept the clothes they received), and during the trance, they will demonstrate whether they accept the clothes or not [12]. Therefore, there are types of clothes, fabrics, more commonly used by some orixás and less used by others (orixás that accept more shine or colors, etc.). Another point in this statement that is corroborated in some costumes is a differentiation of the oboró orixás. In Mother Paula's current phase, there is a greater use of voluminous trousers called "calçolões" (also known as bombachos) without the use of short skirts by the oboró orixás (see Fig. 8, we see both forms: calçolão worn by Xangô on the left of Mother Paula, and skirts worn by Xangô on the right of Mother Paula).

Regarding the materials and techniques currently used for the orixás' tools, it is interesting to note the persistence of certain elements (use of metallic materials such as flandres and iron, use of wood, and fabrics) and the introduction of others (brushed steel, resin, ceramics, chatons, and gold-plated pieces).

[4] Rampazi, Luana. **Interview.** [04 Jul. 2022]. Online. Interview given to the author.

[5] Prado, Georgia. **Interview.** [19 Jul. 2019]. São Paulo. Interview given to the author.

3 Conclusion

In the clothing of candomblé, there are elements that are considered immutable, related to the foundations of the religion and the hierarchical relationship within the terreiro, that must be respected. However, a visual and aesthetic aspect is linked to ephemeral and contemporary tastes, the current textile technologies available, and the aspect of (re)africanization that brings an updated influence from the African continent. As pointed out by Mother Paula in the interview, it is necessary to maintain tradition, as her predecessor did, but the characteristics of the religious leader also manifest in the clothing, such as the use of colors according to the orixá of the ialorixá, for example (see footnote 2).

Changes in clothing occur, even if it is necessary to consult the jogo de búzios (divination system using cowrie shells) to ask for permission from the orixás, as is the case in the terreiro Axé Ilê Obá (as pointed out by Bia of Oxum from Patuá Clothing in an interview (see footnote 3) or in collective meetings with the children of the house, as indicated by Mother Paula, to decide on the use of the lace insertion technique for the attire of iaôs (see footnote 2)). These consultations for permission are necessary because clothing in candomblé is a materiality charged with symbolism and an immateriality, which is axé, the energy and vital force. And since axé is also present in all the individuals of the terreiro, there needs to be a balance between individualities and the collective expressiveness and identity of the house and the religion itself through the clothing of everyone.

In the case of the terreiro Axé Ilê Obá, we have an example that also occurs in other candomblé houses, which is the impact that a babalorixá or an ialorixá has on the collective attire of the entire house, forming "aesthetic phases" during their leadership periods. Nevertheless, we understand that the babalorixá and ialorixás act as dress code guide of axós (clothing), impacting the dressing style of Candomblé.

References

1. Bregolin, D.B.: A moda como linguagem: singularidades e códigos vestíveis no trânsito entre o profano e o sagrado do Candomblé. Dissertation, University of Caxias do Sul (2018)
2. Correa, R.P.: História e memória do terreiro Axé Ilê Obá. Dissertation, Pontifical Catholic University of São Paulo – PUC (2014)
3. Egydio, S.: O perfil do Aché Ile Obá. Edições Populares, São Paulo (1980)
4. Faleiros, A.P.L.M.: Indústria do brilho vai além da folia. O Estado de São Paulo, v. Economy, february 8th, p.B6 (2005). https://www2.senado.leg.br/bdsf/bitstream/handle/id/305372/noticia.htm?sequence=1. Accessed 14 Mar 2021
5. Lody, R.: Moda e história: as indumentárias das mulheres de fé. Editora Senac, São Paulo (2015)
6. Negrão, C.: Processos de comunicação e cultura: oralidade condutora do Axé no terreiro Axé Ilê Obá. Dissertation, Pontifical Catholic University of São Paulo – PUC (2018). https://tede2.pucsp.br/handle/handle/21302. Accessed 15 May 2022
7. Peres, A.: O império das mães de santo. Claudia. Editora Abril, São Paulo, no. 431, p. 17 (1997)
8. Plankensteiner, B.: African Lace: an industrial fabric connecting Austria and Nigeria. Anthrovision **1**(2) (2013). https://doi.org/10.4000/anthrovision.679

9. Prandi, R.: Candomblés de São Paulo: a velha magia na metrópole nova, 2nd edn. Arché Editora, São Paulo (2020)
10. Santos, J.R.L.: Indumentárias de Orixás: arte, mito e moda no rito afro-brasileiro. Dissertation, São Paulo State University - UNESP "Júlio de Mesquita Filho" (2022)
11. de Silva, V.G.: Os orixás da metrópole. Vozes, Petrópolis (1995)
12. Souza de, P.R.: Axós e ilequês: rito, mito e a estética do candomblé University of São Paulo Thesis (2007)

The University of the Mountains. A Historical Comparative Analysis to the Academic Attire of Beira Interior

Rebecca Nantes Silva[1(✉)], Rafaela Norogrando[1], and Alexsander Jorge Duarte[2]

[1] University of Beira Interior, 6201-001 Covilhã, Portugal
rebeccanantes.s@gmail.com
[2] University of Aveiro, 3810-193 Aveiro, Portugal

Abstract. The present article was extracted from the master's dissertation in fashion design titled "Uma análise sobre o Traje Acadêmico Português – Um desenvolvimento Projetual," which focuses on the academic attire of the University of Beira Interior-UBI. The overall objective is to compose a sequence of three articles, with two already published. The first article addressed the history of the attire at the University of Aveiro-UA, and the second article discussed the project that resulted in the creation of an informal summer attire for performances by the members of the Tuna of Aveiro, which was the outcome of the master's dissertation. The specific objective of this third article was to provide the historical background of the UBI attire through oral accounts and to draw a brief comparison between the design of the UBI attire and the design of the UC attire, using Peirce's semiotics as a tool for analysis and understanding of the causality behind the conception of these attires. The methodology for this research consisted of: 1) a literature review, 2) a semi-structured interview, 3) consultation of ethnographic resources for support, and 4) the use of Charles S. Peirce's semiotics, which brings the viewer closer to interpreting the meanings in fashion and tradition.

Keywords: UBI's Academic Attire · Tradition · History · Semiotic · Fashion

1 Introduction

In the Portuguese academic universe, there is a mythical aura surrounding the conception of the attire worn at universities, specifically the UC, which is ingrained in its culture. However, this foundational myth is in conflict with the history of not only the UC attire but also the UBI attire. Chapter one will provide a brief contextualization of the conception of the academic attire at the University of Coimbra-UC to delimit the problem. This will be defined through references that trace back to the roots of the first attire in Portugal, highlighting its differences from the UBI attire. Furthermore, studies and perspectives that offer definitions of its conception will be examined, ultimately determining which definition of attire is relevant to this work.

Chapter two will approach the historical and ethnographic aspects of the conception of the UBI attire through the oral account of its creator, Mr. Lino Torgal. Extracted from a semi-structured interview that values declarative testimony and the personal experience of the individual responsible for this aspect of academic culture in Beira Interior, this chapter will draw upon his personal collection of documentation generated during the period of conceiving the UBI attire.

Chapter three presents a comparative semiotic analysis between the conception of the UBI attire and the traditional and purist perspectives of the UC attire. The aim is to contribute to a new understanding and preservation of academic culture and its tradition in a contemporary context.

2 Understanding the Conception of the Attire

2.1 From the Imaginary of Beliefs

The UC attire[1] was the first attire, considering that it belongs to the first Portuguese institution. However, there is a stigma related to the motivation behind the conception of the UC attire, as this stigma generates a problematic situation where the real reason for the establishment of the attire is obscured by popular beliefs disseminated among students until then. For a brief clarification, it is necessary to examine the UC attire through phases of its existence.

Therefore, the objective of this study is to compare the conception of attires in two institutions, considering an aspect that was once considered a popular belief regarding the creation of the Mater institution's attire but becomes the cause for the idealization of the attire in another contemporary institution, driven by the need for belongingness and a sense of identity. Let us begin with the most repeated popular beliefs among academic communities, and then explore the past.

2.2 From the Realm of the Popular Beliefs

What is recurring among students from different institutions, and especially among those from Coimbra, are sayings that revolve around the motivation behind the creation of academic attire. Among these beliefs, the idea that the attire was intended for social equality among students is one of the most speculated. This factor is often attributed to a misinterpretation, where the attire serves to differentiate students from common civilians, equating them as members of the academy [6] and [8]. There is also a belief that the attire has Jesuit origins. Although this is one of the beliefs, there is some foundation in its origin due to the presence of clergy at the university [6, 8] and [10]. This belief does not go beyond certain limits. However, when it is claimed to belong to a specific order such

[1] Recognized as the uniform worn by students of UC who choose to follow the rituals of Academic Traditions. It is worn both during the period of new student admission, identified as freshmen in university student life, and on solemn occasions. It is also used in daily life if the student chooses to do so according to the rules of the Institution's Code of Praxe, the manual that contains the norms of academic life [8].

as the Jesuits, certainty is firmly placed in an unsubstantiated narrative, disregarding the presence of other monastic orders in the academic environment.

Another belief targeted by students is that the conception of academic attire aims to identify the student according to local culture. Where the institution is located. This notion also does not hold up when considering that the justification for wearing attire in Coimbra actually originates from a body of clergy who attended the university, and later underwent alterations as the institution opened its doors to the bourgeoisie, with statutes being drafted by the monarchy to guide against excesses in order to accommodate this public in classes. Another reason for the confusion and the generation of baseless support for the notion of equalizing social classes among students [13].

The premise that local culture influences the conception of attire and is embedded in the essence of academic attire is established in a contemporary period after 1974 with the Carnation Revolution, where universities that emerged between 1980 and 1990 recognized Coimbra as a reference point for the preservation of academic culture. In the process of asserting their own identity, they began to develop their own codes of practice with their own academic norms.

> "In the 1980s-1990s, the traditions, institutional attire, insignias, and ceremonies of UC enjoyed enormous prestige at the University of Porto (UP), the University of Lisbon (UL), the emerging Portuguese public and private universities, and the majority of Brazilian universities. Many of the Portuguese polytechnic institutes established in the 1970s–1980s looked up to UC's traditions, leading to the assimilation of the Queima das Fitas festivities, serenades, capes and gowns, folders, silk ribbons, fancy top hats, course rings, and almost literal adaptations of the "code of praxe from 1957."" (Nunes, 2013, pp. 16–17).

2.3 To Purist Perspective

Unlike contemporary institutions, UC has a long historical process. Just a few years after its foundation in 1290, specifically in 1309, the University received its statutes, which did not include any clause regarding appropriate attire for attending classes. However, it was the clergy who attended, and consequently, each of them wore their respective monastic order's attire, as shown in Fig. 1 [6] and [10].

Therefore, the influence of attire permeates its essence, both in terms of clerical dressing and bourgeois clothing, which started to be worn within the institution. The bourgeois public, in an effort to integrate into the academic community, attempted to dress similarly to their ecclesiastical counterparts in order to maintain aesthetic harmony within the institution [8].

According to author [8], there was previously the "hábito talar" (long gala habit) prevalent during the liberal and republican period, as a kind of monarchic, Tridentine Catholic, and Baroque heritage, which later transformed into the "hábito curto/hábito privado" (short habit/private habit) or "traje de abatina (abatina attire)." The abatina was promoted by the bourgeois male culture, converging with military and school uniforms, but it did not become a standardized manufactured uniform. A small hypothesis is that the abatina not only has ecclesiastical influence but also a probable legal heritage predating the 19th century, through the tailors' craftsmanship of that period, who used the becas (gowns). And it is not only the legal field but also the attire worn by medical students.

"Detailed comparative exercises between the ancient statutes of UC, the statutes of the Coimbra Seminary, and synodal constituents led me to a quite consistent image of what the university robes would have been like before the mid-19th century. The fact that some tailors in Coimbra still make the old two-layered judicial gown would prove to be a comparative source of paramount importance." (Nunes, 2013, pp. 22).

Fig. 1. Roman cardinal wearing a short habit, private habit, or abatina attire in its earliest morphological formulation, engraving by Buonnani, and short silk habit: roman clergyman, Rome, 1860s, with silver buckle shoes, silk stockings, breeches, black abatina, lined cape, and felt tricorn hat with buckles 1720. Sources: Identidades e Moda, António Manuel Nunes, 2013.

However, cape and cassock were worn differently between 1870–1880: the lente abatina was worn by civilians, while the clergy who attended the University wore the Roman talár cassock according to the statutes and privileges granted to them by their ecclesiastical position [8]. In the modern period, there is a process of feminization of the attire, specifically between the 1940s–1950s, which began at the University of Porto and later spread to Coimbra. After this period of turmoil, the so-called Academic Crisis occurred in 1969 as a result of the Salazarist Regime, leading to the so-called Academic Mourning[2]. The attire ceased to be used and made a comeback between the 1970s and

[2] Period when the attire ceased to be used due to its association with left-wing students, posing a strong threat to the regime of António de Oliveira Salazar [13].

1980s. Alongside this, the rise of universities such as UBI and UA took place in the late 1980s, which, driven by the need for affirmation, developed their own statutes and code of conduct [8].

Therefore, the academic attire of UC encompasses a mixture of specific areas such as ecclesiastical, legal, and medical, which spanned different time periods and influenced its aesthetic, resulting from social and political events in Portugal. Despite disagreements surrounding the myth and student imagination that surround the memory and history of the conception and motivation behind the use and creation of the Coimbra attire, and among them the belief that the local culture is part of this temporal scope that results in the UC attire, it is only validated when it comes to the post-25th of April crisis and new institutions.

To understand the myth of local culture embedded in the student identity of the attires, it is important to complement this perspective with the insight of one of the creators of the attire from one of the contemporary institutions that developed its own attire and code of conduct in the 1990s. To achieve this, the use of a semi-structured interview for the dissertation was employed, as it provided the possibility to contact the creators of the attires from UA and UBI, given the relevance that these oral accounts would have on the overall results of the dissertation, as well as the codes of conduct of these institutions and publications such as periodicals, brochures, etc., dated from that period, which exhibit regional and daily-life characteristics of their inhabitants and workers [5] and [11]. This article will specifically address the attire of Beira Interior.

3 The Journey of the Academic Attire at UBI

UBI originates from IPC - Polytechnic Institute of Covilhã in the 1970s, later becoming IUBI - University Institute of Beira Interior in 1979, and finally transforming into the University of Beira Interior in 1986, situated between the Snow City and the strong Wool Industry. However, what distinguishes the academic attire of UBI from other universities, aesthetically referring to materials, is the region where the institution is located. Located in Covilhã at the foot of Serra da Estrela, UBI had its attire developed in the early 1990s by Lino Torgal, a former member of the UBI Academic Association. In an interview, Torgal stated that there was initially a need to create identity for the university students through the attire. He also noticed that when attending academic events at other institutions with his course colleagues, students from Tuna groups and even regular students were wearing attire specific to their own educational institutions. These events made Torgal realize that each university they visited had a certain status with its own set of conduct rules, similar to the Code of Conduct of Coimbra, known as the "Código da Praxe."

In light of this scenario, Torgal thought that UBI should also have its own attire that would represent them, both for Tuna group members when performing at festivals and for students affiliated with the Academic Association (who were practically the only ones wearing attire at that time, considering the number of university students). According to the testimony, regional characteristics were identified and taken into consideration in order to create a new attire for the University of Beira Interior.

Located in Covilhã, at the foot of Serra da Estrela, UBI had its attire developed in the early 1990s by Lino Torgal (as in Fig. 2 taken from Lino Torgal's collection), a

Fig. 2. Covilhã in a postcard, 19th century. Source: Personal Collection of Mr. Lino Torgal, 2019.

former member of the UBI Academic Association. In an interview, Torgal stated that there was initially a need to create identity for the university students through the attire. He also noticed that when attending academic events at other institutions with his course colleagues, students from Tune groups[3] and even regular students were wearing attire specific to their own educational institutions. These events made Torgal realize that each university they visited had a certain status with its own set of conduct rules, similar to the Code of Conduct of Coimbra, known as the "Código da Praxe."

In light of this scenario, Torgal thought that UBI should also have its own attire that would represent them, both for Tuna group members when performing at festivals and for students affiliated with the Academic Association (who were practically the only ones wearing attire at that time, considering the number of university students). The testimony declares that regional characteristics were identified and taken into consideration in order to create a new attire for the University of Beira Interior.

> "By combining the national attire we were already wearing, the academic identity aspect, and some regional characteristics such as the mountain climate and the cold weather, etc. We mainly attributed to the new design the region's climate and local culture. Then we took the matter to be decided in a Student Assembly. We faced difficulties because it took us over two years to gain everyone's acceptance. At that time, even though we wore the Coimbra attire, we were the members and leaders of the Student Association and members of the Academic Tuna." (Participant, Interview).

To develop the attire, it was necessary to convene a council in the Assembly of Students to justify the needs mentioned by Torgal. Nothing is decided alone in those times and, even today, when it comes to customs and traditions that carry historical and symbolic significance. Therefore, minutes (below in Fig. 3) were taken during these

[3] Musical groups formed by students who attend Portuguese universities. They are part of the experience of traditions and academic culture of each institution [13].

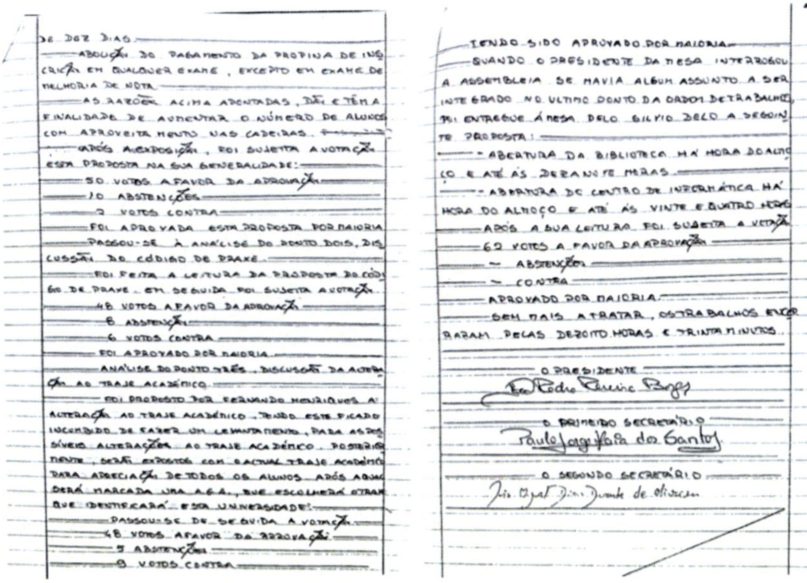
DE DEZ DIAS
ABOLIÇÃO DO PAGAMENTO DA PROPINA DE INSCRIÇÃO EM QUALQUER EXAME, EXCEPTO EM EXAME DE MELHORIA DE NOTA
AS RAZÕES ACIMA APONTADAS, DÃO E TÊM A FINALIDADE DE AUMENTAR O NÚMERO DE ALUNOS COM APROVEITAMENTO NAS CADEIRAS.
APÓS A EXPOSIÇÃO, FOI SUJEITA A VOTAÇÃO ESTA PROPOSTA NA SUA GENERALIDADE:
50 VOTOS A FAVOR DA APROVAÇÃO
10 ABSTENÇÕES
2 VOTOS CONTRA
FOI APROVADA ESTA PROPOSTA POR MAIORIA
PASSOU-SE À ANÁLISE DO PONTO DOIS, DISCUSSÃO DO CÓDIGO DE PRAXE.
FOI FEITA A LEITURA DA PROPOSTA DO CÓDIGO DE PRAXE. EM SEGUIDA FOI SUJEITA A VOTAÇÃO.
48 VOTOS A FAVOR DA APROVAÇÃO
8 ABSTENÇÃO
6 VOTOS CONTRA
FOI APROVADO POR MAIORIA
ANÁLISE DO PONTO TRÊS. DISCUSSÃO DA ALTERAÇÃO AO TRAJE ACADÉMICO.
FOI PROPOSTO POR FERNANDO HENRIQUES A ALTERAÇÃO AO TRAJE ACADÉMICO, TENDO ESTE FICADO INCUMBIDO DE FAZER UM LEVANTAMENTO, PARA AS POSSÍVEIS ALTERAÇÕES AO TRAJE ACADÉMICO. POSTERIORMENTE, SERÃO EXPOSTOS COM O ACTUAL TRAJE ACADÉMICO PARA APRECIAÇÃO DE TODOS OS ALUNOS. APÓS A QUAL SERÁ MARCADA UMA A.G.A., QUE ESCOLHERÁ O TRAJE QUE IDENTIFICARÁ ESTA UNIVERSIDADE.
PASSOU-SE DE SEGUIDA A VOTAÇÃO
48 VOTOS A FAVOR DA APROVAÇÃO
5 ABSTENÇÕES
9 VOTOS CONTRA
TENDO SIDO APROVADO POR MAIORIA
QUANDO O PRESIDENTE DA MESA INTERROGOU A ASSEMBLEIA SE HAVIA ALGUM ASSUNTO A SER INTEGRADO NO ULTIMO PONTO DA ORDEM DE TRABALHOS, FOI ENTREGUE Á MESA PELO SILVIO PELO A SEGUINTE PROPOSTA:
- ABERTURA DA BIBLIOTECA HÁ HORA DO ALMOÇO E ATÉ ÁS DEZANOVE HORAS
- ABERTURA DO CENTRO DE INFORMÁTICA HÁ HORA DO ALMOÇO E ATÉ ÁS VINTE E QUATRO HORAS
APÓS A SUA LEITURA FOI SUJEITA A VOTAÇÃO
62 VOTOS A FAVOR DA APROVAÇÃO
– ABSTENÇÕES
– CONTRA
APROVADO POR MAIORIA.
SEM MAIS A TRATAR, OS TRABALHOS ENCERRARAM PELAS DEZOITO HORAS E TRINTA MINUTOS.
O PRESIDENTE
O PRIMEIRO SECRETÁRIO
O SEGUNDO SECRETÁRIO

Fig. 3. Minutes of the meeting in which Lino Fernandes Torgal and Fernando Henriques Moreira, his colleague, propose the creation of a new attire for UBI on December 6, 1990. Source: Personal collection of Lino Torgal, 2019.

meetings, which contained justifications to prove the need for attire that would truly identify the students of Beira Interior as such.

With the proposal being voted on and approved at the General Assembly of Students, in addition to this occasion, a survey was conducted to promote the attire and a draw (as shown in Fig. 4), with sheets being displayed at the main campus bar to be filled out and placed in an urn.

The new UBI attire wins the elections, with a total of 904 votes from students, where 5 votes were void, 290 votes were against, and 601 votes were in favor, gaining approval by the majority. Subsequently, the event receives immediate news coverage in nearby newspapers (as shown in Figs. 5, 6 and 7 below) such as "Correio da Manhã", "Notícias da Covilhã", "Diário de Notícias", and "Jornal do Fundão".

In addition to the newspapers, banners and posters were displayed to promote the new attire. A specific brochure containing all the information regarding the design of the attire for both males and females was created, highlighting all the changes and features made to it, as well as tables with pricing details for the UBI attire [16].

All these issues were checked to ensure that there were no differences between the male and female attire, such as the number of buttons, the use of a hat in one attire and a hood in the other, etc. And for that reason it was distributed a pamphlet with all these characteristics through the campus (as Fig. 8).

As mentioned earlier, the climate and regional characteristics were taken into great consideration when developing the new design. Torgal explains that a field research was conducted in collaboration with a fashion designer for the creative process. The research

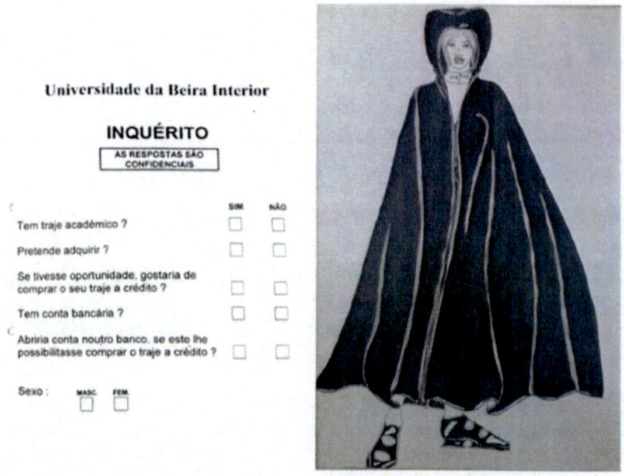

Fig. 4. Survey conducted by Torgal and Moreira to allow students to vote on their approval of the proposed new Academic Attire and the approved sketch at the General Assembly of Students. Source: Personal collection of Lino Torgal, 2019.

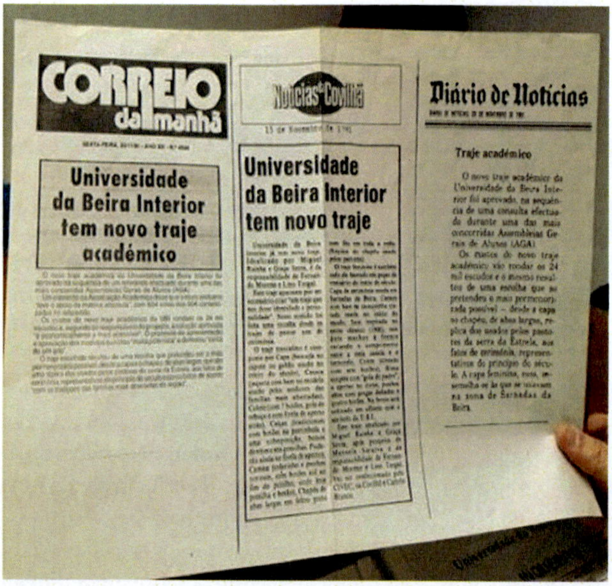

Fig. 5. Newspapers in the region of Covilhã that informed the population about the change regarding the new UBI attire. Source: Personal archive of Lino Torgal, 2019. Source: Personal archive of Lino Torgal, 2019.

was inspired by the peasants who inhabited the Covilhã region several decades ago, their mountainous lifestyle, and the cold weather - particularly regarding the length of the skirts for the ladies, which is why they are long. Other elements include the use of the water-repellent material called burel for the cape, the hat that protects the male, the hood for the female, and the pin on the female attire that replaces the tie.

JORNAL DO FUNDÃO

Novo traje da UBI aprovado por referendo

Fig. 6. Newspaper in the region of Fundão that informed the population about the change regarding the new UBI attire. Source: Personal archive of Lino Torgal, 2019. Source: Personal archive of Lino Torgal, 2019.

Fig. 7. Poster created to promote the new academic attire of UBI. Source: Personal collection of Lino Torgal, 2019.

"We made sure to keep the same number of pieces for both attires. For example, the number of buttons on the vest for the male attire is the same for the female attire, and so on. We proposed this to make them equal despite the gender difference. It was a popular subject at that time: what was in one should also be in the other, and what was not completely equal between them should have something equivalent, such as the tie and the pin. The hat is for boys, just as the hood is for girls, and

O PORQUÊ DE UM TRAJE PRÓPRIO :

Sendo a U.B.I. uma universidade recente, ainda sem tradições, achámos conveniente criar um traje que nos desse identidade e personalidade próprias relacionando-nos com a região da Beira Interior.

Começámos assim com uma recolha que tentou ser o mais pormenorizada possível, desde o traje do pastor aos trajes de cerimónia, tentando conjugar as características mais marcantes de cada um deles.

Chamamos ainda a atenção para algumas peças, como o casaco e as polainas, não terem dado o efeito final pretendido, o que se deve em grande parte ao tempo limitado em que os trajes foram modelados e executados não dando espaço a testes de confecção.

DESCRIÇÃO TÉCNICA

Traje masculino

Capa - Capa de cerimónia baseada no tradicional «capote» ou «gabão» usada no princípio do século. O capote era também usado pelo pastor serrano. Vai até um pouco abaixo do joelho.

Casaco - Jaqueta desenhada a partir do modelo usado pelos senhores das famílias mais abastadas na região da Beira Interior.

Colete - O colete é alto, com sete botões, gola de rebuço e com fivela de aperto atrás. O forro e a traseira são em setinete ou setim.

Calças - Calças tradicionais da zona com botões na portinhola e uma sobreposição, bolsos direitos e seis presilhas. As calças poderão ter ainda fivela de aperto semelhante à do colete na parte de trás.

Camisa - A camisa tem colarinho e punhos normais, aperta com botões até ao fim do «peitilho», onde leva uma presilha e um botão. Esta camisa era também usada no início do século.

Chapéu - O chapéu é de abas largas em feltro preto com uma fita em toda a volta. Este é uma réplica do usado pelos pastores da Serra da Estrela.

DESCRIÇÃO TÉCNICA

Traje feminino

Capa - Capa de cerimónia usada pelas senhoras no princípio do século na zona das Sarnadas da Beira, com capuz e gola levantada.

Casaco - Desenhado com base na casaquinha cintada usada no início do século XX pelas senhoras das famílias mais abastadas.

Saia - Inspirada na época de 1900, de estilo clássico. Tem dois machos à frente variando o comprimento entre a meia canela e o tornozelo.

Colete - Colete cintado com sete botões tentando assim interligá-lo com o colete do traje masculino.

Camisa - Blusa simples com pé de gola (ou «gola de padre»). Na frente levará um alfinete com o símbolo da U.B.I.. Esta camisa aperta atrás e os punhos são altos com pregas deitadas e quatro botões.

Fig. 8. Pamphlet describing the male and female attire and the reason for having a new attire for UBI. Source: Personal collection of Lino Torgal, 2019.

it can be removed from the cape to make them the same. We did this to ensure that there was no discrepancy that would result in disagreements. We designed leggings for the male attire, which were used by shepherds in the Serra da Estrela mountains to keep their legs warm. However, due to aesthetic and fitting reasons, we preferred a classic pair of pants for the male attire and a long skirt for the girls" (Participant, Interview).

In this way, it is possible to perceive the adjustments made in relation to climatic, regional, and traditional cultural factors on a national level (as shown in Figs. 9 and 10 above, referring to the aesthetics of the design and the final costume). Maintaining the use of black in contrast with the white shirt, keeping the closed neckline, the protective cape against cold and rain, and the normal habits that exclude the use of flashy accessories, nail coloring, impactful makeup, among others mentioned in the Praxe Code (Extracted from the Interview with Lino Torgal, January 2019). After the publications, Torgal and Moreira, along with the team that developed and approved the new attire, filed for

Fig. 9. Final sketches of the UBI Academic Attire approved in the G.A.A., 1992. Source: Personal collection of Lino Torgal, 2019.

copyright registration for the attire so that the Academic Association would have the right to a percentage of the profits obtained from clothing companies that produced the attire for UBI. However, they were not successful in their attempt to contribute to AAUBI.

According to Torgal's account in the interview and his personal collection, the influence of local culture and climatic factors in the conception of UBI's attire was evident. However, in accordance with the old Coimbra tradition, these same factors identified in UBI's attire – and in UA's attire, as it also shares these characteristics in its history – do not apply to UC's attire, which is commonly spread among academic communities throughout the country, generating conflicts between purist former students who rigorously preserve the tradition within the context of national attire, and those initiated in the academic culture of both Coimbra and other institutions.

To understand this phenomenon between the attires and their conception, a brief comparative analysis with Peirce's semiotics is proposed.

4 Semiotic Reading for a Perspective on the Presence of Local Culture in Attire

The attire has a proper use with insignias, correct ways of wearing it, suitable occasions to dress in it, and rituals related to the universe to which it belongs (Nunes, 2013). Based on the principle that a sign can be anything [1, 3] and [12], attire is also a sign. It can be

Fig. 10. Students currently wearing the Attire envisioned by Torgal and Moreira in 1990. Source: Personal collection of Rebecca N. Silva, 2019.

read in various ways by different people with different experiences, thus having different interpreters and interpretants.

To provide a better example, we divided it into two groups: Group 1 (G1) consists of UC students, and Group 2 (G2) consists of UBI students. Regarding G1, we have the legisign that operates through a general convention, a certain veiled code that induces the archetype of Academic Praxis, such as students participating in the Queima das Fitas[4]. Students dressed accordingly bring to mind for this group the social and conventional experience that united them during their youth or still unites them if they practice the traditions ([1] and [12]). When the color black is seen in this situation, it represents sobriety and the ancient institutional customs from which academia in Portugal derives.

Therefore, the color black points to the academic universe and traditions, where wearing black demonstrates reverence for academic culture, which designates belonging to a Portuguese higher education institution, whether in Coimbra or elsewhere in the

[4] The "Queima das Fitas", being linked to Praxe, expresses the hierarchical structure imposed on the participants of Praxe, who, based on the group's functioning logic, ascend to new positions through rituals of separation and public consecration in front of the group throughout their formative journey. Only students who received the grelo insignia during the previous year's Festa da Latada are allowed to burn the ribbon (grelo), and from then on, they are referred to as "Candeeiro Fitado" according to the Code of Praxe (2007) (Oliveira da Cruz, 2012, page 7 in Silva, Norogrando & Duarte, 2020).

country, and on this occasion, black attire can be read as a sinsign [12]. Furthermore, it is worth noting what culture means for a reading through lived experiences, which is what Peirce suggests in a simple way. Capa e batina, when seen by foreigners, tend to create an immediate association with fictional characters from movies, such as heroes, villains, etc. And this reading depends on the references and cultural backgrounds of the observers, who interpret the attire through their own cultural lenses. The signification of attire, therefore, is not fixed but subject to interpretation based on cultural and personal perspectives [1] and [12].

Furthermore, it is worth emphasizing what culture means for a reading through lived experiences, as proposed by Peirce in a simple way. When seen by foreigners, the cape and gown tend to immediately evoke a connection to fictional characters from movies, such as heroes, villains, and so on. This interpretation depends on the references experienced by this group of people who have no idea what it means to wear an academic dress as an icon [12]. The author [8] expresses a sense of strangeness upon first encountering the cape and gown, confirming this assumption.

> "Still, in late that same month, I saw a student crossing Praça D. Dinis dressed in black, with a tailcoat just below the knee and what appeared to be a woolen cape draped over the right shoulder. I cannot hide the negative impact this image had on me. So, that was the cape and gown of the UC? I couldn't help but wonder if I was witnessing a student, a chic café employee, a billiards player, or a bourgeois groom. For those familiar with academic culture and who view student experiences in an affectionate and mythological light, the male and female academic dress is imbued with charm. Due to its mythical and symbolic significance, it seems to be much more than its modest aesthetic design suggests. To an alien like me, the ensemble on display was ordinary and inferior in style and fabric compared to most professional attire and traditional Sunday attire that I am familiar with in Portugal." (Nunes, 2013, pp. 21).

Using black just for the sake of it can be seen as a costume or an aesthetic expression for a popular culture event, etc. For this group, it can mean anything according to their own experiences. Therefore, the sign is perceived as an icon and can also be interpreted as a quali-sign in its purest form of being a sign [12].

As a sign, the dress carries a power of representation [3] and [12]. It is the sign itself, the dress as it is, a uniform, a simple and pure set of cape and gown that compose a formal attire and nothing more. It carries its simple meaning as clothing, as garments that cover nudity. According to Peirce's nomenclature, it is the FUNDAMENTAL SIGN, the foundation. It functions as an Immediate Object when it is itself, when it is a SIGN, and this correspondence of its meaning is the act of being clothing, of being a dress and attire. Then we have the SIGN functioning as a Dynamic Object, which gives significance to the object/SIGN as clothing. It only exists as the meaning of something, in this case, as the meaning of a dress that is uniform, culture, and tradition because the immediate object exists as something tangible and physical [13] and [14].

The Immediate Interpretant is the degree of interpretability of the interpreter or receiver of the message conveyed through the Dress/SIGN according to their own references, namely G1 and G2, who have their distinct references. And the Dynamic Interpretant is the effect caused in the minds of these interpreters, and this effect would be the result of the references that both groups have about this sign [3] and [12]. For G1, the references are of Portuguese university tradition, specifically the Coimbra tradition, and for G2, the reference of tradition is that of UBI, which involves the characteristics of the local culture.

When alluding to the use of insignias, one can notice the relationship between the student and their educational institution, their homeland, their country of origin, and their preferences [5] and [11], and it represents the tradition embodied in the cape. By exerting this representative power among practitioners of academic culture, the dress is a sign, an icon, and a symbol at the same time, and its use can be interpreted in various ways.

Still within G1, it is possible to interpret the insignias related to the Coimbra tradition, such as the chamber pot, the club, the scissors, and the spoon[5], as signs. When it comes to G2, the interpretation of local culture translates these elements from Coimbra not only through the attire but also through their insignias. While Coimbra uses course colors on the ribbons, the uniqueness of Beira Interior is reiterated not only in the use of ribbons for the "Bênção das Pastas" (Blessing of the Folders)[6], but also in the application of smaller and thinner ribbons on the cape, forming a star embroidered in the center, representing the Serra da Estrela, the region where UBI is located.

According to Peirce's logic, we can evaluate G1 and G2 through the legi-sign, as both have experienced Portuguese academic tradition and therefore share a common reference regarding the essence of tradition and its social conventions [3] and [12]. However, the prism through which they perceive it is that of identity and belonging to the culture expressed in the attire of each university.

G2, being located in a mountainous region, can be seen wearing garments that require more fabric and display aesthetic differences resulting from their design process. For example, the female dress, with its long skirt, was defined by the cold climate of Serra da Estrela [13], as well as the aesthetics of the cape, which features a hood intended to

[5] The spoon has a strong connection to gastronomy. Some say it is a sign of welcome. A phrase associated with it among students is: "you don't refuse a bowl of soup to anyone." Many consider this to be a myth, but the spoon in this position represents the welcoming of new students. The club refers to the male genital organ and virility due to its anatomy. It represents power and strength, used by senior students as weapons during pranks and initiations. The scissors are traditionally a symbol at the University of Coimbra. They represent the hazing of freshmen, a practice originally carried out by the clergy and passed on to university students for centuries. The chamber pot represents the freshman themselves. According to Nogueira, the freshman is symbolized by the chamber pot because they are like a baby, just starting their academic life and, therefore, need to be taken care of and taught according to the principles and traditions of academic life (Silva, Norogrando & Duarte, 2020).

[6] Held at Portuguese universities, it originated at the University of Coimbra, where the City Bishop blesses the graduation sashes of the final-year students who will enter their professional lives during the celebration of the Burning of the Ribbons (Silva, Norogrando & Duarte, 2020).

reference the hats worn by men, also used by the shepherds of the Serra, and the burel[7] cape that protected them from rain and strong winds.

Therefore, the convergences presented through semiotics indicate a standardization regarding the essence of traditions and their perpetuation, but they also demonstrate the divergences caused by a historical process and the need for affirmation among groups. To understand this, it is important to consider the vibrant culture and sense of belonging within G2 during the period when the UBI dress was conceived. Peirce's semiotic proposal encompasses this sense when it enables the possibility of maintaining a tradition and its essence by adding the perspective of the other and challenging hegemonic standards that did not make G2 students feel represented in some way. Not that it was a problem, but if there was a need for change, there was a lack. There was consideration of identity crisis and affirmation in academic social environments that impacted the academic culture itself. This culture was generated by an initial motivation that did not lose its conventional nature but kept the narrative of tradition alive through the creation of new academic cultures.

5 Results of the Analysis Between Costumes

The commitment of Torgal and Moreira towards creating a distinctive imagery for the students of Covilhã persists to this day, identifying them based on their origins through the insignias embedded within the capes and through the cuts and alterations that refer to the local culture, which is the main foundation for the creation of the official UBI costume.

In comparison to the Cape and Gown, the UBI costume aesthetically incorporates the climatic and regional references perceived both through Torgal's oral account as well as his design archive of the costume. This does not imply any disrespect or mischaracterization of the academic tradition of UC due to its conception or motivation. Although there are clear signs of a new interpretation [7] based on how students from contemporary institutions have included local culture within academic culture, there is a need to understand the sense of lack within the academic community of Beira Interior that led to the conception of its own costume.

Upon analyzing the testimony from Torgal's semi-structured interview, one can perceive the similarity between the identity crisis and cultural creation. Culture, as observed by author [4], is established through ways of doing, the habits of a people that are generated within the society in which this people are part of. It is induced to maintain its academic aspects, as generations of students pass through the university over the years.

> "Every culture, therefore, every society, and even each level of a complex society, has its own imaginary. In other words, the boundary between the real and the imaginary proves to be variable, while the territory traversed by this dividing line is, on the contrary, always identical everywhere, for it is nothing else but the entire universe of human experience, from its most collective and social aspects to its most intimate and personal aspects." (SORCINELLI, 2008, p. 52).

[7] Fabric made of 100% wool from medieval sheep (Bordaleira breed), always associated with Serra da Estrela, the mountains, and the shepherds with their capes. Available at: https://ecolaportugal.com/pt/project/burel-tecido/.

Sorcinelli makes it clear that collective aspects are drivers of culture generation. It is understood that in order for tradition, which is the result of culture, to be maintained, it requires a mechanism that is driven by knowledge so that tradition remains connected to the act of creating for the community. This understanding acknowledges the need for change, not to the extent of losing its character, but to strengthen its identity over time as a participating and constructing community of inherited and new cultures. Consequently, tradition remains alive through the nostalgia of experiences.

The author [2] presents longing for the past as a cultural movement that goes beyond the Portuguese imagination, intersecting with nostalgia. Here, culture moves through nostalgia and longing for the past. Therefore, recognizing that longing for the past and nostalgia are products of experiences within academic culture in general, and especially in the traditions of Coimbra, helps us understand the importance of preserving the memory of the conception of their attire without blemishes added by successive generations.

This does not negate the need for a better understanding of the terms described in the codes, as well as the limits of alterations that can significantly change the true meaning of wearing the attire and its significance within academic communities. However, it is crucial to remember that the passage of time operates and prevails even in some traditional aspects. Culture and tradition survive through repetitive acts of making, recounting, singing, and narrating the past, passed down from parents to children and so on. But it does not mean that they do not undergo changes that may interfere in some way. Considering the evidence and new forms of reading and preserving something sacred within academic communities, it is agreed that in order to propagate and keep the tradition alive, there is a need for some flexibility, change, and caution without radically altering its essence. This proposal aligns with the semiotic perspective. Consequently, the inclusion of local culture assumes this role.

6 Final Considerations

Based on what has been presented, when evaluating both the primary function of student attire and the influence of space and time on it, it is concluded that it undergoes mutations and is subject to interference from generational academic communities. This is a cyclical process also observed in fashion. However, unlike fashion, tradition remains and stays alive through adaptations that bring a sense of belonging. It is a devouring of tradition by a new tradition with the same DNA, but instead of annihilating, it generates vitality and strengthens its roots. In other words, an almost anthropophagic act.

Here, Peirce's argument does not fit in its misused or frivolous sense of burdening real tradition in favor of fads through individual interpretation. Instead, it perpetuates tradition by maintaining its essence through perspectives that reinforce not only the institutional aspect of wearing attire designed for a higher education institution but also emphasize the possibility of enriching and valuing this tradition when unified or, in this specific case, merged with the local culture in which the institution is located. This aims to fill a gap and perpetuate its history through modes of dress.

However, there is still a need for a deeper understanding of academic traditions and their customs involving attire in contemporary Portuguese institutions, where praxe codes were restructured after the Salazarist period. To achieve this, it is necessary to conduct further ethnographic research on the conception of attire in institutions founded after the Regime, as was done in the master's investigation, in order to validate this study.

Annex - Interview Transcript with Mr. Lino Torgal, Former Member of the Student Association of the University of Beira Interior - January 30, 2019, 6:00 PM

Interviewer: Rebecca Nantes Silva, MsC.

Participant: Mr. Lino Torgal – ex-member of AAUBI (1990).

Interviewer – How did you meet Mr. Fernando Henriques, who helped you with the creation of the UBI attire?

Participant – We weren't even in the same program. We knew each other from academic and social life. We thought that, given the specific location of UBI, it made sense to have something different in terms of attire, and because some universities had already made progress in that direction. So, we believed that there was a need for the university to have an attire that would allow us to attend other universities' events, such as the "Queimas das Fitas" in Coimbra and Evora, and be identified by our own attire. We used to attend these academic events quite often. I particularly enjoyed accompanying my friends from the Academic Tuna. What we noticed was that each university had its own status, which made students known for social and aesthetic characteristics specific to that particular institution. Therefore, what we wanted was an attire that would indicate which institution we belonged to, so that we could be identified as students from the University of Beira Interior.

Interviewer – How did the process of developing the new attire unfold? What references were used to create it?

Participant – By combining the national attire we were already wearing, the academic identity aspect, and some regional characteristics such as the mountain climate and the cold weather, etc. We mainly attributed to the new design the region's climate and local culture. Then we took the matter to be decided in a Student Assembly. We faced difficulties because it took us over two years to gain everyone's acceptance. At that time, even though we wore the Coimbra attire, we were the members and leaders of the Student Association and members of the Academic Tuna.

Interviewer – In what year was the project for the UBI attire conceived?

Participant – It started in December 1990. But the process ran between 1990 and 1991. It was proposed by me and my colleague Fernando Henriques Moreira. We faced a challenge: for a university with around two thousand students, there would be about thirty or forty students dressed in attire, who were members of the Male Tuna and the Female Tuna - "As Moçoilas". This entire group would represent forty or fifty people, and later on, around sixty or seventy. When we reached the Student General Assembly (SGA), there were participants who were rivals (these SGA members), who actually had the Coimbra attire, which was known as the national attire, not only used in Coimbra but also in universities that did not yet have their own attire and status. So, it was very difficult for us to explain to people the need for a UBI attire with our identity when the SGA had a maximum of one hundred participants. But only half of those people had the national attire. We wanted to explain to those who had the attire that our identity needed to change and that we would create the new UBI attire. However, every time the votes were taken, the idea could be rejected, and we ended up in a situation where we withdrew the proposal to prevent it from being rejected and, if rejected, not having it recorded, with the possibility of bringing up the same subject and trying to approve it again. First, Miguel and I presented the SGA with drawings we had commissioned from a designer and sketches of people I had known for a few years. In the end, these were the approved sketches, along with more accurate technical drawings of how the new attire would be. When everything was approved and tested, we included these improved drawings in the posters for the first disclosure.

Interviewer – How was the voting process for the new attire, and when did you present it to the Student Assembly?

Participant – There were forty-eight votes, five abstentions, and nine votes against in December. In May 1991, the process returned to the SGA. By this time, we were already gathering information on the costs of the attire and had made contact with fabric manufacturers to find out the ideal fabrics to develop a prototype to present at the next SGA meeting. Importantly, the deadline for presenting the prototype at the SGA was approaching, but as it was not yet ready, with some pins, etc., there was a risk that it would not be approved, and the models wearing the unfinished prototype had difficulty moving around because it was not ready. So, after the SGA, we opened the vote to referendums because it was a trend to open the vote to more people, not just the academic associations that were part of the SGA votes. For this purpose, during the break time, we placed two mannequins, one male and one female, dressed in the new attire near the voting urns in the cafeteria. Shortly after, we asked two students to wear the new attire and walk around the university. So now the mannequins were alive. We proposed this vote first on November 2, 1991, and then on November 7, 1991, near the cafeteria. Thus, out of two thousand students attending UBI, nine hundred and four students voted, with five null votes, two hundred and ninety votes against, and six hundred and one votes in favor of the new attire. We won by majority vote.

Interviewer – Was there any notable differentiation between one attire and the other (e.g., between genders)?

Participant – We made sure to keep the same number of pieces for both attires. For example, the number of buttons on the vest for the male attire is the same for the female attire, and so on. We proposed this to make them equal despite the gender difference. It was a popular subject at that time: what was in one should also be in the other, and what was not completely equal between them should have something equivalent, such as the tie and the pin. The hat is for boys, just as the hood is for girls, and it can be removed from the cape to make them the same. We did this to ensure that there was no discrepancy that would result in disagreements. We designed leggings for the male attire, which were used by shepherds in the Serra da Estrela mountains to keep their legs warm. However, due to aesthetic and fitting reasons, we preferred a classic pair of pants for the male attire and a long skirt for the girls.

Interviewer – Do you hold the copyright for the creation of the attire?

Participant – After the approvals, Fernando and I, along with the designers, registered the copyrights so that both of us and the Student Association would hold the copyrights for the creation. This also meant that only those specific manufacturers could produce the attire, under the condition that they would give us at least between ten and five euros for each attire produced. It was a way for the Student Association to earn money for events. Unfortunately, we never managed to finalize that deal.

References

1. Barthes, R.: Elementos de Semiologia. Publisher, Cultrix (1964)
2. Botelho, A.: Da saudade ao saudosismo. Editora Biblioteca Breve. Instituto de Cultura e Língua Portuguesa, Amadora (1990)
3. D'Oliveira, A.N.: Peirce & Frege, Coleção Os Pensadores. Publisher, Nova Cultural (1989)
4. Eliot, T.S.: Notas para uma Definição de Cultura. Translate by Ernesto Sampaio. Edições Século XXI, Lisbon (1996)
5. Gauthier, D., Carvalhão, D., Correia, J., Ribeiro, N., Salgado, N., Ribeiro, H.: FAINA ACADÉMICA – Ou nem tudo que vem a rede é Praxe, 2nd edn. 3nd. A.A.U.Av., Aveiro (1998). http://conselhosalgado.web.ua.pt/documentos/CodigoFaina.pdf. Accessed 08 July 2023
6. Lamy, A.S.: A Academia de Coimbra: 1537–1990: história, praxe, boémia e estudo, partidas e piadas, organismos académicos. Rei dos Livros, Coimbra (1990)
7. Lipovetzky, G.: O Império do Efêmero. Translate by Maria Lúcia Machado. Companhia de Bolso (1987)
8. Nunes, A.N.: Identidade e Moda. Bubok Publishing S.L, Portugal (2013)
9. Participant, Torgal, L.: Interview conducted January 30, 2019, 6 p.m. [Transcription of interview source, "Audio recording"] (2019)
10. Prata, M.A.C.: A Academia de Coimbra (1880 – 1926): contributo para sua história. Imprensa da Universidade de Coimbra, Coimbra (2002)
11. Praxe, O Código da.: O Código da Praxe – Universidade da Beira Interior. University of Beira Interior. (2014). Neubi. http://neubi.pt/anexos/CodigodePraxeUBI.pdf. Accessed 08 July 2023
12. Santaella, L., Woth, W.: Comunicação e Semiótica. Hacker (2004)
13. Silva, R.N., Norogrando, R., Duarte, A.J.: Uma análise sobre o traje acadêmico português: um desenvolvimento Projetual. (Publication nº 10400.6/10594) (Dissertation of Masters Degree, University of Beira Interior]. Ubibliorum. http://hdl.handle.net/10400.6/10594
14. Silva, R.N., Norogrando, R., Duarte, A.J.A: Universidade dos Mares. In: Raposo, D., Neves, J., Silva, J., Castilho, L.C., Dias. R. (eds.) Investigação e Ensino em Design e Música, vol. II (X-X), pp. 71–79. Coleção Convergências Research Books, Castelo Branco: Edições IPCB (2020)
15. Sorcinelli, P.:Estudar a moda: corpos, vestuário, estratégias. Senac, São Paulo (2008)
16. Torgal, L.R.: A Universidade e o Estado Novo - O caso de Coimbra: 1926–1961. Coimbra: Minerva História, Coimbra (1999)
17. Villelas, J.: Investigação. O Processo de Construção do Conhecimento. Orientações sobre pesquisas em bases de dados científicas. Etapas das revisões sistemáticas e integrativas da literature, 3nd end. Edições Sílabo, Portugal (2020)

Fashion and Product Design

Redesign of the Uniform of the Municipal Guard of Londrina

Patricia Aparecida de Almeida Spaine, Nelio Pinheiro(✉), Raquel Rabelo Andrade, and Livia Marsari Pereira

Universidade Tecnológica Federal Do Paraná, Apucarana, Brazil
{patriciaspaine,nelio,raquelandrade,liviam}@utfpr.edu.br

Abstract. This study is the result of an outreach project between the Civil Municipal Guard of Londrina (GCML) and the Fashion Design undergraduate program UTFPR-AP. The project was created after GCML asked the Fashion Design program to help improve the current uniform model, which has countless wearability, fit, pattern, comfort, and usability problems. To propose a new uniform model, the project team encompassing agents, professors, and students developed various actions over 2 years, especially developing an anthropometric table adequate to the municipal guard agents, studying textile material adequate for the uniform; ergonomic analysis of uniform wearer's needs; developing datasheets adequate to the production of the uniforms; proposing adjustments in the uniform patterns; developing a technical design specifically for uniform production; suggesting models to update the current uniform of agents of the Municipal of Londrina; training agents to identify problems in the future uniforms acquired for the Guard. Hence, the article presents the uniform pattern created in the project, which is only one of the countless actions carried out in the project.

Keywords: redesign · uniform · product design · municipal guard · clothing design · ergonomic design

1 Introduction

The Municipal Guard of Londrina works in that city to "integrate the unified public security system, aiming to ensure public order and the security of people and municipal public patrimony, through the cooperative, coordinated, systemic, and integrated action of federal, state, and municipal security forces" (Official Letter 580/2021SMDS, p. 1).

As Municipal Guards carry out their constitutional and legal duties in Londrina, their uniform stands out as an essential and necessary element to fulfill their routine obligations and for citizens to identify these municipal agents.

Hence, the Municipal Guard of Londrina contacted the Fashion Design undergraduate program to establish a partnership to update their uniforms and solve a series of problems perceived in the current ones – e.g., discomfort as agents carried out their duties and, therefore, the need to rethink their current work uniforms. Thus, various meetings were held between professors, scholarship students, and agents to understand their requirements and needs to improve the current uniforms.

The meetings led to the proposal of redesigning the current uniforms of the Municipal Guard of Londrina. Redesigning refers to reformulating, doing over, reapplying, and redesigning the design of a product. It can be part of a brand's communication or marketing strategy, which seeks to refresh its identification after being present in the market.

The project encompasses clothes design in the following areas of the Fashion Design program: ergonomics, anthropometrics, sewing pattern, production, creation, and sewing – which are all present in the various courses available to students. Hence, the project coordinated teaching, research, and outreach, applying to it the content students acquired in the classroom, encompassing theoretical and practical activities that involved all parts of the project.

1.1 Outreach Project

The outreach project developed to redesign the Guard's uniform had various phases and stages until reaching the final prototype approved by the agents.

The following were the main phases of the project:

1. Weekly in-person and remote meetings between Municipal Guard agents and the UTFPR project team to understand in detail the main problems they had with the current uniform, requiring improvements (FINISHED).
2. Recruiting scholarship students and defining work schedules, dates, and deadlines (FINISHED).
3. Administering questionnaires and research to agents to redesign their uniforms, identifying problems and needs in the current tones (FINISHED).
4. Developing an anthropometric table suitable to the public (municipal agents). This was done by scholarship students, who spent some days with guards in the headquarters of the Municipal Guard of Londrina to measure 30% of the agents (98 men and women). Thus, they found their physical structure and developed a personalized table of measures of the Guard (FINISHED).
5. Researching and selecting adequate textile materials for the uniforms. In this phase, project participants researched the main uniform fabrics available in the Brazilian textile industry. The research encouraged a uniform fabric factory to donate material, which was used to develop the prototypes (FINISHED).
6. Ergonomic analysis of the users' needs regarding their uniforms (FINISHED).
7. Identifying and proposing adjustments in uniform patterns. Having developed the table of measures, the team created patterns according to the agents' bodies to avoid wearability errors in the new uniform (FINISHED).
8. Proposing redesigned uniforms for the agents of the Municipal Guard of Londrina. After analyzing the questionnaires, holding meetings with the agents, creating the table of measures, and making patterns, the team reached a result, proposing a new uniform design (FINISHED).
9. Developing datasheets of the uniform and its production (FINISHED).
10. Producing samples of the proposed uniforms. Throughout this 2-year process, the project team made some partners specialists in uniforms. After much research, they decided on an ideal fabric for the new uniform. The company that produced this

raw material provided fabric samples and indicated another company certified to produce high-complexity uniforms. Hence, it was possible to produce adequate uniform samples (FINISHED).

11. Helping develop reference terms/basic project and detailed technical study for the bidding process (FINISHED).
12. Training agents to identify problems in the future uniforms acquired by the city government of Londrina through the bidding process (ONGOING).

2 Proposed Uniforms

2.1 Current Uniform

Figure 1 shows the Municipal Guard's current uniform design, as the new one is undergoing the bidding process. The current pattern was developed without any technical study regarding agents' needs to carry out their duties. Thus, it has various usability problems. All difficulties found in the current pattern were pointed out by agents in meetings and in a questionnaire developed by the project team and administered to the agents of the Municipal Guard of Londrina.

Fig. 1. Current uniform of the Municipal Guard of Londrina (Source: GCML, LONDRINA, 2023).

The questionnaire administered to the agents and the team meetings brought up various problems with their current uniforms:

- **Tactical pants**: men's and women's pattern (it is the same for both), too short and straight crotch (it is painful on the agents' private parts), shallow pockets (it has little volume, unfit for holding objects), little flexibility on the knees (causing tears), thin and short belt loops (they do not hold the weight of the agents' gun belts).
- **Combat shirts**: short sleeves (agents get sunburned on sunny days), pockets (small and useless), front opening (unfit for wearing with a bulletproof vest), shirt pattern (inadequate to their duties).

Furthermore, all of them reported problems with the current sizes of the uniforms, due to the lack of an adequate table – the company that made the current pieces used the same table for male and female agents.

2.2 The Proposed Uniform

This project and its resulting uniform were conceived based on principles of clothing design and ergonomics. Clothing design combines social, anthropological, ecological, ergonomic, technological, and economic factors coherent with the consumers' needs and expectations. This process fits the design problem-solving creative procedure, thus establishing a relationship between fashion and design (SPAINE, 2016).

Ergonomics aims at users' security, satisfaction, well-being, and relationship with the products and production systems. Ergonomics must be applied throughout the clothes production process, as the clothing industry aims to develop products that ensure greater interaction with users. Using ergonomic factors goes beyond investigations to improve work and its organization. It must also gather elements and procedures to better adjust products to people's tastes and shapes (IIDA, 2005).

Therefore, agents of the Municipal Guard of Londrina participated in all parts of the project to conceive the new uniform. They are the ones who know what functions agents carry out in their everyday work and can clearly point out the problems in the current uniform regarding wearability, pattern, fit, and comfort, also making suggestions for the updated uniform. Thus, they helped the project team solve the existing problems.

The projects teams of both the Municipal Guard of Londrina and UTFPR developed a new uniform model that met the agents' needs and solved countless problems perceive in the current one.

The prototype was manufactured by a private company that agreed to help make the model created in the outreach project. The raw material for the prototype was provided by another partner private company, which donated the specific fabric used for tactical purposes.

The process of creating the new uniform required various actions over almost 2 years, in a combined effort between all project partners, leading to the proposed uniform model. Figures 2 and 3 show the new combat shirt and tactical pants models approved by the Municipal Guard of Londrina. They are currently undergoing the municipal bidding process.

Sections 2.3 and 2.4 present the main innovations in the prototype.

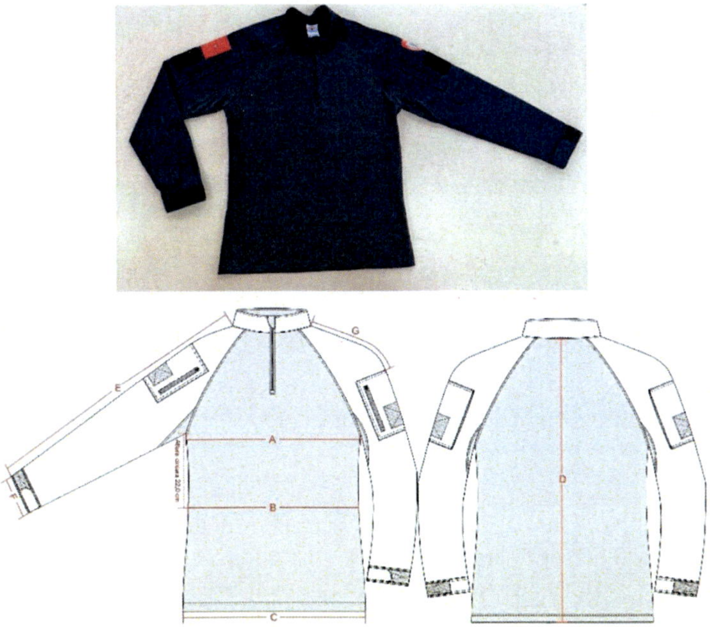

Fig. 2. Proposed Combat shirt – front and back (Source: GCML, LONDRINA, 2023).

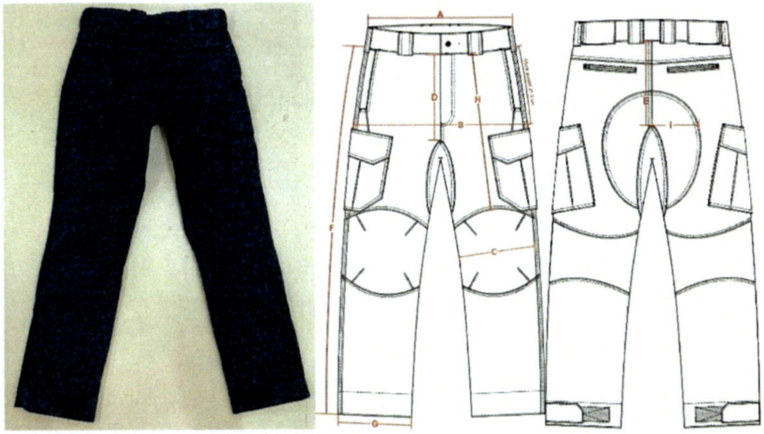

Fig. 3. Proposed tactical pants – front and back (Source: GCML, LONDRINA, 2023).

2.3 Combat Shirt: Specifications and Characteristics

The new combat shirt has the following differential characteristics:

- **Body:** jersey front and back (designed to facilitate transpiration).
- **Front opening (Fig. 4):** partial front opening with a zipper to the upper part of the collar, making it easier to put it on and adapt it to different neck sizes. Its lower end

is outlined with jersey piping. The zipper end sewn on the inner left side forms a covering to protect the user's neck.

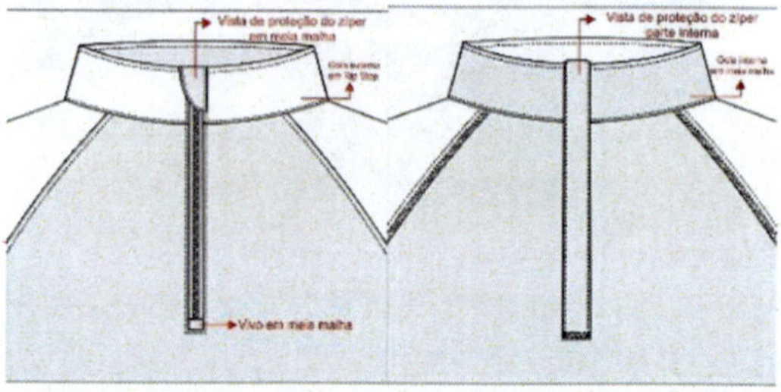

Fig. 4. Detailed outer front/detailed inner front (Source: GCML, LONDRINA, 2023).

- **Collar:** interfaced inside. The outside was made with the main fabric, and the inside was made with jersey to protect the neck.
- **Cuff (Fig. 5):** with adjustable folds to fit different wrist sizes.

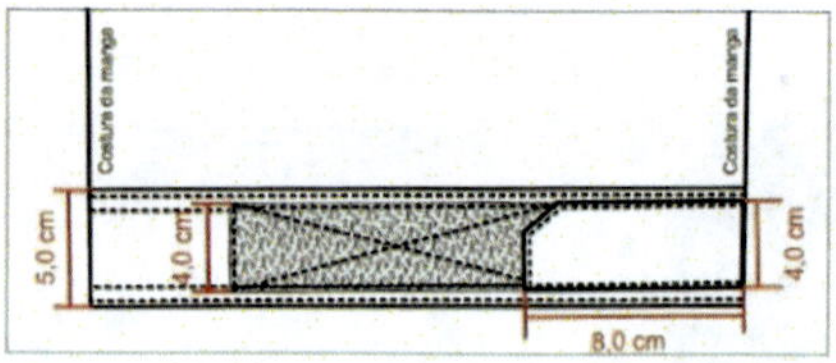

Fig. 5. Illustrated open cuff. (Source: GCML, LONDRINA, 2023).

- **Sleeve pocket (Fig. 6):** Two pockets will be placed over the sleeves (one on each), made with the main fabric, and opened with a zipper (positioned to make it easy to put objects in it and handle in everyday work).

- **Sleeves:** raglan sleeves made with the main fabric, proportional to the dress form. The lower part of the armhole was made with a strip of jersey to facilitate transpiration and mobility.

Overall analysis: The combat shirt was designed with long sleeves to protect the agent in different weather conditions, especially sunny days. It was made with fabric designed specifically for tactical uniforms, ensuring greater durability and comfort when doing activities.

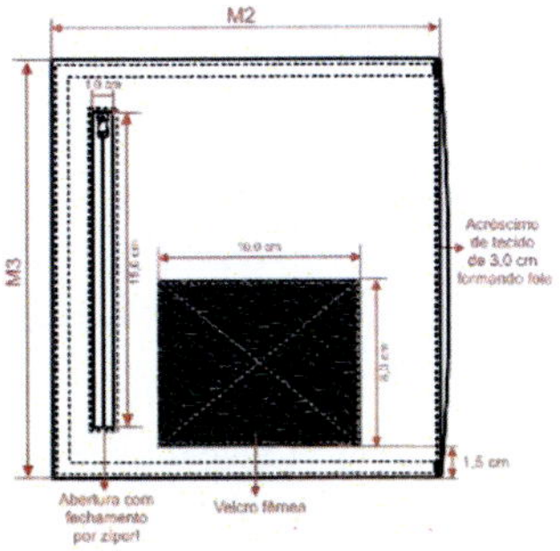

Fig. 6. Sleeve pocket (Source: GCML, LONDRINA, 2023).

2.4 Tactical Pants: Specifications and Characteristics

The tactical pants had most of the agents' complaints and difficulties for use at work. The main changes were as follows:

- **Waistband (Fig. 7):** self-adjustable elastic waistband, making it possible to enlarge the waist. The inner part is covered with an interface (self-adjustment enables people with the same hip sizes and different waist sizes to wear the same-sized pants, as the adjustment enlarges up to 8 cm in the waist).

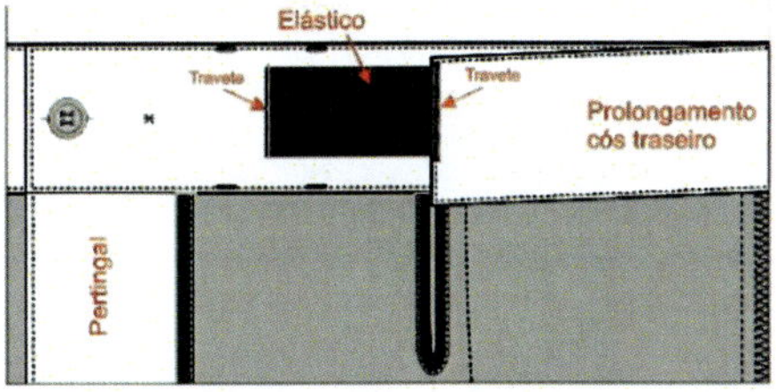

Fig. 7. Inner front waistband

- **Waistband belt loops (Fig. 8):** six 4-cm wide belt loops were made with the main fabric (larger to hold the agents' belts).

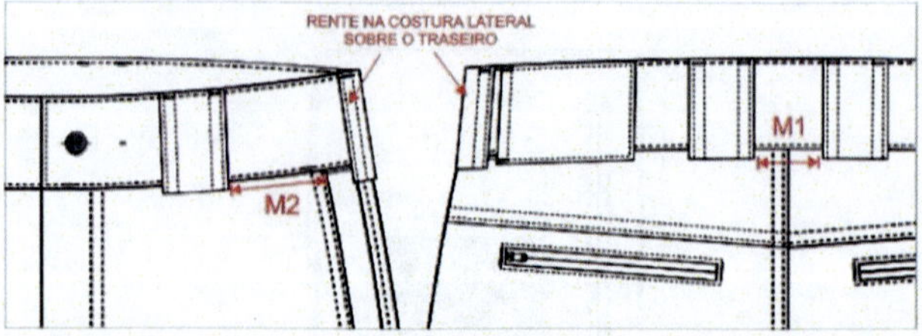

Fig. 8. Location of the belt loops (Source: GCML, LONDRINA, 2023).

Front pockets (Fig. 9): one pocket was placed on each side, with a vertical opening and an angle in the lower end to make it easier for agents to put their hands in.

Front pocket reinforcements: overlapping the lower end of the pocket opening, made in the main fabric, to make it stronger as agents place objects in the pocket.

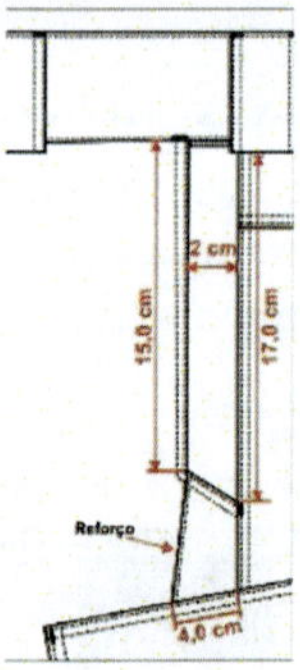

Fig. 9. Front pocket view with reinforcements (Source: GCML, LONDRINA, 2023).

- **Back (Fig. 10):** with a simple yoke and two pockets, one on each side, built-in 2 cm down from the yoke, opened with zippers running from the side to the crotch, making it easier for agents to use them.

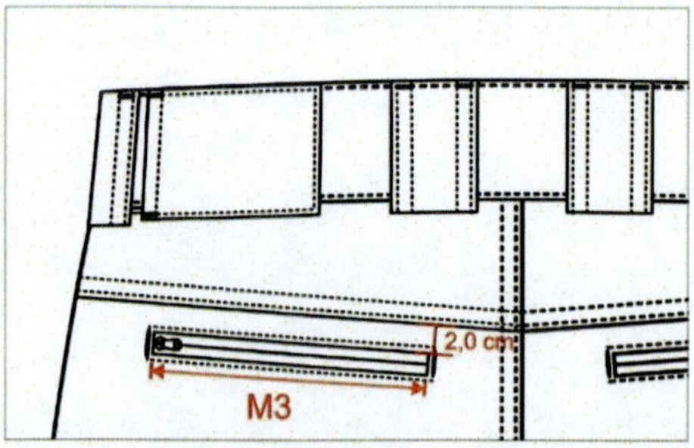

Fig. 10. Back yoke and back pocket (Source: GCML, LONDRINA, 2023).

- **Side leg pockets (Fig. 11):** two pockets, one on each leg, placed over the side seam, slightly inclined to make opening easier, with two vertical folds to enlarge the pocket and place objects.

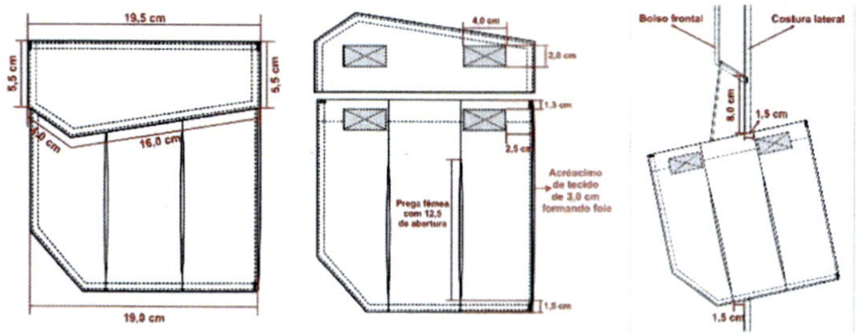

Fig. 11. Side leg pocket (Source: GCML, LONDRINA, 2023).

- **Inseam between the legs:** double-fabric strip sewn between the front and back crotch between the legs to ensure mobility and prevent friction in the agents' private parts against the crotch (Fig. 12).

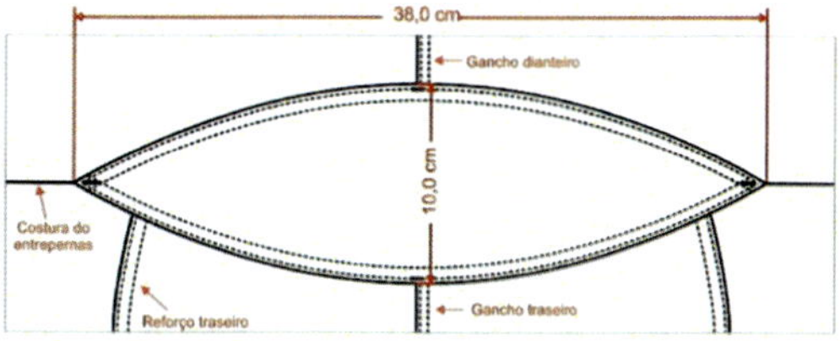

Fig. 12. Measures of inseam between the legs (Source: GCML, LONDRINA, 2023).

Gusseted crotch (Fig. 13): gusset made with the same fabric overlapping the crotch to ensure greater comfort when sitting down.

Knee (Fig. 14): double-fabric knee patches on the front, with folds to enlarge them and improve mobility. Back side of the knee whose cut must be made with yarn at 45° (bias) to ensure greater mobility.

Bottom: with the adjustable hem to fit different ankle sizes (Fig. 15).

The uniform presented here in detail is the result of the 2-year joint effort between the UTFPR project team (students and professors) and guard agents.

It was approved by the guards who participated in the project team. It was also tested by various agents who received evaluation prototypes. The uniform is undergoing the final stage of the bidding process (available from: http://portaltransparencia.londrina.pr.gov.br:8080/transparencia/licitacoes/verLicitacao?formulario.codEntidade=480&formulario.exercicio=2023&formulario.codLicitacao=95&formulario.codTipoLicitacao=6).

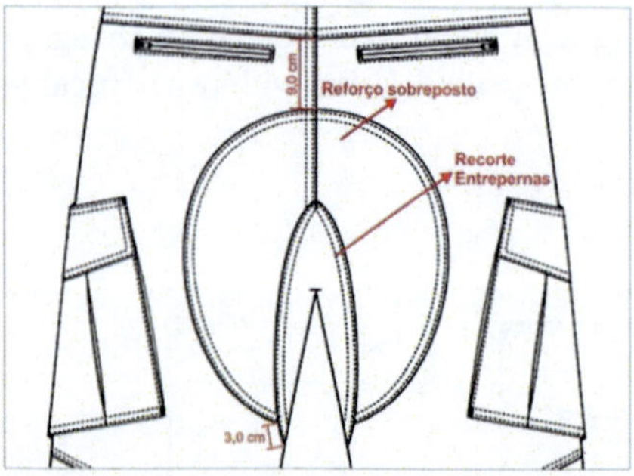

Fig. 13. Back gusseted crotch (Source: GCML, LONDRINA, 2023).

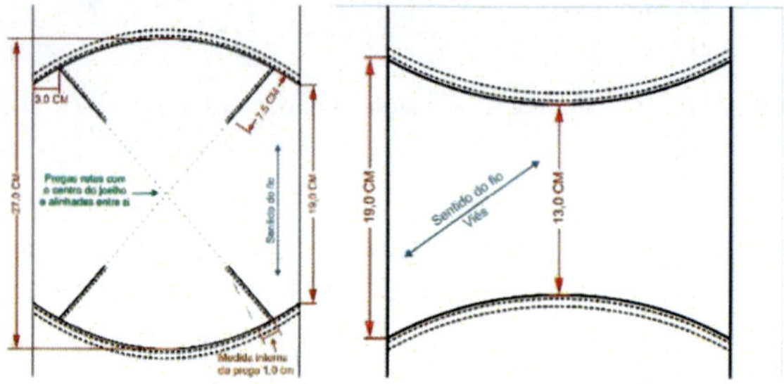

Fig. 14. Front knee patch/back knee cut (Source: GCML, LONDRINA, 2023).

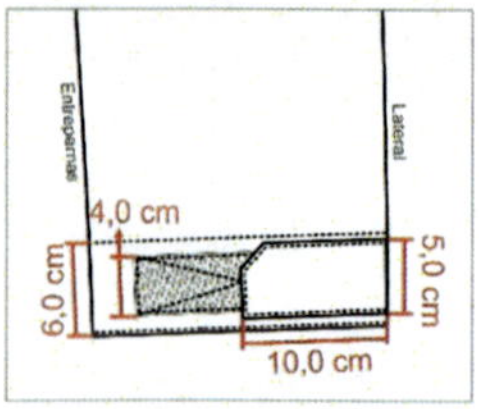

Fig. 15. Bottom hem (Source: GCML, LONDRINA, 2023).

3 Considerations

The partnership between the Fashion Design undergraduate program and the Municipal Guard of Londrina allowed the practical use of various principles of design, resulting in a product. The prototype approved for bidding was defined after countess samples and tests with the agents over almost 2 years of activity.

The main results for the Municipal Guard of Londrina were their redesigned uniform, development of the reference terms/basic project and detailed technical study for the bidding process, training agents to identify problems in the future uniforms acquired by the municipal government of Londrina through the bidding process, and particularly the improved performance of work activities.

As for the students involved in the project, the main contributions were seeing in practical terms the process of solving real product problems (in this case, a uniform), understanding the importance of studying fashion design to develop products, and the relevance of projects to support the community, understanding how clothes design can improve the lives of those who wear them, and developing material to provide related scientific publications and make them available to the academia.

The main results for the professors involved in the project were helping students and agents solve real problems based on the principles of design, visualizing real situations of topics approached in the classroom, experiencing the relationship between the UTFPR community and the outside community, and developing material to provide related scientific publications and make them available to the academia.

The main positive results for UTFPR are presenting to their community and the outside community the importance of the university to society and allowing students and professors to help solve society's real problems.

Thus, the study of the prototype helped the whole project team understand the relevance of design to improve people's quality of life, specifically that of guard agents as they carry out their everyday activities – as the current uniform has various product development problems

Acknowledgments. The authors of this paper extend their gratitude to all people and entities that contributed directly or indirectly to its development, especially the agents of the Municipal Guard of Londrina, scholarship students, the UTFPPR board of directors, Apucarana campus, the private companies partnered with the Municipal Guard, Araucaria Foudantion, and Seti (Secretary of Science and Technology of Paraná).

References

1. Iida, I.: Ergonomia: projeto e produção. São Paulo: Edgard Blucher (2005)
2. Ofício 580/2021-SMDS: Parceria para desenvolvimento de novo uniforme destinado a Guarda Municipal de Londrina. Londrina (2021)
3. Spaine, P.A.A.: Diretrizes para o ensino e construção da modelagem: um processo híbrido [thesis]. Bauru: UNESP – Universidade Estadual Paulista "Júlio de Mesquita Filho", FAAC – Faculdade de Arquitetura, Artes e Comunicação (2016)
4. Tair, R.S.: Uniformes militares: o emprego de novas tecnologias e padrões de camuflagem operacionais [senior project]. Escola de Saúde do Exército, Rio de Janeiro (2021)
5. Prefeitura de Londrina. Legislação/Competências – Defesa Social. [updated September 29, 2021; cited April 12, 2023]. https://portal.londrina.pr.gov.br/menu-oculto-defesa-social/legislacao-defesa

The Resilience of Made in Italy Productions: Research Experience at Officina Vanvitelli

Patrizia Ranzo(✉), Roberto Liberti, and Rosanna Veneziano

Department of Architecture and Industrial Design, University of Campania Luigi Vanvitelli, 81031 Aversa, CE, Italy
{patrizia.ranzo,roberto.liberti,
rosanna.veneziano}@unicampania.it

Abstract. The paper aims to describe the research on design for the enhancement of territorial tangible and intangible capital through cases study that experiment sustainable processes, shared with different actors, for the development of new products in the fashion field.

The resilience of Made in Italy productions, such a strategic sector for the country system, must be supported through strategies capable of combining technologies, advanced innovation (both financial and technological) and humanities (social and cultural values and innovations).

The latter characterize Made in Italy in a meaningful way, which conveys the complex Italian identity condensed in the high quality of its productions to the world.

The paper reports cases study, developed in OFFICINA Vanvitelli HUB - research infrastructure dedicated to made in Italy - which are part of the sustainable transition context of the tanning sector and of the cosmetic sector. The first experimentation, starting from free chrome tanning, was carried out through the collaboration of Made in Italy companies and with the projects of 3D printing processing techniques. The second experimentation, moves in circular economy, using bio-manufacturing of materials deriving from by-products or agricultural and biological products for define skincare cosmetic lines.

Keywords: Sustainable Value Chains · Industrial symbiosis · Traceability · Biofabrication

1 A Place of Research and Experimentation of New Industrial Visions for Made in Italy

The resilience of Made in Italy productions, such a strategic sector for the country system, must be supported through strategies capable of combining technologies, advanced innovation (both financial and technological) and humanities (social and cultural values and innovations). The latter characterize Made in Italy in a meaningful way, which conveys the complex Italian identity condensed in the high quality of its productions to the world.

The environmental sustainability of products and processes is now an unavoidable factor in contemporary industrial scenarios; a fact unknown to most is that Italy presents excellent data from this point of view: it is the European country with the highest percentage of recycled waste (79% of the total); it is first among the big Europeans in the energy sector renewables on total energy consumption (18.3%); the Italian furniture sector has been at the forefront for years due to the high content of recycled materials in its products; over 95% of waste deriving from wood processing becomes raw material for the production of semi-finished products (Source: FederlegnoArredo Research Office on UN COMTRADE data); in 2018 Italy is among the first countries in the world for the number of organic crops with 70,000 organic agricultural producers (Source: Symbola, Processed by Coldiretti on Sinab/Eurostat/FIBL-IFOAM data); our agriculture is among the most sustainable, with just 7.2% of all national emissions (Source: Symbola, Processed by Coldiretti on Eurostat 2017 data). As far as data on the agri-food and food and wine sector are concerned, Made in Italy products are world leaders in quality and safety, with the lowest amount of chemical residues beyond the legal limits (Source: Symbola, Bonifiche Ferraresi, Coldiretti on EFSA data 2019).

Within the Italian fashion sector, more and more companies are sensitized in this sense, also through studies and actions by trade associations (See "Manifesto on Sustainability", Camera Nazionale della Moda Italiana; "Viaggio verso la sostenibilità della filiera", Sistema Moda Italia). These results are above all the expression of small and medium-sized Italian enterprises, which transform their apparent fragility of the system into flexibility to change. The same phenomenon also occurs in the advanced technologies sector: 18.6% of companies operating in the 4.0 environment are small, demonstrating that size is not always the cause of the failure to adopt new technologies, but rather the absence of corporate strategies and visions (Source: from a survey by Marco Bettiol and Eleonora di Maria, Fabbrica Futuro, 2019). An important figure is the growth of robot installations in Italy, which between 2017 and 2018, marked + 27.3%, more than in other European countries: in the last three years robots in the company have increased by + 48% in the food industry, + 27% in fashion; + 21% in wood-furniture, + 23% in engineering (Source: Symbola, IFR, World Robotics, 2019). These data demonstrate that, in order to outline solutions and strategic lines for the competitiveness of the sector, the general and systemic vision in which they are inserted is fundamental, especially with respect to the many opportunities that are already present at a European and Italian level at the moment (PNR, PNR 2021–27).

Precisely the systemic and collaborative action is the basis of the vision "Officina Vanvitelli" of the University of Campania "Luigi Vanvitelli". In line with the European challenges of the 2030 Agenda and the research programs for strengthening the strategic supply chains of the Campania region, the Officina Vanvitelli project - research infrastructure dedicated to made in Italy, D.M. no. 1082, 10/09/21 - represents a case of particular importance at a national level and an opportunity at the same time for the regional territory aimed at consolidating its excellence (Fig. 1). The Hub, entirely dedicated to design and fashion-oriented research, is located in the monumental complex within a multifunctional space equipped with machinery for advanced production.

Exposed on the very wide views of the historic site, with terraces of citrus groves behind it, Officina Vanvitelli promotes collaborative and integrated productions, trans-disciplinary research, and a planning of processes and products oriented towards a perspective of circular ecology, inside and beyond the reference territory.

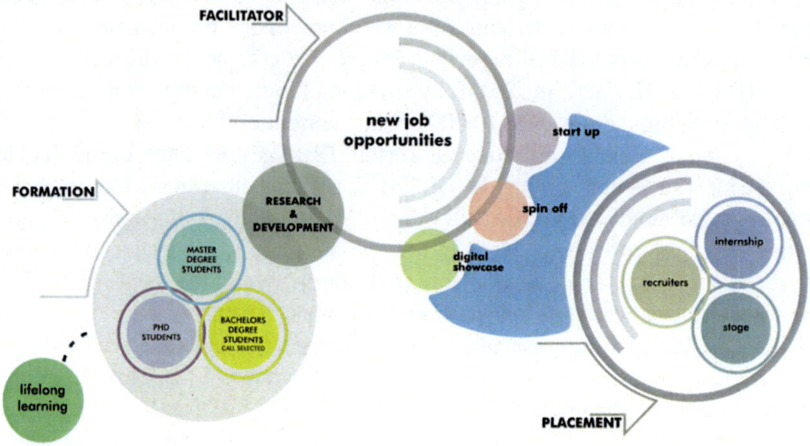

Fig. 1. The Oganizational chart research infrastructure.

2 Research Between Nature, Sustainability and Digital Technologies

This essay starts from the analysis of the research experiments that are being developed at the Officina Vanvitelli Hub in the "conscious" area of the Italian tanning industry, starting from the agreement developed with the Lineapelle trade association signed in 2022. Starting from the partnerships that we had on the local and national territory with the tanning companies, the research activities carried out with the researchers and students of the master's degree in "Design for Innovation" of the Fashion Ecodesign curriculum are described below.

In particular, the "Mutations" project will be analysed, which deepens the research conducted on the innovative and sustainable scenarios of leather and its processing. The contemporary trends analyzed are inclusiveness, sustainability, slow lifestyles and productions as opposed to the fast fashion scenarios. These trends led to the deepening of three themes defined as follows: hyperhumanism, hypernatural, hyperminimal. The three scenarios have deepened the analysis of trendy moodboards which correspond to material research on leather and biological and composite materials, up to a stylistic project of capsule collections of accessories, footwear and leather garments.

The experimentation, starting from free chrome tanning, was carried out through the collaboration of Made in Italy companies and with the experimentation of 3D printing processing techniques.

The research is part of the context of the sustainable transition of the tanning sector, which is already providing good results: compared to 2003, there was a sharp drop in energy consumption in the tanning sector, around 37%, the consumption of chemical products decreased by 8% and those of water 15%. The waste known as SAO, Animal Subproducts of Origin, is recovered and valorized, as a percentage of 77.4% of waste elements per square meter of leather produced, while the remaining part, 22.6%, is destined for disposal; these are sludge, paint residues or polymaterials that cannot be recovered [1]. The companies of the Italian leather system, especially in the Campania area, have developed a high degree of innovation which has allowed the development of new products completely free from traditional concepts. The development of "free metal" processes that use synthetic tannins is a process increasingly used in leather tanning. Another essential aspect is traceability, as suggested by the UNECE Traceability of Sustainable Value Chains project, with respect to the necessary traceability in the clothing and footwear sector for an informed and responsible choice.

Traceability helps us to understand the history of the product, and therefore its origin from the farm to the final transformation in the food sector. The current standard requires us to provide information only on meat and products intended for human consumption, but not to provide information on other kind of products; the task of our research in the "conscious leather design" sector is to propose, among the various focuses, a traceability of leathers used in fashion products [2].

Within the "Mutations" project, in addition to interesting experiments on the visual and structural language of the leather surface, we started from an approach to design that hybridizes nature and digital manufacturing technologies. Among the most interesting results is a collection that starts from the study of the interaction between sound waves and water; the resulting digital images, at different frequencies, generated extraordinary surfaces printed on leather, substantiating the concept of "hypernatural" for leathers destined to the luxury market. The experiments described, carried out in collaboration with some of the tanning companies of the Solofra District, including the vegetable-tanned leather garment made with "Officina Chimica" and with the "Hyperhumanism" project team, led to the "Mutations" exhibition presented at the 2022 edition of Lineapelle Milan (Fig. 2).

The companies and researchers involved will converge in the advanced training project in September 2023 "Leather Conscious Design Academy" which will train, on the job, conscious designers for the luxury fashion sector.

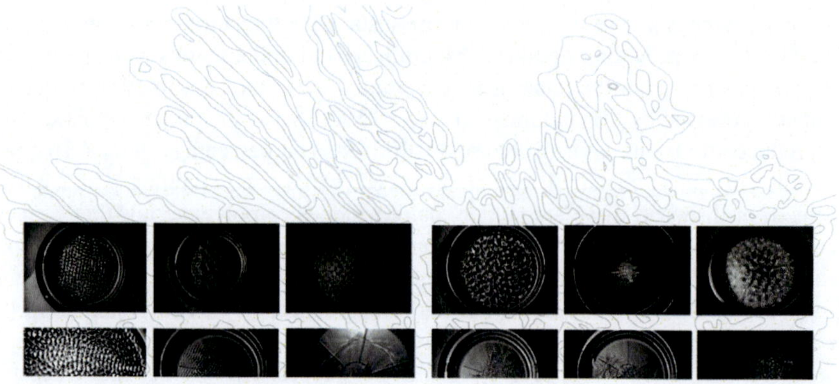

Fig. 2. Cymatics experiments from the Hypernatural research team.

3 Industrial Symbiosis and Biofabrication: Sustainable Materials for New Design Experiments

The processes of industrial symbiosis allow, through the construction of collaborative systems, the strengthening of territorial production chains and the experimentation of new processes and products in a systemic perspective.

In particular, through the systemic design methodology which facilitates the construction of relationships between different actors and stakeholders, the outputs of resources become inputs for other local realities and, the connections are consolidated, generating productive alliances and collaborations between companies [3, 4, 5]. Among the projects that move in circular economy and sustainability direction, the bio-manufacturing of materials deriving from by-products or agricultural and biological products has been used in various production sectors.

Bacterial nano cellulose, known by the name Scoby (Symbiotic Colony of Bacteria and Yeast), has been the subject of design trials of eco-oriented products, which confirm the interest in the material and the high environmental performance of the production processes and phases of disposal as well as the potential and technical and functional characteristics of the support. The research on BioCouture carried out since 2010 by Suzanne Lee, which takes the form of collections with fabrics based on bacterial nano cellulose based on caffeine and kombucha, conducted by Theanne Schiros, in collaboration with the streetwear brand "Public School" with the creation of sneakers in bio leather, grown from the production waste of the local brewery "Kombucha Om Champagne Tea" are just some of the experiences conducted starting from the collaboration of closed-loop supply chains and research centers (Fig. 3), (Fig. 4).

Since 2016, the MakeGrowLab laboratory (www.makegrowlab.com) has launched a project funded by the European community, which intends to disseminate an efficient and scalable model for transforming local organic waste into natural and compostable materials for domestic use.

The first experiments on the cellulose-based material created by the Kombucha fermentation process were not considered suitable for commercial production due to the poor mechanical resistance characteristics, therefore a series of tests on the material,

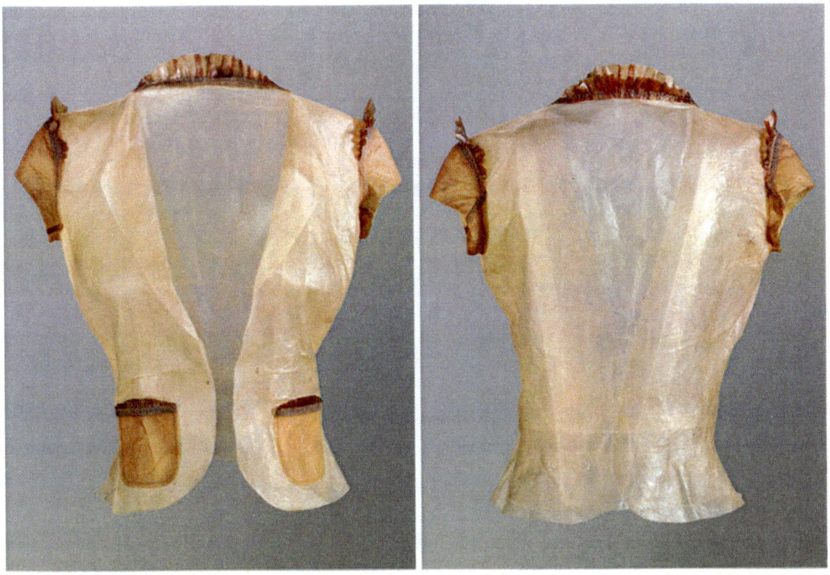

Fig. 3. BioCouture growind textile by Suzanne Lee.

the founders Roza Janusz and Josh Brito created Scoby Packaging Materials, a durable, compostable and 100% plastic-free material. Today, the research lab has active partnerships to make a variety of water-soluble or impermeable materials and films that act as a barrier to oxygen, grease, and water vapor. The sector in which MakeGrowLab concentrates its research is that of food packaging, experimenting with edible solutions and alternatives to laminated paper, but also in the cosmetic sector the laboratory has consolidated strategic partnerships with companies such as Palmolive.

In the cosmetic sector, the bio-material Scoby, due to the water and microbial barrier properties of the film, has found a space for experimentation both as a support for active ingredients, as a reference or patch, and as an alternative material to traditional packaging, offering significant benefits in the post-consumption disposal.

The best-known examples of the use of bacterial nano cellulose are the Yeabucha line, created to spread the production of Scoby from Kombucha and transfer production protocols and methods of use to the community. "Yeabucha" provides kits, guides, and ingredients for making material from kombucha probiotic culture. The "From Peel to Peel" project is microbial cellulose-based films made with fruit and vegetable leftovers. Using different plants during the cultivation of the material allows obtaining films with a chromatic variety (Fig. 5).

The work starts from the reuse of food waste which, combined with bacteria and yeasts, generate films for disposable packaging creation. Even in the "Ponto" project by designer Elena Amato, the bacterial cellulose sheets are similar to paper in which natural pigments such as spirulina, hibiscus, saffron, and charcoal, added to the compound during the mixing process, allow to obtain different colors (Fig. 6).

Fig. 4. Public School NY x Dr. Theanne Schiros' microbial leather sneakers. Photo by Jon Brown courtesy of One x One.

Fig. 5. From Peel to peel, eco-friendly food packaging.

In the cosmetics sector, where R&D investments are higher than the manufacturing industry average, the use of sustainable materials, the valorization of local resources,

Fig. 6. Ponto, packaging for cosmetics.

and the building of connections with universities and research centers are some of the competitive levers essential for success in national and international markets. For some years at Officina Vanvitelli a series of educational and research initiatives have been taking place in close collaboration with companies in the Campania region of the sector. Workshops, thematic seminars, and research interactions with the Department of Economics and the Department of Environmental, Biological, and Pharmaceutical Sciences and Technologies, make it possible to build a network of skills at the service of the ecosystem of companies in Campania, to build a supply chain of cosmetics, as happens in other Italian regions.

The research group, coordinated by Rosanna Veneziano, deals with defining concepts of cosmetic lines, implementing design-oriented strategies, aimed at optimizing supply chain processes, and spreading sustainable practices. The design experiments used bacterial nano cellulose as a support for highly functional formulations and as packaging, exploiting the water and microbial barrier performance of the films. The research groups collaborated with the Kforbusiness - Biologic company which followed the development process of the concepts of the cosmetic lines and guided the workshops for the definition of the scenarios and the functionality of the material. Another important collaboration was carried out with the Biancamore company which for years has been experimenting with the integration of the agri-food chain, in particular the dairy one, with that of cosmetics, with a view to the circularity of the flows of production and waste materials. The enhancement of territorial resources, the partnership with local producers, and the desire to experiment with a local collaboration model make Biancamore closely connected to the Paestum area, its community, and its historical identity. The concept of the Arteria cosmetic line was developed to consolidate connections with local producers of ingredients for formulations by integrating manufacturing processes and optimizing the inputs and outputs of the supply chains (Fig. 7).

Fig. 7. Arteria, capsule collection for Biancamore.

These experiments show that through an integrated and collaborative approach to the production chains, it is possible to support and consolidate new opportunities for the development of sustainable entrepreneurship and to build an ever-wider collaborative ecosystem between institutions, companies, and communities [5, 6].

4 Conclusions

Officina Vanvitelli identifies its main reference in the cultural and creative industry. In this context it hosts a continuous dialogue and feeds dialectical orientation scenarios for the formulation of new training paths, for internationalization and promotion, through innovation in design and fashion sectors and in the profiles of its students.

The activities carried out at Officina Vanvitelli are oriented to promote the enhancement of territories and support the local realities defining the network of values underlying the material and immaterial landscapes of Italian culture and project.

It is the place for comparison with production companies, institutions and bodies that represent the complex reality of the territory and its geographies, not only physical, but cultural and social, with the aim of continuously deriving new scenarios, knowledge and relationships.

Author Contributions. Patrizia Ranzo is mainly responsible for Sect. 1. Roberto Liberti is mainly responsible for Sect. 2. Rosanna Veneziano is mainly responsible for Sect. 3. All authors wrote the Abstract and Conclusions collaboratively.

References

1. Fletcher, K.: Moda, Design e Sostenibilità. Postmedia, Milano (2018)
2. Roberto, L.: Men's tailoring, neapolitan essence of made in Italy. Firenze: Altralinea ed (2017)
3. Bistagnino, L.: microMACRO - Micro relazioni come rete vitale del sistema economico e produttivo, Edizioni Ambiente (2014)

4. Silvia Barbero e Cristian Campagnaro. Dai sistemi viventi ai sistemi industriali aperti, Slow Food Review (2008)
5. Battistoni, C., Nohra, C.G., Barbero, S.: A systemic design method to approach future complex scenarios and research towards sustainability: a holistic diagnosis tool. Sustainability **11**(16), 4458 (2019). https://doi.org/10.3390/su11164458
6. MdDonough, W., Braungart, M.: The Upcycle: Beyond Sustainability, Designing for Abundance, Unabridged Edizione (2012)

Ergonomics and Materials Technology: The Case of the Uniform of the Cycle Couriers in the City of Recife–Pernambuco/Brazil

Thuanne R. F. Teixeira[1(✉)], Etienne A. A. Silva Martins[2], and Germannya D'Garcia Silva[1]

[1] Universidade Federal de Pernambuco, Recife, Pernambuco, Brazil
{thuanne.teixeira,germannya.asilva}@ufpe.br
[2] Universidade Federal Rural de Pernambuco, Recife, Pernambuco, Brazil
etienne.silva@ufrpe.br

Abstract. This article brings the partial result of the dissertation of the Postgraduate Program in Design of Federal University of Pernambuco, in the line of research Design, Ergonomics and Technology, which aims to propose improvements for professional clothing in outdoor environments, in order to promote comfort and well-being to the cycle couriers of the Brazilian Company of Posts and Telegraphs (ECT/Correios) in the city of Recife - Pernambuco. The problem analyzed was the relationship between the mechanical properties of the textile material selected for the uniform and the ergonomic problems related to the task of the cycle couriers of the postal company, through ergonomic appreciation and laboratory textile analyses. The research procedures were based on some steps of the OIKOS (Martins, 2019) and SHTM (Moraes e Mont'Alvão, 2012) methodologies. As a result, a strong relationship was verified between the mechanical properties of the fabric used in the uniform and the ergonomic problems regarding displacement and interface, as well as physical-environmental, natural and chemical problems, and problems related to occupational accidents, configuring itself as a possible point of contribution of design regarding the selection of materials for wearable pieces, specifically for uniforms for use in outdoor environments with active mobility of the users.

Keywords: Uniforms · Workers in Outdoor Environments · Ergodesign · Usability

1 Introduction

Garments are present in our lives, mediating the relationship between humans and the physical environment through their functions, manifesting themselves as use values during their use [1]. Unlike natural objects, articles of clothing are created by humans, with different purposes and roles. In the case of a piece of clothing, it presents itself as a wearable item, with other functions in addition to the function of solving issues related to social decency: in addition to human survival, through coverage and body protection, it manages to communicate information about users and their preferences, about a collective or even about a social group, as well as influence the perception of comfort and well-being of the user, affecting that user both physiologically and psychosocially [2–4].

The main material used in wearable pieces are fabrics, textile materials that have been present in human life for millennia, accompanying individuals from their birth to the end of their lives [5]. Such materials have the ability to provide physical or psychological comfort or discomfort to users, as they are in direct contact with human skin, acting as a second skin in the person-environment interface and interfering with safety and well-being [6]. With regard to uniforms, they are a type of clothing that provides protection in the work environment and that must meet the practical issues of use [7]; they also constitute the corporate identity, carrying with them symbolic values linked to the profession or institution they represent [8], they must also attend to the subjective questions of use, aiming at the satisfaction of their users, the workers. In the context of work in an outdoor environment, and with the use of active mobility to perform tasks, also considering the geographical location in which activities take place, it is important that the selection of materials for the uniforms is consistent with the work demands. Therefore, there needs to be an adaptation to the user's profile and expectations, as well as to the situation of use [9], such as the work context (tasks performed and environment where they are performed).

Fig. 1. Postal cycle courier of the company Correios (ECT). Available on: < www.correios.com.br >.

In this context, there is the possibility of contributing to uniform projects, or to the selection of materials for them. In an interdisciplinary way with other areas of knowledge, such as ergonomics and materials technology, for example, Design enriches fabrics with new concepts and better performance, obtaining results that go beyond simple dressing, with finishes and functionalities that add value to wearable pieces, such as intelligent fabrics [10, 11]. In view of the exposed scenario, the research problem analyzed in this work is presented, namely the relationship between the mechanical properties of the textile material selected for the uniforms and the ergonomic problems associated with the task performed by the cycle couriers of the Brazilian Company of Posts and Telegraphs (ECT/Correios) in the city of Recife–Pernambuco (see Fig. 1).

This research builds on the results of the dissertation entitled "Analysis of the technical and emotional attributes of uniforms for outdoor environments: the case of Correio's cycle couriers in Recife–PE", linked to the Postgraduate Program in Design at the Federal University of Pernambuco. The uniform of the Company of Posts and Telegraphs was chosen for this investigation because it contemplates the requirements of symbolic factors, as it is a public, federal, national and secular institution, and very present in the repertoire of Brazilian society, which impacts on the user's perception of this clothing; in addition to these, practical factors also stand out, enabling the analysis of uniforms intended for external environments, related to the work with cycle delivery, which, currently, is configured as a mostly precarious job and on the rise since the period of the pandemic caused by the SARS-CoV-2.

2 Ergonomic and Usability Perspectives

The activities carried out at work, as well as the physical and social environments in which they are carried out, expose the worker to loads and human costs, with physical, mental, emotional and affective wear [12]. Several professional outdoor activities carry significant health risks, so deploying prevention efforts across different professions based on their individual needs and activities may be key to reducing workloads, ergonomic problems and risks to the health and well-being of the worker [11, 13]. In this context, a uniform is configured as a wearable artifact that operates at the interface person-task-machine-environment, therefore, it must function in a way that contributes to the good performance of work activities, without bringing constraints to its user. With this in mind, there must be technical quality in the proper choice of materials, finishes, operation and maintenance of this product; psycho-aesthetic quality with regard to the combination of colors, shapes, textures, the pleasure provided to the user and their symbolic and significance values; and ergonomic quality, such as comfort when interacting with the product, anthropometric adaptation, compatibility with the activity performed, and user safety and satisfaction [14, 15].

The process of selecting the materials, an important phase in the design process, influences the feeling of comfort in the user of the final product, and this comfort can be psycho-aesthetic or psychosocial, sensory or tactile, ergonomic and thermophysiological [16, 17]. In this context, the technical properties of textile materials are essential in the production of clothing, as they are decisive for the acceptance of the product [18], where the mechanical and superficial characteristics of the fabrics play an important role in the evaluation and perception of the product in the process of use [4]. Therefore, aspects such as weight, elasticity, surface characteristics and other properties of the fabric can be associated with an assessment of the workers' comfort, relating it to the use of the uniform during the activities, for a better understanding of its relationship with the execution of tasks, uniting the perspectives of usability and ergonomics for this analysis [18].

The analysis of the uniform from the perspective of the Human-Task-Machine-System (SHTM), considering the environment where the tasks are carried out, makes it possible to identify existing ergonomic problems in carrying out the task. On the other hand, the analysis of the properties of the fabric used in the uniform makes it possible to identify which textile characteristics are related to these problems. In this multidisciplinary way, the steps and tools of the SHTM [12] and OIKOS [18] methodologies were used to assess the uniform of the cycle couriers of the company Correios (ECT) in Recife, Pernambuco, Brazil.

3 Procedures and Tools for Data Collection

Table 1. Methodological stages and procedures.

Methodological stages	Research procedures and tools
Characterization of the pieces of clothing that make up the uniform of the cycle courier	Identification of the composition, trims and finishes; laboratory tests of grammage (ABNT NBR 10591), color fastness to rubbing (ISO 105) and tensile strength (ASTM D5034), at the Textile Engineering Laboratory of the Federal University of Rio Grande do Norte (Brazil), with new uniforms, provided by ECT for analysis
Survey of problems related to the worker's job (with emphasis on uniform)	Document analysis; photographic record and unsystematic observation of the task; semi-structured interview with volunteers from the administrative sectors (seven people) and the delivery sector (six people), during the activities. Such techniques were applied at the Home Distribution Center (CDD) of the ECT and also outdoors, in the city of Recife (Pernambuco/Brazil), after the formal commitment of the research with the institution's Code of Ethics, signed between the research entity and the ECT

This investigation was based on the OIKOS methodology [18] for conducting the semi-structured interviews and for observing the activity with regard to points related to usability criteria (compatibility with the user and with the tasks performed) and to comfort criteria (subjective perception and investigation of fabric aspects such as touch, fit, weight and elasticity). In order to understand the relationship between the textile characteristics of the uniform and the ergonomic problems of the task, techniques and tools of ergonomic appreciation were used, part of the SHTM method [12], analyzing the system problems (see Table 1).

4 Results of the Analyses

4.1 Characterization of the Garments that Make up the Uniform of the Cycle Courier

For the evaluation of usability, the characterization of the garments that make up the uniform was carried out, in order to understand their behavior in a real situation of use (see Table 2). It was possible to notice that the upper part of the uniform is lighter than the lower part, corroborating the report of the users, who perceive the uniform shirt as being "light" and the pants as being "heavy". With regard to the test of color fastness to rubbing, it was observed that there was no transfer of color from the fabric of the upper part of the uniform. However, this does not exempt the dyeing from fading due to exposure to ultraviolet rays (UVR). Its structure, knitted and with high stretching capacity, as well as the percentage of cotton in its composition, contribute to the feeling of freedom of movement and to the alleviation of sensory and thermophysiological discomfort. Both pieces are made mostly of synthetic fiber (PES), contributing to the feeling of heat and stuffiness.

Table 2. Characterization of the garments that make up the uniform of the ECT.

Piece	Composition	Trimming	Finishes
Shirt	67% polyester (PES) and 33% cotton (CO)	Elastic collar and cuffs	Plasticized internal instructional label. Knitted fabric, yellow dyeing, with straight seams and inner overlock finish
Pants	67% polyester (PES) and 33% viscose (CV)	Zipper, button, clasp.	Plasticized internal instructional label. Plain twill fabric, navy blue dye, with straight seams and inner overlock finish

The lower part of the uniform, with a twill structure, showed low stretching capacity in both directions, weft and warp. On the other hand, the fabric has good mechanical strength. In the test of color fastness to rubbing, it showed transfer to the control body, mainly in the test with wet fabric, demonstrating that, in contact with water or sweat, the loss or transfer of the dye occurs more quickly.

This indicates that, during the worker's activity, the fabric used in the pants of the uniform may show discoloration and wear after a short time of use, with the risk of transferring color to another fabric in friction, as well as being absorbed by the skin during the movements performed, presenting a risk to the health of the user. In addition to these risks mentioned, there is also the risk of environmental contamination, as the uniforms are subjected to water-based domestic washing. The data obtained from the analyzes can be seen in table 3.

Table 3. Laboratory tests aiming at the textile characterization of the uniform of the ECT

Piece	Grammage (g/m^2)	Color fastness to rubbing	Tensile strength and elasticity
Shirt	171.7	Dry samples: 1–1; Wet samples: 1–1	Weft: Stretching capacity of 100.6% and a strength level of 457.6N Warp: stretching capacity of 165.8% and a strength level of 305N
Pants	235.2	Dry samples: 2–2 Wet samples: 3–3/4	Weft: stretching capacity of 26.9% and a strength level of 680N Warp: stretching capacity of 21.8% and a strength level of 1.210N

4.2 Problems Related to the Task of the Cycle Courier with Emphasis on the Uniform

Although the approach of this investigation is focused on garments, under the premise that the characteristics of the fabrics influence the comfort and well-being of the user, footwear (it is part of the uniform, but it is not a research object of this investigation) stood out, pointing to the need for studies focused on this specific component (see Fig. 2).

Fig. 2. footwear that forms part of the uniform.

Among the problems identified in the system through the ergonomic assessment, those that are related to the uniform (and the fabric used in it) used in the activities are presented here. Therefore, ergonomic problems were identified related to: interface, displacement, physical-environmental, natural, biological, occupational accidents and chemical (see Table 4).

Problems related to the interface were the most cited by the participants. The embarrassment caused by the inadequate contact between the lower part of the uniform and the body causes several points of discomfort in the workers. The most prominent points of pain were the legs and the foot, with the pants and footwear being identified as not suitable for the activity, being qualified by the workers as "heavy", "tight", "too hard" (shoes), "bad" (pants fabric) and "painful". The shoes stood out, becoming the worst rated item in terms of the users' perception of comfort.

Table 4. Survey of system problems.

Nature	Ergonomic problems
Interface	Biomechanical constraints in carrying out the activity caused by repetitive movements and pain/discomfort caused by parts of the uniform, such as the bag strap, pants and shoes
Displacement	Workers move an average of 13 km/day through active mobility (pedaling and walking) while performing their outdoor tasks
Physical-environmental	Activity in an external environment, therefore without noise, temperature or light control
Natural	Postmen are exposed to intense ultraviolet radiation (UVR) (from 2 pm to 5 pm), and exposed to the weather, which causes interruption of activity during rains, or exposure of the worker to them
Biological	Postmen are exposed to urban waste such as garbage, sewage or contaminated mud
Occupational accidents	In addition to the urban risks inherent to the nature of an activity that involves movement from one place to the another, there are also risks of canine attacks, which have already occurred among volunteers
Chemical	Risk of absorption of uniform dyeing by the wearer's skin

To ease the discomfort caused by displacement and the feeling of heat, ECT offers the possibility of choosing the uniform model (where the different options have the same textile composition). Therefore, the user can choose a shirt with short or long sleeves; shorts or pants. Due to the thermophysiological discomfort resulting from the environmental context in which the work is carried out, and also from the movements necessary for active mobility, some users prefer to wear shorts instead of pants, exposing, however, the lower part of the body to a greater risk of occupational accidents (abrasions and canine attacks) and natural accidents (increased exposure to UVR). In this sense, workers have to choose which risk they will be more exposed to: whether to the risks related to interface and displacement, or whether to the risks concerning the natural and occupational accidents related to the task.

5 The Relationship Between Fabric and Ergonomic Problems

By crossing data from ergonomic and laboratory analyzes (see Table 5), it was observed that the uniform is presented as a tool that could reduce the identified ergonomic risks if made with suitable materials. The current uniform does not provide such protection, workers are more exposed to risks and are also more susceptible to destandardize the image and identity of the company, deliberately changing the composition of their professional clothing.

In the lower part, changes were made to the pants and shoes, in which regular jeans were added to the uniform, as well as sneakers, as part of the users' quest for greater comfort while performing their tasks (see Fig. 3).

Table 5. Data crossing.

Uniform	Textile characterization	Problems in the system
Shirt	67% PES and 33% CO; Low tensile strength	Susceptible to natural, biological and physical agents
Pants (or shorts)	Plain twill fabric; 67% PES and 33% CV; Heavy material; Low elasticity; Low color fastness	Does not allow freedom of movement (pants); cause of heaviness and heat (pants); exposes the legs to canine attacks and abrasions (shorts); susceptible to natural and biological agents
Shoes	Main cause of discomfort in users: pain, tiredness, contact with contaminated material	

Fig. 3. Changes to the lower part of the uniform.

In the upper part of the uniform, in the search for protection against natural ergonomic problems, there was the addition of a long-sleeved shirt, knitted, with improved UV protection (see Fig. 4).

Fig. 4. Modification made to the upper part of the uniform.

As much as the user has sought to use colors that are close to the visual identity of the institution, this change, as well as those mentioned above, goes against the institution's own internal manuals, accessed during document analysis in the ergonomic assessment. Even though the uniform of the company ECT has undergone countless updates throughout its long history, the technological update on the product was not enough. In this sense, the lack of conformity between the clothing and the user can lead to risks during the execution of the tasks, so this product must meet the specificities of the tasks performed, the diversity of biotypes existing in the group, the characteristics of the users and must also take into account technical, environmental and systemic restrictions, in order to favor a good user-product-task-environment relationship [9, 15]. In the context of clothing for the professional segment, such needs are accentuated, as it is a scenario where there are goals, expected results, methods and specific contexts of use. It is the responsibility of the clothing design to think about the entire cycle of the garment, including modeling, materials, finishes, as well as the manufacturing, use and disposal processes [19, 20].

However, according to what was observed in this investigation, professional clothing did not prove to be the most problematic part and the cause of discomfort in the user. In this regard, it is worth emphasizing the important role that the user's repertoire plays in judging the garment [21, 22]. Volunteer users have a profile of stability in terms of employment, with long experience within the company (some of them for more than twenty years). The experience with previous uniforms, which presented greater inadequacies, can contribute to the attention now being focused on solving more urgent problems, such as pain caused by the shoes.

Since the uniform carries with it not only technical values, but also symbolic ones [8] and that the surroundings and garments are judged through affections [23, 24], this will influence the perception of users, causing the focus of complaints to be taken to the footwear, which is the piece that most negatively affects them at the moment. In view of the above, interdisciplinarity and a holistic look at the analyzes of this piece are essential.

6 Final Considerations

The material used in the uniform of the cycle courier who works for the ECT/Correios showed characteristics incompatible with the task, especially the lower part, which causes bodily constraints in relation to movement, to the exchange of body heat with the environment and to the risks that arise during the activity, such as dog attacks, for example. It can be seen that the uniform is the first barrier between the body and the environment, and can be used as a tool to protect health and enhance the well-being of the worker.

In addition to laboratory tests on the mechanical nature of the fabric, studies are needed regarding modeling and absorption capacity of ultraviolet radiation, as this is a job in an external and daytime environment. In addition, studies are also needed on intelligent textile materials that protect users from biological and physical-environmental risks and from occupational accidents, as well as on the use of fibers that allow the exchange of heat body-environment, and that allow body movements, so as not to constrain the movements necessary for the activity, also providing thermophysiological comfort. In this scenario, being attentive to the innovations in the textile industry is essential for projects that consider materials that are in accordance with the real situations of use of the piece of clothing, without failing to consider the expectations of the user, thus meeting the objective and subjective needs related to the use of the product.

This research focused on the garments of the professional uniform. However, despite the perceived need for investment in more suitable materials for making the uniform for this activity, the point of intervention that proved to be the most urgent was footwear. This is the item that most negatively affects users at the moment, being a source of physical and psychological discomfort for workers, as it causes pain and dissatisfaction, and may impair the performance of the activity itself, which is not desirable either for the user or for the institution that employs him.

Finally, when verifying, during the semi-structured interviews (self-report), the important role that affections (sensations, feelings and emotions) play in the judgment of the garments by their users, the relevance of interdisciplinary studies for a better understanding of the user's relationship with his uniform stands out, especially because it is a nationally recognized institution and present in the repertoire of the Brazilian population, adding symbolic values to the uniform analyzed in this work. In this sense, investigating this relationship from the perspective of emotional design can contribute to the understanding of the practical and affective needs of users.

Acknowledgements. This research was carried out with the support of the Coordination of Superior Level Staff Improvement–Brazil / Coordenação de Aperfeiçoamento de Pessoal de Nível Superior–Brasil (CAPES) and with the authorization and collaboration of the Brazilian Company of Posts and Telegraphs / Empresa Brasileira de Correios e Telégrafos (ECT / Correios, Brazil).

References

1. Löbach, B.: Design Industrial: Base para a configuração dos produtos industriais. Blucher, São Paulo (2001)
2. Rocha M.A.V.: Algumas reflexões sobre o consumo de moda. In: Silva MZT Saraiva JM, Santana DESL (Orgs.) O consumo no contexto da família, da infância e da adolescência. Editora Universitária da UFRPE, Recife (2016)
3. Moura, L.L.: Moda como expressão de identidade no mundo contemporâneo. Dissertação (Mestrado em Psicologia Social), Universidade Federal de Sergipe, São Cristóvão (2018)
4. Broega, A.C, Cunha, J.L.L., Cabeço-Silva, M.E.: O conforto no vestuário, seus aspectos conceituais e subjetivos. In: Martins, S. B. (Org.). Ergonomia, usabilidade e conforto no Design de Moda: a metodologia Oikos, Estação das Letras e Cores, São Paulo, pp. 14–32 (2019)
5. Pezzolo, D.B.: Tecidos: história, tramas, tipos e usos. Editora Senac, São Paulo (2021)
6. Martins, S.B.: Ergonomia, usabilidade e conforto em projeto de produto de moda e vestuário. In: Martins SB (Org.). Ergonomia, usabilidade e conforto no Design de Moda: a metodologia Oikos. Estação das Letras e Cores, São Paulo, pp. 56–77 (2019)
7. Dias, S.F.P.: O fardamento na indústria: desenvolvimento do fardamento na empresa Celtejo. Dissertação (Mestrado em Design de Vestuário e Têxtil), Instituto Politécnico de Castelo Branco, Portugal (2017)
8. Farias, R.C.P.: Transubstanciação simbólica do uniforme de trabalho em signo de prestígio. Anais do Museu Paulista: história e cultura material, São Paulo, **18**(2), 263–284 (2010)

9. Rosa, L.: Vestuário industrializado: inserção da ergonomia na concepção e desenvolvimento de produto. In: Martins SB (Org.). Ergonomia, usabilidade e conforto no Design de Moda: a metodologia Oikos. Estação das Letras e Cores, São Paulo, pp. 116–138 (2019)
10. Laschuck, T.: Aplicação de têxteis inteligentes a produtos de design de moda. Dissertação [Mestrado em Engenharia Têxtil], Universidade do Minho, Braga, Portugal (2008)
11. Martins, E.A.A.S.: O estudo da aplicação de acabamentos funcionais de barreira UV em fibras previamente ativadas por plasma. Tese (Doutorado em Engenharia Têxtil), Departamento de Engenharia Têxtil, Universidade do Minho, Portugal (2015)
12. Moraes, A., Mont'Alvão, C.: Ergonomia: conceitos e aplicações. 2AB, Rio de Janeiro (2012)
13. Zink, A., Tizek, L., Schielein, M., Böhner, A., Biedermann, T., Wildner, M.: Different outdoor professions have different risks–a cross-sectional study comparing non-melanoma skin cancer risk among farmers, gardeners and mountain guides. J. Eur. Acad. Dermatol. Venereology **32**, 1695–1701 (2018)
14. Hancock, P.A., Pepe, A.A., Murphy, L.L.: Hedonomics: the power of positive and pleasurable ergonomics. Ergonomics Des. Q. Hum. Factors Appl. **13**(1), 8–14 (2005)
15. Iida, I., Guimarães, L.B.M.: Ergonomia: projeto e produção. Edgard Blücher, São Paulo (2016)
16. Slater, K.: Subjective textile testing. J. Textile Institute, Part 1 **88**(2), 79–91 (1997)
17. Broega, A.C., Silva, M.E.C.: O conforto total do vestuário: design para os cinco sentidos (2010). https://fido.palermo.edu/servicios_dyc/encuentro2007/02_auspicios_publicaciones/actas_diseno/articulos_pdf/A6012.pdf. Accessed 25 June 2023
18. Martins, S.B.L.: OIKOS: metodologia de avaliação da usabilidade e conforto de produtos de moda e vestuário. In: Martins SB (Org.). Ergonomia, usabilidade e conforto no Design de Moda: a metodologia Oikos. Estação das Letras e Cores, São Paulo, pp. 92–113 (2019)
19. Santos, J., Veloso, I.: Características ergonômicas de calçados femininos para tomada de decisão e comportamento das usuárias. In: Ergodesign e HCI, número 1, vol. 5, Rio de Janeiro (2017)
20. Vianna, C., Quaresma, M.: Ergonomia do vestuário para idosas de 60 a 75 anos. In: Ergodesign e HCI, número especial, vol. 7, Rio de Janeiro (2019)
21. Norman, D.: Emotional Design: Why We Love (or Hate) Everyday Things. Basic Books (2005)
22. Cardoso, R.: Design para um mundo complexo. Cosac Naify, São Paulo (2012)
23. Desmet, P., Hekkert, P.: Design and Emotion. Int. J. Des. **3**(2), 1–6 (2009)
24. Damásio, A.: Sentir & saber: as origens da consciência. Companhia das Letras, São Paulo (2022)

The Importance of Comfort in Sport Caps Through Subjective Assessment in Real Weather Conditions

Rosimeiri Naomi Nagamatsu[1(✉)], Maria José Abreu[2], Cosmo Damião Santiago[3], Debora Mizubuti Brito[1], and Marcio Roberto Ghizzo[4]

[1] Fashion Design Department, Federal University of Technology – Paraná, Campus Apucarana, Apucarana, Brazil
naomi@utfpr.edu.br
[2] 2C2T, Textile Engineering Department, Minho University, Braga, Portugal
[3] Mathematics Department, Federal University of Technology – Paraná, Campus Apucarana, Apucarana, Brazil
cosmo@utfpr.edu.br
[4] Humanities Department, Federal University of Technology – Paraná, Campus Apucarana, Apucarana, Brazil
marcioghizzo@utfpr.edu.br

Abstract. This article aims to present the results of the investigation of the perception of the comfort properties of sports caps through wear trial in real weather conditions in an uncontrolled environment. The sensations of general comfort, temperature, humidity and the pressure that ten samples of sports caps exert on the head of ten volunteers during subjective assessment were evaluated. The two-way ANOVA statistical tool was used for statistical data analysis. Volunteers assigned very low averages to the affective ratings of the sports cap samples. This may indicate the volunteers' difficulty in judging the attributes when the caps are worn on the head. In future research, it is intended to compare the data obtained in this study with the data obtained in tests of use carried out in an environment of controlled temperature and humidity.

Keywords: Textile Comfort · Wear Trial · Sports Caps · Textile Attribute · Subjective Evaluation

1 Introduction

The clothing microenvironment is the space between the clothing and the skin. If clothing is a dynamic microenvironment, then how do you make that microclimate pleasant?

In large urban centers in southern Brazil, the occurrence of thermal amplitudes is very common, which can vary between 10° and 20 °C. Brazilians usually call this phenomenon "the four seasons of the day" and, in order to regulate the temperature of their microenvironment, in the morning they are wrapped up and, as the temperature rises, they take off the layers of clothing in an attempt to provide comfort to the body. Controlling the thermo-physiological properties of clothing during changing weather conditions is one of the factors that can contribute to thermos-physiological balance.

When studying microclimate issues related to clothing comfort, [1, 2] and [3] cite that, for [4], it is necessary to meet four fundamental aspects to achieve the total clothing comfort: ergonomic comfort, psychological comfort, thermo-physiological comfort and sensory comfort. Each of these aspects is formed by a set of factors/attributes that influence the performance of the garment.

Given the above, this article is focused on investigating the perceptions of the comfort properties of sports caps through subjective evaluation of use in real weather conditions in an uncontrolled environment.

1.1 Thermos Physiological Comfort

One of the functions of clothing is to protect the user from cold and heat, thus, heat transfer is one of the factors that help maintain the thermos-physiological balance, or, in some cases, that make it difficult for sweat to evaporate from the skin [5, 6] confirmed in their studies that a large part of total comfort is related to thermos physiological comfort.

Thermo-physiological comfort is defined by the American Society of Heating Refrigeration and Air Conditioning Engineers (ASHRAE) Standard 55–66 and ISO 7730:2005 as "the mental condition in which man expresses satisfaction with the thermal environment" [7]. It is also defined as a state of comfort in thermal and humidity terms, which involves the transport of heat and water vapor through clothing. People reach this comfort state when they don't need to put on or take off clothes to be satisfied with the temperature.

The microclimate created by clothing encompasses the temperature, humidity and air velocity between the body and clothing. For [8], the microclimate considered comfortable for the user is 32 ± 1°C for skin surface temperature, $50 \pm 10\%$ for relative humidity and 0.25 ± 0.15 m/s for air velocity.

Many tests are available to measure the thermo-physiological comfort of clothing and can be performed using instruments (objective evaluation) or through wear trials with volunteers (subjective evaluation). For [9] measuring thermal comfort is highly complex, as the perception of thermal comfort by the user is affected by several parameters such as temperature and air movement, humidity, types of clothing, activity levels, radiant temperature, among others.

Investigations on thermo-physiological comfort are focused on subjective assessments of clothing, in which researchers usually use data from wear trial carried out in climate-controlled indoor environments. In open spaces, however, the difficulties lie in the "low control of temperature and humidity variables due to the transport of mass and energy provided by the action of the winds" [10].

2 Materials and Methods

As a way of approaching the research question and the proposed objectives, the methodological procedures used are explanatory research, based on experimental procedures and applications of qualitative and quantitative analysis.

In this investigation, in evaluations of aspects of thermo-physiological comfort, the explanatory method experimentally collects data from subjective and instrumental evaluations of the relationship between the user and the clothing on the properties of thermo-physiological comfort during wear trial.

The application of surveys to obtain subjective and instrumental responses is carried out from a protocol, where the comfort of the cap is evaluated by the user through tests of use in real weather conditions. For this, a group of 10 evaluators were recruited in Brazil, who evaluated 10 different models of caps in 10 different tests in an external area of the University Campus of the Federal Technological University of Paraná, in real and uncontrolled climatic conditions, that is, in real conditions: temperature at 25.6 °C (±2 °C) and humidity at 74% (±5%) during the month of December.

The raw materials used to make the caps to be tested are five types of fabric used by the cap industries in Apucarana – Brazil, plus three in wool, in addition to the inclusion of a sample from Portugal (laminated cork), totaling nine samples. Of these nine samples, ten different caps were produced, five in the baseball model (Table 1), namely B1 (100% CO), B2 (100% CO), B4 (100% WO), B6 (65% PES/ 35% WO) and B8 (100% CORK and 100% PES), and five in the snapback model (Table 2), caps B3 (100% CO), B5 (65% PES and 35% WO), B7 (100% CORK), B9 (100% CORK and 100% PES) and B10 (100% PES).

Table 1. Baseball caps samples

	Samples	B1	B2	B4	B6	B8
	Composition	100% CO	100% CO	100% WO	65% PES 35% WO	Cork 100% PES
	Mass per unit area (g/m²)	294	197	248	215	Cork 321 PES102
Basebol Model	Thickness (mm)	0,718	0,456	0,696	0,6	0,546
	Water vapour resistance (Pa/m²/W⁻¹)	4,2	2,88	4,48	3,22	Cork 36,5 PES -0,8
	Air Permeability (l/m²/s)	72,4	81,8	215,6	0,6	-

It is not common to use 100% polyester warp knitting on the front of the caps, so in order to explore cork, a material rarely used in the development of caps in Apucarana/Brazil, two samples were made with two different raw materials - the front in cork and the sides and back in 100% polyester warp knitting, namely sample B8 in the baseball model and B9 in the snapback model.

The wear trial lasted 30 min and was divided into 3 phases, as shown in Table 3. In the first phase, the volunteers sat down for 5 min at rest to adapt to the weather conditions. Phase 2 was divided into two parts: first, the volunteer pedaled the ergometric bicycle at a moderate speed of 10 to 14 km/h for 15 min; and then at an intense speed of 15 to 30 km/h for 5 min, with the speed being monitored every minute. In the last phase, the volunteers

Table 2. Snapback caps samples

	Sample	B3	B5	B7	B9	B10
 Snapback Model	Composition	100% CO	65% PES 35% WO	100% Cork	100% Cork 100% PES	100% PES
	Mass per unit area (g/m2)	378	197	321	Cork 321 PES 102	430
	Thickness (mm)	0,756	0,448	0,748	Cork 0,748 PES 0,546	0,708
	Water vapour resistance (Pa/m2/W-1)	5,38	2,44	36,5	Cork 36,5 PES 0,8	5,6
	Air Permeability (l/m2/s)	45,67	178,6	-	45,67	138,2

remain seated at rest for another 5 min. At the end of each phase, the volunteers are photographed with an infrared camera and answer the questionnaire. Every 5 min, the heart rate and Peripheral Oxygen Saturation are recorded, with the aid of the oximeter, and the pressure that the cap exerts on the volunteer's head, with the pressure sensor.

Table 3. Phases of wear trial test.

Phases	Position	Activity	Time	Descriptions
Pre-test		Resting	5 minutes	At this stage you will be seated resting
Exercise		Low velocity	15 minutes 10 a 14 km/h	At this stage you will exercise on the bike.
		High velocity	5 minutes 15 a 20 km/h	
After-Exercise		Resting	5 minutes	At this stage you will be seated resting

The ISO 10551 standard – Ergonomics of the thermal environment–Evaluation of the influence of the thermal environment using subjective judgment scales [11] presents 3 types of scales for subjective judgment: perception scale, affective evaluation scale and preference scale. These scales were adapted for data collection.

The perception scale was used in this work to assess thermal comfort, as shown in Table 4.

Affective rating scales were used to rate sensations of humidity, general comfort, the pressure that the cap exerts on the head and the attributes: weight, thickness, roughness,

Table 4. Perception scale to assess thermal comfort

Attribute	Scale						
Thermal Sensation	–3 Very cold	–2 Cold	–1 Cool	0 Natural	1 Warm	2 Hot	3 Very Hot

stiffness and itching, as presented in Table 5. The attributes were extracted from the list of attributes developed by [12] and [13].

Table 5. Attribute affective assessment scale

Attribute	Scale				
Humid sensation	0 Dry	1 Less dry	2 Slightly wet	3 wet	4 Very wet
Comfort sensation	0 Comfortable	1 Slightly comfortable	2 uncomfortable	3 Very uncomfortable	4 Extremely uncomfortable
Pressure sensation	0 No	1 Slightly	2 Neutral	3 Very	4 Extremely
Weight sensation	0 No	1 Slightly	2 Neutral	3 Very	4 Extremely
Thickness sensation	0 No	1 Slightly	2 Neutral	3 Very	4 Extremely
Rough sensation	0 No	1 Slightly	2 Neutral	3 Very	4 Extremely
Stiffness sensation	0 No	1 Slightly	2 Neutral	3 Very	4 Extremely
Itches Sensation	0 No	1 Slightly	2 Neutral	3 Very	4 Extremely

Two-way ANOVA with repetitions was conducted for statistical analysis of the data, with the aim of identifying differences between the samples, and using the Pairwise Comparison Method, the sample averages were compared pairwise in relation to each attribute.

The SIDAK correction was used as a complement to the analysis of variance, as it performs each test within a limited significance level to ensure that the false-positive rate applied to the test set does not exceed the specified value [14, 15 and 16]. The mean difference is significant at the 0.05 level, that is, if $p < 0.05$ there are differences between samples and if $p > 0.05$ there are no differences between samples. For the treatment of the data collected, the SPSS software was used.

Because it is a research involving human beings, the experimental protocols of this project were submitted to the Ethics Committee in Research involving Human Beings of UTFPR - Federal University of Technology of Paraná, with registration CAAE 45651115.5.0000.5547.

3 Results

Tests for use in real weather conditions were conducted in the external areas of the Federal University of Technology of Paraná Campus Apucarana during the month of December. The temperatures recorded during the tests were 21 °C minimum and maximum 29 °C, relative air humidity between 72% and 95% and air speed from 0 to 4 m/s.

Volunteers were recruited on UTFPR premises and are students and people from the local community. The tests were scheduled according to the volunteers' availability, so that the tests would not compromise the volunteer's routine.

In tests in external environments under real weather conditions, oxygen levels among volunteers did not differ significantly, and heart rate, thermal sensation and humidity sensation accompanied the rhythm of physical activities from wear trial.

Regarding the thermal sensation, the samples did not show significant differences in their averages, but some showed differences between the phases of the wear trial. It was possible to verify that some samples presented the same behavior, having been, in this way, grouped. Samples B1 (100% CO), B2 (100% CO), B3 (100% CO), B4 (100% WO) and B10 (100% PES) did not show significant differences between phases. The means of samples B5 (65% PES/ 35% WO) and B8 (100% CORK and 100% PES) show that the thermal sensations suffered a gradual increase until the phase of intense activity and, in the last phase, a rapid cooling, not differing from the means of the pre-test. In samples B6 (65% PES/ 35% WO) and B9 (100% CORK and 100% PES), the thermal sensation had a gradual increase until the phase of intense activity and, in the last phase, the volunteers judged that the thermal sensations were similar to those of the moderate activity phase. With regard to sample B7 (100% CORK), the volunteers felt that the head temperature rose more quickly during moderate physical activity, which remained stable at the end of intense physical activity, and felt a slight cooling in the post-test.

For the purposes of a better understanding of the data, it was verified that the equality between the variances of the samples demonstrates that there is no sudden change in the thermal sensation. Considering that the averages closest to the zero axis are ideal for the sensation of thermal comfort, it is assumed that samples B3 (100% CO) and B4 (100% WO) are the ones with the best thermal performance, suffering an increase in the sensation of heat at the beginning of the physical activities and remaining homogeneous until the end of the test, while the worst are samples B5 (65% PES / 35% WO), B6 (65% PES / 35% WO) and B8 (100% CORK and 100% PES), as shown in Graph 1.

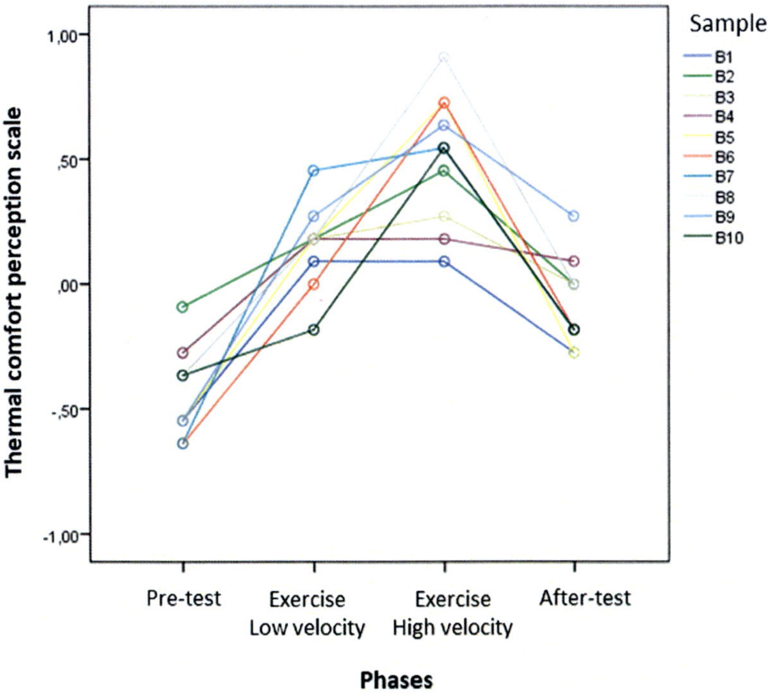

Graph 1. Averages of thermal sensations in real weather conditions.

Graph 2 illustrates the averages of the sensations of humidity perceived by the volunteers. In the pre-test, samples B4 (100% WO), B5 (65% PES/ 35% WO), B6 (65% PES/ 35% WO), B7 (100% CORK) and B9 (100% CORK/100% PES) were judged by the volunteers as dry and sample B3 (100% CO) as the least dry. In the moderate physical activity phase, sample B1 (100% CO) was the driest and B10 (100% PES) the least dry. In the phase of intense physical activity, samples B1 (100% CO), B4 (100% WO) and B5 were considered by the volunteers as the least humid and samples B6 (65% PES and 35% WO) and B10 (100% PES) as the most humid.

In the post-test, samples B1 (100% CO) and B7 (100% CORK) were the least humid and samples B6 (65% PES and 35% WO), B9 (100% CORK and 100% PES) and B10 (100% PES) were the wettest. In general, sample B1 (100% CO) was judged by the volunteers as the driest and sample B10 (100% PES) as slightly wet, according to the affective scale.

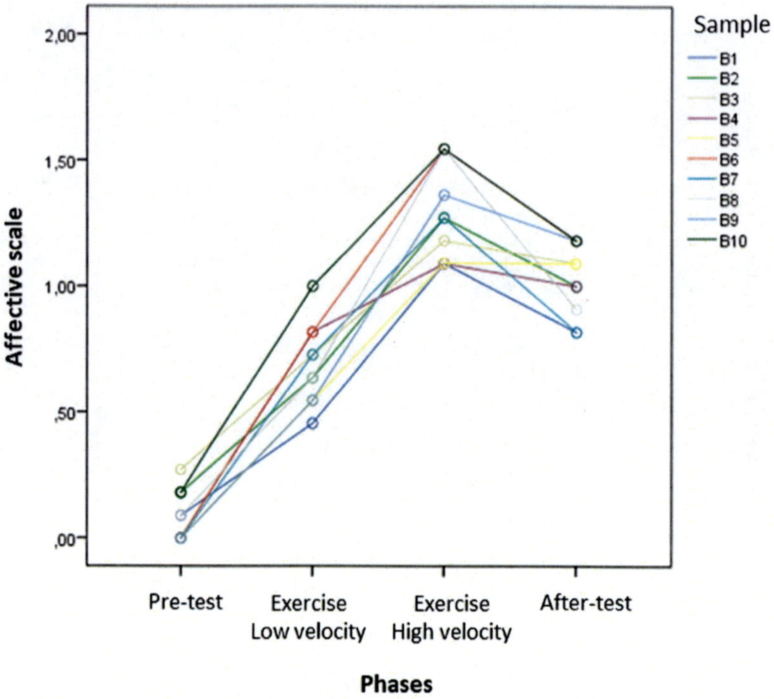

Graph 2. Average humidity sensations in real weather conditions.

In Graph 3, the means of the affective evaluation of the attributes did not show significant differences. In the weight attribute, the volunteers did not feel the weight of sample B8 (100% CORK and 100% PES), in turn, sample B3 (100% CO) was judged as the heaviest. In the thickness attribute, samples B1 (100% CO), B2 (100% CO), B4 (100% WO), B5 (65% PES/ 35% WO) and B10 (100% PES) were considered the thinnest and B3 (100%CO) is the thickest. The volunteers did not feel rough samples B2 (100% CO), B9 (100% CORK and 100% PES) and B10 (100% PES), samples B4 (100% WO), B7 (100% CORK) and B8 were considered the roughest and B2 (100% CO), B9 (100% CORK and 100% PES) and B10 (100% PES) the least rough. Regarding the stiffness attribute, sample B8 was considered by the volunteers to be the least stiff and B3 (100% CO) to be the stiffest. The volunteers felt no itching in sample B3 (100% CO) and sample B8 (100% CORK and 100% PES) felt itchier.

Sample B1 (100% CO) performed better in thermal sensation and humidity, sample B3 (100% CO) in general comfort and sample B6 (65% PES/ 35% WO) in pressure sensation. The evaluators pointed to sample B8 (100% CORK and 100% PES) as the warmest, B10 (100% PES) as the most humid and B7 (100% CORK) with the worst performance in the feeling of general comfort and pressure.

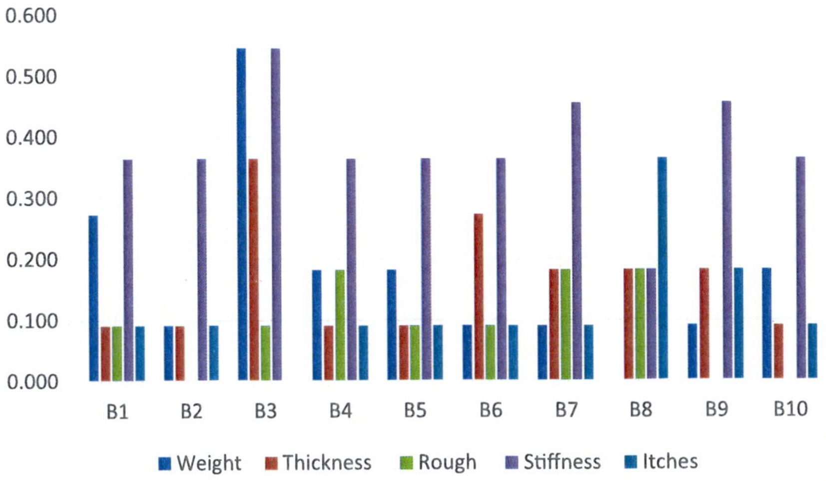

Graph 3. Sensory profile of cap samples in real weather conditions.

4 Final Considerations

The main objective of this article was to present the results of the investigation on the perceptions of the comfort properties of sports caps through a test of their use in real climatic conditions in an uncontrolled environment.

It was observed that the volunteers assigned very low averages to the cap samples (0 to 0.6). This may indicate difficulty in judging these attributes when worn.

The application of the evaluation of the comfort of sports caps was carried out with volunteers. In this way, other investigations can be carried out with users of caps, as well as with athletes who wear caps during competitions, such as: skateboarders and baseball players during sports and during long periods of time.

Acknowledgments. To the Araucária Foundation (Support for Scientific and Technological Development of Paraná) through CP20/2013, Doctoral Program in Textile Engineering in Partnership with a Foreign Institution (UMINHO/UEM/UTFPR).

To the Federal Technological University of Paraná (UTFPR), and Directorate of Research and Graduate Studies (DIRPPG).

This work was funded by FEDER through - COMPETE and by national funds through FCT - Foundation for Science and Technology under the project UID/CTM/00264/2021.

References

1. Abreu, M.J.A.M.:. Contribuição para o estudo da parametrização de têxteis hospitalares. Universidade do Minho (2004)
2. Broega, A.C. da L.: Contribuição para a Definição de Padrões de Conforto de Tecidos Finos de Lã. Universidade do Minho (2007)

3. Broega, A., Silva, M., Silva, R.: A neural network approach in the sensorial comfort of wool light fabrics by subjective and objective evaluation (2010). http://repositorium.sdum.uminho.pt/handle/1822/19255
4. Slater, K.: Subjective textile testing. J. Text. Inst. **88**(2), 79–91 (1997). https://doi.org/10.1080/004050009708658532
5. Das, A., Alagirusamy, R.: Science in Clothing Comfort. New Delhi: Wooldhead Plubishing India Pvt. Ltd. (2010b)
6. Stanković, S.B., Popović, D., Poparić, G.B.: Thermal properties of textile fabrics made of natural and regenerated cellulose fibers. Polym. Testing **27**(1), 41–48 (2008). https://doi.org/10.1016/j.polymertesting.2007.08.003
7. ISO, (2005). ISO 7730 - Ergonomics of the thermal environmet - Analytical deermination an interpretation of thermal comfort using calculation of the PMV an PPD indices and local thermal comfort criteria
8. Das, A., Alagirusamy, R.: Introduction of clothing comfort. In: A. Das and R. Alagirusamy, eds. science in clothing comfort. [online] New Delhi: Wooldhead Plubishing India Pvt. Ltd, pp.1–12 (2010a). https://doi.org/10.1533/9780857092830.1
9. Abreu, M.J., Catarino, A.P., Cardoso, C., Martin, E.: Effects of sportswear design on thermal comfort. AUTEX 2011 World Textile Conference, [online] (June), pp.50–55 (2011). http://repositorium.sdum.uminho.pt/handle/1822/15155
10. Monteiro, L.M.: Modelos preditivos de conforto térmico: quantificação de relações entre variáveis microclimáticas e de sensação térmica para avaliação e projeto de espaços abertos. Universidade de São Paulo (2008)
11. ISO, (1995). ISO 10551- Egonomic of thermal evironment - Assessment of the influence of the thermal evironment using subjective judgement scales
12. Nagamatsu, R.N., Abreu, M.J., Santiago, C.D.: Tactile feeling of textile: a comparative study between textiles comfortable attributes of France, Portugal and Brazil. In: FIBER SOCIETY, ed. FIBER SOCIETY (2017)
13. Nagamatsu, R.N., Abreu, M.J., Santiago, C.D., Braga, I.: Selection and training of a panel of evaluators for sensory analysis of tactile comfort in Brazil. Textiles, Identity and Innovation: Design the Future, pp.9–14 (2018). https://doi.org/10.1201/9781315100210-4
14. Conagin, A., Barbin, D.: Bonferroni's and Sidak's modified tests. Scientia Agricola, **63**(1), 70–76 (2006). https://doi.org/10.1590/S0103-90162006000100011
15. Abdi, H.: The Bonferonni and Šidák Corrections for Multiple Comparisons. Encyclopedia of Measurement and Statistics, pp.103–107 (2007). https://doi.org/10.4135/9781412952644
16. Filingeri, D., Redortier, B., Hodder, S., Havenith, G.: Thermal and tactile interactions in the perception of local skin wetness at rest and during exercise in thermo-neutral and warm environments. Neuroscience, **258**,121–30 (2014). https://doi.org/10.1016/j.neuroscience.2013.11.019

Design Methodologies for Sustainable Products from Mixed Waste Plastics and Foundry Sands

Daniel Vieira[1,2], Fernando Castro[1], Helder Carvalho[2], and Bernardo Providência[3](✉)

[1] Mechanical Engineering and Resource Sustainability Center, University of Minho, Guimarães, Portugal
mail@danielvieira.pt, fcastro@dem.uminho.pt

[2] Centre for Textile Science and Technology, University of Minho, Guimarães, Portugal
helder@det.uminho.pt

[3] Landscape, Heritage and Territory Laboratory, University of Minho, Guimarães, Portugal
providencia@eaad.uminho.pt

Abstract. This design research proposes, through cooperation between the private sector (industry) and the public sector (university), new furniture and cladding solutions to integrate the urban environment. It seeks to raise awareness on the reuse of industrial waste originating from mixed plastics and foundry sands, promoting collection, valorisation, and transformation. It potentiates sustainable advances that allow contributing to a Circular Economy, through the premise of safeguarding natural resources, as well as environmental ones. For the development of the proposal, a design methodology that combines the Double Diamond model with the Circular Design model is applied. From this conjugation, sustainability is explored as an intrinsic condition of the product, capable of producing added value through design and innovation, minimising the impact of industrial waste on the environment. Simultaneously, the modular concept was explored, that is the application of a single component in the development of several solutions, thus reducing production costs and increasing the range of design possibilities. With this approach, we intend to contribute to a change in the design paradigms, through sustainable and environmentally friendly production techniques. The results obtained contributed to the awareness of designers, architects, or engineers, allowing the identification of new ways and contributions of design in industrial customs and traditions, as well as, transferring technical-productive knowledge about the positive impact of the reuse of industrial waste in the design of new products.

Keywords: Circular Design · Circular Economy · Eco Design · Foundry Sand · Plastic Waste · Product Design

1 Introduction

Following the Stockholm Declaration of 1972, the debate for the conservation of the human environment began, a warning for present and future generations. Twenty years later, following the Rio Declaration, in the environmental and development field, a course was set in terms of environmental care, assuming from then on, the relevance of sustainability, having as main axes the social, economic, and environmental dimension [1]. Despite these warnings, contemporary society consumes raw materials at an unsustainable rate, and it is common knowledge that many of these raw materials, available in nature are finite and non-renewable, can be obtained with the desired characteristics and applied in the manufacture of new products. On the other hand, with the current unbridled consumption of renewable raw materials, nature's regeneration capacity is exceeded, emerging the need to apply in design recycled materials, from the end of the product life cycle, as well as the replacement of these raw materials by secondary raw materials [2].

Today it is crucial, both in environmental and economic terms, to consider Circular Design, from the beginning of the design process, as well as at all stages of product creation. Supporting sustainable strategies and targeting the fundamental principles for the Circular Economy, promote product maintenance for the maximum lifetime, decoupling consumption from finite resources and help companies in the transition towards the consumption of renewable resources [3]. These fundamentals have been promoted by the Ellen MacArthur Foundation [4], through the slogan "It's time for a circular economy", which considers "Through design, we can eliminate waste and pollution, circulate products and materials, and regenerate nature, creating an economy that benefits people, business and the natural world." With the aim of alerting to the fact that the production cycle of a product can be more sustainable. Considering that the current economic model (take → make → use → dispose → pollute) is reaching its limits both environmentally and socially. The principles of Circular Economy seek to redefine society, support a transition to a circular model (recycle → make → use → reuse → remake → recycle), build economic, natural, and social capital, building on three principles: 1) eliminate waste and pollution from the start; 2) keep products and materials in use; lastly, 3) regenerate natural systems.

Within the scope of the project "D'ECO–Design Economic Circular Opportunities" [5], this design research aims to strengthen sustainable technological development and innovation in design. Through the application of composite material–developed by the Department of Mechanical Engineering (DEM) of the University of Minho –resulting from the union of industrial waste from the industrial and construction sectors. With origin in mixed plastics and foundry sands, a quality composite was obtained, with functional properties of technical and economic interest. A viable alternative to a serious environmental problem, it promotes the reduction, reuse, and recycling of this waste.

2 State of the Art

We live asphyxiated by increasingly diversified products that uncontrollably consume the planet's natural resources. Design is diving into the era of the 10Rs of sustainability (Respect, Refuse, Reduce, Reuse, Renew, Recycle, Responsibility, Rethink, Replant, Restore) as a policy to achieve environmental conservation goals and social identity. In this sense, design does not prefigure tomorrow, but participates in its formation, being able to improve the future through the instruments made available by it, offering new solutions, revealing new possible and different paths [6]. On the other hand, designers, architects, and engineers, play an important role in choosing and applying materials, as well as making decisions that ensure sustainability, a choice that should be understood as a systematic approach [7].

The current economy is largely linear. Traditional companies procure raw materials to produce products that end their life cycle after short and medium-term use. The application of sustainable policies offers unique opportunities for change, both in design and in the multidisciplinary development of technologies that enable the reduction, reuse, and recycling of waste from industry and the construction sector. It is possible to identify in the literature products (Fig. 1) from the recycling of various types of waste: (1) plastic waste and sand "Silica Plastic Block" by Rhino Machines; (2) plastic waste "Sea Chair" by Studio Swine, "Pots Plus" by The New Taw, "Jää Bench" by And New, "Substantial Chair" by Alexander Schul, "h-Bench" by Studio Segers, "Alex Chaise Longue" by Alessandro Mendini; (3) styrofoam waste "Refoam" by We Plus; (4) clay waste "Breathing Wall" by Celestin Tanari; lastly, (5) paper waste "PaperBricks" by Studio Woojai, "Paper Paper Bin" by Fritz Jakob, "Recycling Reject" by Tim Teven.

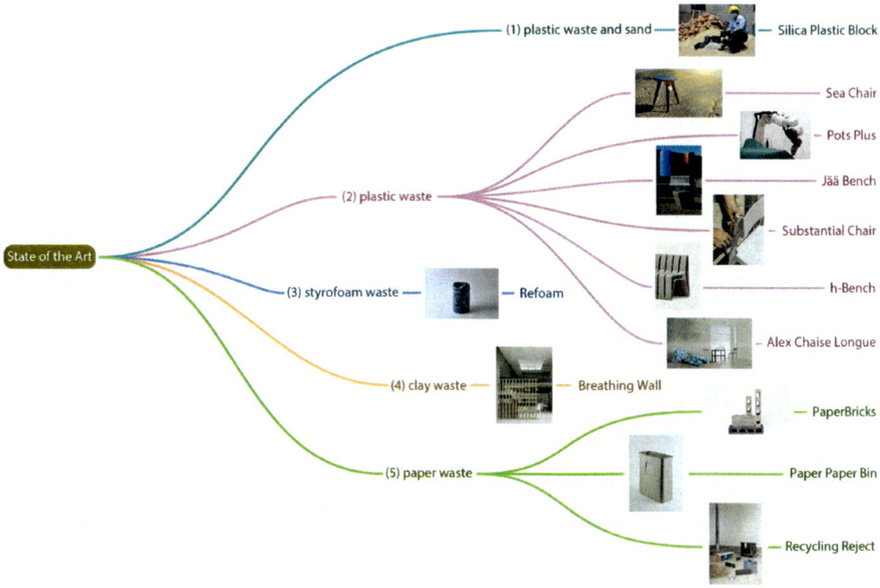

Fig. 1. Diagram showing products from waste recycling.

In a Circular Economy, products and/or their materials return to the industrial ecosystem, through a conscious process of sustainable production, and this return may occur at different stages of the production process.

2.1 Mixed Waste Plastics with Foundry Sands

During the activities promoted by the project "D'ECO–Design Economic Circular Opportunities", the Department of Mechanical Engineering of the University of Minho in cooperation with Resifluxo, Lda., a company responsible for the triage of much of the waste from the area of Vale do Ave, Portugal, developed a proposal for a composite material through the recycling of industrial waste, with 50% mixed plastics and 50% foundry sands.

It was possible to test, cutting and machining characteristics, through the transformation of a sample specimen produced in the scope of the project, into a sample with 90 × 40 × 25 mm, using a simple mechanical saw normally applied in the cutting of solid wood, without compromising the chemical and mechanical properties of the composite, creating only a smooth surface finish (Fig. 2).

Fig. 2. Composite sample 90 × 40 × 25 mm, weight/mass 108 g, density 1200 kg/m3.

2.2 Existing Products on the Market

Some research, especially for the construction sector, has been studying this type of composite material, as well as, demonstrated scientific interest [8, 9]. Above mentioned, the product "Silica Plastic Block" created by Rhino Machines, is an example of this, aspires to sustainable construction, through the application of a brick produced from the recycling of 80% of foundry dust/sand waste and 20% of mixed plastic waste [10]. Outside the remit of the "D'ECO–Design Economic Circular Opportunities" project and to date, the application of this composite typology in urban furniture design and related, does not seem to exist yet, or at least, we have not found it in the literature.

3 Materials and Methods

The research presents a design methodology (Fig. 3) applied to a case study, where the problem is addressed through research, synthesizing the findings and the idea development, proposing a unique solution, not encompassing the final testing phase. This is achieved by combining the Double Diamond model [11] with the Circular Design model [3].

Each interaction of this design process approach involves the creation of four distinct phases: 1) "discover—understand" corresponds to the research and discovery of the purpose, through sustainability and innovation criteria; 2) "define—define" corresponds to the determination and synthesis of the theme/problem, through intuition and opportunity criteria; 3) "develop—make" corresponds to the development of ideas, designs and prototypes, through context and technology criteria; lastly, 4) "deliver—release" corresponds to the delivery of a unique solution. To this end, it is understood: i) the purpose and legacy of the study (why); ii) the research method applied (how); iii) the solution implemented (what).

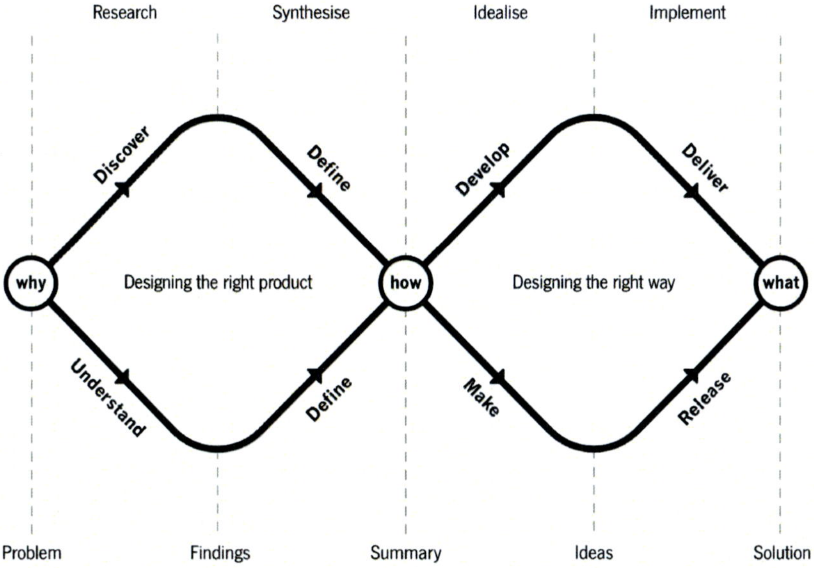

Fig. 3. Design methodology (the Double Diamond model with the Circular Design model).

3.1 Design Process

Through a visit to Resifluxo, Lda. Facilities, it was possible to understand both the manufacturing processes and the company's pretension to design a new product. The first phase of the "discover—understand" process, aimed to research and identify ideas, shapes, and morphologies. The theme was defined as the design of road and pedestrian barriers. Through initial research it was possible to select 24 safety barriers (Fig. 4).

Fig. 4. Systematisation of 24 road/pedestrian safety barriers.

Their mapping using the web application "Coogle" exposed a morphological homogeneity, restricted to the volumetry of a triangular prism, which may present a grey pigmentation provided by the concrete, but also a brown, red, orange, or yellow pigmentation provided by the plastic.

After this result new research was carried out outside the scope of the defined archetype, as a result 12 design products were selected (Fig. 5). Although this selection did not present solutions with origin in industrial waste, the identification of these provided an analysis, even if superficial, for possible morphologies and manufacturing processes by pressing of the composite material. This information revealed itself important to glimpse the possibility of integration of the composite in a design proposal.

Fig. 5. Selection of 12 design products.

The second phase, "define—define", aimed to systematize, and summarize the interpretation of the research of the identified ideas, shapes, and morphologies. For a redefinition of the program, new design aspects were identified. After this result, new research was conducted with the keyword "zig-zag" in several design blogs, as a result 12 design products were summarized (Fig. 6).

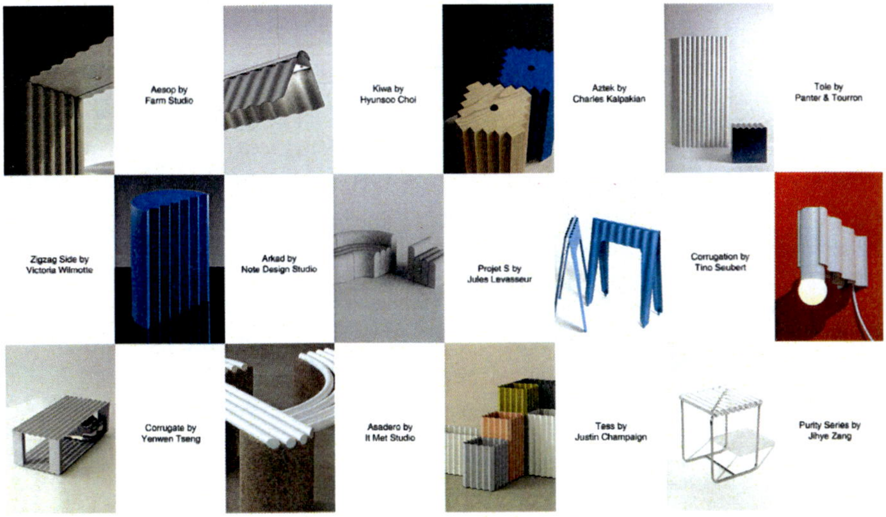

Fig. 6. Summary of 12 design products.

Although this selection also did not present solutions originating from industrial waste, their identification provided a deeper analysis of possible morphologies and technical-productive processes by extrusion. This information guided the definition, design, and development of the solution extruded with composite material, originating from industrial waste, composed of 50% mixed plastics and 50% foundry sands.

3.2 Concept Definition and Prototype in Polystyrene Foam

The third phase, "develop—make", aimed to develop the idea through the construction of volumetric studies in polystyrene foam using CNC (Computer Numeric Control) machining. The first study corresponded to the mock-up of a profiled texture with a hexagonal base, developed using the software "Rhinoceros 3D". It had as a starting point the technical-productive capacity for extrusion provided by Resifluxo, Lda., always considering the alternative by pressing (Fig. 7).

The creation of this profiled component had as a differentiating factor the modular design, allowing the conception of a set of design hypothesis. Through its development we intended to communicate, through modularity and aesthetics, the premises of the Circular Economy and Circular Design, protecting the environment through the resilience of sustainability. Creating the opportunity to respond to diversified sectors, for example, the street furniture or exterior cladding sector.

A second study made it possible to explore the modularity of the profiled component with circular base (Fig. 8a). The development of several volumetric studies in polystyrene foam allowed us to envisage various morphologies and hypothetical solutions for the

Fig. 7. Technical-productive process and mock-up for hexagonal profiled texture.

urban furniture sector using the software 3D, such as bollards, chairs, benches, litter bins, barriers, or defences. The volumetric studies also made it possible to explore the surface textures of the profiled component with octagonal base (Fig. 8b). Through these it was possible to glimpse various aesthetic morphologies for the exterior cladding sector. In this way we respond to a hypothetical list of products that fit the Resifluxo guidelines, as a means of applying the composite developed by the D'ECO project, thus valorising industrial waste based on mixed plastics and sand resulting from foundry waste.

Fig. 8. Study for a profiled component: a) with circular base, and b) with octagonal base.

4 Results and Discussion

The fourth phase "deliver—release", aimed to present a unique solution, thus, it was proposed a profiled component with hexagonal base and circular ends, with a width of 420 mm and a maximum length of 3000 mm (Fig. 9).

Fig. 9. Solution for profiled component applied to a pedestrian safety barrier project.

Through the solution developed for the profiled component, the aim is to reduce resources and production costs, in a Circular Economy perspective, which favours the process chain to reduce energy consumption, increasing the product's lifetime and minimizing the impact of industrial waste. Simultaneously, it responds through a wide range of possible sustainable solutions (Fig. 10).

This approach underpins the creation of a future prototype that supports the materialisation of the possible solutions, achieving a solution closer to reality, with a higher degree of confidence in both the design and the integration of the composite. The resulting material can be re-crushed and extruded, maintaining the circularity of the recycled raw material, although some mechanical characteristics may progressively lose qualities. This research also allowed the debate and presentation of ideas for urban environment, potentiating a Circular Economy, through the safeguarding of natural resources, and environmental, with the application of design methodologies, life cycle extension and green production process [12].

Fig. 10. Possible sustainable solutions for the design of benches.

5 Conclusion

Scientific and technological research plays an important role in the path towards a Circular Economy. Through a design methodology, the Double Diamond model is combined with the Circular Design model to identify new paths, promoting research, discovery, and design, as well as waste collection, recovery, and transformation.

The aim of this study was to reuse industrial waste from the manufacturing and construction sectors in the design of new, more environmentally friendly products. Aiming to safeguard natural resources through the reuse of industrial waste, a case study was presented that values the application of mixed plastic waste and foundry sands, whose destination would be landfill. Based on this assumption, it is proposed as a solution a profiled component produced by extrusion of composite material, resulting from the union of industrial waste with 50% mixed plastics and 50% foundry sands. A modular solution that intends to contribute to new design solutions, without defining an exact product morphology, covering a wide spectrum of possible archetypes, making it possible to reach a sustainable solution, capable of producing added value. Morphologies oriented to the urban environment, products that normally end their life cycle after a long use. Contributing in this way to issues such as sustainability, reuse and modularity of the hypotheses presented. Promoting through product design the collection, valorisation, and transformation of waste.

Possible applications of this waste may need technical-productive adaptations, to adapt this waste to the characteristics of the final product. Future advances in a prototype, both for the component and the product, will enable tests to be carried out to understand the viability and resistance of the proposed connections between the modular components, to improve design issues and technical-productive knowledge. Finally, this study alerts designers, architects, and engineers about the importance of reusing industrial waste in the design of new products.

Acknowledgements. This work was developed within the project "D'ECO – Design Economic Circular Opportunities", with ref. NORTE-01–0247-FEDER-049824, financed by the Competitiveness and Internationalisation Operational Programme and by the Operational Programme of the North supported by the European Regional Development Fund (FEDER).

References

1. Handl, G.: Declaration of the United Nations conference on the human environment (Stockholm Declaration), 1972 and the Rio Declaration on Environment and Development, 1992. United Nations Audiovisual Library of International Law 11:1-11 (2012)
2. Castro, F.: Resíduos como Recursos: Barreiras e Aspetos Técnicos. Indústria e Ambiente **98**, 14–16 (2016)
3. MacArthur, E.: Circular Design for Fashion. Ellen MacArthur Foundation Publishing, Isle of Wight (2021)
4. Ellen Macarthur Foundation (2023). https://ellenmacarthurfoundation.org. Accessed 30 Apr 2023
5. Gonçalves, R., Castro, F., Providência, F., et al.: D'ECO - design economic circular opportunities. Revista de Fundição **301**, 36–40 (2022)
6. Manzini, E., Jégou, F.: Sustainable everyday - Scenarios of urban life. Edizioni Ambiente, Milano (2003)
7. Kandachar, P.: Chapter 7 - Materials and social sustainability. In: Karana E, Pedgley O, Rognoli V (ed) Materials Experience - Fundamentals of Materials and Design. Butterworth-Heinemann, Oxford, pp. 91–103 (2014). https://doi.org/10.1016/B978-0-08-099359-1.00007-2
8. Aneke, F.I., Shabangu, C.: Green-efficient masonry bricks produced from scrap plastic waste and foundry sand. Case Stud. Constr. Mater. **14**, e00515 (2021). https://doi.org/10.1016/j.cscm.2021.e00515
9. Rajesh, K.N., Raju, P.M.: Performance of recycled plastic waste and used foundry sand as a replacement of fine aggregate in concrete. In: Das, B.B., Gomez, C.P., Mohapatra, B.G., (eds.) Recent Developments in Sustainable Infrastructure (ICRDSI-2020)–Structure and Construction Management, Springer, Singapore, p 735–747 (2022). https://doi.org/10.1007/978-981-16-8433-3_61
10. Katsikopoulou, M.: 'Silica plastic block' is a sustainable brick made from recycled sand + plastic waste (2020). https://www.designboom.com/design/silica-plastic-block-sustainable-brick-from-recycled-sand-plastic-waste-06-15-2020. Accessed 30 Apr 2023
11. Council, D.: Eleven Lessons: Managing Design in Eleven Global Companies – Desk Research Report. Design Council, London (2007)
12. Bakker, C.A., den Hollander, M.C., van Hinte, E., et al.: Products that Last: Product Design for Circular Business Models. BIS Publishers, Amsterdam (2019)

Standardization of Classification of Terms for Clothing Pattern Cutting: A Need Detected from Bibliographic Investigation in the Brazilian Context

Sophia Codato Ferreira, Luana da Costa Honda, Maria Eduarda Maragno Luiz, Lucimar de Fátima Bilmaia Emídio, Thassiana de Almeida Miotto Barbosa(✉), and Franciele Menegucci

State University of Londrina, Londrina, Brazil
{sophia.codato.ferreira,luana.costa.honda,eduarda.maragno, lucimar,thassi,franmenegucci}@uel.br

Abstract. Considering the need for building a system of classification that considers the different domains of knowledge in the clothing pattern cutting field, this qualitative and exploratory research aimed to map the universe of elements that are related and their level of extension in the context of pattern cutting books. We present an analysis regarding how contents from this field of knowledge are being addressed in the bibliographies used in pattern cutting courses in Brazil, either related to their topics, focus, or content structure. The results evidence the need of applying classification methods, proposed by theoretical basis from the Information Science field, aimed at processes of organization and recovery of information in pattern cutting, above all, to facilitate the process of knowledge building in this field of fashion design.

Keywords: Fashion Design · Clothing Pattern cutting · Classification and Standardization of Terms · Basic Bibliography

1 Introduction

Clothing pattern cutting teaching, within creative and productive dimensions, contributes to training fashion designers capable of comprehending the whole process in the clothing productive chain, as well as developing new products with excellence and assertiveness. However, as observed by Oliveira [8] educational institutions introduce courses on pattern cutting in diverse moments and ways in their pedagogical projects, demonstrating the autonomy each institution has towards their courses and syllabus proposed in their programs.

Yet, despite the gradual increase of scientific literature produced in the field of pattern cutting, in Brazil, it is observed that basic bibliographies used in this field remain similar, severely focused on how to make it not on why. Moreover, the traditional teaching of pattern cutting courses presents itself with a focus on convergent thinking. Mostly approaching behaviorist reinforcement principles of stimulus-response associations that, according to Jonassen [5] do not sustain the complex thinking needed in meaningful learning toward problem-solving within the real context, nor develop the ability that enables students in transferring competencies acquired to new situations, aside from not stimulating the creation of new ideas.

Pattern cutting knowledge-building might be derived from the classroom, but not having technical and/or theoretical-practical supporting material might difficult the student learning process. Thus, choosing accessible and complete bibliographies is fundamental to support courses. These can be used by professors and students as a guide during the teaching-learning process. Still, such literature is highly important as a form of registry and knowledge documentation, to the extent of perceiving how teaching this modality has changed and improved throughout the years.

Hence, this research presents an analysis of the way contents are approached in the bibliographies of pattern cutting courses in Brazil, related to their themes, focus, or content structure. This analysis is one of the phases of a research project, coordinated by one of this research's author, that mainly aims to: "identify problems and produce solutions for pattern cutting under the MODThink model", proposed by Emídio [3], which seeks to contribute to break teaching paradigms predominantly traditional.

From that, this study's purpose was to raise data referenced to the content presented in pattern cutting literature summaries and, from this verification, identify the main aspects related to the lack of a pattern of classification of nomenclature in pattern cutting book, especially the ones focused on flat pattern cutting, whose terms and sequences adopted by authors tend to not follow the same logic. This complicates students' search for content and, consequently, their knowledge-building in pattern cutting, once contents presented in a book are not related to others, even though they approach the same matter.

2 Development and Methodological Aspects Adopted to this Study

The analysis emerged from hard copies of books available at the State University of Londrina (UEL) Sectorial Humanities Library, complementary, we investigated basic bibliographies from syllabus courses in the key educational institutions that offer Fashion Design programs in Brazil. The selection of these institutions occurred by applying pre-established criteria during the first phase of the research project in question, being "[…] 1) the register of a Fashion Design program; 2) having a bachelor program with an ENADE and e-MEC score greater than 3.0, considering the period in which the research took place; 3) contemplate at least one program from each region of the country". [8].

From that, the selected institutions were: Anhembi Morumbi University (UAM), National Service for Industrial Training–Technology Center of Chemical and Textile Industry (SENAI CETIQT), State University of Minas Gerais (UEMG), and Federal University of Minas Gerais (UFMG)– representing the southeast; Santa Catarina State University (UDESC) and State University of Maringá (UEM)– represent the south; Federal University of Goiás (UFG)–representing the west center; and Federal University of Piauí (UFPI)–representing the northeast. It is important to highlight that during this study, we could not find the Pedagogical Projects of fashion programs in the north region, nor the bibliographies used by UAM, which is the reason why both were suppressed in the analysis.

Cataloguing and data analysis were used through qualitative bibliographic research, which consists in searching already published works relevant to knowing and analyzing the thematic problem of the research undertaken [11]. According to Severino [10], the bibliographic approach is undertaken "from the available registry, derived from prior research, in printed documents, such as books, articles, theses, among others". Hence, bibliographic research was fundamental to identifying relevant references in the pattern cutting field, contributing to the study's theoretical foundation.

Based on Emídio [3], who stresses how "it is important to gather as much relevant information as possible about the topic and search for different perspectives to understand it better", the research opted to obtain the greatest number of data possible, from phase 2 of MODThink Model [3], proposed by the author, that focuses on investigating through theoretical-methodological allowances for the construction of pattern cutting. Using the referred methodology contributed to the organization and analysis of the summaries in the works related to pattern cutting. The application of tools, suggested to phase 2 of the Model, backed up the identification of gaps or areas of greater relevance in the literature related to pattern cutting.

Thus, we gathered 83 books, from which it was possible to access the summary of 48 of them. The contents from the mapped books were divided into items of analysis, from a checklist, a tool suggested by Emídio [3]. Being: 1) body's physical structure–feminine, masculine, or infant; 2) books' content; 3) pattern cutting approach; 4) types of products; 5) pattern cutting contents; 6) specific components of products; and 7) constructive resources. The definition of such criteria was based on the summaries as an object of study, and the nomenclatures used in each item related to the way they were named in the bibliographies accessed.

3 Results and Discussions

In the first item analyzed, we sought to understand the general aspects of the books, identifying among the body's physical structures–feminine, masculine, and infant which were approached the most by pattern cutting books.

Through this analysis we observed an expressive number of bibliographies addressing feminine pattern cutting, once out of the 48 books analyzed, 36 addressed the construction of pattern cutting directed to the feminine body, 14 to the masculine, and 10 to the infant. We infer that this happens because the feminine body possesses more curvatures and particularities to be met through pattern cutting.

However, it is important to stress that infant pattern cutting also presents its singularities, and despite countless safety, ergonomic, and comfort questions that need to be met, above all, for this audience, this group is still the least worked by authors. It is relevant to highlight that most of the infant clothing content is included in dressmaking and sewing books, where only the interpretation of the model is present, or even as a short chapter in pattern cutting books that addresses more than one type of physical structure. Works of literature specifically driven to infant pattern cutting are still very limited, this is due to the unawareness of this field in the historical context, in which just modifications were made by the view that the child was a mini adult [15].

Regarding the masculine, its presence also takes place in great amounts, as inclusions in books that jointly address feminine pattern cutting, as in dressmaking and sewing books. It is important to point out that books that specifically work pattern cutting in this segment, expressively appear in content directed to tailoring.

Referring to the second item analysis, types of books' content, we delimited the following criteria: practical, theoretical-practical, theoretical, and aimed at teaching-learning from the design standpoint. Related to this item of investigation, 24 books addressed practical content; 20 theoretical-practical; 3 theoretical, and only 1 covered teaching-learning from the design standpoint. It is observed that practical books, that mainly focus on the teaching of model constructions, are the majority. On the other hand, books with just a theoretical approach are the minority, and in this, the contents of ergonomics are covered.

As for the theoretical-practical books, they cover these two contents together, which enables a more complete and straightforward reading, given that the construction of the model englobes several anthropometric and ergonomic knowledge [6]. Moreover, questions such as body measurement, materials, fabric choice, and historical context are common in books with this type of approach to assist the process of relating pattern cutting content to project thinking, from initial phases.

The third item for analysis regards what concerns the pattern cutting approaches that correspond to the main groups of contents present in the books analyzed. We noticed that the majority of them approach flat pattern cutting (34); followed by knitwear (8); moulage (7); underclothing (6); tailoring (4); and 'moda festa'[1] (3), in which many approaches tend to be present in the same book.

According to Sabrá [9], flat pattern cutting is "a building technique used to transform a shape of a bi-dimensional flat drawing [...] into a tridimensional item, obeying the spatial geometry principles". Such a technique is broadly used in industry, once it allows the standardization of measurement tables and basic pattern, making the production process faster with the help of software.

It is worth highlighting that, despite knit products being expressively used by Brazilian consumers, the number of books that address or explain the pattern cutting particularities of this type of material is greatly limited. As Sabrá [9] stresses "the product also has to be adapted according to the most varied types of fabrics. It is not recommended, for instance, to use programmed molds to flat fabrics in knit fabrics or [fabrics] with elasticity".

Another parameter delimited contemplates the type of products present in the analyzed bibliographies, the fourth item for analysis. The products were named according to the present nomenclatures in the summary of the analyzed books, being blouse, dress, pants, skirt, coat, panties, bra and bikini, briefs, corset, and bodysuit. Because of it, the term "blouse", for example, englobes items such as shirts and t-shirts. On the other hand, what could have been classified just as "underclothing" ended up being sorted into several categories. This happens mainly because each author categorizes the contents in their books the way they understand them as being the most appropriate.

The products that appeared the most in these bibliographies are the "blouse", "dress", "pants", and "skirt". The category "coat" shows up after them, with a difference of 13 books. The occurrence in a great number of these four products: blouse (35), dress

[1] "Moda festa" is a specific segment in the Brazilian fashion market, aimed at manufacturing products for formal events, usually made in sewing studios.

(29), pants (28), and skirt (28), simultaneously in different books, can be explained by the fact that most of the books bring, at least, the construction of the basic pattern for them. The interpretation of models, when contemplated, is part of what can be named "basic products". As Treptow [14] explains, these models are "functional and usually have guaranteed sale", which means, by knowing these models, in theory, it would be possible the insertion of the modeler in the industry, a field which the majority of bibliographies bring as focus. Moreover, the reason why the category "coat" appears in lesser numbers can be explained by the fact that the interpretation of this clothing piece might emerge from the base of a shirt, for instance.

The other categories appear with less frequency (panties, bras and bikinis {8}, briefs {6}, corsets {4}, and bodysuits {3}), being normally reserved for specific books, such as those designed for underclothing or knitwear studies, for example.

Another aspect important to be observed refers to the interpretations of the model. Treptow [14] also cites the category of "fashion" products, which the author defines as clothing pieces that are in accordance with the current aesthetic. Thus, it is possible noticing the books which, at that time, were intended for housewives, such as the ones distributed by Singer (1945, 1978), the collection "Faça você mesma" (Do it yourself) by Brandão (1980–1988), and the Sewing Course by Sudbrack (1983). They bring an interpretation of "fashion" of the basic pattern. Dated to a time when pattern cutting practice had a greater bond with homemade production, it was common that these magazines brought what was popular at that time. However, once the modeling process starts gaining space in the industries, the content of these books also starts using a more technical language, as seen in the book Brazilian Industry Pattern cutting, published in 2009, a reference to the study in this field [13].

Regarding book contents, the fifth item analyzed in this study, it is interesting perceiving that the instructions for the interpretation of these contents take place in 28 summaries from the bibliographies analyzed. This happens mainly due to the freedom and method each author uses in their explanation, being needed such explanation once a book might have its teaching methods very different from others.

The main contents contemplated in the books analyzed were: firstly, about the construction of basic patterns; followed by instructions for body measurement and measurement table; instructions about the interpretation of the book; interpretation and adaptation of models; constructive resources; finishings and trims; dart studies; grading; adequacy of basic patterns according to body conformations, ending with the different techniques of pattern cutting.

As Souza and Pereira [13] reported, pattern cutting learning, especially flat, has a greater focus on the "how" instead of the "why" of processes. This makes many students just follow the orders given by the material, most of the time without comprehending the logic embedded in such steps, making some future professionals not hold enough pattern cutting understanding, which can be harmful to their careers. Yet, even though they pursue the field of creation, the pattern cutting knowledge cannot be separated in this step, once it helps the understanding of how that product will suit a body [7].

Despite many of the books bringing an interpretation and adaptation of models in their summaries, their contents boil down to step-by-step, without necessarily explaining the logic of each step. The same happens with darts, to which the measurements and places are given to the reader, who learns by just reproducing the transpositions. The constructive resources, in turn, mostly appeared tied to products (such as the skirt with slits). It is observed the lack of specific chapters that present each one of them.

This makes us notice, as Souza and Pereira [13] mention, the gap of material that deals with different bodies, as for the flat pattern cutting perspective as for the tridimensional. From the analyzed books, just 10 of them bring adequations in the basic pattern according to body conformations as content. This becomes a problem when the tables widely used by industries (and also by a great part of pattern cutting books) bring a body standard present in just 5% of Brazilian women, represented by broad hips and small waist. In a survey conducted by the Brazilian National Standards Organization (ABNT in Portuguese), the body of Brazilian women is, in general, rectangular (76%), which the waist is slightly accentuated and there is a minimal difference between the chest and the hip [4]. Thus, it is possible to notice that the tables used by these materials do not necessarily address the needs of their audience.

Beyond that, body conformation may also be derived from sequels or pathologies. It is possible noticing an expressive lack regarding materials that study these specificities. Among the 48 books analyzed, just one shared one of these conditions, hemiplegia, that cites all its characteristics and what should be done in pattern cutting so this body ends up comfortable wearing a piece of clothing. Elderly bodies are also not addressed, even though there is the knowledge that they might need specific adaptations, with special regard to finishing and trims.

Lastly, grading, which is a process very usual within the industrial process, is approached in just 9 of the analyzed books. This may be due to the use of CAD software, which has this process almost automated in its system. However, the importance of learning this process remains once the information needs to the inputted into the system and errors in the grading of molds are capable of jeopardizing an entire production.

Regarding the sixth item in this study, specific components, sleeve, collar, waistband, pocket, yoke, cuff, neckline, trims, types of hemlines, sleeve placket, lining cloth, e hood, are approached by most of the authors as specific techniques to be used in the interpretation of distinct clothing products.

It is evident the number of books that address sleeves and collars, due to the great variety of types that can be studied in these two components. Different processes of amplification and reduction of sleeves, lengths, and finishings, for example, illustrate very well all this diversity, especially because they are techniques that, with a systemic thinking of pattern cutting, can be applied to other types of patterns cutting, such as flared skirts. About the collars, an entire historical background derived from tailors and years of techniques is studied and compressed in a step-by-step.

It was possible to observe that pockets studies are also very present in pattern cutting literature, both related to the upper and lower parts of clothing. Nonetheless, in contrast to sleeves and collars, these studies are generally addressed in a wider context, most of the time being modeled with other clothing pieces, as for the case of the step-by-step of a pair of pants. This way, the study of pocket is not an isolated pattern cutting technique, but a fundamental element in the creation of complete and well-finished clothing pieces.

It is worth stressing that the components that focus more on clothing pieces finishing appear with less frequency, as for the components for opening and closing and the lining.

According to Alves [1], "finishing is the most delicate step of confection. The opening and closing finishings are fundamental parts of clothing and, without a doubt, the ones that demand more ability and attention". As a component, the hood rarely appears, even though it is an item with great use in current fashion, mainly in street style and sportswear styles.

Subsequently, we analyzed the seventh item indicated in this study, the constructive resources present in the bibliographies. These constructive resources assist the suitability of the clothing piece to the body, which can be related to giving volume, adjustment, or even aesthetic purposes.

According to Souza and Menezes [12], from the development of constructive resources in textile structures, several alternatives for manipulating the fitting emerge and their properties to contemplate what is desired in the pattern cutting.

From the resources listed in this research, there are cutouts, darts, ruched, peats, A-line styles, godet, transpass, draping, flounce, nesga[2]/taco, pleated, slit, and ribbing, among others. These were considered from the products present in the books, once they always find themselves tied to the same items, such as the darted skirts.

Among the mapped constructive resources, the most recurrent one is the cutout. This happens especially due to different model interpretations that use this resource aiming to increase movement or even make the pattern cutting adjust to the body shape without the need for darts, a resource that appeared right after as the second most addressed one. However, cutouts that focused on the aesthetic function, especially the hollows, receive very little attention in this literature.

Some resources like nesga, ribbings, and basque are almost not present in this literature, maybe due to their level of use in current society being less recurrent than compared to previous eras.

According to Barbosa and Emídio [2], knowledge about pattern cutting contributes to designers so they can explore hypotheses of formal, structural, constructive, and productive solutions; to build a product assertively; so, there is no divergence in the materialization of ideas, most of the time resulted from the lack of technical knowledge, referred to the building mechanisms and production processes involved.

The authors corroborate the understanding of the lack of a chapter that addresses this topic only due to its great importance to the exploration of the potentialities of pattern cutting in the project process in fashion design.

4 Final Considerations

After undertaking this study, a very visible aspect was the absence of a classification system for the pattern cutting field, above all to contents related to flat pattern cutting. Their terms and indications tend to not follow the same logic in books that are not from the same author, this makes it difficult to search for specific contents as well as their understanding. The change of terms or orientations to the construction of molds, for instance, can also be a problem, due to the spread of a learning that is focused on the "how" and not on the "why". Because of these variations, there may be some confusion

[2] Nesga is a triangular piece of fabric usually used on the side of a pair of pants to enlarge it.

by the reader and/or student in their comprehension process about a determined content being studied.

It was observed, however, that even though it comes from different authors, books that focus on moulage contain contents and summaries similarly organized, which does not happen in books with a focus on bidimensional pattern cutting.

A constant issue in all the bibliographies analyzed is the lack of diversity in body structure or even questions more related to aesthetics. Infant pattern cutting is little addressed in the books and, when it happens, it is generally not followed by ABNT safety regulations for this type of clothing. Beyond that, the lack of existing body types or even conformations or deficiencies is concerning once diversity represents the world. The lack of chapters that cover pattern cutting adequations to different bodies also corroborates for the absence of this topic in pattern cutting books.

We understand that the diversity of formats of content presentation and the lack of a pattern of classification might be considered huge issues to knowledge building in the field. Hence, the results presented in this study demonstrate a need for the application of methods of classification, proposed by theoretical basis from the Information Science field, directed to the process of organization and recovery of information in pattern cutting, above all, to facilitate the process of knowledge building in this field of fashion design. So, a second phase of the research was structured and is aimed at investigating theories and/or methods of classification, proposed by the Information Science theoretical basis, to identify the one that fits better to cover the contents of pattern cutting and propose a system of categorization of terms and nomenclatures used in this field of expertise.

References

1. Alves, P.: Modelagem masculina. Editora Senac, São Paulo (2009)
2. Barbosa, T.A.M., Emídio, L.F.B.: Contribuições do conhecimento de modelagem para a construção do pensamento de projeto no design de moda. Revista DOBRAS **11**, 272–287 (2018)
3. Emídio, L.F.B.: MODThink: Projetando a Modelagem do Vestuário. Estação das Letras e Cores: São Paulo (2021)
4. Gomes, D.: Maioria das brasileiras não tem o chamado corpo ´violão´, diz pesquisa. In: Revista Veja. Available via DIALOG (2023). https://veja.abril.com.br/comportamento/maioria-das-brasileiras-nao-tem-o-chamado-corpo-violao-diz-pesquisa/. Accessed 23 jan 2023
5. Jonassen, D.: Computadores, Ferramentas Cognitivas: desenvolvendo o pensamento crítico nas escolas. Porto Editora: Porto (2007)
6. Menezes, M.S., Spaine, P.A.A.: Modelagem plana industrial do vestuário: diretrizes para a indústria do vestuário e o ensino-aprendizado. Projética **1**(1), 82–100 (2010)
7. Nishimura, M.D.L., Vergara, L.G.L., Gontijo, L.A.: Modelagem do vestuário a partir das percepções dos professores em Santa Catarina. DAPesquisa **12**(18), 68–78 (2017)
8. Oliveira, S.F., Stersa, V.R., Oliveira, K.R., e Emídio, LF.B.: A tradicionalidade das ementas de modelagem nos cursos de Design de Moda e sua problematização para o ensino sob a ótica do design. In: XVII Colóquio de Moda, online, Brasil, September 2022 (2022)
9. Sabrá, F.: Modelagem: Tecnologia em produção de vestuário. Anhembi Morumbi: São Paulo (2009)
10. Severino, A.J.: Metodologia do trabalho científico. Cortez: São Paulo (2014)

11. Sousa, A.S., Oliveira, G.S., Alves, L.H.: A pesquisa bibliográfica: princípios e fundamentos. Cadernos da Fucamp **20**(43), 64–83 (2021)
12. Souza, P.M., Menezes, M.S.: Estratégias construtivas para a configuração do produto de moda. Projética **2**(1), 82–94 (2011)
13. Souza, B.P., Pereira, A.C.: Nem todo trajeto é reto: limites e possibilidades para a sensibilização de estudantes de Design de Moda por meio do ensino de modelagem. Ensinarmode **4**(2), 11–29 (2020)
14. Treptow, D.: Inventando moda: planejamento de coleção. Author Edition: Brusque (2003)
15. Zanatta, T.A.F.: Modelagem infantil: dificuldades antropométricas atuais. Monografia (2014). Available via DIALOG. http://repositorio.unesc.net/handle/1/2507. Accessed 10 may 2023

The ADeQMat Model Facilitates Material Selection to Project the Product in Fashion

Patrícia de Mello Souza[1](✉) and Isabel Cristina Italiano[2]

[1] State University of Londrina, Rodovia Celso Garcia Cid | PR 445 Km 380, Londrina 86057-970, Brazil
patriciademellosouza@gmail.com

[2] University of Sao Paulo, Rua Arlindo Bettio, 1000, São Paulo 03828-100, Brazil
isabel.italiano@usp.br

Abstract. The research discusses aspects linked to the configuration of the shape and to textile analysis in the context of fashion clothing project, in the academic scope. It contemplates the study of fitting alterations generated by various materials, in which drapeability, evaluated by visual estimative, is identified as a characteristic that relevantly influences the textile behavior in silhouette configuration. As final result, we propose the ADeQMat Model that, inserted in the selection of materials phase, allows guiding the sequencing and interaction of project actions linked to tridimensional modeling to enable a more assertive choice of the material to be defined for the development of products.

Keywords: Fashion Design · ADeQMat Model · Product Development · Material Selection · Tridimensional Modeling

1 Introduction

This work presents a framing from a research developed during the post-doctoral program in Textile and Fashion at the School of Arts, Sciences, and Humanities at the University of São Paulo (USP-Brazil), which resulted in the deposit of the patent named "ADeQMat: an adequacy model of textile materials to the configuration of clothing products linked to tridimensional modeling" (BR 102020024988–6).

The world of work scenario, where the project of fashion clothing is framed, portrays professionals who directly deal with the complex task of selecting material to develop their products. Many cannot visualize and predict fabric's behavior before seeing them applied in the manufactured pieces, which, frequently, entails results that do not meet the project's demands, especially regarding the fitting and the reliability of the proposed silhouette.

Those who teach in the formation of future professionals in the field must warn them about the challenges involving the complexity of systems and contemporary society, that point out to an integrated and contextualized knowledge-building process, in a way it links teaching to work's reality. The emphasized growth in research in the field of

J. Cunha et al. (Eds.): CIMODE 2023, *Advances in Fashion and Design Research II*, pp. 203–212, 2024.
https://doi.org/10.1007/978-3-031-43937-7_18

material has been conducted to a situation of frequent releases in the textile industry, increasing the urgency of updating such knowledge.

In an attempt to contribute to decision-making, regardless it is in academic project, in industry, or in similar scenarios, the present work prioritizes the study of aspects that interfere with and condition textile behavior, searching for solutions to be applied, initially, in the project teaching-learning context, that promote improvements in the selection of materials process, in a way that the choices become more adequate each time.

To do so, we intend to answer the following research question: how can we enhance the possibility of selecting materials correctly to design the shape of the clothing, in the project process context in the design field?

To systematize the research investigation, we established as a general objective: to propose a process model to optimize the adequation of material to the configuration of products in a fashion clothing project. As specific objectives, we seek to: identify characteristics that influence the most formal aspects of the products; investigate methods and models related to the configuration of the shape and the selection of textile material; employ tridimensional modeling techniques to evaluate the construction of structures and the behavior of materials in prototypes configuration; elaborate instruments for tests; validate the proposed model.

2 The Importance of the Winifred Aldrich Method

Initially, it is needed to present the Aldrich method [1]–define it and report its procedures–due to its importance in the context of, not only this research's theoretical foundation, the researcher's own trajectory, who discusses and applies it throughout the last 10 years. The method provides a way of determining how specific characteristics of a material might affect the behavior of a model. The evaluation criterion must also consider the effect produced regarding the user o this product, being the body dressed and in movement.

The process starts with a thorough fabric analysis, referring to the investigation of five characteristics pointed as crucial to obtain products with an adequate integration among modeling, material, and shape. They are weight, thickness, distortion, drapeability, and elasticity. By distortion, it is understood the deformation that might occur between the warp yarns and the weft; drapeability refers to the fabric's fitting; and elasticity, the ability to stretch provided by the fiber and the structure itself. Thickness and drapeability are considered by the author as visual characteristics, being, then, evaluated by visual estimation.

It is defined a scale of five points to evaluate the fabrics according to such characteristics, in which index 1 indicates lighter and thinner materials, with great ability for distortion and drapeability, and with great elasticity. As far as the index grows, the materials present changes in their properties and behavior until they reach the maximum opposite value, index 5, which represents heavier and thicker materials, with a low ability for distortion and drapeability, and with low elasticity.

Once there are countless materials, the scale was divided into five categories as an attempt to contemplate the maximum number of possibilities possible. However, this is a flexible method that can be adapted to several needs, that is, it can be used both for investigating the huge variety of fabrics and to verify one of their specific groups.

Aldrich [1] stipulates simplified procedures, in special, at the students' level, held to quickly enable a practical classification of textile materials characteristics. The method was primarily thought for students' practices.

To evaluate, regardless of the specific procedures employed in verifying each characteristic, it was adopted a squared sample with each side of 20 cm, precisely cut throughout warp and weft yarns. Such fact denotes coherency with the context in which the method is applied, once it assumes that, most of the time, it is just a sample of the fabric that the designer has at the moment of deciding large quantities purchase of material to develop a series production; and, on the other side, it is also feasible in an academic context.

This research limits presenting the procedures to proceed, exclusively, with the evaluation of drapeability characteristic, due to prior researches from the author [9] have proved that despite the influence exerted by the other characteristics, drapeability is the one that conditions the most formal aspects, considering the context being studied.

Hence, for the falling measurement, it proceeds as follows (Fig. 1): on a blank thick card a central point at the top is marked, tracing a central line down and, from it, other two lines at 45°. The area from each side is divided into five sections of 9° each, marking points from 1 to 5. A nail or big pin is put at the superior central point of the drawing and there the fabric to be analyzed is hung, that is, the squared sample with 20cm on each side. The drapeability is recorded. In Fig. 1, it is observed that the fabric on the left presents index 1, which means, a high ability of drapeability, while the fabric on the right presents index 4, which means, a very lower ability (average-low), according to the fabric evaluation scale.

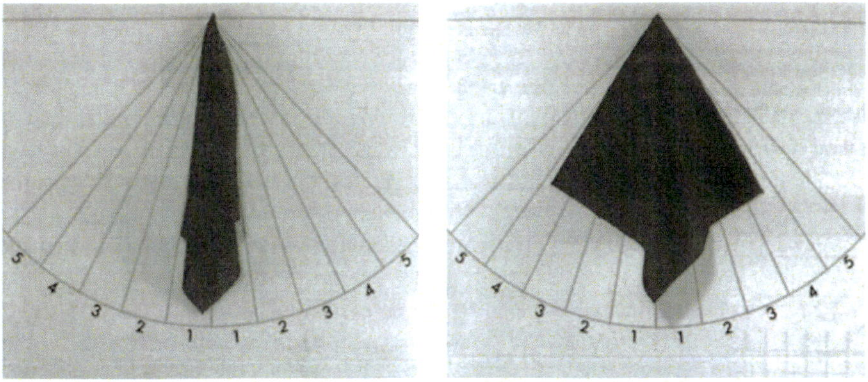

Fig. 1. Falling measurement; method for student's practice. Source: Aldrich [1], p.26.

Throughout the research, beyond Aldrich's method [1], which focuses on falling and fitting as the dominant aspect–and also from Sanad et al. [8]–it was identified others who discuss fabric selection and who can be grouped in three different types of approach: the one that describes requirements and dimensions to be considered for the selection; the one that prioritizes the sensorial analysis as relevant for choosing; and the one that provides information and specifications about the materials per se. We restrict to quoting the main authors investigated, and not the description of respective methods or models, due to the limitation of this framing and the fact that they do not exert direct influence for the development of the ADeQMat Model. They are the following: Andrade [2]; Petreca, Baurley, and Bianchi-Berthouze [6]; Petreca et al. [7]; Baraunaa, Razerab, and Heemannb [3]; and Brehm [4].

3 Methodological Design

This research is qualitative, characterized as exploratory and descriptive, with data collected through field survey, experiment, and interventions. The bibliographic investigation occurred concomitantly with the experimentation that took place at the State University of Londrina (UEL-Brazil)–at the modeling laboratory in the Design department. Tridimensional modeling techniques, through the use of technical manikin in a 1:2 scale, enabled the process of experimentation and facilitated the subjective evaluation of results, in a way that the falling perception of the materials was accomplished through sensorial analysis, based on comparisons by visual estimative, according to Lerma, Di Giorgi, and Allione [5] directions.

The study continued an investigation that started in the context of an applied research project, coordinated by the author, that investigated the application of different building resources in distinct textile materials, to evaluate the applicability of the resource as a structuring element, and the material behavior through the countless interventions.

Then, it started from a relevant number of fabric samples already measured through Winifred Aldrich's method [1] following the characteristics of weight, thickness, distortion, drapeability, and elasticity; and from a set of prototypes manufactured with such materials, which contemplates a diversity of models and building resources. All of the materials and prototypes derived from prior investigations, as well as the present one, resulted from project activities developed with students at the modeling laboratory, in course contexts, or as part of actions linked to scientific or technological initiation.

In the specific case of the ADeQMat Model proposed, its validation took place through the application in the context of the practical course Laboratory of Advanced Shape, conducted by the author, with 15 students per group on average, that were in their third year of the Bachelor Program in Fashion Design at the State University of Londrina. Regarding the contents, in determined periods the projects exclusively happen in the context of this course; in others, they are experimental projects that integrate more courses and respective professors and contemplate interdisciplinary activities. In both situations, the objective of the course is that the students gain the ability to create alternatives to formal configurations through tridimensional modeling, employing the textile materials and building resources in a coherent aesthetic, constructive, and ergonomic composition, that addresses the project requirements.

4 Exploring Drapeability

The function of drapeability has been identified as a relevant characteristic on what refers to interfering in the formal aspects, it is needed to investigate it deeply, before starting the analysis of the materials applied in diverse configurations.

At the beginning of the measurement, the samples were hung by any of their edges to, then, measure the drapeability. By continuing with the measurements, it started being noticed that the sample felt in different ways, depending on the edge chosen to be hung. So, it was established–and it was not part of Aldrich's method procedures–numbering each one of the square edges, differentiating the right and wrong side of the fabric. Thus, the samples started being photographed measures in 8 positions: 4 on the right side–identified as 1R, 2R, 3R, and 4D; and 4 on the wrong side–identified as 1W, 2W, 3W, and 4W.

It was observed that the drapeability indexes in several positions kept nearly unchanged, considering that small variations occurred inside the same "slice" of the same index. On the other side, there was relevant differentiation in shapes configured by samples in some of the registered positions, even by having one of the sides leaned against the meter–as Aldrich's method indicates. It was noticed, then, if they were placed further from the measurement apparatus, in a way they would freely fall, the samples would start creating volumes. Hence, it started being considered the possibility of measuring such volumes as a way of expanding the drapeability evaluation.

The visualization of the volume filled could be more easily noted if there was an incidence of light above the sample. To do so, the use of a LED lamp–light distributed in its encapsulation, in which the incidence is dispersed–placed above the samples was thought, in a way it projects its shadow on the inferior basis. This way, it was developed an equipment capable of addressing these new requests, repositioning even the place where the sample was hung to free it from any contact: this is the electric meter with LED light.

During the process of measuring, photographing, analyzing, and discussing–what occurred during project meetings with scholarship-students–it was noticed that it could be interesting measuring the shadows and not just observe their conformation captured by a digital camera. Hence, it was decided to insert in the inferior basis of the meter, where the shadow was projected, a blank squared paper of 1cm, so the number of filled squares could be counted from the projected volume.

This new way of measuring, with the samples freely hung and the volume visualization, revealed itself as relevant to contribute to material choice learning. The thorough observation of this context referred, by analogy, to the vision of a dressed body being seen from the front and, at the same time, allowing to notice the configured volume around it. So, what was missing was observing it from the side.

From this, to contemplate the "from the side" observations, another measurement was established, which was named as side view; the others were named as front view and projected view on the inferior surface. To enable the alterations another support to the electric meter with LED light was added. A retractable tripod also started being used to facilitate the cellphone's digital camera positioning. Figure 2 presents the scenario that illustrates how the measurements were conducted, to proceed with the drapeability measurements, considering the three views.

Figure 3 shows how they were registered, with a basis on the images and indexes.

Throughout the research, by analyzing the procedures adopted, it was noted that the light could be replaced for the use of a cellphone digital camera that would be placed on the inferior surface–where the shadow was projected–and from there the image of the

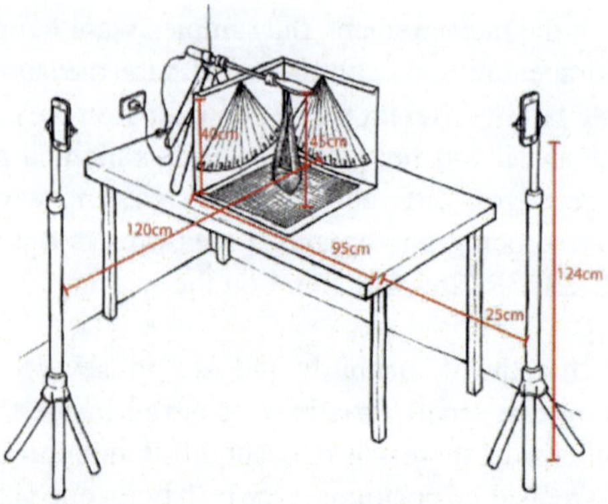

Fig. 2. Scenario for measurements. Source: the authors, 2019.

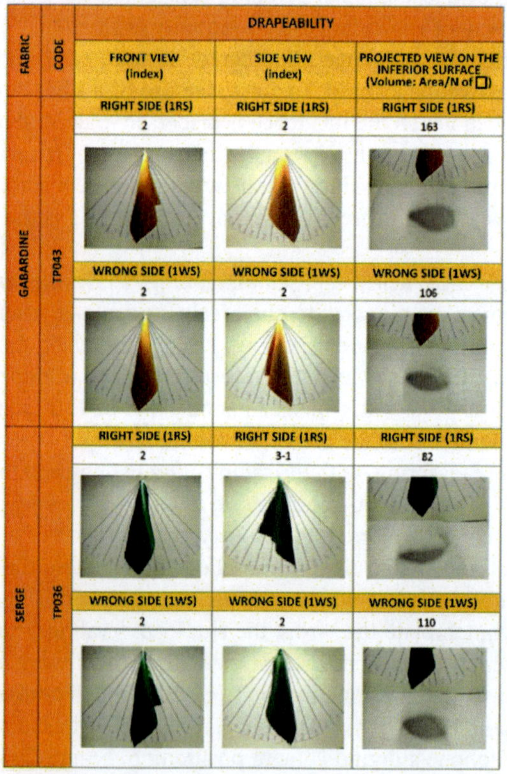

Fig. 3. Drapeability measurement of 3 views. Source : the authors, 2019.

hung sample could be directly captured. This way, the image obtained, also enables the understanding of what is the sample volume, as Fig. 4 shows.

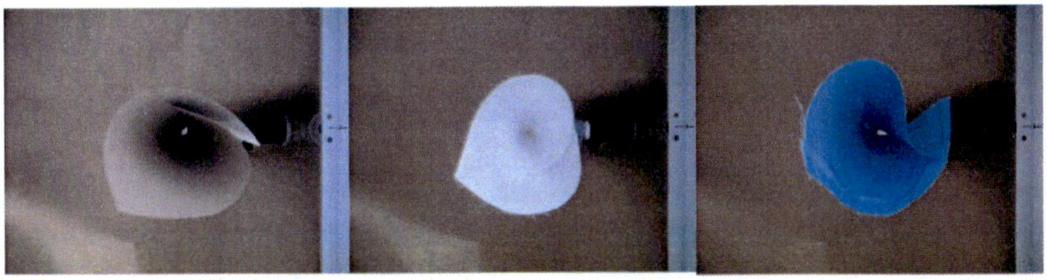

Fig. 4. Views of the samples, captured from the bottom-up. Source: the authors, 2019.

At last, regardless of how the measurement of data was processed, they were always registered in tables or charts, that were also changing, as to reflect the constant updates. In this sense, more than the tables and charts that were presented, another type was used, that grouped and synthesized numeric data, dividing them into images, to facilitate certain analyses. Sometimes, the recordings were made by decoupling the measurements of the right and wrong sides; but not other times, depending on what was intended to verify.

The most important finding in the deep investigation of drapeability and its unfoldings, which makes a huge difference to this entire research, was that many samples that presented the same frontal and/or lateral drapeability indexes, recurrently presented projected views absolutely distinct. This is evident in the recordings of almost all the material investigated. They were an essential data for the objectives of this study because they validated the importance of volume measurement filled by the sample, as a differential to evaluate the behavior of material in shape configuration.

5 Configuring Silhouette

Almost all of the materials measured were implemented by the students in the development of the prototypes on a 1:2 scale, in the context of project activities of the Laboratory of Advanced Shape course. Divided into groups, they would choose a referential model and reproduced it through tridimensional modeling techniques, with many different fabrics, so they could compare them.

Many activities like this one were carried out. The purpose was to determine, in the evaluation of the prototypes, if the importance of measuring the volume filled by the sample was truly confirmed, as a differential to evaluate the behavior of materials when applied in silhouette configuration. We limit presenting a single model, manufactured from four different fabrics, all white, to facilitate the visualization and perception of the shape for considering it enough to exemplify the procedure and point out the aspects observed, that attest to the results priorly obtained.

Figure 5 shows the skirts' prototypes, in which the samples refer to their front view measurement. Among the items manufactured, organza is the one that creates the most structured and wider silhouette and has the greatest distance from the body. Shantung is a little less structurally configured, yet, with the most assembled fitting. About crêpe and serge, the silhouettes get closer to the body, and the fitting loses the stiffness from the previous ones.

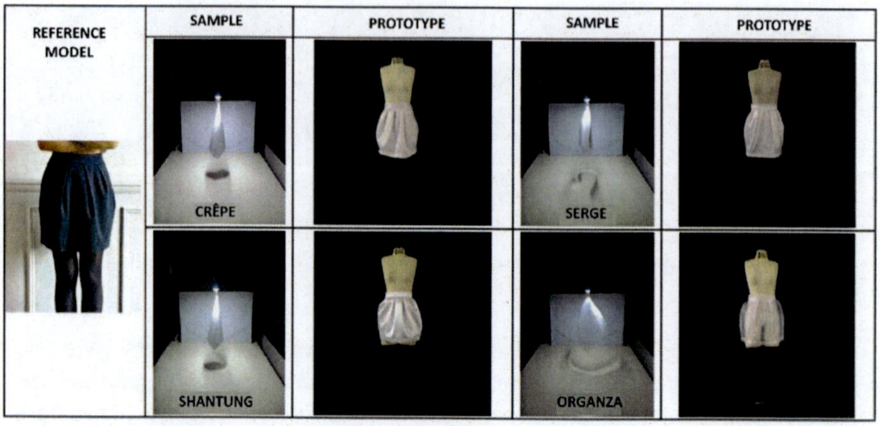

Fig. 5. Equal models in different fabrics. Source: the authors, 2019.

Serge is the material that configures the silhouette most similar to the referential model. By comparing the drapeability data of these materials, organza is the one that presents the greatest index in the front view, only partially reducing in the side view. Even though the projected area in numbers of squares does not appear with precision–due to the difficulty of perceiving the shadow nuances, especially in this case of very "open" samples–it is possible to notice from the image how wide it is, agreeing then with the analysis of the prototype. Shantung and serge present the same drapeability index in the front and side view, but in the comparison of the projected views, the former occupies a greater volume. Such an aspect is confirmed in the observation of the prototypes once the skirt in shantung presents itself way more structured and far from the body than the skirt in serge. In the verification of serge and crêpe drapeability, it is possible to notice that they registered the same index in the front view; and the latter, a little smaller than the former in the side view–despite the shape of the projected shadow being very distinct–however, it did not show itself as relevant in the comparison of the prototypes that are very similar.

Hence, it is seen a considerable coherence between the behavior noticed in the configuration of the prototypes–not only these ones but in the sampling totality–and the relation verified between the drapeability indexes of the materials used to manufacture them. From this, and considering the vast variety of formal configurations possible, we conclude that the most coherent way of inserting the silhouettes in the proposition of the Model would be thinking about them in terms of proximity or body distance, relating them to each material and contour lines suggested by them.

6 The ADeQMat Model

The investigations allowed the proposition of the ADeQMat Model–a model of adequation of materials to product configuration. Its validation occurred practically, and concomitant to its development, once the advances in the trajectory of the research were always linked to the experimentations and findings that occurred in the course context, in the direct contact with students and scholarship-students. The Model englobes a set of eight sequential actions, among which, some do not need to be broken down once they were already presented and discussed throughout this work; others will be detailed hereinafter.

1. Gather the available materials susceptible to being used in the product.
2. Prepare samples of the materials with 20 by 20 cm of measurement.
3. Measure, by visual estimative, frontal, side, and projected on the inferior surface drapeability, of each one of the samples.
4. Elaborate a table containing drapeability data of each material and other meaningful information for the selection phase (Table 1).

Table 1. Data for the selection of material. Source: the authors, 2019.

		Drapeability			**Silhouette** (contour line)		
Material Name/Reference	**Sample** Material physics	**Front View** (index)	**Side View** (index)	**Projected view on the inferior surface** (Volume: Area/N of□ + projected form draft)	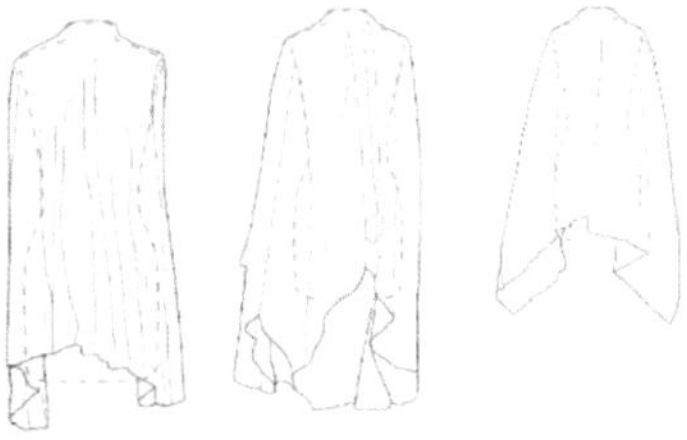 Close	Intermediate	Distant

5. Investigate the correlation among the three views measured of each material.
6. Identify patterns of silhouettes that can be linked to each material and register in the data table for the selection of material, in the area designed to do so.

   The silhouette presents itself with a double characteristic: silhouette as formal configuration, that defines the limits of the garment; and the silhouette as contour line, that defines the limits around the body. A same formal configuration can define distinct contour lines, depending on the material that is being used. For example: a manufactured garment in rigid fabric that presents a trapezoidal silhouette, will possibly define a voluminous contour line, and keep distance from the body; in malleable fabric it will define a more fluid contour line, keeping itself closer to the body. In the table, signs of (+) or (-) can be attributed to indicate that the silhouette is a little closer or a little further from the determined pattern.
7. Compare the data of the materials in the table filled.

It is recommended that the materials are placed in sequence regarding the silhouettes: first the closer ones, then the intermediates, and, lastly, the further ones. If there are (+) or (-) signs, consider them in the same way.

7 Final Considerations

It is proved that the model developed–ADeQMat–expands the possibilities of correctly selecting materials. To the extent that, it guides the sequencing and interaction of actions that involve the configuration of the garment shape more directly, being possible to apply it to any one of the project process phases, according to each need. The table, recommended as one of the actions of the model is constituted as an essential and indispensable document, to be continually supplemented when new evaluation of materials takes place, being of great value to assist, not only what concerns the specificities of each product, but in the decision of materials purchase. It is believed that from the spread of the ADeQMat and its systematic application in projects, it will be possible to expand the quality of the prototypes and minimize the loss of materials due to wrong choices, not only in the academic scope but also in the clothing manufacturing industry.

References

1. Aldrich, W.: Tejido, forma y patronaje plano. Gustavo Gili, Barcelona (2010)
2. Andrade, R.B.: Uma ferramenta para a seleção de tecidos no desenvolvimento de produtos de moda. 2016. Tese (Doutorado em Design)–Faculdade de Arquitetura, Artes e Comunicação, Universidade Estadual Paulista, Bauru (2016)
3. Barauna, D., Razera, D.L., Heemann, A.: Seleção de materiais no design: informações necessárias ao designer na tomada de decisão para a conceituação do produto. Revista Design & Tecnologia **5**(10), 1–9 (2015)
4. Brehm, L.S.: Contribuição para classificação e descrição do caimento dos tecidos de seda 100% empregados no vestuário. 2011. Master Thesis–Universidade Federal do Rio Grande do Sul, Escola de Engenharia, Faculdade de Arquitetura, Porto Alegre (2011)
5. Lerma, B., De Giorgi, C., Allione, C.: Design e materiali: sensorialità_sostenibilità_progetto. Milano: Francoangeli 2011
6. Petreca, B., Baurley, S., Bianchi-Berthouze, N.: How do designers feel textiles? In. International conference on affective computing and intelligent interaction, 6., 2015, Xi'an. Proceeding[…].Washington: IEEE Computer Society, pp. 982–987 2015
7. Petreca, B., Baurley, S., Bianchi-Berthouze, N., Tajadura-Jiménez, A.: Investigating nuanced sensory experiences in textiles selection. In. International Joint Conference on Pervasive and Ubiquitous Computing, 2016, Heidelberg. Proceedings[…]. New York: ACM, p. 989–994 2016
8. Sanad, R., Cassidy, T., Cheung, V., Evans, E.: Fabric and Garment Drape Measurement - Part 2. Journal of Fiber Bioengineering and Informatics, 6(1), 2013 (2013). http://manu27.magtech.com.cn/Jwk_JFBI/EN/Y2013/V6/I1/1. Accessed 15 Feb 2019
9. Souza, P.M.: Estratégias de construção para estruturas têxteis vestíveis. 2013. Design doctoral thesis–Faculdade de Arquitetura, Artes e Comunicação, Universidade Estadual Paulista, Bauru (2013)

Marketing and Consumption

The Effect of Religiosity on Purchase Intentions of Fashion Products: Is Faith an Important Factor in Consumer Behaviour?

Vitória Bernardo(✉)

Universidade da Beira Interior, 6201-001 Covilhã, Portugal
vitoriarbernardo@gmail.com

Abstract. As a result of the projected growth of the global fashion industry, and the fact that the majority of the world's population is affiliated with a religious faith, the present study aimed to assess the state of the art on the intersections of fashion, religion and consumer behaviour. This study is done in order to help guide the strategies of marketers and theorists on this subject, which despite its significance is still relatively new in scientific literature, therefore including limited research. This literature review touches on the difference between religion and religiosity, and how it can be a moderating factor in consumer behaviour, focusing on the influence of Christian religiosity in apparel shopping behaviour and the purchase intentions of fashion products by Muslim consumers. This study also investigates the impact of religiosity on pro-environmental and sustainable fashion attitudes and how different faiths see this relationship. Similarly, it was also concluded that while most faiths give great emphasis to modesty and, therefore, its incompatibility with status consumption, religiosity has very little influence on the consumption of luxury fashion, as consumers did not see a connection between consuming luxury goods and their faith.

Keywords: Religion · Religiosity · Fashion · Consumption · Consumer Behaviour · Modesty

1 Introduction

It is commonly believed that as society progresses the population tends to become more secularised, however, 84% of the world self-identifies as being part of a religious group (Sherwood, 2018). Additionally, this religious demographic is often younger and prone to procreate more than non-religious cohorts, meaning that the world is very likely to become more religious rather than the contrary (Sherwood, 2018). According to 2015 data (Hackett and McClendon, 2017), Christianity is the most numerous religion, accounting for 2.3 billion members (31.2% of the world's total population), followed by Islam which currently has 1.8 billion followers, 1.1 billion Hindus and 500 million Buddhists. As reported by Pew Research Center (2017) and Statista (Jenik, 2021), if current trends in fertility and religious conversion continue until 2060, although Christianity will persist to be the most popular religion, Islam will expand the most, rising from 24% to 31% of the worldwide population in just 40 years.

Due to spirituality's omnipresence and importance in society, numerous studies have focused on the effects of religion on Consumer Behaviour (CB), more specifically on topics such as the influence of religiosity on consumer responses to sexual appeals and religious cues/symbols in advertising, religiously motivated consumer boycotts (Kalliny, Minton and Benmamoun, 2018), Islamic marketing, counterfeit product consumption as a "sin", as well as themes related to the effect of religion in partaking in sustainable/ecological consumption. But does religion play an important factor in women's consumption of fashion?

Previous research has established that concerns affecting social, cultural, economic, and political dimensions impact the consumption of fashion products (Lundblad and Davies, 2016; Ashaduzzaman, et al. 2021; Cavusoglu and Atik, 2021; Cruz, Seo, and Buchanan-Oliver, 2018). It then becomes clear that religion can be used as a determining factor among these other aspects, as faith is intimately tied to a person's behaviour and frequently recommends proper methods of attire to the followers of the doctrine in question.

Scholars such as Guindi (1999), Hume (2013), Arthur (1999), Lewis (2013), Tarlo and Moors (2013) and Almila (2018) have dedicated their work to researching the interfaces of fashion and religion, concluding that the core principle expressed by religious writings from the main Abrahamic religions concerning women's fashion is that they should not flaunt their bodies or earthly possessions, therefore putting forward the concept of modesty. According to Modest Fashion journalist and author, Hafsa Lodi, "though the terms may differ, *haya*, *modesty* and *tzniut* are all intended to safeguard the chastity of women" (Lodi, 2020:59). Nevertheless, the religious concept of modesty means a lot more than rejecting the female display of sex appeal that has become so prevalent in current fashion trends, and instead, strongly focuses on humility, moderation and self-effacement (Hartman, 2015; Ambrosio, 2019) which seem incompatible with the overconsumption, materialism, ethical issues and vanity associated with the fashion industry.

This paper aims to critically review the state of the art (Snyder, 2019; Grant and Booth, 2009) by thoroughly investigating and critically integrating, analysing and summarising information from several sources of current literature on the intersection of fashion, religion and CB. The theoretical originality of this literature review will stem from the unique views and areas for additional research that will be identified.

2 Methodology

The research was conducted between November 2021 and January 2022, updated in April 2023, and consisted of detailed article searching based on academic peer-reviewed journals written in English or Portuguese between the years 2012 and 2022, focusing on the keywords: "religion", "religiosity", "clothing/apparel", "fashion", "consumption" and "consumer behaviour". After thoroughly screening through the titles and keywords of various publications across Google Scholar, ProQuest, Taylor & Francis Online, Wiley Online Library, Oxford University Press, Emerald Insight, Bloomsbury Fashion Central, and ResearchGate, 314 articles supposedly correlated all the previously identified keywords. After examining the abstracts of those papers, 75 were considered to be linked directly or indirectly to the aims of this study.

Table 1. Number of analysed articles in various journals (Source: Author's research)

Journal	Number of articles
International Journal of Consumer Studies	13
Journal of Islamic Marketing	9
Religions (by MDPI)	5
Journal of Consumer Marketing	4
Journal of Business Ethics	3
Journal of Consumer Psychology	2
Journal of Consumer Behaviour	2
Other Journals	37
Total	75

Table 1 shows the number of publications that are in some way related to the influence of religion on fashion CB in various journals within the 10-year range considered for this literature review. Additionally, Fig. 1 presents an annual distribution of the selected articles, and Table 2 shows an overview of these papers, in which it is possible to conclude that the primary research for these publications was conducted in 25 different countries (grouped in geopolitical regions). The geographical distribution of the articles perfectly illustrates the focus on the Islamic faith (35 articles) followed by Christianity (22 articles), because, although the single country which carried the greatest share of research was the USA, most contributions to this theoretical framework are from predominantly Muslim populations.

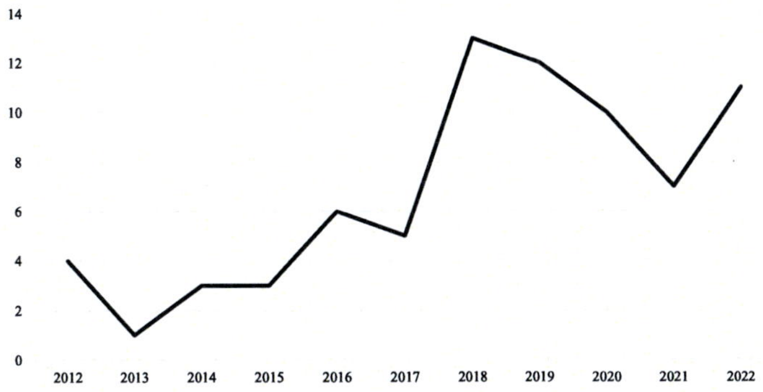

Fig. 1. Annual Distribution of the analysed articles (n = 75) (Source: Author's research)

While this literature review does not include articles that belong solely to one of the main research areas in question (Fashion, Religion and/or CB), the Venn diagram in Fig. 2 demonstrates the interactions of the areas of study.

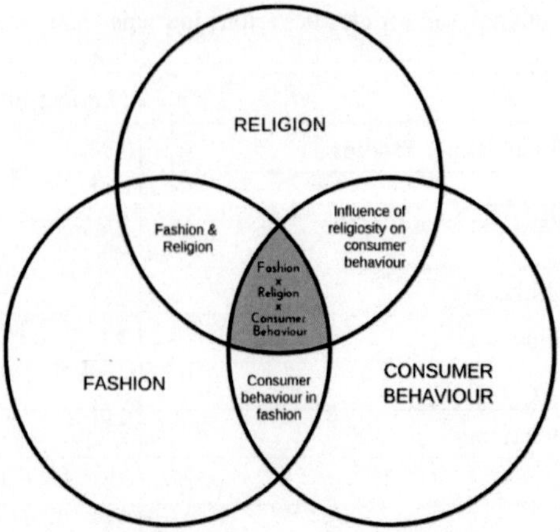

Fig. 2. Venn diagram of the research topics (Source: Author's research)

Table 2. Overview of the analysed articles (n = 75) (Source: Author's research)

Parameters		No. of articles
Regions and Countries	Southeast Asia/Australia (Malaysia, Indonesia, Australia, and New Zealand)	17
	North America (USA)	13
	Europe (UK, Italy, Luxembourg, Spain, Netherlands, Poland, and Portugal)	12
	Middle East/North Africa (Turkey, Iran, Egypt, Tunisia, Morocco, UAE, and Qatar)	12
	South/Central America (Brazil and Chile)	9
	South Asia (India and Pakistan)	9
	Sub-Saharan Africa (Nigeria and South Africa)	2
	East Asia (Korea)	1
Religion	Islam	35
	Christianity	22
	Hinduism	1
	Interreligious/All	14
	Not Applicable	3
Research Area	Fashion x Religion x Consumer Behaviour	37

(*continued*)

Table 2. *(continued)*

Parameters		No. of articles
	Religion x Consumer Behaviour	29
	Fashion x Religion	6
	Fashion x Consumer Behaviour	3
Methodology	Survey	36
	Analysis of in-depth interviews	12
	Literature Review/Theoretical Framework	9
	Multifactor analysis	4
	Editorial/Conceptual Paper	2
	Other	12
Sample Size	≤ 100	14
	100–499	34
	≥ 500	12
	Not Applicable	15

3 Literature Review

3.1 Religiosity as a Moderating Factor in Consumer Behaviour

Religion is seen as a significant factor that influences CB due to the set of beliefs, rituals, values, and collective devotion that its members share and commit to (Sardana, Gupta, and Sharma, 2018; Mathras et al. 2015). Therefore, it is believed that:

> *The integration of religion with brand personality will help to influence the devoted believers to stay loyal to their brand because devout consumers will strongly follow their religious principles and it will impact their behaviour in the market, level of confidence in a brand and most importantly their commitment or loyalty towards the brand.* (Zainudin, et al. 2018:990)

Spirituality has always been considered the foundation of all faiths, however, throughout recent decades, the public conception of religiosity shifted dramatically, and people started to distinguish between "being religious" and "being spiritual". Simultaneously, there is a rising recognition that people engage in religious activities for a variety of purposes, not all of which are necessarily spiritual (Sardana, Gupta, and Sharma, 2018). Mathras, et al. (2015) wrote one of the most cited papers on this topic, which creates a conceptual framework on the effects of religion on CB, discussing previously discovered links between degree of religiosity and many behavioural qualities, including agreeability, conscientiousness, decreased impulsivity, conservativeness/moderation, and independence - which would imply that non-religious customers would be more vulnerable to a range of marketing methods that religious consumers would not be so susceptible to.

At the crux of the dispute about religion and consumption is the case of Muslim consumers' distinctiveness as a result of their particular life philosophy and purchasing behaviours that are intrinsically linked to their religious convictions (Floren, Rasul, and Gani, 2019). As the Muslim population grows, there is a significant increase in demand for Halal (allowed/Sharia-compliant) services and goods among Muslim consumers. More particularly, Halal items should not include Haram (forbidden) components (Islam and Chandrasekaran, 2019), and Muslims are instructed to scrutinise each product and refrain from consuming it if they are unsure. Furthermore, Floren, Rasul, and Gani (2019) define Islamic Marketing as a "God-conscious approach to marketing" and systematically review the existing literature on its influence on CB. Specifically regarding Islam and fashion consumption, Hassan, and Ara (2022) examined how - from the perspective of individuals involved directly in the Islamic fashion industry, such as Muslim clothing retailers - "Hijab fashion" symbolises the current ideal Muslim women's dress that promotes an image in accordance with Islamic principles. Therefore, all around the world, most Muslim women follow a dress code known as "hijab" (meaning "covering" in Arabic) which requires women to cover their bodies in public, with the exception of their face and hands, often wearing clothes that are loose and modest, with the aim of not drawing attention to the shape of the body.

Finally, the undeniable influence of religiosity on CB led to various studies on the presence of religious cues in advertising or packaging for different products within groups of Christian, Muslim and Hindu consumers, concluding that in some cases these religious messages increase purchase intention in people with high levels of religiosity (Ustaahmetoğlu, 2020) but that devout consumers can quickly perceive them as commercialization of their faith, manipulation or proselytising attempts (Agarwala, Mishra, and Singh, 2021; Zehra and Minton, 2020; Taylor, Halstead, and Moal-Ulvoas, 2017), which lead them to develop negative feelings about the brand in question. Some other studies also focused on religion as a product and people as consumers of religion (Pinelli and Einstein, 2019), specifically when it comes to attitudes toward church retailing (Bundwini and Lappeman, 2018) or the detraditionalization of sacred occasions like Christmas and Hanukkah in order to marketize these holidays (McAlexander, et al. 2014). Furthermore, other papers bring up religion in a completely different way, explaining the concept of brand religiosity (Wang, Sarkar, and Sarkar, 2018).

3.2 Muslim Consumers' Attitudes Towards Islamic Fashion

As previously mentioned, people with high levels of religiosity perceive higher danger, therefore they are expected to be more brand conscious and less novelty-fashion-conscious as a way to lessen the risk involved. In the case of Islam, although faith does not encourage a full rejection of earthly desires and leisure, it does condemn an overemphasis on hedonistic pleasures such as the excessive use of luxuries or materialism (Avcı and Hacıkeleşoğlu, 2021). Therefore, in a 2019 study (Islam and Chandrasekaran, 2019) focusing on Muslim women in India, it was concluded that more religious individuals were less likely to be both novelty-fashion-conscious and brand conscious, similar to a previous paper (Rahman, Albaity, and Maruf, 2017) which argued that highly religious consumers were, overall, less interested in fashion.

Regarding Islamic fashion retail store attributes in Malaysia, Zainal, Yen-Nee, and Ahmad (2022) demonstrated that the physical environment has no considerable influence on customer satisfaction, perhaps because retail businesses produce comparable physical settings, and shoppers are unable to tell the difference between those with Islamic influence and those without. Moreover, Kusumawati, et al. (2019) who aimed to examine the impact of religiosity on fashion knowledge, Customer Perceived Value (CPV) and patronage intention of fashion brands among Muslim women in Indonesia, concluded that a customer's level of religiosity was not the sole reason for patronage intention and that therefore, an individual's level of religiosity does not explain their inclination to return (or not) to a particular store. When it comes to CPV, the study's findings indicate that religiosity has an essential role in determining CPV because consumers have a set of beliefs that influence their evaluation of a particular product or brand:

> *Religious consumers will consider purchasing decisions because they conform to their religion; thus, consumers will consider their intention to return to the store that sells products in line with their religiosity.* (Kusumawati, et al. 2019:274)

In the same country, Aruan and Wirdania (2020) conducted a specially interesting study on the dimensionality of religiosity, more specifically, about the differences in the CB of Sharia versus non-Sharia modest dressers. The research discovered that religion has a major impact on consumer purchasing decisions, but that the mediating effects of emotional attitude and self-presentation are only detected for some types of clothing: while Sharia responders were unlikely to purchase fashionable clothing, non-Sharia respondents were equally likely to purchase both fashioned and Sharia types of clothing. In consequence, this article makes valuable suggestions for marketing Islamic fashion to different types of women by focusing their communication on modest qualities (e.g. fully covered body, non-see-through materials) to Sharia dressers and reinforcing aesthetic attributes when targeting non-Sharia consumers.

Indonesian Muslim women had previously been segmented into six groups based on their fashion lifestyle, those being "hijab fashionist, aspirant Sharia oriented, religious moderate dressing, economic fashion follower, Sharia fashion follower, and pragmatic hijabers" (Kartajaya, et al. 2019:319). This paper also considered that "hijab fashionistas" were the most important and potentially lucrative target demographic for Islamic fashion as these consumers tend to be early adopters of fashion trends, especially within luxury brands. Similarly, it was found in a different study (Hassan and Harun, 2016) that Muslim women who are more fashion aware have a higher proclivity than other women to engage in distinctive and trendy Muslim fashion consumption as a way of differentiating themselves from others and demonstrating their personality. These fashionistas, who get their fashion inspiration from "catalogues, fashion magazines, friends, fashion fairs and the internet" (Hassan and Harun, 2016) are also more inclined to have their clothing handmade to order and tend to choose Islamic/Modest Fashion for reasons related to comfort, individuality, peer acceptability, and compliance to particular events.

One of the most popular papers about fashion and religion focuses on the CB of young, Muslim, and Egyptian consumers (Farrag and Hassan, 2015) and argues that if religion was given greater consideration in marketing and communications, society would see a prosperous expansion of the fashion sector. Muslims have firm convictions rooted in the religious concept of modesty that is opposed to the "flamboyant display and show off of prestige and status" (Farrag and Hassan, 2015; O'Cass, Lee, and Siahtiri, 2013), and therefore, Farrag and Hassan (2015) advise marketers to create communication strategies based on the fundamental ideals of moderation and simplicity, in contrast to the common uses of sex appeals (Mitterfellner, 2020:147–149), which lower purchase intentions in highly religious individuals who find this type of communication to be unethical (Putrevu and Swimberghek, 2013). Comparably, the findings in Baran and Barutçu (2023) indicate that hijab clothing mobile apps are generally well-received by conservative Muslim customers, who express a preference for visual content and advertising that aligns with Islamic principles.

Lastly, Sobh, Belk, and Gressel (2012) centred their research on the Muslim dress of women in the Arab Gold Region, where strict religious and cultural traditions and legislations prevent young women from completely adopting the current global fashion trends in public, leading them to adapt by engaging in more of what they can exhibit and purchase, such as the extensive usage of global luxury brands and cosmetics. In contrast, a study with a multi-ethnic perspective by Ashraf, Williams, and Bray (2022) showed that although Muslim women living in the United Kingdom made fashion choices compatible with their identity, they were concerned about a lack of variety and availability, and that, therefore:

> *(...) ranges must be developed that can both maintain modesty but enable Muslim women to be up to date with latest trends and convey their individual personalities. Consumers are seeking to be represented based on their beliefs and would respond strongly to brands who create clothing that adheres to their needs. This consumer response is likely to include significant word-of-mouth marketing and avid attitudinal and behavioural loyalty to the brand.* (Ashraf, Williams and Bray, 2022)

3.3 Christian Religiosity in Apparel Consumer Behaviour

Although most of the academic and political attention has been almost exclusively on the associations between Islam and fashion, people of other faiths also make decisions concerning fashion, and according to Dr Anna-Mari Almila, "these choices are worth researching just as much as the more politicised phenomenon of Islamic veiling" (Almila, 2020). The literature on the interfaces of fashion and Christianity is very scant, nonetheless, some publications have explored the topic (Neal, 2019; Covolo, 2020).

In Brazil, Caixeta, et al. (2012) found the differences between different denominations of Christianity regarding CB of fashion products and concluded that Evangelical Christian women were more influenced by their religion when purchasing clothing items than followers of a Catholic doctrine. Furthermore, in the same country, Albuquerque, Duque-Arrazola and Rocha (2018) investigated the consumption of "gospel fashion" by members of a conservative Evangelical church and found that these women shop from Christian clothing brands to be legitimised and considered by the secular world as being modern and fashionable even if not able to wear trousers or shorter dresses/skirts. Other articles on this topic in Brazil are also Cezar (2010) and Gonçalo (2016), which address the aesthetic productions of women's clothing as proof of devotion; Becheri, et al. (2022)

which studies Instagram's impact on the Modest Fashion market; Silva et al. (2020) and Becheri et al. (2020), who focus on the influence of Christian religiosity on consumer behaviour in fashion; as well as Almeida et al. (2021) and Albuquerque (2021), who explore the negotiation of religious dress rules on social media.

Furthermore, an American study (Davis, 2014) showed that religion has major influence on Christian customers' clothing preferences, particularly on quality awareness, fashion consciousness, and price sensitivity, which means these individuals are more inclined to look for discounts and reduced prices. According to the study's findings, more devoted Christians are not brand conscious or loyal, possibly because religious consumers are more motivated to demonstrate their identity via their faith instead of a brand name. In a more recent paper (Davis, 2016), the researcher explored the effect of religion on the CPV of clothing store features among Christian customers in the United States. According to the data, more devout Christians are more engaged and demanding fashion consumers, putting a lot of importance on store attributes, reputation and social status as well as increased brand loyalty to companies whom they perceive to respect the same religious values.

Another interesting fashion subcategory in which it is possible to see the differences between Christianity and Islam is the case of religious sportswear. While Muslim consumers prioritise sports apparel that covers the body but that is not necessarily from Modest Fashion collections or brands (Baber, 2019; Leonnard, et al. 2019), Christian consumers are more likely to use athletic wear as a "profession of faith" (Ornella, 2019), flaunting t-shirts with biblical verses and/or religious symbols (Neal, 2017).

3.4 Influence of Religion on Sustainable Consumption

Perera and Hewege (2018) explained the relationship of various religions with nature. Christians tend to believe that God intends that mankind has dominion over nature (Minton, et al. 2018) and uses it for their own purposes, therefore having lesser correlations between religious and ethical CB (Raggiotto, Mason, and Moretti, 2018), another common explanation for this is "the idea of dispensationalism, belief in the 'end of time' and renewal of the earth in eternity" (Graafland, 2017:132). Differently, Muslims and Hindus believe in the concept of responsible stewardship of the environment. Lastly, Buddhism is often regarded as the most ecologically conscious religious ideology due to its belief that all life forms are fundamentally equal and its teachings that recommend "a gentle, inclusive approach to nature and environment, minimising individual presence and impact on the environment" (Raggiotto, Mason, and Moretti, 2018:622).

In a recent case study, Sadowski (2021) analysed how the Sunday shopping restrictions implemented in Poland contributed to the development of a culture of communion by reinforcing interpersonal ties and changes in consumption patterns, as well as laying the groundwork for the development of a culture of fair balance in consumption. The researcher put forward that consumerism has been identified "as one of the greatest challenges to man's integral development" (Sadowski, 2021:12) and that because the overwhelming majority of the population is Roman Catholic, Christian denominations could help decrease consumerist culture and promote sustainable behaviours regarding fashion consumption.

In a different study (Hwang, 2018), it was concluded that religious consumers were more willing to pay higher prices and reduce their standard of living than secular consumers, showing that religious engagement can encourage individuals to prioritise public interests over personal interests. It is crucially important to keep in mind geographical and cultural differences when it comes to using religiosity as a mediating factor in issues related to sustainability because there are strong differences between Western and Eastern religions when it comes to ethical/moral consumption (Minton, Bret, and Upadhyaya, 2018). For instance, care for others motivates Eastern religious customers, whilst self-concern motivates Western religious consumers.

Razzaq, et al. (2018) explored the involvement of Islam in the consumption of sustainable fashion and concluded that individuals who follow the religious concept of modesty will resist the temptation to overindulge in all areas of life, including fashion, therefore becoming increasingly interested in the consumption of sustainable items, which support their religious ideals of being less selfish and more altruistic.

Lastly, because it is known that "a person's religiosity does influence their decision making in situations that involve ethical issues" (Vitell, Ramos-Hidalgo, and Rodríguez-Rad, 2018:677), some other studies are based on religion and consumer ethics, such as Souiden, Ladhari, and Amri (2018) as well as Kasber, El-Bassiouny, and Hamed (2022), who explored Muslim consumers' attitudes towards counterfeit products; Aziz and Bakar (2020) who studied ethical apparel consumption in Malaysia; Sandıkcı (2021) who presents a moral economy approach to everyday consumption ethics of products such as Halal nail polish; and Marhana, et al. (2012) who concluded that highly religious Malaysian Muslims have higher intentions of purchasing cause-related products.

3.5 Effects of Religiosity on Luxury and Status Consumption (SC)

SC can be explained as a "consumer's overt or ostentatious display of consumer products that symbolise status, with the hope of gaining social approval or recognition from surrounding others" (O'Cass, Lee, and Siahtiri, 2013:442).

Arli, Gil, and Esch (2020) tested the effect of Christianity/Catholicism on perceptions of luxury goods among young consumers in Chile. According to the study's findings, shoppers regarded materialism and luxury products as two distinct conceptions. Religious consumers reject materialism as a connection to worldly belongings, yet they retain an emotional affinity to luxury items. In short, there are no substantial variations in the acceptability of luxury items between religious and non-religious consumers, leading the researchers to state that "religious youth consumers love God, but they also love Gucci" (Arli, Gil, and Esch, 2020:181). Similarly, when it comes to the impact of religion and culture on Nigerian Muslim consumers' intention to acquire luxury items, Aksoy and Abdulfatai (2019) found that Islamic morals had little effect on consumer purchase intentions of these products and that consumers did not see a connection between consuming luxury goods and their religiosity. This view is shared in Varma, et al. (2022) and Ramazani and Kermani (2022), where the authors found a positive relationship between personal spirituality (specifically Islamic religious commitment, in the case of the latter) and conspicuous consumption of fashion.

In parallel, when investigating Muslim consumers in Turkey - also researched by Erden (2019), Nazlı (2018), and Sandikci and Ger (2010) - and the relationship between religion, materialism and SC, Yeniaras and Akkemik (2017) concluded that consumption of luxurious brands like Dior, Hermes and Gucci is a tool for Muslims to display their wealth and status as well as their moral characteristics. Similar results were shown by Rehman, Al Shammari and Al-Mamary (2022), where it was established that Saudi and Indian consumers highly valued the interpersonal aspect of luxury, primarily viewing luxury goods as symbols of status and as "social enablers".

Contrastingly, Geiger-Oneto and Minton (2019) explored the influence of Western religions in the purchasing of luxury products and concluded that consumers with high religiosity had the poorest evaluations of high-end goods. Nonetheless, the researchers consider that it is also likely that customers' views about luxury consumption are changing due to the increased exposure to luxury items on social media and the existence of products that merge religion and prestige (such as designer hijabs). Furthermore, Arli, Cherrier, and Tjiptono (2016) state that "religious consumers are not necessarily anti-materialism and often opt for luxury brands over purely utilitarian possession". Overall, as concluded by Alanadoly and Salem (2022), Muslim women who consume fashion are willing to pay premium prices for hijab fashion brands that fulfil their religious and social commitments and offer modern designs with high product quality.

4 Conclusion and Directions for Further Research

Although religion affects consumers' decision-making, influences their CB and instils values related to modesty and altruism, this influence may vary according to the religion being followed or to the individual's degree of religiosity, therefore it is crucial to explore the multidimensionality of religiosity (Mathras, et al. 2015) instead of treating religious consumers as a homogeneous group. It was also concluded that one's level of religious commitment "cannot be determined simply by what she wears, with age, marital status and education found to have far greater impact on a woman's choice of clothing than religiosity" (Bachleda, Hamelin, and Benachour, 2014).

This literature review concluded that Christian consumers tend to be more price-sensitive and look for discounts, and when it came to clothing purchases, Evangelical Christians were more impacted by their faith than Catholics (Caixeta, et al. 2012). Similarly, these consumers enjoy using their clothing as an expression of their religious beliefs, wearing garments and accessories with slogans and symbols related to their faith (Ornella, 2019; Neal, 2017), while more fashion-conscious Muslim women are more likely to participate in unique apparel consumption because they want to distinguish themselves from their peers and display their individuality (Aruan and Wirdania, 2020; Kartajaya, et al. 2019; Hassan and Harun, 2016).

Regarding the impact of Religiosity on pro-environmental and sustainable fashion attitudes (Razzaq, et al. 2018), it was found that Christians tend to believe that it is God's plan for humans to have sovereignty over nature, but Muslims believe in the notion of environmental responsibility (Perera and Hewege, 2018) because they are expected to follow the religious values of generosity, sincerity (Marhana, et al. 2012) and prioritise collective welfare over personal benefit (Hwang, 2018). Nevertheless, it is still considered that "when companies invest in sustainability, they will be able to attract consumers who have higher spirituality as well as religiosity" (Iqbal, and Khan, 2020).

Subsequently, while marketing actions must be different for different religious groups, most faiths maintain strong beliefs about modesty and aversion to flaunting/bragging, status consumption and display of sexuality, therefore reacting better to communication strategies based on brand values, quality and simplicity (Mitterfellner, 2020; Farrag and Hassan, 2015; Putrevu and Swimberghek, 2013). Nonetheless, as mentioned in Rodrigues, Sousa, and Torres (2022) brands should adopt a diverse mindset and try to deliver targeted messages that accommodate for the varying spiritual backgrounds of their customers.

Finally, the main theoretical contributions of the present work are in the presentation of various explorations about the effects of religious commitment on the consumption of fashion, as well as in the mapping of previous discussions about the intersection of religiosity and CB. In terms of future research, it is suggested to determine the degree of influence of religious digital influencers on the fashion CB of religious consumers, study the intersectionality of Modest Fashion, as well as a focus on Christian consumers - to understand the differences between these Western religious women and visibly Muslim (Tarlo, 2010) veiled women who tend to be the protagonists of studies on the interfaces of fashion and religion. This will bring theoretical contributions, as these women are underexplored within the business of fashion, as well as being able to guide researchers in the field of fashion, culture and society.

References

Agarwala, R., Mishra, P., Singh, R.: Evaluating the impact of religious icons and symbols on consumer's brand evaluation: context of Hindu religion. J. Advert. **50**(4), 372–390 (2021). https://doi.org/10.1080/00913367.2021.1940394

Aksoy, H., Abdulfatai, O.: Exploring the impact of religiousness and culture on luxury fashion goods purchasing intention: a behavioural study on Nigerian Muslim consumers. J. Islamic Mark. 10, 768–789 (2019). https://doi.org/10.1108/JIMA-01-2018-0022

Alanadoly, A.B., Salem, S.F.: Hijabista willingness to accept premium pricing: an analytical study of the effect of social and self-identity on hijab fashion brands satisfaction. J. Islamic Mark. **13**(1), 227–245 (2022). https://doi.org/10.1108/JIMA-02-2020-0041

Albuquerque, H.C., Duque-Arrazola, L.S., Rocha, M.A.V.: Santas e Estilosas": o consumo de moda gospel por mulheres pentecostais. Oikos, **28**(2), 316–331 (2018). https://periodicos.ufv.br/oikos/article/view/3757

Albuquerque, H.C., et al.: Fé, consumo e gênero: reflexões sobre os discursos de vlogueiras gospel. Braz. J. Dev. **7**(4), 33477–33489 (2021). https://doi.org/10.34117/bjdv7n4-009

Almeida, R.D., Casotti, L.M., Miranda, A.P.C.: Negociações simbólicas das regras religiosas do vestir em postagens no Instagram. dObra[s] - revista da Associação Brasileira de Estudos de Pesquisas em Moda. **31**, 237–261 (2021)

Almila, A.M.: Veiling in Fashion: Space and the Hijab in Minority Communities. [e-book] London: Bloomsbury Publishing (2018)

Almila, A.M.: Introduction: fashion/religion interfaces. Religions **11**, 133 (2020). https://doi.org/10.3390/rel11030133

Ambrosio, A.F.: Invisible dress: weaving a theology of fashion. Religions **10**(7), 419 (2019). https://doi.org/10.3390/rel10070419

Arli, D., Cherrier, H., Tjiptono, F.: God Blesses those who wear prada: exploring the impact of religiousness on attitudes toward luxury among the youth of Indonesia. Mark. Intell. Plan. **34**(1), 1–33 (2016)

Arli, D., Gil, L.A., Esch, P.: The effect of religiosity on luxury goods: The case of Chilean youths. Int. J. Consum. Stud. **44**, 181–190 (2020). https://doi.org/10.1111/ijcs.12559

Arthur, L.B.: (ed.) (1999) Religion, Dress and the Body, Oxford: Berg

Aruan, D.T.H., Wirdania, I.: You are what you wear: examining the multidimensionality of religiosity and its influence on attitudes and intention to buy Muslim fashion clothing. J. Fash. Mark. Manag. **24**(1), 121–136 (2020). https://doi.org/10.1108/JFMM-04-2019-0069

Ashaduzzaman, M., et al.: Acculturation and apparel store loyalty among immigrants in Western countries. J. Mark. Manag. **37**(5–6), 488–519 (2021). https://doi.org/10.1080/0267257X.2020.1833963

Ashraf, S., Williams, A.M., Bray, J.: Female Muslim identity and modest clothing consumption in the UK. J. Islamic Mark. (2022). https://doi.org/10.1108/JIMA-05-2021-0167

Avcı, İ., Hacıkeleşoğlu, H.: The Effect of Religiosity on Hedonic Consumption and Wasteful Consumption Behavior. İlahiyat Tetkikleri Dergisi (2021). https://doi.org/10.29288/ilted.871339

Aziz, N.A., Bakar, R.A.: Ethical apparel consumption: a study on Muslim consumers in Malaysia. Asian J. Res. Bus. Manage. **2**(2), 86–95 (2020)

Baber, H.: Influence of religiosity on the behavior of buying sports apparel: a study of the Muslim market segment in India. Organ. Markets Emerg. Econ. **10**(2), 212–226 (2019). https://doi.org/10.15388/omee.2019.10.11

Bachleda, C., Hamelin, N., Benachour, O.: Does religiosity impact Moroccan Muslim women's clothing choice? J. Islamic Mark. **5**(2), 210–226 (2014). https://doi.org/10.1108/JIMA-05-2013-0038

Baran, T., Barutçu, S.: Attitudes toward the design of Islamic mobile shopping apps in Turkey: the case of hijab clothing mobile apps from the perspective of conservative consumers. J. Islamic Account. Bus. Res. **14**(3), 398–415 (2023). https://doi.org/10.1108/JIABR-12-2020-0380

Becheri, J., et al.: Efeitos da religião no mercado da moda: uma revisão integrativa [Conference paper]. In: XLIV Encontro da ANPAD, pp. 2177–2576. (2020) Available through: https://www.researchgate.net/publication/347090491_Efeitos_da_religiao_no_mercado_da_moda_uma_revisao_integrativa

Becheri, J., et al.: Instagram no Mercado da Moda Modesta: uma Análise sobre a Ótica das Estruturas Sociotécnicas [Paper presentation]. In: XLVI Encontro da ANPAD - EnANPAD 2022 On-line. Brazil, September 21–23, 2022 (2022)

Bundwini, N., Lappeman, J.: Attitudes toward Christian church retailing: a study in Cape Town, South Africa. Int. J. Consum. Stud. **42**, 744–754 (2018). https://doi.org/10.1111/IJCS.12442

Caixeta, R.P., et al.: Religião E Comportamento De Consumo De Vestuário. Pretexto. **13**(4) (2012). https://doi.org/10.21714/pretexto.v13i4.1048

Cavusoglu, L., Atik, D.: Accumulating capital through social media: the transformative power of underrepresented fashion consumers. J. Consum. Mark. **38**(5), 514–524 (2021). https://doi.org/10.1108/JCM-08-2020-4074

Cezar, M.S.: A estética como comprovação da devoção. dObra[s] - revista da Associação Brasileira de Estudos de Pesquisas em Moda. **4**(10), 97–105 (2010). https://doi.org/10.26563/dobras.v4i10.190

Covolo, R.: Fashion Theology. Baylor University Press (2020)

Cruz, A.G.B., Seo, Y., Buchanan-Oliver, M.: Religion as a field of transcultural practices in multicultural marketplaces. J. Bus. Res. **91**, 317–325 (2018). https://doi.org/10.1016/j.jbusres.2017.07.022

Davis, L.: Effects of religiosity on apparel shopping orientation: an exploratory study. Int. J. Bus. Anthropol. **5**, 24–36 (2014). http://hdl.handle.net/10211.3/164842

Davis, L.: Religiosity and retail store choices: exploring US Christian consumers' apparel shopping. Int. J. Bus. Anthropol. **6**, 47–86 (2016)

Erden, O.: The new religion-based work ethic and cultural consumption patterns of religiously conservative groups in Turkey. Religions **10**(10), 541 (2019). https://doi.org/10.3390/rel10100541

Farrag, D.A., Hassan, M.: The influence of religiosity on Egyptian Muslim youths' attitude towards fashion. J. Islamic Mark. **6**(1), 95–108 (2015). https://doi.org/10.1108/JIMA-04-2014-0030

Floren, J., Rasul, T., Gani, A.: Islamic marketing and consumer behaviour: a systematic literature review. J. Islamic Mark. **11**(6), 1557–1578 (2019). https://doi.org/10.1108/JIMA-05-2019-0100

Geiger-Oneto, S., Minton, E.A.: How religiosity influences the consumption of luxury goods: exploration of the moral halo effect. Eur. J. Mark. **53**(12), 2530–2555 (2019). https://doi.org/10.1108/EJM-01-2018-0016

Gonçalo, R.: Moda Church – Performances e produções estéticas do vestir feminino em igrejas evangélicas cariocas. Mosaico **7**(11), 9–31. (2016). https://doi.org/10.12660/rm.v7n11.2016.64775

Graafland, J.: Religiosity, attitude, and the demand for socially responsible products. J. Bus. Ethics **144**(1), 121–138 (2017). https://doi.org/10.1007/s10551-015-2796-9

Grant, M.J., Booth, A.: A typology of reviews: An analysis of 14 review types and associated methodologies. Health Info. Libr. J. **26**(2), 91–108 (2009). https://doi.org/10.1111/j.1471-1842.2009.00848.x

Guindi, F.: Veil: Modesty, Privacy and Resistance. Berg, Oxford (1999)

Hackett, C., McClendon, D.: Christians remain world's largest religious group, but they are declining in Europe. Pew Research Center (2017). www.pewresearch.org/fact-tank/2017/04/05/christians-remain-worlds-largest-religious-group-but-they-are-declining-in-europe/ Accessed 22 Dec 2021

Hartman, L.M.: Environmental modesty: reclaiming an ancient virtue. J. Religious Ethics, **43**(3), 475–492 (2015). http://www.jstor.org/stable/24586129

Hassan, S.H., Harun, H.: Factors influencing fashion consciousness in hijab fashion consumption among Hijabistas. J. Islamic Mark. **7**(4), 476–494 (2016). https://doi.org/10.1108/JIMA-10-2014-0064

Hassan, S.H., Ara, H.: Thematic analysis of hijab fashion from Muslim clothing retailers perspective. J. Islamic Mark. **13**(11), 2462–2479 (2022). https://doi.org/10.1108/JIMA-03-2020-0066

Hume, L.: The Religious Life of dress: Global Fashion and Faith. Bloomsbury, London (2013)

Hwang, H.: Do religion and religiosity affect consumers' intentions to adopt pro-environmental behaviours? Int. J. Consum. Stud. **42**, 664–674 (2018). https://doi.org/10.1111/ijcs.12488

Iqbal, S., Khan, M.I.: Spirituality as a predictor of psychological well-being: an explanatory mechanism of religiosity and sustainable consumption. Religions **11**(12), 634 (2020). https://doi.org/10.3390/rel11120634

Islam, T., Chandrasekaran, U.: Religiosity, values and consumer behaviour: a study of young Indian Muslim consumers. J. Consum. Mark. **36**(7), 948–961 (2019). https://doi.org/10.1108/JCM-10-2017-2415

Jenik, C.: The World's Largest Religious Groups Over Time. Statista (2021). https://www.statista.com/chart/25572/worlds-largest-religious-groups-over-time/. Accessed 22 Dec 2021

Kalliny, M., Minton, E.A., Benmamoun, M.: Affect as a driver to religious-based consumer boycotts: evidence from qualitative and quantitative research in the United States. Int. J. Consum. Stud. **42**, 840–853 (2018). https://doi.org/10.1111/ijcs.12450

Kartajaya, H., et al.: Segmenting Islamic fashion lifestyle on Indonesian woman. Res. J. Text. Appar. **23**(4), 306–322 (2019). https://doi.org/10.1108/RJTA-02-2019-0003

Kasber, A., El-Bassiouny, N., Hamed, S.: Can religiosity alter luxury and counterfeit consumption? An empirical study in an emerging market. J. Islamic Mark. **7**, 1768–1769 (2022). https://doi.org/10.1108/JIMA-02-2022-0058

Kusumawati, A., et al.: The impact of religiosity on fashion knowledge, consumer-perceived value and patronage intention. Res. J. Text. Appar. **23**, 269–290 (2019). https://doi.org/10.1108/RJTA-04-2019-0014

Leonnard, L., et al.: Young Indonesian Muslim women consumer behavior towards modest sportswear. J. Bus. Retail Manage. Res. **13**(4), 277–288 (2019)

Lewis, R. (ed.): Modest Fashion: Styling Bodies, Mediating Faith. I.B, London (2013)

Lodi, H.: Modesty: A Fashion Paradox, Illustrated Neem Tree Press, London (2020)

Lundblad, L., Davies, I.: The values and motivations behind sustainable fashion consumption. J. Consum. Behav. **15** 149–162 (2016)https://doi.org/10.1002/cb.1559

Marhana, M., et al.: The role of religiosity in socially responsible consumption. Int. J. Asian Soc. Sci. **2**, 1467–1476 (2012)

Mathras, D., et al.: The Effects of religion on consumer behavior: a conceptual framework and research agenda. J. Consum. Psychol. **26**, 298–311 (2015). https://doi.org/10.1016/j.jcps.2015.08.001

McAlexander, J.H., et al.: The marketization of religion: field, capital, and consumer identity. J. Consum. Res. **41**(3), 858–875 (2014). https://doi.org/10.1086/677894

Minton, E.A., Bret, L.R., Upadhyaya, S.: Religion's influence on consumer response to moral vs. justice message appeals. Int. J. Consum. Stud. **42**, 768–778 (2018). https://doi.org/10.1111/ijcs.12453

Minton, E.A., et al.: Greening up because of god: the relations among religion, sustainable consumption and subjective well-being. Int. J. Consum. Stud. **42**, 655–663 (2018). https://doi.org/10.1111/ijcs.12449

Mitterfellner, O.: Fashion Marketing and Communication: Theory and Practice Across the Fashion Industry. Routledge, New York (2020)

Nazlı, A.: Fashion and Consumption. In Faith and Fashion in Turkey: Consumption, Politics and Islamic Identities. New York: I.B. Tauris, pp, 187–226. Bloomsbury Fashion Central (2018)

Neal, L.S.: OMG: authenticity, parody, and evangelical Christian fashion. Fash. Theory **21**(3), 223–244 (2017). https://doi.org/10.1080/1362704X.2016.1143574

Neal, L.S.: Religion in Vogue: Christianity and Fashion in America. New York University Press(2019)

O'Cass, A., Lee, W., Siahtiri, V.: Can Islam and status consumption live together in the house of fashion clothing?. J. Fashion Mark. Manage. 17, 440–459 (2013)https://doi.org/10.1108/JFMM-03-2013-0023

Ornella, A.D.: 'Jesus saves' and 'clothed in Christ': athletic religious apparel in the Christian crossfit community. Sport Soc. **22**(2), 266–280 (2019). https://doi.org/10.1080/17430437.2017.1360580

Perera, C., Hewege, C.: Religiosity and environmentally concerned consumer behaviour: ' becoming one with god (nature)' through surrendering environmental identities. Int. J. Consum. Stud. **42**, 627–638 (2018). https://doi.org/10.1111/ijcs.12469

Pew Research Center. The Changing Global Religious Landscape (2017). https://www.pewforum.org/2017/04/05/the-changing-global-religious-landscape/. Accessed 22 Dec 2021

Pinelli, M., Einstein, M.: Religion, science and secularization: a consumer-centric analysis of religion's functional obsolescence. J. Consum. Mark. **36**, 582–591 (2019). https://doi.org/10.1108/JCM-11-2017-2451

Putrevu, S., Swimberghek, K.: The influence of religiosity on consumer ethical judgments and responses toward sexual appeals. J. Bus. Ethics **115**(2), 351–365 (2013). https://doi.org/10.1007/s10551-012-1399-y

Raggiotto, F., Mason, M.C., Moretti, A.: Religiosity, materialism, consumer environmental predisposition. Some insights on vegan purchasing intentions in Italy. Int. J. Consum. Stud. **42**, 613–626 (2018). https://doi.org/10.1111/ijcs.12478

Rahman, M., Albaity, M., Maruf, B.: The role of religiosity on the relationship between materialism and fashion clothing consumption among Malaysian generation Y consumers. Soc. Indic. Res. **132**, 757–783 (2017)

Ramazani, A., Kermani, M.: Spiritualism versus materialism: can religiosity reduce conspicuous consumption? J. Islamic Mark. **13**(8), 1639–1655 (2022). https://doi.org/10.1108/JIMA-09-2019-0184

Razzaq, A., et al.: The impact of fashion involvement and pro-environmental attitude on sustainable clothing consumption: the moderating role of Islamic religiosity. SAGE Open **8**, 2158244018774611 (2018). https://doi.org/10.1177/2158244018774611

Rehman, A.U., Al Shammari, S., Al-Mamary, Y.H.: Role of religiosity and the mediating effect of luxury value perception in luxury purchase intention: a cross-cultural examination. J. Islamic Mark. **13**(4), 975–995 (2022). https://doi.org/10.1108/JIMA-03-2021-0091

Rodrigues, P., Sousa, A., Torres, I.: Generation Y and brand love: understanding the effects of individual cultural values and religious commitment. J. Creative Commun. (2022). https://doi.org/10.1177/09732586221092637

Sadowski, R.F.: The role of Catholicism in shaping a culture of sustainable consumption. Religions, **12**, 598 (2021). https://doi.org/10.3390/rel12080598

Sandikci, O., Ger, G.: Veiling in style: how does a stigmatized practice become fashionable? J. Consum. Res. **37**(1), 15–36 (2010). https://doi.org/10.1086/649910

Sandıkcı, O.: Religion and everyday consumption ethics: a moral economy approach. J. Bus. Ethics **168**, 277–293 (2021). https://doi.org/10.1007/s10551-019-04422-2

Sardana, D., Gupta, N., Sharma, P.: Spirituality and religiosity at the junction of consumerism: exploring consumer preference for spiritual brands. Int. J. Consum. Stud. **42**(6), 724–735 (2018). https://doi.org/10.1111/ijcs.12467

Sherwood, H.: Religion: why faith is becoming more and more popular. The Guardian (2018). https://www.theguardian.com/news/2018/aug/27/religion-why-is-faith-growing-and-what-happens-next. Accessed 22 Dec 2021

Silva, R.S.V., et al.: Em um Mundo com Tantas Opções, por que Mulheres Evangélicas Optam pela Restritiva Moda Comportada? [Conference paper]. In: 14º Congresso Latino-Americano de Varejo e Consumo (CLAV 2021) (2020). http://bibliotecadigital.fgv.br/ocs/index.php/clav/clav2020/paper/view/7522/2137

Snyder, H.: Literature review as a research methodology: an overview and guidelines. J. Bus. Res. **104**, 333–339 (2019). https://doi.org/10.1016/j.jbusres.2019.07.039

Sobh, R., Belk, R., Gressel, J.: Modest seductiveness: reconciling modesty and vanity by reverse assimilation and double resistance. J. Consum. Behav. **11**, 357–367 (2012)

Souiden, N., Ladhari, R., Amri, A.: Is buying counterfeit sinful? Investigation of consumers' attitudes and purchase intentions of counterfeit products in a Muslim country. Int. J. Consum. Stud. **42**, 687–703 (2018). https://doi.org/10.1111/ijcs.12466

Tarlo, E.: Visibly Muslim Fashion, Politics, Faith. Berg, Oxford (2010)

Tarlo, E., Moors, A. (eds.): Islamic Fashion and Anti-Fashion: New Perspectives from Europe and North America. Bloomsbury Academic, London (2013)

Taylor, V., Halstead, D., Moal-Ulvoas, G.: Millennial consumer responses to Christian religious symbols in advertising: a replication study. J. Advert. **39**(2), 79–92 (2017). https://doi.org/10.2753/JOA0091-3367390206

Ustaahmetoğlu, E.: The influence of different advertisement messages and levels of religiosity on attitude and purchase intention. International Journal of Islamic and Middle Eastern Finance and Management. Ahead-of-print (2020). https://doi.org/10.1108/IMEFM-02-2019-0064

Varma, I.G., et al.: Impact of spirituality on the conspicuous consumption of fashion consumers of generation Z: moderating role of dispositional positive emotions. Int. J. Emerg. Markets (2022). https://doi.org/10.1108/IJOEM-01-2022-0159

Vitell, S., Ramos-Hidalgo, E., Rodríguez-Rad, C.: A Spanish perspective on the impact on religiosity and spirituality on consumer ethics. Int. J. Consum. Stud. **42**, 675–686 (2018). https://doi.org/10.1111/ijcs.12438

Wang, C.L., Sarkar, A., Sarkar, J.G.: Building the holy brand: towards a theoretical model of brand religiosity. Int. J. Consum. Stud. **42**, 736–743 (2018). https://doi.org/10.1111/ijcs.12430

Yeniaras, V., Akkemik, K.A.: Materialism and fashion consciousness: the moderating role of status consumption tendencies and religiosity. J. Sci. Study Relig. **56**, 498–513 (2017). https://doi.org/10.1111/jssr.12363

Zainal, N.N.B., Yen-Neem G., Ahmadm T.B.: The Effects of Islamic fashion retail store attributes on customer satisfaction and patronage behaviour in Malaysia: a preliminary study. Global Bus. Manage. Res. Int. J. **14**(3), 133–147 (2022). http://www.gbmrjournal.com/pdf/v14n3s/V14N3s-10.pdf

Zainudin, M.I.B., et al.: Halal brand personality and brand loyalty among millenials modest fashion consumers in Malaysia: a conceptual paper. Int. J. Asian Soc. Sci. **8**(11), 985–994 (2018). https://doi.org/10.18488/journal.1.2018.811.985.994

Zehra, S., Minton, E.: Should businesses use religious cues in advertising? A comparison of consumer perceptions across Christianity and Islam. Int. J. Consum. Stud. **44**, 393–406 (2020). https://doi.org/10.1111/ijcs.12512

The Values Perceived by Fashion Consumers in Luxury Brands

Madalena Barata[1(✉)] and Paulo Duarte[2]

[1] Universidade da Beira Interior, Covilhã, Portugal
madalenaba@gmail.com
[2] Universidade da Beira Interior, NECE – Research Center for Business Sciences, Av. Marquês D'Ávila e Bolama, 6201-001 Covilhã, Portugal
pduarte@ubi.pt

Abstract. This study analyses the values perceived by consumers of luxury fashion before the purchase option based on a literature review. To frame this subject, the study begins by exploring the concept of luxury from the perspective of authors with works of reference in the area to offer distinct or complementary approaches to each other. It is followed by an analysis of the luxury fashion target audience through the analysis of its main characteristics, whose focus is based on three consumer groups: millennials, Generation Z, and Generation X. Finally, this study develops a flowchart based on luxury fashion literature and created from several theoretical approaches, thus contributing to the comprehension of consumer motivations and perceptions of value in luxury fashion.

Keywords: Luxury fashion · Generation X · Generation Z · Millennials

1 Introduction

The luxury market is very wide and competitive, and its target audience has been changing as it has acquired specific characteristics, becoming increasingly well-informed and with access to more channels and resources. Therefore, luxury brands must understand and adapt to the changes by focusing on studies that bring them closer to customers and defining strategies more consistent with new behavioral patterns of different generational segments.

For this purpose, a literature review was conducted using as criteria for identifying relevant articles the keywords: consumer behavior, luxury fashion, Millennials, Generation Z, and Generation X, allowing us to confront the points of view of several authors.

This study is structured in five sections, starting with the definition of luxury fashion and its main characteristics, continuing by analyzing how generations born in different years behave at the time of purchase, and finishing with the values perceived by this consumer during the buying process.

The conclusions of this generational analysis in this work allow different audiences to decide the direction brands should take in strategic terms of adequacy to the specificities of their audience. Future research can use the findings in this study to contextualize research questions and findings regarding the values of luxury products across generations.

2 Concept and Value Proposition of Luxury

Despite suggesting a recent phenomenon, luxury consumption has always existed. According to Allérès (2000, p. 100), "In all eras, a class or an elite of the population has indulged in the pleasures of luxury, for religious, tribal or exclusively profane purposes". Luxury was primarily responsible for demarcating an unbridgeable boundary between the privileged classes and the rest of the population.

"Luxury" has become a commonly used word, encompassing various products, services, and even a specific way of life; Despite being an ambiguous concept with a still unclear definition, it takes different forms according to different types of people and depends on the consumer's state of mind and experience (Wiedmann and Henigs, 2013). Table 1 summarizes several authors' perspectives on luxury fashion brands' main characteristics.

Table 1 shows that most authors agree and associate luxury fashion in terms of premium prices and exceptional quality. The exclusive character is also considered extremely important in this market. The feeling of high status and belonging to a select or exclusive group in society, who can buy certain expensive goods, is another factor associated with luxury brands highlighted by several authors.

Thus, it is observed that luxury brands share certain characteristics, namely (Nueno and Uelch, 1998): the consistent delivery of superior quality in all product lines, the heritage of handmade/craftsmanship, a distinguishable style or design, a limited edition that ensures exclusiveness, associated with marketing strategies that combine an emotional appeal to a superior product of global reputation. In some cases, also the association with a home country that has a solid reputation in a specific industry (e.g., Italian fashion and footwear or Swiss watches). Therefore, luxury fashion consumers are assumed to buy more than products or services. They buy a complete service of experiences, emotions, and identities formed by the product, the service, and the characteristics associated with the brand.

Sheth, Newman, and Gross (1991) describe the functional aspect of luxury as being connected with essential benefits and fundamental utilities, such as usability, exclusivity, quality, reliability, and durability. For Park, Jaworski, and McInnis (1986), usability can be assessed based on the ease of use and can be recognized through the physical, technical, concrete, or abstract features of the product or service, which may be based on product properties as well as on consumer needs. The functions, appearance, durability, and performance of the product consumers purchase are all things they expect from it (Fennel, 1978).

Gentry et al. (2001) suggest that consumers are drawn to purchase luxury brands because of the superior quality presented in the brand name, which is seen as an indispensable attribute of any luxury brand (Quelch, 1987; Garfein, 1989; Roux, 1995).

Table 1. Luxury brands' key characteristics

Authors	Characteristics of luxury brands
Barnier et al. (2012)	Exceptional quality; Hedonism; High price; Selective distribution; Personalized service; Exclusive character
Jackson (2004)	Exclusivity; Premium prices; Image and status
Kapferer (2012)	Innovation and culture; Quality of craftsmanship; Symbols; Glamour; High price indiscretion
Sjostrom et al. (2016)	Exceptional quality; Aesthetic uniqueness; Rarity and symbolism; Long history and reputation; High price level
Wordiq (2004)	Product positioned at the top of the market in quality and price
Vigernon and Johnson (2004)	Abstract concept; Meaning determined by interpersonal and personal motives; Built on consumer perception
Phau and Prendergast, (2000) Vikers and Renand, (2003)	Well-structured brand identity; High brand perception; High sales levels; Perceived quality; Customer loyalty
Mansharamani and Khanna (2013)	A mixture of product attributes with additional psychological factors added, which are perceived as benefits by the consumer as they happen with added value, esteem, and prestige
Okonkwo (2009)	Innovation; Creativity; Unique and attractive products; Consistent delivery of premium quality; Exclusivity in the production of goods; Firm distribution control; Heritage of craftsmanship; Distinctive brand identity; Global reputation; Emotional appeal; Premium price; High visibility

This is congruent with the perception that luxury brands offer higher product quality and a greater assurance of safety (Garfein, 1989; Roux, 1995; Quelch, 1987; Nia and Zaichkowsky, 2000; O'Cass and Frost, 2002; Vigneron and Johnson, 2004).

Uniqueness is another crucial aspect to consider. It is believed that the perceived exclusivity and rarity of the product enhance consumer appeal or preference (Verhallen, 1982; Lynn, 1991; Pantzalis, 1995). Additionally, the desire for a brand increases in tandem with its price (Groth and McDaniel, 1993). Therefore, the more exclusive and expensive a brand is, the more valuable it becomes (Verhallen and Robben, 1994). Moreover, the functional value of uniqueness reinforces individuals' need for exclusivity (Snyder and Fromkin, 1977). Similarly, individuals' desire for differentiation and individuality can solely be fulfilled by consuming and using a particular brand that is provided to an exclusive niche of clients (Leibenstein, 1950; Vigneron and Johnson, 1999, 2004).

The individual dimension of luxury value centers around the customer's positioning towards luxury consumption and addresses factors like materialism (Richins and Dawson, 1992), hedonism, and self-identity (Vigneron and Johnson, 2004; Hirschman and

Holbrook, 1982). Materialism refers to the extent to which individuals perceive products as playing a central role in their lives (Chang and Arkin, 2002). The more materialistic consumers are, the more likely they are to hold positive attitudes towards acquiring and possessing material goods. Highly materialistic individuals consider products extremely desirable and have the tendency to invest more time and energy in activities related to buying and using them (Belk, 1985). Consequently, the use of luxury goods allows consumers to incorporate symbolic meaning into their own identity (Holt, 1995; Vigneron and Johnson, 2004). Brands may also serve as a means for consumers to express and develop their self-identity (Hirschman, 1988; Dittmar, 1994), since certain products and services hold emotional value beyond their functional utility (Hirschman and Holbrook, 1982; Sheth, Newman, and Gross, 1991, Westbrook and Oliver, 1991). As a result, luxury products are likely to create certain subjective, intangible benefits (Dubois and Laurent, 1994).

Research has revealed that luxury consumption is associated with emotional responses, such as sensory pleasure, aesthetic beauty, and excitement (Benarrosh-Dahan, 1991; Fauchois and Krieg, 1991; Roux and Floch, 1996; Vigneron and Johnson, 2004). This pursuit of pleasure, referred to as hedonism, involves the personal satisfaction and inherent appeal derived from acquiring and using luxury brands, evoking feelings and emotional states related to personal rewarding, accomplishment, or social recognition (Sheth, Newman, and Gross, 1991; Westbrook and Oliver, 1991). The social aspect of luxury value revolves around noticeable and prestigious value within social groups, which can impact the assessment and tendency to buy or use luxury brands (Vigneron and Johnson, 1999; Bearden and Etzel, 1982; Brinberg and Plimpton, 1986). Bearden and Etzel (1982) propose that recognizable products tend to be consumed publicly, while recognizable luxury brands are more likely to be consumed privately. Recognizable consumption continues to play a significant role in determining preferences in many cases, particularly for luxury brands (Braun and Wicklund, Vigneron and Johnson, 2004). Thus, the noticeability of luxury brands may have relevance to individuals seeking status and social representation and play an important role in purchasing.

3 Generation-Based Luxury Brands Consumption: Generation X, Millennials, and Generation Z

With the rapid expansion of luxury brands, the behavior of the luxury consumer has undergone significant changes. Today, the global luxury market is ultra-competitive, experiencing exponential growth in recent decades, and is expected to reach a value of €320–350 billion by 2025 (Bain and Company, 2018).

Generation X, also known as "baby busters", is a generation that holds less significance in the luxury consumption market. Born between 1961 and 1980 (Howe and Strauss, 2007), Generation X individuals are characterized as savvy and cynical shoppers who rely on the Internet to reinforce their opinions about products and brands (Solomon 2010). This consumer segment consists of approximately 30 million people across Europe spanning various fields like fashion, politics, marketing, and culture. Mitchell et al. (2015) emphasize that members of Generation X are practical and skeptical of misleading advertising, demonstrating caution when it comes to 'marketing'.

Foulkes (2016) notes that Generation X tends to be brand-oriented, often purchasing products that contribute to a youthful appearance and exhibiting higher brand loyalty.

According to a study carried out by True-Luxury Global Consumer Insight 2021, Millennials (born between the early 1980s and mid-1990s) and Gen Z (born after the second half of the 1990s) are particularly relevant for the market, as they are expected to represent more than 60% of Global Personal Luxury Market by 2025, with an estimated market value ranging from 235 to 265 billion euros. These new generations prove to be very influential on the overall market, thanks to their digital engagement creation and trendsetting ability. Millennials share common values, beliefs, and experiences (Howe and Strauss, 2009). They are eager to exert social influence (Butcher, Phau, and Shimul, 2017), which impacts their purchase intentions of luxury products (Soh, Rezaei, and Gu, 2017).

Several studies show that millennials are more sophisticated buyers, influenced by the symbolic aspects of luxury brands (Shin et al. 2017), and more motivated to consume for status than older customers (Eastman and Liu, 2012). Although the millennial market segment is particularly attractive, they pose a challenge for marketers because it is difficult to create a strong emotional and psychological connection with them (Adkins, 2016) since this generation disconnects very easily from brands compared to older generations. They use digital media both to expose their dissatisfaction with a brand and to praise it. They are knowledgeable about materials and their sources. They travel a lot and are interculturally competent.

Millennials develop an independent vision throughout their adolescence, and as they reach adulthood, they become permeable to new experiences (Ryff, 1989). As such, they seek new opportunities that allow them to enrich their knowledge and bring new emotions, thus expressing all facets of their personality (Mattingly and Lewandowski, 2014), namely through investment in the experience associated with the brands they own (Mundel et al., 2017).

Generation Z, classified as the digital native generation, represents only 4% of the luxury consumer market, and they are entering that market with a very different way of consuming than their elders. Thus, the luxury industry is interested in preparing for these future consumers by knowing more about their behavior and values, explains (Pianon (2016). Among these young consumers, 95% interact with brands and their products almost exclusively through the Internet and social media. They appear as the segment that is most attracted to collaborations created by brands and the purchase of second-hand products.

According to the True Luxury Consumer Insight Survey (2019), generation Z has a unique set of behavior and values that brands must begin to understand better:

- They buy a greater number of items from collaborations with other brands or artists.
- They attach importance to the resale value when buying luxury goods.
- They are influenced by sustainability when shopping.
- They use social media to interact with luxury brands, bloggers, or digital influencers.

According to the comparison presented in Table 2, the three generations present significant differences. However, quality is a major concern for all consumers, as well as cost, which emerges as a factor that cuts across all generations. The younger generations are concerned with the fun and entertainment factor, while the older ones maintain a

Table 2. Top five value elements by generation (Generation X, Millennials, and Generation Z) *(Bain and Company, 2021)*

Generation x	Millennials	Generation z
Quality	Quality/Fun and entertainment	Fun/Entertainment
Design and aesthetics	Reducing the cost	Quality/self-updating
Reducing the cost	Reduce anxiety/uninterrupted experience	Design and aesthetics/Reduce cost
Heritage	Self-updating	Reduce anxiety
Fun and entertainment/ relic	Design and aesthetics	Self-transcendence

priority over heritage. It can be concluded that older consumers prioritize more traditional values such as heritage. However, self-actualization is also important, trying to keep up with trends and what is currently being worn. Millennials continue to care about quality and value the emotional experience and inspiration brands can give them. For Generation Z, fun and entertainment are key, along with achieving self-fulfillment.

4 Luxury Brand Index

Vigneron and Johnson (2004) create a "luxury brand index" to assess luxury brands. They propose that the decision-making process of luxury consumers' can be described by five key factors forming a semantic network. These factors include personal perceptions, (1) perception of the extended self, (2) perception of hedonism and the more usual non-personal perceptions (3) perception of conspicuousness, (4) uniqueness, and (5) quality. Additionally, what adds luxury value to consumer perception is based on four latent dimensions: financial, functional, individual, and social (Wiedmann, Hennigs, and Siebels, 2007). Although these four dimensions operate independently, they work together and influence the perceptions and behaviors of individuals with luxury preferences. These behaviors can be used to differentiate and segment luxury consumers.

The model presented in Fig. 1 is based on the analysis of luxury research and was created from various theoretical approaches, which improves the understanding of consumer motives and value perception in luxury consumption.

Figure 1 illustrates distinct but correlated dimensions of value. Each dimension contributes to the creation of a luxury value that consumers inevitably recognize when making a purchase. The financial dimension of luxury value pertains to direct financial aspects like price, resale value, price reductions and discounts, or investment potential. It involves both the intrinsic and extrinsic value of the product and the sacrifices required to obtain it (Ahtola, 1984; Monroe and Krishnan, 1985). Status-conscious consumers often consider price as an indicator of prestige (Berkowitz et al., 1992; Groth and McDaniel, 1993). Thus, as premium prices imply high quality or status, products or services with these prices may become more desirable (Groth and McDaniel, 1993). However, it is important to note that a product or service does not necessarily have to be expensive to be considered a luxury good, nor is it automatically luxurious solely based on its price.

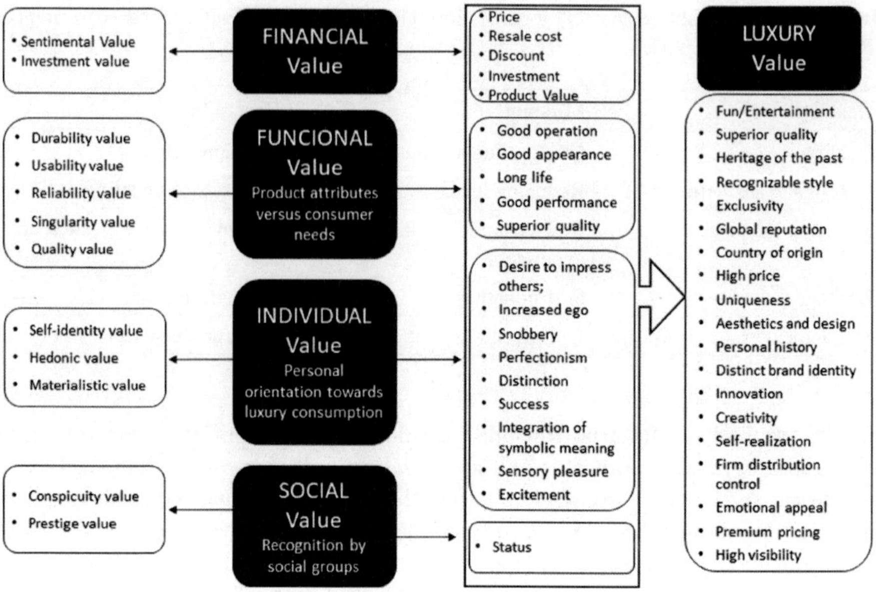

Fig. 1. Brand luxury index. Analysis of Consumers' Perception of Values (Adapted from Vigneron and Johnson, 2004, p. 5).

5 Conclusion

Luxury fashion has been the subject of extensive research, but the conclusions are ambiguous, and the various authors' perspectives vary according to the consumer's experience. The main characteristics of luxury brands coined by different authors argue that exceptional quality, high price, value for money, exclusivity, and innovation. Luxury fashion consumers buy more than products or services. They buy a full service of experiences, emotions, and identities formed by the product and characteristics associated with the brand.

Due to the rapid expansion of luxury brands, consumer behavior has changed, and the main consumer groups may vary according to individual preferences as they evolve: Generation X, known for having an age group that values luxury fashion, individuals of this generation have financial stability and have the willingness to buy these items. They tend to look for established brands and exclusive products, although they gather information before purchasing. Generation Y is characterized by including more sophisticated buyers, influenced by the symbolic aspects of the brands, and uses digital media to praise or show dissatisfaction with a product. Although they have less purchasing power than the previous generation, they use luxury fashion for personal expression and social status. They value authenticity, exclusivity, and the brand's social responsibility. The Z generation is attracted by the associated values, namely, sustainability and collaborations with brands or artists. Social media highly influence the youngest generation, which tends to look for brands that offer a unique and authentic style, showing a strong interest in luxury brands.

The present analysis of consumer behavior, the main values associated with luxury brands, and the characteristics of each generation allow researchers to better frame future studies. On the commercial side, allow companies to define more adjusted marketing actions based on creating and preserving the most important values for each generation and more appropriate marketing strategies to stimulate the purchase and make these consumers more loyal to the brand.

References

Adkins, A.: Brands aren't winning millennial consumers. Gallup Bus. J. (2016). https://news.gallup.com/businessjournal/192710/brands-aren-winning-millennial-consumers.aspx

Ahtola, O.T.: Price as a "give" component in an exchange theoretic multi-component model. Adv. Consum. Res. **11**, 623–636 (1984)

Danielle, A.: Luxo: Estratégias, marketing. Rio de Janeiro: Editora FGV, p. 262 (2000). 852250329X, 9788522503292

Bearden, W.O., Etzel, M.J.: Reference group influence on product and brand purchase decisions. J. Consum. Res. **9**, 183–194 (1982)

Belk, R.W.: Materialism: traits aspects of living in the material world. J. Consum. Res. **12**, 265–280 (1985)

Berkowitz, E.N., Kerin, R.A., Hartley, S.W., Rudelius, W.: Marketing, 3rd ed.Homewood, IL: Irwin (1992)

Brinberg, D., Plimpton, L.: Self-monitoring and product conspicuousness on reference group influence. In R. J. Lutz (Ed.), Advances in consumer research, Vol. 13. p. 297–300). Provo, UT: Association for Consumer Research (1986)

Butcher, L., Phau, I., Shimul, A.: Uniqueness and status consumption in Generation Y consumers: does moderation exist? Mark. Intell. Plan. **35**, 673–687 (2017). https://doi.org/10.1108/MIP-12-2016-0216

Chang, L., Arkin, R.: Materialism and an attempt to cope with uncertainty. Psychol. Mark. **19**, 389–406 (2002). https://doi.org/10.1002/mar.10016

Dittmar, H.: Material possessions as stereotypes: material images of different socioeconomic groups. J. Econ. Psychol. **15**(4), 561–585 (1994)

Dubois, B., Laurent, G.:. Attitudes toward the concept of luxury: An exploratory analysis. In S. Leong and J. Cote (Eds.), Asia Pacific advances in consumer research, vol 1, pp. 273–278. Provo, UT: Association for Consumer Research (1994)

Eastman, J., Liu, J.: The impact of generational cohorts on status consumption: an exploratory look at generational cohort and demographics on status consumption. J. Consum. Mark. **29**, 93–102 (2012). https://doi.org/10.1108/07363761211206348

Abby, F.:. A comparison of luxury perceptions of Generation X and Y consumers in the United Kingdom. Thesis for BSC (Hons) Business Management (2016). https://doi.org/10.13140/RG.2.2.35857.22888

Garfein, R.T.: Cross-cultural perspectives on the dynamics of prestige. J. Serv. Mark. **3**, 17–24 (1989)

Gentry, J.W., Putrevu, S., Shultz, C., Commuri, S.: How now Ralph Lauren? The separation of brand and product in a counterfeit culture. Adv. Consum. Res. **28**, 258–265 (2001)

Groth, J., McDaniel, S.W.: The exclusive value principle: the basis for prestige pricing. J. Consum. Mark. **10**, 10–16 (1993). https://doi.org/10.1108/07363769310026539

Hirschman, E.C., Holbrook, M.B.: Hedonic consumption: emerging concepts, methods and propositions. J. Market. **22**, 1–16 (1982). https://doi.org/10.1086/209431

Howe, N., Strauss, W.: Millennials Rising. The Next Great Generation. Vintage Books, New York (2009)

Howe, N., Strauss, W.: Millennials go to college. Great Falls, Va.: LifeCourse Associates (2007)

Jackson, T.B.: International retail marketing. Elsevier Butterworth Heinemann, Offorf (2004)

Kapferer, J.N.: Abundant rarity: the key to luxury growth. Bus. Horiz. **5**(5), 453–462 (2012)

Leibenstein, H.: Bandwagon, snob, and Veblen effects in the theory of consumers' demand. Quart. J. Econ. **64**, 183–207 (1950)

Lynn, M.: Scarcity effects on value: a quantitative review of the commodity theory literature. Psychol. Mark. **8**, 45–57 (1991). https://doi.org/10.1002/mar.4220080105

Mansharamani, A., Khanna, S.: Marketing of Luxury Brands, Faculty of Management Studies, North Campus, FMS Delhi (2013). http://www.brandchannel.com/images/papers/297_luxury_branding_india.pdf. Accessed 11 Feb 2014

Mattingly, B.A., Lewandowski, G.W.: Expanding the self brick by brick. Soc. Psychol. Pers. Sci. **5**(4), 484–490 (2014). https://doi.org/10.1177/1948550613503886

Mitchell, V., Bakewell, C., Jackson, P., Heslin, C.: How message framing affects consumer attitudes in food crises. Br. Food J. **117**(8), 2200–2211 (2015). https://doi.org/10.1108/BFJ-01-2015-0015

Monroe, K.B., Krishnan, R.: The effect of price on subjective product evaluations. In: Jacoby, J., Olsen, J. (eds.), Perceived Quality, pp. 209–232. Lexington, MA:Lexington Books (1985)

Mundel, J., Huddleston, P., Vodermeier, M.: An exploratory study of consumers' perceptions: what are affordable luxuries? J. Retail. Consum. Serv. **35**(March), 68–75 (2017)

Nia, A., Zaichkowsky, J.L.: Do counterfeits devalue the ownership of luxury brands? J. Prod. Brand Manage. **9**(7), 485–497 (2000). https://doi.org/10.1108/10610420010351402

O'Cass, A., Frost, H.: Status brands: Examining the effects of non-product brand associations (2002)

Okonkwo, U.: Luxury fashion branding – trends, tactics, techniques. J. Brand Manage. 413–415 (2009). https://doi.org/10.1057/bm.2008.34

Phau, I., Prendergast, G.: Consuming luxury brands: the relevance of the rarity principle. J. Brand Manage. **8**(2), 122–138 (2000). https://doi.org/10.1057/palgrave.bm.2540013

Pantzalis, I.: Exclusivity strategies in pricing and brand extension. Unpublished doctoral dissertation, University of Arizona, Tucson, AZ (1995)

Park, C.W., Jaworski, B.J., McInnis, D.J.: Strategic brand concept-image management. J. Mark. **50**, 135–145 (1986)

Quelch, J.A.: Marketing the premium product. Bus. Horiz. **30**, 38–45 (1987)

Richins, M., Dawson, S.: A consumer values orientation for materialism and its measurement. J. Consum. Res. **19**, 303–316 (1992). https://doi.org/10.1086/209304

Ryff, C.D.: Happiness is everything, or is it? J. Pers. Soc. Psychol. **57**(6), 1069–1081 (1989). https://doi.org/10.1037/0022-3514.57.6.1069

Roux, E.: Consumer evaluation of luxury brand extensions. EMAC Conference, May,ESSEC, Paris, France (1995)

Shin, H., Eastman, J.K., Mothersbaugh, D.: The effect of a limited-edition offer following brand dilution on consumer attitudes toward a luxury brand. J. Retail. Consum. Serv. **38**, 59–70 (2017)

Sheth, J.N., Newman, B.I., Gross, B.L.: Why we buy what we buy: a theory of consumption values. J. Bus. Res. **22**(2), 159–170 (1991). https://doi.org/10.1016/0148-2963(91)90050-8

Solomon, M.R.: Consumer Behavior: Buying, Having, and Being, 9th edn. Prentice Hall (2010)

Snyder, C.R., Fromkin, H.L.: Abnormality as a positive characteristic: the development and validation of a scale measuring need for uniqueness. J. Abnorm. Psychol. **86**, 518–527 (1977). https://doi.org/10.1037/0021-843X.86.5.518

Verhallen, T.M.: Scarcity and consumer choice behavior. J. Econ. Psychol. **8**(4), 299–321 (1982)

Verhallen, T., Robben, H.: Scarcity and preference: an experiment on unavailability and product evaluation. J. Econ. Psychol. **15**(2), 315–331 (1994). https://doi.org/10.1016/0167-4870(94)90007-8

Vigneron, F., Johnson, L.W.: A review and a conceptual framework of prestige-seeking consumer behavior. Acad. Market. Sci. Rev. **2**(1), 15 (1999)

Vigneron, F., Johnson, L.W.: Measuring perceptions of brand luxury. J. Brand Manag. **11**(6), 484–506 (2004)

Westbrook, R.A., Oliver, R.L.: The dimensionality of consumption emotion patterns and consumer satisfaction. J. Consum. Res. **18**(1), 84–91 (1991). https://doi.org/10.1086/209243

Wiedmann, K.-P., Hennigs, N., Siebels, A.: Measuring consumers' luxury value perception: a cross-cultural framework. Acad. Mark. Sci. **7**(7) (2007)

The Ghanaian Wedding Industry: Review of Factors that Influences the Consumption of a Bridal Gown

Haruna Ibrahim[1], Mónica Romãozinho[2(✉)], and Fernando Moreira da Silva[3]

[1] Universidade da Beira Interior, R. Marquês de Ávila e Bolama, 6201-001 Covilhã, Portugal
[2] Labcom-UBI, Universidade da Beira Interior, Rua Marquês de Ávila e Bolama, 6201-001 Covilhã, Portugal
monica.romaoz@gmail.com
[3] Universidade de Lisboa, Rua Sá Nogueira, Polo Universitário do Alto da Ajuda, 1349-063 Lisbon, Portugal

Abstract. The bridal gown is one of the most highly symbolic objects in the contemporary wedding market that is all-encompasses with key issues influencing consumer behaviour. The study reviewed factors that influences the consumption of bridal gowns in Ghana. A quantitative research approach and analysis of results is employed as depicted in my Ph.D. thesis. This study is an extract of the explorative phase of my Ph.D. thesis that extensively reviewed related literature on practices and consumption of bridal gowns to assess and affirm the current evidence that pertains in designing for longevity in the bridal gown industry. The population for the study is 300 married and yet-to-be married couples. Convenient sampling technique was used to select 230 out of 300 representing 76.7% response rate. The study concluded that influencing factors were apparent and deciding factors that drive consumers' consumption of bridal gown in Ghana. This implying that, manufacturers and marketers could introduce product differentiation using these influencing factors. This sociological approach was fundamental for understanding Ghanaian women opinion and needs to identify sustainable strategies that fashion designers can considerer when designing a bridal gown for this specific territory with strong traditions.

Keywords: Bridal gown · Consumption · Design · Ghanaian culture

1 Introduction

Fashion in the light of globalization, is a $1.75 trillion industry with the sales of many major fashion brands growing rapidly worldwide of which bridal gown industry is not an exception. Bridal gowns are more than simply just a style. According to (Cecil, 2014), the modern study of weddings and bridal gown and it associated influences in the modern society in terms of consumption was first depicted in Ingraham's White Weddings "Romancing Heterosexuality in Pop Culture" in 1999. His study was more about how the acceptable parameters of wedding ceremonies are established in popular culture through wedding themes that permeate entertainment and advertising. Ingraham also points to the mind-boggling revenue generated by the wedding industry ranging from gown, diamond, and wedding décor sub-industries. In a similar vein, Kalmijn (2004) confirms that the expression of happiness that newly wedding couples and families feel about the marriage affect consumption plan by socializing the bride and groom into their marriage in an elaborate fashion.

In Ghana, one can find fashion in bridal gown, either through the style of dress, type of fabric, or style of accessories (Alabi, 2020). Brides are not only wearing world fashions, but they are also actively revising world fashions to suite their own tastes and preferences. One such example was featured in the 1956 article from the Sunday Mirror "New Styles in Frocks for Brides". As Takyiwa described, "in a society wedding in Accra, the bride sprang a welcome surprise on fashion-mongers when she turned up in a short white wedding gown. The dressmaker used 65 yards of embroidery in designing the gown (Takyiwa, 1956: 4). As Takyiwa asserted, "A discriminating bride and a competent dressmaker combined to give the fashionable white wedding gown".

The prominence given to world's fashion by brides suggests that they directly influenced by numerous factors encouraging brides to emulate the styles of world fashion that is distant from Ghanaian traditions. There are different styles of bridal gown made to embody the any figure type of Ghanaian woman; thus, there is a lot of fashionable bridal gown to choose from as a Ghanaian woman who is getting married. The fluidity of the Ghanaian wedding gown fabric makes the voluptuous Ghanaian bride a sight to behold (Alabi, 2020). While the bride is the main view, the wedding gowns are made to match the bride.

Ghana is a third world country faced with a very low socio-economic problem, many families live in poor conditions and fashion solutions should be more sustainable, versatile and connected to their culture and economic conditions.

This study is an extract of my Ph.D. thesis as earlier on indicated and the thesis seeks to contribute knowledge in responsible consumption and production in the Fashion Design field, as stipulated in the seventeen Sustainable Development Goals formulated by United Nations (Agenda, 2030). The primary research question is: how can the involvement (co-design) of the wearers of clothes, particularly in the bridal gown conception be a source of design thinking techniques that will influence the multiplicity of usage?

2 Literature Review

The business of fashion and luxury goods and services has always been on the sport light in consumer marketing. According to (Taylor & Costello, 2017) recent years have seen considerable increase in the desire of customers to acquire luxury goods, including those that can be described as fashion products. One important motivating force that influences a wide range of consumer behavior towards the selection and buying of bridal gowns is the desire to gain status or social prestige from the acquisition and consumption. More often not, yet-to be couples purchase expensive and luxury goods and services to display their social standing during their ceremony. In addition, individuals purchase status products to fulfil their material desires and reinforce their group identity (Madinga, Maziriri, &

Lose, 2016). Bridal gown and its consumption can be influenced by psychological, personal, economic, social and cultural factors.

2.1 Influence of Psychological Factors in Bridal Consumption

Bridal gown consumption is influenced by psychological factors such as motivation, perception, learning, beliefs, and attitudes (Guan, Luoi & Tangi, 2015). Individuals have many needs at any given point in time. These needs can be physical or psycho-logical. The need for recognition, esteem, and belongingness are examples of psy-chological needs. Madinga, Maziriri and Lose (2016) mentioned that the perception, feelings, beliefs, and attitudes influence a person consumption of bridal gown.

2.2 Influence of Personal Factors in Bridal Consumption

Personal factors are psychological characteristics of a person that are different from others that cause relatively consistent and enduring responses to the environment. Consumers buying decisions are also influenced by personal characteristics, namely age and life cycle stage, occupation, economic situation, lifestyle, personality, and self-concept. Min, Ceballos and Yurchisin (2018) viewed that personal factors are influential in the consumption of wedding gown. According to Santosa (2021) the preferences of the brides on the use of bridal gown is influenced by their characteristics such as age, occupation, lifestyle, and personality.

2.3 Influence of Economic Factors in Bridal Consumption

The buying tendency of an individual is directly proportional to his income/earnings per month. How much an individual brings home decides how much he spends and on which products? As indicated by Semesta, Fahmi and Jahroh (2020), individuals with high income would buy expensive and premium products as compared to individuals from middle- and lower-income group who would spend mostly on necessary items. It is hardly finding an individual from a low-income group spending money on designer bridal gown. The bride would be more interested in buying a gown necessary for her survival.

2.4 Influence of Social Factors in Bridal Consumption

Consumer behaviour is largely influenced by social factors, such as family, reference groups, and social roles and status. Reference groups expose individuals to new behaviours and lifestyles and shape their attitudes and self-concept (Kotler & Keller, 2016). This formation of attitude and self-concept is triggered by direct or indirect points of comparison or reference. The task of marketers, in this case, is to identify the reference groups for their target markets. The importance of reference groups varies across products and services.

2.5 Influence of Cultural Factors in Bridal Consumption

Culture plays an important role in the decision making of consumers. Every individual consumer belongs to a social group and such have the potentials in influencing consumption of bridal gown. Expressing his concerns about Weddings and cultural related issues and their impact on consumption (Bell 1997) cemented his thoughts by saying that wedding consumptions is an unchanging ritual where individuals believe they must transmit and respond to local traditions in the past. Boden (2003) stated that family traditions and culture is an influential factor in the consumption of wedding goods and services. Western fashion has influenced the evolution of the bridal replacing the traditional dresses mostly after the Independence of the country (1957), such as the *abosoɔ*.

3 Methodology

The study employed quantitative methods and utilized primary data. The primary data was collected through field survey and relied on google form of questionnaire as a data collection instrument (Creswell, 2003). Taking into consideration other constraints (such as financial and difficulty in selecting all married couples and yet to be coupled) to answer the research question, 300 married couples and yet to be coupled in the Kumasi Metropolis of Ghana were conveniently selected. All the married couples and yet to be coupled in Kumasi Metropolis were used as the sample frame.

The main instrument used was a structured questionnaire *sent to the e-mail and WhatsApp number of the sampled respondents and a total of 230 out of 300 were received attaining 76.7% response rate. I*tems on respondents' demography includes gender, age, educational level, and marital status of the respondents. A 7-point Likert Scale was also used ranging from strongly disagree (1) to strongly agree (7) to solicit responses on the factors influencing the consumption of bridal gown. *The study discovered psychological factors (mean = 4.5884); personal factors" (mean = 5.0925); economic factors (mean = 4.6243); and cultural factors (mean = 4.4211) influences the consumption of bridal gown. The finding showed psychological factors ($F = 20.714, p < 0.01$); personal factors ($F = 47.594, p < 0.01$); economic factors ($F = 17.247, p < 0.01$); social factors ($r = 39.298, p < 0.01$); cultural factors ($F = 36.052, p < 0.01$) have statistically significant relationship with bridal gown consumption.*

The data collected was refined, coded, and fed into SPSS version 21 for both descriptive and inferential data analysis. Frequencies, Percentages, Means, and Standard deviations were used to summarize the data. A Pearson correlation and regression analysis were performed to ascertain the relationship between influencing factors, and consumption of bridal gown at a 5% level of significance.

4 Results

4.1 Descriptive Statistics of the Survey Participants

This section of the paper presents the result and analysis of the data obtained from the questionnaires distributed. Table 1 presents the background information of the respondents.

Table 1. Descriptive statistics of the survey participants.

Characteristics	Frequency (%)
Gender	
Male	43 (18.7)
Female	187 (81.3)
Age (years)	
Below 20 years	28 (12.2)
20–29 years	128 (55.7)
30–39 years	55 (23.9)
40–49 years	15 (6.5)
50–59 years	4 (1.7)
Educational Background	
No Formal Education	–
First degree	213 (92.6)
Masters degree	15 (6.5)
PhD	2 (0.9)

As shown in Table 1, 18.7% of the respondents were males while 81.3% were females with most of the respondents falling between the ages of 20–29 years. The least age group of the respondents included in the study was 50–59 years represented by 1.7%. A closer look at Table 1 reveals that almost 92.6% of the respondents have at least first-degree education with the rest attaining master's degree and PhD qualifications. Demographic information of the respondents was sought; the researcher deemed it necessary to look into demographic information of respondents because they make a person who he or she is.

4.2 Influencing Factors on Bridal Gown Consumption

The factors that influence consumers' consumption of bridal gown in Ghana were analyzed. The internal consistency of the questionnaire variables was initially assessed. In this study, Cronbach's alpha value was 0.849 indicating a higher internal consistency and the reliability of the questionnaire. In this section, respondents were asked to rate the factors that influence their consumption of bridal gown using a seven-point Likert scale. From Table 2, the influencing factors were grouped into psychological, personal, economic, social, and cultural factors.

From Table 2, five influencing factors presented to the respondents were reported to influence the consumption of bridal gown. It is interesting to note that, " Psychological factors" (mean = 4.5884, SD = 1.27397); "Personal factors" (mean = 5.0925, SD = 1.27303); "Economic factors" (mean = 4.6243, SD = 1.19211); "Social factors" (mean = 3.9671, SD = 1.40718); and "Cultural factors" (mean = 4.4211, SD = 1.63686) are

Table 2. Influencing factors and bridal gown consumption

Construct	No. of items	Mean	Std. Dev.	1	2	3	4	5	6
Psychological factors	5	4.5884	1.27397	1	.490**	.417**	.317**	.242**	.296**
Personal factors	5	5.0925	1.27303		1	.495**	.358**	.367**	.423**
Economic factors	5	4.6243	1.19211			1	.440**	.438**	.274**
Social factors	5	3.9671	1.40718				1	.616**	.392**
Cultural factors	5	4.4211	1.63686					1	.376**
Bridal Gown Consumption	4	4.0735	1.01448						1
R^2 Adjusted				.088	.179	.071	.149	.137	.206
F-Value				20.714	47.594	17.247	39.298	36.052	11.576

**. Correlation is significant at the 0.01 level (2-tailed).
Cell entries are the standardized beta coefficient

the five (5) most influencing factors. However, "bridal gown consumption" (mean = 4.0735, SD = 1.01448) had acceptable average score. All these factors met the predetermined cut-off point of 4.0 if the exception of social factors. According to the result, personal factors with the highest mean score are the dominant factors in influencing the consumption of bridal gown. It became evident that at a 5% level of significance, the mean ratings of consumer preference of cement brands were significant (p-value > 0.01). The result shows significant association between psychological factors ($r = 0.296$, $p < 0.01$); personal factors ($r = 0.423$, $p < 0.01$); economic factors ($r = 0.274$, $p < 0.01$); social factors ($r = 0.392$, $p < 0.01$); cultural factors ($r = 0.376$, $p < 0.01$), and bridal gown consumption. This is an indication that these influencing factors affect the consumption and selection criteria of bridal gown. The significant effect of the influencing factors (independent variable) and bridal gown consumption (dependent variables) were measured using a 7-point Likert scale (1 = strongly disagree to 7 = strongly agree). The explained variance for factors increased from 7.1 to 14.9% (Model 1, 2, 3, 4 and 5) with all variables except for Model 6 when it entered the equation. Thus, the consumption of bridal gown is largely influenced by these factors. Among the influencing factors, psychological factors ($F = 20.714$, $p < 0.01$); personal factors ($F = 47.594$, $p < 0.01$); economic factors ($F = 17.247$, $p < 0.01$); social factors ($r = 39.298$, $p < 0.01$); cultural factors ($F = 36.052$, $p < 0.01$) were found to be statistically significant on bridal gown consumption. Furthermore, the change in $R^2 = 20.6\%$ (Model 6) was statistically significant indicating that influencing factors were in and among themselves important factors that drive consumers' consumption of bridal gown in Ghana.

5 Discussion

The study aimed at contributing to knowledge of factors that influences the consumption of a bridal gown in Ghanaian perspective. Psychological factors, personal factors, economic factors, and cultural factors have considerable influence and provide a cue for the bridal gown consumption. The mean ratings for the brand awareness were between (Mean = 5.09 – 3.96). Furthermore, the results indicated that consumers of bridal gown are mostly influenced by personal factors such as age and life cycle stage, occupation, economic situation, lifestyle, personality, and self-concept. This is supported by Semesta, Fahmi and Jahroh (2020) who found that personal factors like occupation, age and life cycle stage, lifestyle, and personality play a major role in buying behavior of a consumer. The findings agree with other researchers who asserted that psychological, personal, economic, social and cultural are important factors considered by consumers when making purchasing decisions on bridal gown (Madinga et al., 2016; Boden, 2003). There was a substantial significant relationship between the influencing factors and consumption of bridal gown. This is in agreement with Santosa (2021), which established three factors as main determinants of wedding dress: psychological, cultural and social factors.

6 Conclusion and Recommendation

As a result of globalization, wedding ceremonies over the years has gone through tremendous changes, thus: from romantic event to commercialized event. In fact, Bridal gowns are more than simply just a style. In Ghana, one can find fashion in bridal gown, either through the style of dress, type of fabric, or style of accessories (Alabi, 2020). Brides are not only wearing world fashion; they are actively revising world fashions to suite their own tastes and preferences. The prominence given to world's fashion by brides suggests that they are directly influenced by numerous factors encouraging brides to emulate the styles of world fashion. According to the present study, psychological factors, personal factors, economic factors, and cultural factors have considerable influence and provide a cue for bridal gown consumption. Drawing from the result, one can conclude that psychological, personal, economic, social and cultural factors are influencing factors that have a direct relationship with the consumption of bridal gown in Ghana. It can also be concluded that understanding the roles and behavioural characteristics of consumers particularly married couples and yet-to be couples are very crucial and cannot be over emphases in the bridal market, wedding ceremonies over the years has gone through tremendous changes, thus: from romantic event to commercialized event. Weddings are normally perceived to be emotional event, but the narrative has change as a result of socio-economic activities such as income, industrialization educational and other preferences. Hence, manufacturers and marketers should focus their strategies on understanding consumers of all ages, gender, and educational backgrounds as their purchasing decisions will go a long way in influencing production and consumption of bridal gown. This approach is the first phase of the application of facilitation process, a practice of structuring and guiding dialogues, meetings, events, decision-making processes, and other activities using intentional strategies that help groups converse and collaborate more respectfully and productively (Hanson & Hanson, 2011).

References

Alabi, J.: Ghanaian traditional wedding dresses: different styles, brands, lengths, ages and how to wear them (2020). https://yen.com.gh/105773-ghanaian-traditional-wedding-dresses-different-styles-brands-lengths-ages-how-wear-them.html

Bell, C.: Ritual Perspective and Dimensions. Oxford University Press, New York (1997)

Boden, S.: Consumerism, Romance and the Wedding Experience. Palgrave, London (2003)

Cecil, M.: Empowered ambivalence: an analysis of feminist wedding planning (2014). http://mars.gmu.edu/jspui/bitstream/handle/1920/9027/Cecil_thesis_2014.pdf;jsessionid=9DF9315C2C4B9BD068DF8C3E5FA16784?sequence=1

Creswell, J.W.: Research Design: Qualitative, Quantitative, and Mixed Methods Approaches, 2nd edn. Sage, Thousand Oaks (2003)

Guan, L., Luoi, Y., Tangi, L.R.: An exploratory study of decision makers for choosing wedding banquet venues: push and pull motivations. Int. J. Tour. Cities **1**(2), 162–174 (2015)

Kalmijn, M.: Marriage rituals as reinforcers of role transitions: an analysis of weddings in the Netherlands. J. Marriage Fam. **66**(3), 582–594 (2004)

Kotler, P., Keller, K.L.: Marketing Management, 5th edn. Pearson Education South Asia Ltd., New Jersey (2016)

Madinga, N.W., Maziriri, E.T., Lose, T.: Exploring status consumption in South Africa: a literature review. Invest. Manag. Financ. Innov. **13**(3), 131–136 (2016)

Teaching and Education

Accessibility in Digital Games

Regina de Oliveira Heidrich(✉), Sheisa Amaral Bittencourt, Nathália Gomes Corrêa da Silva, Maria Eduarda Rotilli Eibs, and Emanuel Obino Wiest

Feevale University, Novo Hamburgo, RS 93525-075, Brazil
RHeidrich@feevale.br, sheisa.court@gmail.com, nathaliagomescsilva@gmail.com, mrotillieibs@gmail.com, 12e3qsdasd@gmail.com

Abstract. According to data extracted from the first report on disability and development carried out in 2018 by the UN, there are one billion people with disabilities in the world, which represents at least one-eighth of the world's population and, consequently, this represents a large portion of possible consumers of digital games that the big industry has overlooked. According to Prodanov and Freitas (2013) the methodology used in this work is qualitative and the design methodology is Scrum. This article proposes a methodology for analyzing accessibility in digital games using the development of the digital game Planeta ODS as a case study. The goal is to contribute to the inclusion of individuals who rely on accessibility features in digital games and highlight the fundamental criteria for promoting this inclusion.

Keywords: accessibility · digital games · evaluation criteria

1 Introduction

According to the World Report on Disability (2012), surveys conducted in over 100 countries revealed that disability is a universal challenge with social and economic costs for individuals, families, communities, and nations. The estimates suggest that there are approximately 785 million (15.6% according to the World Health Survey) to 975 million (19.4% according to the Global Burden of Disease) people aged 15 and over living with a disability, based on the 2010 population estimates of 6.9 billion inhabitants, with 1.86 billion people under 15 years of age. The same report (2012) indicates that among them, the World Health Survey estimates that 110 million people (2.2%) face significant functional difficulties, while the Global Burden of Disease estimates that 190 million people (3.8%) have "severe impairments" - equivalent to inferred impairments for conditions such as quadriplegia, severe depression, or blindness. Considering children, it is estimated that more than one billion people (or about 15% of the world's population) are living with a disability. Disability varies according to a complex combination of factors, including age, gender, life stage, exposure to environmental hazards, socioeconomic status, culture, and available resources, which significantly vary between regions. Thus, there are one billion people with disabilities in the world, representing at least one-eighth of the world's population and, consequently, a significant portion of potential consumers of digital games that the gaming industry has overlooked.

Bittencourt et al. (2020) state that, in the Latin American region, the impact on education until the beginning of 2021 and the ten-month closure of schools led to an increase from 55% to 71% of the population classified as having poor learning outcomes.

This article aims to contribute to addressing the learning gap through gamification by developing an accessible digital game for the web, where students in Brazil can learn about the UN's sustainable development goals. It also provides teachers with materials to incorporate the content of these goals into the classroom, promoting interdisciplinary connections. Whether in face-to-face or remote settings, the game is structured in chapters. The initial phase conceptualizes the problem, while subsequent phases, released periodically, delve into each of the 17 UN sustainable development goals in detail, according to Fig. 1. At times, multiple objectives are present within the same phase, as they often complement each other and facilitate discussions on the topic. This methodology allows for the game's launch well before completion. We also hope that the game provides support for teachers to explore themes such as the environment, technology, inequality, and cultural diversity. Alongside fostering independent thinking, creativity, and problem-solving skills, the game aims to encourage critical thinking about the long-term consequences of small actions and each individual's responsibility towards the planet.

Fig. 1. The UN's 17 Sustainable Development Goals.

2 Inclusion and Accessibility

One cannot think about inclusion and sustainability without ensuring that games are accessible to everyone. Assistive technologies, which encompass resources, methods, and strategies that promote the autonomy and inclusion of the elderly and people with disabilities (PwD), are based on universal design concepts. This worldwide movement emphasizes the importance of designing products and environments that meet the needs of the majority of users (NSCU, 1998). In recent years, there has been growing interest in incorporating assistive technologies in educational settings to enable students with disabilities to fully participate in the classroom. However, with the rise of technology and the expansion of digital learning environments, it is crucial to consider assistive technologies that ensure equivalent educational experiences for PwDs in digital environments and provide inclusive leisure experiences.

The concept of inclusion emerged from the advocacy efforts of disabled individuals, which led to their integration into society and classrooms. In Brazil, the Law of Inclusion (LB, BRAZIL, 2015), specifically Article 74, guarantees that "persons with disabilities have guaranteed access to assistive technology products, resources, strategies, practices, processes, methods, and services that maximize their autonomy, personal mobility, and quality of life". This legislation aims to facilitate the social inclusion and educational participation of individuals with disabilities.

In order to include people with disabilities, it is necessary to adapt various aspects of society. The challenges posed by visual, motor, and other impairments have long been recognized in the entertainment industry. Nowadays, cinemas have wheelchair-accessible seating and subtitles for the deaf, among other practices. However, digital games present several barriers for this population, such as complex controls and reliance on visual and auditory elements. The demands of controlling a character, following the plot, and simultaneously perceiving objects and objectives can make it difficult for everyone to engage with this industry. However, a game has gained recognition for its focus on accessibility: The Last of Us II (2020).

Produced by Naughty Dog (2020) and published by Sony Interactive Entertainment, The Last of Us II focuses on narrative and is set in a post-apocalyptic world, serving as a direct sequel to the previous game. The game offers more than 60 accessibility options, catering to people with difficulties in following subtitles and even those who are completely blind.

Bittencourt et al. (2020) analyze the available options in the game, considering three criteria: Kilpp Frames, Universal Design (UD), and Nielsen Heuristics.

Kilpp's frame method (2010) involves dissecting and dividing images into groups based on proximity or similarity, extracting video fragments from the flow and analyzing them separately.

Universal Design (UD), introduced by Ron Mace (1987), suggests developing products and environmental designs that can be used by all individuals without the need for adaptation for people with disabilities (Bassani et al.). According to Carletto and Cambiaghi (2008) (cited in Bassani et al.), UD is based on seven fundamental principles: equitable use, flexibility in use, simple and intuitive use, perceptible information, tolerance for error, low physical effort, and appropriate size and space for use.

Finally, Nielsen Heuristics is a method aligned with Universal Design, aiming to create an analysis methodology specifically for audio aspects in digital games. Bittencourt (2020) points out that The Last of Us II has auditory accessibility features, and while some criteria, such as "equitable use", are fully satisfied by the available options, others like "user control and freedom" are not adequately addressed. Nonetheless, the game still offers numerous accessibility options.

The Game Awards (TGA) is an annual awards ceremony that recognizes the best electronic games of each year and celebrates significant achievements in the industry. It has been referred to as the "Oscars of the Video Game industry". In recognition of the increasing importance of inclusion and accessibility, the awards introduced a category called "Innovation and Accessibility". This category highlights the advancements and efforts made in creating inclusive and accessible gaming experiences. Additionally, the ceremony also serves as a platform for announcing new game titles.

3 Microsoft Accessibility Guidelines

Microsoft has developed its own accessibility guide known as XAG (Xbox Accessibility Guidelines) (2023). This guide aims to assist developers in creating inclusive games that break down barriers for people with disabilities. Microsoft provides explanatory articles and videos to support and educate individuals interested in working in the field of accessibility. These resources offer insights into the best practices for creating enjoyable environments and projects that cater to a wide range of users. Table 1 presents the accessibility guidelines indicated pela Microsoft. By reaching a larger percentage of helpers and allies, we can foster a more accessible society.

An interesting development is that developers now have the opportunity to submit their games to Microsoft's accessibility team for evaluation against the requirements outlined in the guide. This process ensures that games meet the necessary accessibility standards. As an integral part of contemporary culture, games have the ability to unite people as a community and connect them through shared experiences. With a global gaming population of nearly 400 million individuals, it is crucial to prioritize the creation of inclusive and accessible gaming experiences. Through thoughtful design, we can make gaming accessible to as many players as possible.

Table 1. Presents the accessibility guidelines indicated pela Microsoft

Cognitive Accessibility: Cognition and disability	Cognitive, learning, or other impairments that contribute to a player's ability to perform the actions mentioned above. They can affect player experiences in different ways
Attention	Varieties of sensory stimuli can affect attention
Memory	Often, navigating gaming experiences requires players to retain past information in order to make future decisions
Learning, cognitive processing	Players are often given complex objectives or tasks that contain the game's progress until resolved. The way to complete some of these tasks can be easily understood. However, other solutions may require players to piece together subtle game clues that are not blatantly obvious

(continued)

Table 1. (*continued*)

Learning, cognitive processing	The way to complete some of these tasks can be easily understood. However, other solutions may require players to piece together subtle game clues that are not blatantly obvious
Input Accessibility	Typically, completing gaming tasks requires players to move their fingers with speed, precision, and motor coordination while interacting with physical input devices
Mobility and disability	Players with disabilities that affect arm and hand movements may also experience barriers in the game
Strength	Players must be able to operate input devices supported by their desired platforms and games
Dexterity	Dexterity, in the context of games, refers to the rapid movement of a player's fingers to activate controls within a specific period
Coordination	The ability to perform smooth and precise movements with the arms, hands, and fingers is vital in gaming experiences
Resistance	Determined how long someone can perform an activity without getting tired
Visual Accessibility: Games - no vision	These players must interpret all information in a game only through audio channels or tactile indications
Games and low vision -	Nowadays, many gaming experiences use visual cues or pointers to guide player success
Vision and disability -	When creating accessible visual experiences, one must be aware of visual impairments and their potential impact on player experiences
Hearing Accessibility	Hearing and Impairment—Players who cannot hear any audio provided by a game, platform, or hardware device will have different experiences than players with varying hearing levels
Games without audio	These players may miss important game information when presented via audio only
Games and partial hearing	Many gaming experiences offer multiple sounds simultaneously

4 Planeta ODS Game

The game was designed to belong to the "Point and Click" genre (Fig. 2), with a strong emphasis on problem-solving through observation. As such, it encourages players to explore various scenarios using mouse, keyboard, or controller inputs to control the main character's movement. The player interacts with objects, animals, and NPCs (non-player characters) along the way. Set in a dystopian future, the game follows the journey of our heroine who travels back in time to aid citizens and overcome challenges while traversing between the past and the future. Moreover, the game incorporates knowledge related to humanitarian and ecological actions.

To ensure accessibility, the game employs distinct color schemes for the past and future scenarios, taking into consideration individuals with visual impairments. Accessibility options are easily accessible from the main menu. The choice of font for the in-game texts was also a matter of concern, opting for a sans-serif font with distinguishable upper- and lower-case characters.

The narrative commences by introducing a central problem that addresses one or more themes from the UN's sustainable development goals. Initially, a character suggests a possible solution to an obstacle, which triggers dialogue with other NPCs. The player progresses by collecting items and verifying if the problem is resolved in the future. The core mechanics of the game revolve around exploring the environment, controlling the player character, manipulating the surroundings, and interacting with objects and NPCs. These mechanics are designed with inclusivity in mind, aiming to accommodate players with disabilities (Fig. 3).

The Table 2 below presents the accessibility options categorized into three classifications that are either already included or will be implemented in the game.

Table 2. Accessibility options

VISUAL	MOTOR COORDINATION	AUDITORY
Minimum text size	Being able to play with one hand	Additional channels for visual and audio hints
(Customizable) High contrast on interactive objects	Do not create an action that requires holding, pressing two buttons simultaneously, or clicking several times in sequence	Events such as shots, damage to the enemy, the presence of interactive objects, and a new objective, among others, cannot be indicated visually or audibly only, as players with vision or hearing loss may lose information

(continued)

Table 2. (*continued*)

VISUAL	MOTOR COORDINATION	AUDITORY
Minimal contrast in non-textual elements	Possibility for the player to readjust all controls	The game provides hints through various sensory methods
Customizable high contrast on Ul (User Interface)	Show the appropriate shortcut on the screen	
External software, in-game narration, texts, UI, menus, and interactive objects	Caution with the drag action. Enable activation for when the player clicks again or drops the "item"	
Visual, auditory, and tactile (vibration) feedback		

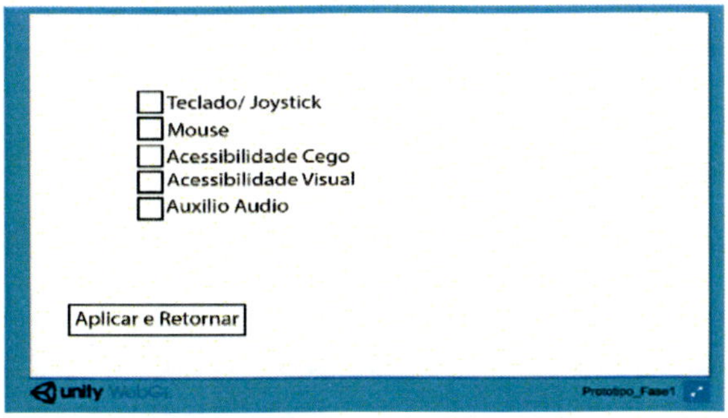

Fig. 2. Included accessibility options

5 The Importance of Accessibility in Digital Games

According to SBGames (2019), digital games play a significant role in the socialization of children, adolescents, and young adults. However, blind and visually impaired individuals face numerous barriers in accessing various game titles, which limits their participation and often excludes them from certain social groups. To make games accessible to a wide range of players, it is crucial to have useful, simple, and easy-to-understand tools.

One of the key improvements demanded for better accessibility is the use of sound cues to indicate subtle variations in movement, jumping, and proximity to hazards. An effective solution to this issue involves implementing more accurate and realistic sound effects commonly used in games. Another interesting approach is to provide players with an introduction to the game's sound effects before they start playing, allowing them to familiarize themselves with the mechanics while becoming accustomed to the auditory cues.

Accessibility, especially in the realm of digital games, is still a challenge for many companies. Ubisoft has been attempting to address this issue in their recent titles, such as Assassin's Creed: Valhalla (2020). Although their accessibility features are still somewhat limited, it represents a significant step forward, particularly for a renowned company like Ubisoft. Additionally, other companies like Obsidian Entertainment and Finji Games, among others, are increasingly surprising the gaming market with their accessibility options and bringing a fresh perspective to the world of digital games.

According to Pereira et al. (2018), certain games incorporate filters that allow users to adjust the color palette. This serves as an accessibility feature, ensuring adequate contrast between similar tones. The authors also suggest that such filters enable players with color blindness to discern information more effectively, thereby enhancing their understanding of the game's intended message conveyed through the use of colors. However, the appropriate accessibility resource may vary depending on the specific type of color blindness exhibited by the individual, such as deuteranomaly, tritanomaly, protanomaly, or monochromacy. Figure 3 presents the type of deuteranomaly visualization, Fig. 4 presents the type of visualization of protanomaly, Fig. 5 presents the type of visualization of tritanomaly and Fig. 6 presents the character in the original version. Consequently, for digital games to cater to different types of color blindness and ensure accessibility, they should offer four distinct color filter options in their configurations.

Images Showing the Different Types of Color Blindness

Fig. 3. Deuteranomaly

Fig. 4. Protanomaly

Fig. 5. Tritanomaly

Fig. 6. Original Image

6 Final Considerations

The present study addresses a methodology for analyzing accessibility in digital games based on existing methodologies, such as universal design principles and Nielsen's heuristics. Its purpose is to contribute to the inclusion of individuals who will utilize these functionalities in digital games, as well as to outline the fundamental criteria that should be observed to promote such inclusion. Given the growing market for accessible digital games, it is of paramount importance for designers to incorporate these considerations into their digital products. Our project has been testing the accessibility features throughout the game's development process for a year.

Acknowledgements. This project was financed by FAPERGS (Foundation for Research Support of the State of Rio Grande do Sul) under the PUBLIC NOTICE FAPERGS 07/2021 - PROGRAMA PESQUISADOR GAÚCHO – PQG.

References

Bassini, P.B.S.: Usabilidade e acessibilidade no desenvolvimento de interfaces para ambientes de educação à distância (2010). https://seer.ufrgs.br/renote/article/view/15180/8947. Accessed 30 Mar 2023

Bittencourt, S., Heidrich, R., Bittencourt, A.: A verificação da acessibilidade ligada à audição nojogo Last Of Us 2 a partir dos princípios do design universal e das heurísticas de Nielsen. Inovamundi (SPG v.13), pp 3498–3507 (2020). http://www.feevale.br/divulgacao/2020/sitefeevale/SPG/Seminario%20de%20Pos-graduacao%202020.pdf. Accessed 30 Mar 2023

Carletto, A.C., Cambiaghi, S.: Desenho universal: um conceito para todos. Mara Gabrilli, São Paulo (2008). https://www.maragabrilli.com.br/wp-content/uploads/2016/01/universal_web-1.pdf. Accessed 30 Mar 2023

Microsoft. Xbox Accessibility guideline V3.1 (2023). https://learn.microsoft.com/en-us/gaming/accessibility/guidelines. Accessed 15 Mar 2023

Pereira F., et al.: Game accessibility guidelines for people with sequelae from macular chorioretinitis, vol 28. Elsevier, Minas Gerais (2018). https://reader.elsevier.com/reader/sd/pii/S1875952118300144?token=2090D08CE2A62E2D8A088B51546DDE3D42A15F93E8A142588C4187438048A6DC30C750F6FC71B95C049E63035E2D44D0&originRegion=us-east-1&originCreation=20230412183315. Accessed 12 Apr 2023

UN. Disability and Development Report (2018). https://social.un.org/publications/UN-Flagship-Report-Disability-Final.pdf. Accessed 5 Apr 2023

World Health Organization, The World Bank. Relatório mundial sobre a deficiência; tradução Lexicus Serviços Linguísticos. - São Paulo: SEDPcD, 334 p. Título original: World report on disability 2011 ISBN 978-85-64047-02-0 1. Pessoas com deficiência - estatística e dados numéricos. 2. Pessoas com deficiência - reabilitação. 3. Prestação de serviços de saúde. 4. Crianças com deficiência. 5. Educação Especial. 6. Emprego apoiado. 7. Política de Saúde. I Organização Mundial da Saúde. Banco Mundial. II. Título (2012). https://apps.who.int/iris/bitstream/handle/10665/44575/9788564047020_por.pdf. Accessed 10 Apr 2023

A New Life for Textile Waste – Upcycling in a Fashion Collection

Sofia Moreira(✉) and António Dinis Marques

Universidade do Minho, Campus de Azurém, 4800-058 Guimarães, Portugal
soracomo@gmail.com

Abstract. The fashion industry is one of the most important industries in the world from an economic and social point of view, but it is also one of the most polluting, being a larger consumer of natural resources and a considerable producer of waste throughout its value chain. It is imperative to change the way the fashion industry operates, the fast fashion model has to be rethought and the circular economy emerges as an alternative. These changes should start from the awareness of young fashion designers for the impact fashion has on the environment. Design schools, in general, have the function of opening the horizons of future designers to this problem and to the role they play with the choices they make.

The present case study intends, in this way, to present an academic exercise, from the bachelor's degree in Fashion Design and Marketing, from the University of Minho (2018/19 academic year) that consisted in the idealization and implementation of a fashion collection whose basis was a sustainable and circular perspective. "Signs of Life" collection took this stance by incorporating the reuse of pre-consumer textile waste, giving it an added value – upcycling.

Keywords: Fashion Design · Circular Economy · Upcycling · Education

1 Background

The impact of the fashion industry is tremendous, not only in an economic and social level but specially on an environmental scale, mainly because we are dealing with "the world's third biggest manufacturing industry after automotive and technology industries" [1]. Therefor the effect that this industry has on the environment has repercussions in several spheres:

- Water pollution and microplastics – "Washing, solvents, and dyes used in manufacturing are responsible for one-fifth of industrial water pollution" [2]; "Each year, around half a million tonnes of plastic microfibers resulting from the washing of textiles are estimated to be released into the ocean" [3] and "By 2025, there will be 1 ton of plastic for every 3 tons of fish in the oceans, and by 2050 the weight of plastic will overtake that of fish" [4];
- Natural resources – "Textiles production (including cotton farming) uses around 93 billion cubic metres of water annually, representing 4% of global freshwater withdrawal" [5];

- Biodiversity – "The apparel industry is a significant contributor to biodiversity loss. Apparel supply chains are directly linked to soil degradation, conversion of natural ecosystems, and waterway pollution" [6];
- Emission of greenhouse gas – "under a business-as-usual growth scenario, sector emissions will grow to 1,588 Gt by 2030 - well off pace to deliver 45% absolute reduction" [7] and even more far from the net zero by 2050;
- Pesticides and insecticides – "cotton cultivation currently uses lots of chemicals – 4% of all world pesticides and 10% of insecticides are used in cotton-growing" [8];
- Waste - "three out of five of the 100 billion garments made in 2018 end up in landfill within a year" [9] and "one garbage truck of textiles is landfilled or incinerated every second" [5].

Currently, the fashion industry is based on a business model that encourages the over-consumption which generates all the environment issues described above, and unfortunately even many more. Based on a linear economy, "take-make-dispose" [10], the commonly known fast fashion model "involves increased numbers of new fashion collections every year, quick turnarounds and often lower prices. Reacting rapidly to offer new products to meet consumer demand is crucial to this business model" [1]. Braungart and McDonough [11] describe it as an industrial linear model "from cradle to grave", characterized by the extraction, manufacture, use and consequent disposal.

For all this, it is essential to bring sustainability to the fashion industry, and the circular economy (CE) presents itself as an indisputable guideline because is more than gradually reducing the harm of our current model. According to Ellen MacArthur Foundation (EMF) [12] "it tackles the root causes of global challenges such as climate change, biodiversity loss, and pollution, while creating opportunities for better growth". Unlike the fast fashion model, "is intended to be restorative and regenerative by having better product design and processes to promote maximum reuse of resources and prevention of waste" [10].

The Cradle-to-Cradle model (also known as C2C), idealized by the authors William McDonough e Michael Braungart in 2002 [11], already presents a very innovative vision of this theme, because instead of proposing the prevention of the waste, this model creates value, goes beyond the notion of having recycling as the final step (and consequent beginning) in the product's life cycle and, as an alternative, it is based on the idea that waste doesn´t need to exist, like in Nature. They argue that life and products can be designed around the notion of nutrition, a model in which everything is a resource for something else, a circular path.

Based on this model, the CE promoted by the EMF [12] goes a little deeper, by presenting three principles that must be undeniable for all fashion products - used more (through durability, reuse and repair), made to be made again (the products enable composting, design for disassembly, recycling and remaking), and made from safe and recycled or renewable inputs (be aware of hazardous substances, microfibers, recycled material, regenerative production practices, renewable material and waste). Like the C2C model, it also considers a biological and a technical cycle (butterfly diagram) and in each cycle all the products, components and materials are keep in the highest utility and value. For this matter, the concept of pre-consumer waste no longer has a definition in a circular economy, waste in general "is 'designed-out' by intention" [12]. This way, the technical materials must be designed to go back to technical cycle, allowing them to be ""upcycled" rather than recycled – to retain their high quality in a closed-loop industrial cycle" [11]. The process of upcycling is "converting materials into new materials of higher quality and increased functionality" [13], the opposite of downcycling where the created products have less value than the initial one (which normally happens with recycling).

Therefore, it is extremely important to look at waste as a resource, whatever its category [14], as there is still a large amount of pre- and post-consumer textile waste and, with the current model, most of this textile material is lost in the supply chain during fabric and clothing production. For this very reason, it is imperative to rethink (re)manufacturing before recycling, and Runnel and other authors [15] approach the problem through three different design strategies: "invisible remanufacturing", where production leftovers invisibly, in areas of the parts that are not visible; "visible remanufacturing", where textile structures are reused in small details on the outer part of the garment; "design-led remanufacturing" is all about designing the entire piece with a view to fully utilizing waste.

Taking this background into account, while we are in a transition phase towards a circular production process, education, especially of young fashion designers, is seen as a driver of a peaceful transition, students will be taught to think and act in a circular way. "Bringing circular economy principles into education (…) will equip learners with the systems-thinking skills and mindsets needed to become active shapers of a circular economy in general, and a new textiles economy in particular" [5]. The labor market will surely be changed by the CE, and it is "vital for product designers to take circular ideas from practice to reality. In order to apply these principles in practice, education incorporate and teach these principles as well, across specializations" [16].

2 Goals

The following study case intends, in a more general way, to highlight the possibility of the design process to be focused on sustainability, not just taking in account the concern with textile waste but having in consideration the entire cycle of the product created. It proposes to answer the research question "Is it possible for a fashion collection and/or brand (that takes into account fashion trends and follows the steps of design process) to incorporate the principles of circular economy?". Taking this research question into account, the most convenient option of research has the action-research, with the researcher being part of the design process and its analysis.

Despite being just an academic exercise, this collection shows that it is possible for the entire process of creating a brand and idealizing collections to be thought of to have a lesser impact on the environment, without neglecting the design, the brand's identity, and the target audience.

3 Design Process

The academic exercise proposed, in the discipline of "Interdisciplinary Project in Fashion Design V", began with a briefing to present the main theme – "Living in the virtual world" - on which the collection would have to be conceptualized. The design process then proceeded according to the fashion design methods proposed by McKelvey and Munslow [17], starting with the identification of the problem and ending with the fashion products, which together forms the "Signs of Life" collection. As soon as the problem was identified, the need for the brand and the collection created to have a sustainable core, through the upcycling of mostly pre-consumer waste, was imperative. Fashion design that reuses these materials can be an interesting approach to environmental sustainability [18], and the pyramid model presented by Hawley [19] has the "conversion to new products" of waste (textiles and apparel) as an important textile recycling tool.

Prior to the design process itself, it was necessary to conceptualize the brand, its identity, marketing plan and image book. These steps were fundamental as they also allowed the definition of the target audience, the future consumer of the brand, without which the entire design process is also devoid of economic purpose, which was also one of the objectives of the project (financial sustainability). Thus, the clothing and accessories brand called "Redefined" was conceived (Fig. 1), whose motto "Waiting in the line to be new again" reflects the ecological component that the brand incorporates. The reuse of textile structures appears as the core of the brand (upcycling), seeking in this way that the brand has a positive impact on its consumers and reduces its (and its consumers') ecological footprint.

Fig. 1. Redefined brand logo.

Based on this assumption, which significantly alters the design process and even the production process, close contact with partners in the textile industry was essential to understand what kind of textile waste normally results from the entire production process. It was quickly found that there are many fabrics and knits that are surplus from previous collections (a few meters), as well as that the amount of textile waste resulting from cutting is also significant. The main source of raw material thus focused on these two alternatives, for larger areas of the pieces and smaller details, respectively.

Having the briefing as a starting point, the design process itself began with the analysis and defragmentation of the theme, through the brainstorming tool, building ramifications based on more concrete concepts, starting from the general to the particularization (brainstorming web). "Brainstorming has traditionally been used to spur group creativity with the intention of generating concepts and ideas regarding a specific challenge" [20]. From the three broadest terms – Technology, Energy and Virtual Communication – the duality between the "real self" and the "virtual self" emerged, which ended up becoming the concept for the "Signs of Life" collection, which is assumed as a satire on lonely society that looks for signs of life not only in the real world but also in the virtual one. The concept of the collection was described in a moodboard where the "I" seeks in virtual communication new experiences and relationships, the illusion of a new life, moments of fun, freedom to express and be accepted or even this communication is just a refuge for the real world. The creation of alter-egos in this virtual world is assumed as a "new" way of living, as if there is an imaginary world that is opposed to the real world.

After devising the concept of the collection and contacting future partners, and because the primary raw material for this collection were old and cutting remnants from more recent collections, it was necessary to select and catalog everything that was collected. Only with this information was it possible to define the colors and later shapes, which resulted in the first sketches of the collection (Fig. 2). The choice of colors, materials and in the definition of the design of the pieces, had in consideration not only the target audience, but also the trends dictated by the WGSN style office. Inspired by street culture, in a young, urban, and bold environment, the "Glocal Connection" trend emerges, for the autumn/winter 18/19 season. Focusing on the globalization of urban culture, with references to various places, where multiculturalism is reflected, not only through the palette of cores, but also in contrasting patterns.

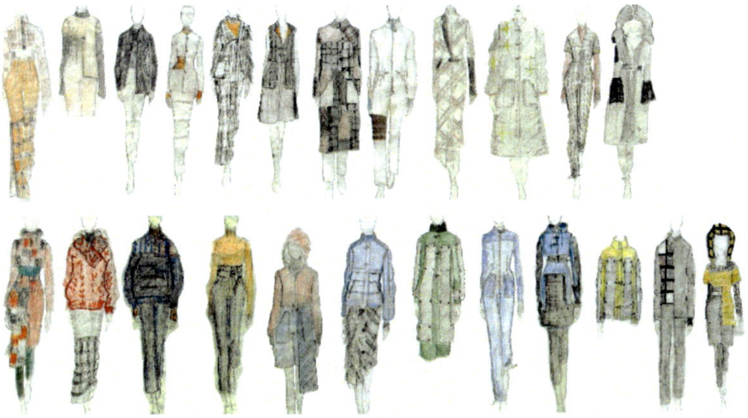

Fig. 2. Sketches of "Signs of Life" collection.

As mentioned before, one of the pillars on which the "Signs of Life" collection is based largely influences the color palette - textile structures and patterns available. Upcycling thus presents itself as a key element in the way the entire collection was designed, because although there was a focus on colors and materials that reflect the spirit of the trend, this choice was conditioned by the textile waste that was available.

Still regarding the upcycling process, it should also be mentioned that all accessories (ribbons, springs, fasteners, ...) were also reused and came from various sources. From the partnership with a haberdashery, it was possible to "rescue" fasteners and lines that had been in stock for several years and that, most likely, ended up in the common rubbish bin (and consequently in landfills). With a junkyard it was possible to collect carabiners and plastic parts from the seat belts of very old cars that were going to be destroyed and whose new purpose was to be transform into belt buckles and fasteners for accessories.

With a passementerie company snap button and plaques were created, the material used resulted from the surplus of previous productions and proceeding in the same company to laser engraving these accessories. Finally, using the resources that the University itself has, filling paste and jacquard ribbons with the brand's logo were developed (with a mixture of photoluminescent yarn and recycled polyester yarn).Traditional printing was only used for the labels, not only because it is a technique that uses a lot of natural resources (mainly water), but also because the "patterns" created resulted mostly from the assembly of various fabric and knitted remnants (patchwork).

While the production process of all accessories was taking place, the necessary patterns for the production of all pieces (clothing and accessories) were developed. The cutting of the fabrics and knits was a manual process and the cutting plans developed tried to produce as little waste as possible. These surpluses resulting from the cutting process were used as labels for garments and accessories.

The prototype footwear resulted from a combination of fabric and leather waste (surpluses from the cut of previous footwear production), and the same company that supplied these waste also dealt with the pattern making and production of the boots. It should also be noted that the sole used in the prototypes also resulted from the recycling of rubber, provided by another company specializing in soles.

With the entire collection completed - accessories, footwear, and clothing (Fig. 3) - the brand would develop a series of initiatives to promote the collection. The photo shoot with catalogue development and the presentation of the entire collection in a runway show were carried out. The continuation of the marketing process did not take place as this was not required in the academic exercise. Although there was no continuity, the entire process of marketing and positioning of the brand in the market was studied, namely a very important aspect for the brand concept was defined - the pieces of the

Fig. 3. "Signs of Life" collection – some examples

collection sold, used and whose end had arrived would be collected and, if possible, the raw materials and accessories would be reused. If this option were not viable, recycling of the various materials would be attempted, that is, the brand's objective was to close the production cycle (circular economy).

4 Results

The "Signs of Life" collection was presented in a joint fashion runway with the other students of the bachelor's degree in Fashion Design and Marketing, in an iconic place in the city of Guimarães - IDEGUI - Instituto de Design de Guimarães. This show was based on the "Wasteland" theme, a post-apocalyptic world where the fragility of planet Earth reflects extreme consumption, from which not only textiles but also technological waste result. This presentation to the main public was the culmination of the entire design process and it presents itself as the first time the pieces created were seen. This runway show had not only coverage by the local media and the University of Minho's own communication vehicles, but the main result obtained was the projection that the collection had among the various partners – it was very fruitful in raising awareness of the issue of reuse of textile waste.

It can be said that the result of this case study was all the pieces created, with the transformation of waste into value-added fashion products, however it was more than that - it was an awareness of the problem of textile waste, not only of the students who created the collection and their peers, but also of the partners who contributed and allowed its development, and even, we dare to say, the public that came across it.

The exhibition of the collection did not end with the runway, as there was also a presentation of a more academic nature, with the teachers of the bachelor's degree present, as well as students from the University. In addition to this presentation, in 2019, some pieces of the collection were exhibited at the 8th Encuentro Bid_Enseñanza y Diseño, in Madrid. This last exhibition became another opportunity to make known not only the collection with its aesthetic sense, but also through it the problem of the impact of the fashion industry on the environment, more specifically with the surplus of textile waste, presenting itself this (or other) collection(s) as a solution. Another opportunity was envisioned to, in some way, bring this issue to the public eye, in this case more specifically to international university students and professors.

Finally, this article can also be seen as a result of the academic work developed, which once again generates reflection on the impact of textile waste on the environment.

5 Conclusion

The fashion design process, despite being relatively structured, with fundamental steps that are usually unavoidable, is based mostly on the designer's creativity and freedom, from conceptual development to the making of clothes. This design process, in the present case study, was triggered by the desire of the designers to embrace the sustainable component, using upcycling.

This approach brought several challenges, the first being the limitation of available raw materials, not only in terms of textile structures (the textile structure is not always suitable for the intended purpose), colors and even the size of the textile waste. The creativity and freedom characteristic of the process ended up being somewhat conditioned, and the process became a little more analytical and structured. Thus, in the conceptualization phase of the garments, the size of the textile materials influenced the design, cutting and sewing area and consequently in the pattern making.

Even the pattern making process itself was very fluctuating, because when cutting, and to create the least amount of waste possible, small design changes were made that allowed the incorporation of smaller pieces of fabric, resulting from the cutting process itself.

Despite all the challenges and being a more arduous task than purchasing raw materials and accessories without any limitation, it proved to be a possible process that really brings added value to the garments. This is not only visible in the transformation of textile waste into wearable pieces, but also on the impact, the awareness that projects of this magnitude have on the general population, and also on the industry itself, with all partners getting involved spontaneously and even emotional in the process. As previously mentioned, initiatives such as these, in the case of an academic nature, can be replicated by the fashion industry, if there is awareness of the impacts, including financial ones, that these and other measures that focus on the circular economy can have on the industry itself and the world we all live in.

Acknowledgements. It is mandatory to give a special thanks to the group of designers that developed this collection – Beatriz Moreira, Carla Mendes, Cristiana Pacheco, Sofia Moreira, and Tânia Augusto. Also a special thanks to all partners without whom this project would not materialize – AFF, Inarbel, CombinedTex, Jooze, Albano Morgado, Triple Marfel, Retrosaria Glória, Guimanos, Nesmatex, MoreTextile Group, and Fiverlarte.

This work is financed by Project UID/CTM/00264/2021 of 2C2T – Centro de Ciência e Tecnologia Têxtil, funded by National Founds through FCT PhD Studentship with reference 2022.13244.BD.

References

1. Creagh, M., et al.: Fixing fashion: clothing consumption and sustainability. Sixteenth Report of Session 2017–19. In: House of Commons: Environmental Audit Committee. UK Parliament (2019). https://publications.parliament.uk/pa/cm201719/cmselect/cmenvaud/1952/1952.pdf. Accessed 20 Feb 2023
2. Amed, I., André, S., Balchandani, A., Berg, A., Rölkens, A.: The state of fashion 2023: holding onto growth as global clouds gather. McKinsey & Company (2022). https://www.mckinsey.com/industries/retail/our-insights/state-of-fashion. Accessed 10 Nov 2022
3. Boucher, J., Friot, D.: Primary microplastics in the oceans: a global evaluation of sources. International Union for Conservation of Nature and Natural Resources (2017). https://doi.org/10.2305/IUCN.CH.2017.01.en. Accessed 11 Nov 2022
4. Mermaids Consortium, et al.: Microfiber release from clothes after washing: Hard facts, figures and promising solutions. Plastic Soup (2017). https://www.plasticsoupfoundation.org/wp-content/uploads/2017/07/Position-Paper.Microfiber-release-from-clothes-after-washing.PSF_.pdf. Accessed 03 Oct 2022

5. Morlet, A., Opsomer, R., Herrmann, S., Balmond, L., Gillet, C., Fuchs, L.: A new textiles economy: redesigning fashion's future. Ellen MacArthur Foundation and Circular Fibres Initiative (2017). https://emf.thirdlight.com/file/24/uiwtaHvud8YIG_uiSTauTlJH74/A%20New%20Textiles%20Economy%3A%20Redesigning%20fashion%E2%80%99s%20future.pdf. Accessed 12 Oct 2022
6. Granskog, A., Laizet, F., Lobis, M., Sawers, C.: Biodiversity: The next frontier in sustainable fashion. McKinsey & Company (2020). https://www.mckinsey.com/industries/retail/our-insights/biodiversity-the-next-frontier-in-sustainable-fashion. Accessed 17 Nov 2022
7. Sadowski, M., Perkins, L., McGarvey, E.: Roadmap to net-zero: delivering science-based targets in the apparel sector. working paper. in: sustainable and ethical apparel. Resources Institute (2021). https://doi.org/10.46830/wriwp.20.00004. Accessed 20 Jan 2023
8. Common Objective: Fibre Briefing: Cotton. Ethical Fashion Group Ltd. (2021). https://www.commonobjective.co/article/fibre-briefing-cotton. Accessed 23 Jan 2023
9. Clean Clothes Campaign: Waste and pollution. Clean Clothes Campaign (2019). https://cleanclothes.org/fashions-problems/waste-and-pollution. Accessed 30 Oct 2022
10. Bukhari, M.A., Carrasco-Gallego, R., Ponce-Cueto, E.: Developing a national programme for textiles and clothing recovery. Waste Manage. Res. **36**(4), 321–331 (2018). https://doi.org/10.1177/0734242X18759190
11. Braungart, M., McDonough, W.: Cradle to Cradle: Remaking the Way We Make Things. North Point Press, New York (2002)
12. Ellen MacArthur Foundation: Vision of a circular economy for fashion (2020). https://emf.thirdlight.com/link/nbwff6ugh01m-y15u3p/@/preview/1?. Accessed 15 Oct 2022
13. Ellen MacArthur Foundation: Towards the Circular Economy: Economic and Business Rationale for an Accelerated Transition. (2013). https://emf.thirdlight.com/file/24/xTyQj3oxiYNMO1xTFs9xT5LF3C/Towards%20the%20circular%20economy%20Vol%201%3A%20an%20economic%20and%20business%20rationale%20for%20an%20accelerated%20transition.pdf. Accessed 19 Oct 2022
14. Lacy, P., Rutqvist, J.: Waste to Wealth. The Circular Economy Advantage. Palgrave Macmillan, Hampshire (2015)
15. Runnel, A., Raihan, K., Castel, N., Oja, D., Bhuiya, H.: Creating digitally enhanced circular economy. Reverse Resources (2017). http://www.reverseresources.net/about/white-paper. Accessed 30 Jun 2021
16. Het Groene Brein: Knowledge Map Circular Economy (2020). https://kenniskaarten.hetgroenebrein.nl/en/kenniskaart/circular-economy/. Accessed 12 Feb 2023
17. McKelvey, K., Munslow, J.: Fashion Design: Process, Innovation & Practice. Blackwell Publishing, Hoboken (2003)
18. Vezzoli, C., Manzini, E.: Design for Environmental Sustainability. Springer, London (2008). https://doi.org/10.1007/978-1-84800-163-3
19. Hawley, J.M.: Textile recycling: a system perspective. In: Wang, Y. (ed.) Recycling in Textiles. Woodhead Publishing Limited, UK (2006). http://hdl.handle.net/2097/595. Accessed 12 Jun 2021
20. Martin, B., Hannington, B.: Universal Methods of Design. 100 Ways to Research Complex Problems, Develop Innovative Ideas and Design Effective Solutions. Rockport Publishers, Beverly (2012)

Co-creating Case Studies to Teach and Learn Fashion at a Business School

Silvia Pérez-Bou(✉), María Ángeles Burguera, and Teresa Sádaba

ISEM Fashion Business School, University of Navarra, Marquesado de Santa Marta, 3, 28027 Madrid, Spain
sperezb@unav.es

Abstract. Case studies are a recognized method to teach and learn in a business school, as they provide the students with the opportunity to analyze real situations in companies in an in-depth manner.

The goal of this paper is to describe a novel methodology of co-creating fashion case studies among students, professors, and companies in a fashion business school, and to establish learnings and improvements in this teaching methodology.

Two are the methods used to create the cases: one top-down, triggered by fashion companies, and the other bottom-up, based on the students' interests. The analysis of a sample of 133 cases elaborated in ISEM Fashion Business School of the University of Navarra (Spain) since 2005 shows both the evolution of the challenges faced by the Spanish fashion industry and the permanence of some topics across time. Moreover, it highlights the contribution of students and academics to the companies' development.

Keywords: Case Study · Fashion Companies · Sustainability · Teaching and Learning Experiences · Fashion Studies · Learning by Doing

1 Introduction

Business Schools are focused on providing students with the skills and knowledge to be able to manage business organizations and make decisions by analyzing situations, evaluating alternatives, and assessing the effectiveness of those decisions.

One of the most common methods of teaching in Business Schools is case studies and their use in the classroom has been on the rise over the past few decades [1]. The case study method of teaching and learning is a powerful tool for students to apply theoretical course concepts to real-life situations, dilemmas, or opportunities faced by professionals and companies [2].

Although case studies in different sectors may serve for teaching specific disciplines (general management, ethics, accounting, strategy, entrepreneurship, etc.) in all industries, the closer the sector is to the area of expertise of the students, the easier for them to be engaged with the case. Fashion management students are especially interested in facing - during their learning period - real situations that occurred in the industry, to acquire specific knowledge and skills that may serve their future professional life.

The fashion industry has not been, until recently, considered by the academy as a field of research [3, 4]. This explains why it is difficult to find cases addressing the specific issues faced by fashion companies' managers. This motivated ISEM, the Fashion Business School of the University of Navarra in Spain, to start using and writing fashion case studies as a means to improve the teaching and learning processes in their fashion postgraduate programs. Two programs, the Executive Master in Fashion Business Administration or FBA (an MBA focused on fashion), and The Right Fashion (a sustainability-focused program), are two scenarios to co-create different case studies: top-bottom or bottom-up, generalist, or sustainability-focused.

The goal of this research is to analyze the development of this practice, and the results for the different stakeholders participating in the creation of the cases: industry, academia, and students. In doing so, other aspects arise: the maintenance and evolution of the different topics across this period -the connection between the past and present-, allowing us to learn from the past and from the real situations in the industry. It is also evident how the academy is leading and being one step ahead of industry challenges, thus able to offer paths for a better fashion future.

2 Theoretical Framework

Universities are preparing new professional profiles for the fashion industry, and the learning process has to be managed according to the economy derived from information and social media [5]. Bridging academia with both the community and the profession in a transdisciplinary manner gives students the opportunity to act as mediators, lead, negotiate, work collaboratively, and develop transversal skills and competencies [6].

Business cases have long been regarded as a very effective tool in higher education [7]. Their use in the classroom allows students to develop an in-depth understanding [8] and appreciation of complex, real-life phenomena in contemporary settings [9–11]. Live case studies take those advantages one step further by bringing direct interaction with business organizations into the classroom and presenting industry challenges in real-time, dynamic solutions requiring cross-disciplinary approaches, and knowledge applications [12]. Cases are best understood within the context of a particular disciplinary tradition [13]. For fashion business students, the interest in the cases increases when they are related to that sector. To enlarge the available teaching and learning material, our students elaborate cases playing the role of consultants interacting with real complex situations and providing solutions to the organization gaining a whole system perspective and exhibiting cooperative behavior [12].

3 Methodology

To analyze the case studies, the authors have performed a double methodology with a qualitative and quantitative analysis. This allows them to enrich the discussion and conclusions, as data are seen from different perspectives. Also, it provides arguments for future work in the case study as a learning process for students.

First, a qualitative approach to the case studies elaboration on the fashion programs has been performed. Data are collected from ISEM Fashion Business School projects, from the academic years 2004–2005 to 2022–2023, with a total of 123 generalist projects, and 10 sustainability-focused from 2019 on.

Second, a quantitative methodology is used, analyzing the list of case studies through the STATA IC/15.1 program, in order to explore the number, topics, and evolution of case studies with a univariate analysis of frequencies and bivariate analysis of frequencies and years.

With these analyses, in the last section, there is a discussion of the results, highlighting the roles of each of the actors involved in the processes, and comparing the outcomes of both approaches.

4 Findings and Analysis

4.1 Qualitative Discussion

ISEM Fashion Business School has created an ecosystem among fashion companies, postgraduate students, and academia, to provide a range of case studies tackling relevant aspects of the industry.

The FBA program has been offered since 2004 as an official executive master in business administration (MBA), focused on fashion business. The students take all the typical courses of an MBA (finance, general management, ethics, analysis of business problems, etc., and the ones specific to fashion: luxury, retail, buying, internationalization, trends, culture and fashion, marketing, communication, sustainability, etc.). This program is unique in Spain, as there are no more masters with that specific approach to fashion (an integrated perspective well connected to the industry), and because it also allows access to Doctorate studies. According to El Mundo ranking, it has been considered the best specialized Master in Spain for the last 5 years. Around 50–55 students from different countries of the world enroll each year in the program.

The FBA case generation process starts with the Call for Consultancy from the academic institution (ISEM) to selected fashion companies. They appreciate the added value from three perspectives: they are looking for a young point of view of the problems (students are on average 26 years old), they want a solution and they don't have time or resources to find it out, and they want to build bonding with the academic institution based on trust and growing relationship.

Most of the companies are Spanish brands or fashion groups. A small percentage of the cases come from internationally expanded brands or groups whose headquarters are in other countries but bring specific topics that complement the content for the fashion students (cosmetics, fur, watches, etc.). The process usually follows the same patterns: small companies propose high cases (oriented to business growth or strategy), while big companies tend to offer medium cases (an area of the business). The result is a top-bottom pattern, as the companies propose the topics to the academia and the students.

ISEM encourages the company to think of some problem areas and launch a brief, based on a template. The projects are randomly assigned to the students' groups (six-to-eight people each), who will have between eight-to-twelve weeks to provide a solution. The professor and the mentors (part of the faculty or the ISEM's alumni network) help the students to monitor the projects and provide them with inspiring solutions. From the fashion company, there is a designated team to facilitate information and the company DNA to the students during the whole process.

The delivery of the case is a real proposal for their company, based on an in-depth external analysis of the market, an internal analysis of the company, the goals definition, and the action plan. The resulting dossier is sent to the company and defended before the board in a public presentation (with other companies and students). Company members ask questions acting as clients, and the students act as consultants.

The Right Fashion (TRF) is a shorter program (around 100 sessions during three or six months, depending on the format: executive or part-time) exclusively focused on a technical vision 360 of the sustainability issues in a fashion company: design strategies, sourcing, human rights across the entire value chain, social audits, certifications, the problems of logistics and delivery, health and safety of the products, how to meet different regulations in different countries in the world, etc. It is a niche program, followed by ten-to twenty students each year, with different backgrounds and positions.

The TRF case generation process follows a bottom-up pattern. The students (professionals working in fashion companies) choose a real problem they have to face in their work and try to solve it by applying the concepts learned during the course, with the help of professors. As a result, ten case studies were presented in the three past editions. To name just a few: how to resolve the crisis caused by the fire at a supplier factory in Bangladesh; how to handle the complaint of a customer who has received an online order with clothes having bloodstains; which was the reputational impact on fashion brands after the collapse of Rana Plaza building in 2013, how to escalate a technology to absorb carbon emissions, or how to improve traceability with blockchain.

In both programs, the outcome, the case study, is a valuable working document for the company: useful, realistic, implementable, and eye-opener: Usually, the students bring disruptive thinking, ambitious goals, and the asset of the intergenerational point of view.

4.2 Quantitative Discussion

From 2004–2005, there have always been case studies (consultancy projects) at the Fashion Business School. In those 18 years, with 133 projects in total, the average and median are 7 projects per year. However, over two years, the number increases due to the sustainability projects in the TRF program (2018–19, 2021–22).

About the topics of ISEM case studies analyzed (Table 1), communication and marketing issues are the most relevant (together, more than 37% of the projects), followed by the general strategy of the company (13%) and internationalization and sustainability at the same level (11%). Retail, e-commerce, or customer topics, although very important and more changeable in this sector, are present with a lower number of projects.

Among the companies involved in the projects, it is interesting to highlight how the Spanish fashion group Tendam has launched 13 projects with the ISEM students, and the Finnish company Saga Furs, 10. Both groups started to propose studies from the very first beginning.

Over 18 years, as Table 2 shows, the importance of topics has evolved in a discontinuous way. It is remarkable that in 2020–21, 5 projects out of 7 were on communication

Table 1. Frequency of cases' topics and percentage

Topics	Frequency	Percentage
Communication	28	21.05
Consumer	8	6.02
Creativity	2	1.50
Customer Services	1	0.75
E-commerce	8	6.02
Internationalization	15	11.28
Marketing	22	16.54
Retail	5	3.76
Sales	10	7.52
Strategy	18	13.53
Sustainability	15	11.28
Visual Merchandising	1	0.75
Total	133	100

issues. The year 2013/14 was very important for internationalization, as many Spanish companies were trying to diversify their markets because of the economic crisis.

Retail topics have experienced two waves of interest. Another interesting thing is that sustainability is not considered the protagonist one could think about, except for the years when the TRF projects appear.

Table 2. Number of cases by topics per year

Academic Year	CM	UX	CR	CS	EC	IN	MK	RT	SL	ST	SU	VM	Total
2004–05	1	0	1	0	0	1	1	0	0	1	0	1	6
2005–06	3	0	0	0	0	0	0	0	0	2	0	0	5
2006–07	2	1	0	0	0	1	1	0	1	1	0	0	7
2007–08	1	0	0	0	0	1	2	0	2	0	1	0	7
2008–09	2	2	0	0	1	1	1	0	0	1	0	0	8
2009–10	1	1	0	1	1	0	1	1	0	1	0	0	7
2010–11	1	0	0	0	1	1	2	1	0	1	0	0	7
2011–12	1	1	0	0	0	1	4	0	0	0	0	0	7
2012–13	1	0	0	0	1	2	0	0	0	2	0	0	6

(*continued*)

Table 2. *(continued)*

Academic Year	CM	UX	CR	CS	EC	IN	MK	RT	SL	ST	SU	VM	Total
2013–14	3	0	0	0	0	3	0	0	0	0	0	0	6
2014–15	0	1	0	0	1	0	3	0	0	0	1	0	6
2015–16	0	1	0	0	1	1	0	0	1	2	0	0	6
2016–17	0	0	0	0	0	1	2	0	1	2	0	0	6
2017–18	2	0	0	0	0	0	2	0	1	0	1	0	6
2018–19	2	1	0	0	0	0	0	0	2	0	4	0	9
2019–20	1	0	0	0	0	1	1	1	1	1	0	0	6
2020–21	5	0	0	0	0	0	0	1	0	1	0	0	7
2021–22	2	0	0	0	0	1	1	1	1	1	7	0	14
2022–23	0	0	1	0	2	0	1	0	0	2	1	0	7
Total	28	8	2	1	8	15	22	5	10	18	15	1	133

CM: Communication; UX: User Experience; CR: Creativity; CS: Customer Service; EC: E-Commerce; IN: Internationalization; MK: Marketing; RT: Retail; SL: Sales; ST: Strategy; SU: Sustainability; VM: Visual Merchandising.

5 Conclusions

The live project pedagogy can enrich future graduates' ability to deal with the complexity of their future professional area. Students have the opportunity to gain valuable experiences by actively being involved in real-world fashion conditions and by interacting with different actors and disciplines, as well as with the practice. Bridging academia with both the students and the profession in a transdisciplinary manner gives students the opportunity to act as mediators, lead, negotiate, work collaboratively, and develop transversal skills and competencies [14].

The goal of this paper was to analyse the practice of fashion case studies across time, to establish learnings and improvements in this teaching methodology. After the quantitative and qualitative study of cases, there are some interesting points to consider.

Supporting the reviewed literature on case studies, the analysis shows how a live case study can be an opportunity to unite academia, students, and industry for a common goal. For all of them it is a) a learning experience, b) a collaborative work where new relationships are established, and c) a better understanding of the real situation of fashion. Then, and going a step further on the previous literature, it can be said that this process gives not just learning benefits for the students, but also for all stakeholders (industry and faculty) involved in the process. There is no specific way to implement this approach, but what is more relevant is a) the involvement of the three stakeholders (faculty/mentors, students, and industry); b) the selection of a relevant topic c) the planning of timing and delivery of the consultancy case.

In regard to topic selection, it is observed a repetition in the issues studied. The most frequent topics have been Communication and Marketing (37.6%), Internationalization (11.3%), and Sustainability (11.3%). On the one hand, this selection might relate to the evolution of the main challenges of the industry over the years. In addition, a decision of avoiding sharing major issues with students can be playing a role in the limited topics offered by the companies. On the other hand, the continual presence of some of our sponsors, like Tendam-Cortefiel or Saga Furs, could be somehow influencing these topics' selection (i.e., if the company manager involved comes precisely from one of these frequent areas, or if the mentor from the side of the business school is more familiarized with the matter proposed).

From the point of view of the issues, ISEM Fashion Business School should reflect on how to guarantee a wider array of options, by introducing new companies or, if it is not possible, trying to attract other department managers different from the present ones.

It is very interesting to see how the bottom-up process boosts sustainability in recent years, meaning that academia is observing and taking the lead. A similar process was noticed with e-commerce years ago, with increasing attention from academia when not even the industry was interested in it. This idea requires further research but could change the established thinking about academia: always being behind reality.

To improve the process, the use of live cases, as a methodology, suggests the need to implement more reflexive spaces along the process to follow and strengthen the professional skills and competencies learned by students. Although teamwork on a real case solution is, without any doubt, a great advantage, sometimes students might not be fully aware of the benefits, because of the stress of preparing and presenting the dossier before the companies. To continue improving the teaching and learning experience, ISEM Fashion Business School should pay more attention to providing the students with more advice in this direction.

References

1. Sádaba, T., Pérez-Bou, S.,Kadiyska, R., Steffek, V.: Teaching fashion with the case study methodology: comparison between Spain and Canada. In: Communication in 25th IFFTI Conference, Connecting Pathways. Otago Polytechnic, Dunedin (2023)
2. Davis Burns, L.: Fashion Business Cases. A student guide to learning with case studies. Fairchild Books, New York (2021)
3. Calia, R.M.: The Imaginary dress. An interdisciplinary fashion approach among sociological, anthropological, and psychological orientations. Fashion Theory **25**(2), 243–268 (2021). https://doi.org/10.1080/1362704X.2020.1837526
4. Kawamura, Y.: Doing Research in Fashion and Dress: An Introduction to Qualitative Methods. Berg, Oxford (2011)
5. Marques, A.D., Moschatou, A.: Learning process in fashion design students: link with industry and social media. In: IOP Conference Series: Materials Science and Engineering, vol. 254, p. 232005 (2017). https://doi.org/10.1088/1757-899X/254/23/232005
6. Panayi, C., Charalambous, N.: Community-engaged design studios: learning through "live" projects. In: 8th International Conference on Higher Education Advances (HEAd'22) Universitat Politècnica de València, València (2022). https://doi.org/10.4995/HEAd22.2022.14683
7. Barnes, L.B., Hansen, A.J., Christensen, C.R.: Teaching and the Case Method: Text, Cases, and Readings, 3rd edn. Harvard Business School Press, Boston (1994)
8. Stake, R.E.: The Art of Case Study Research. SAGE Publications, Thousand Oaks (1995)

9. Yin, R.K.: Case Study Research: Design and Methods. SAGE Publications, Thousand Oaks (2009)
10. Crowe, S., Cresswell, K., Robertson, A., et al.: The case study approach. BMC Med. Res. Methodol. **11** (2011). https://doi.org/10.1186/1471-2288-11-100
11. Larsen-Freeman, D.: Saying what we mean: Making a case for 'language acquisition' to become 'language development.' Lang. Teach. **48**(4), 491–505 (2015). https://doi.org/10.1017/S0261444814000019
12. Daly, P.: Business apprenticeship: a viable business model in management education. J. Manage. Dev. **36**(6), 734–742 (2017). https://doi.org/10.1108/JMD-10-2015-0148
13. Piekkari, R., Welch, C., Paavilainen, E.: The case study as a disciplinary convention: evidence from international business journals. Organ. Res. Methods **12**(3), 567–589 (2009). https://doi.org/10.1177/1094428108319905
14. Charalambous, N.: Responding to uncertainty through the housing design studio pedagogy: implications and opportunities for architectural education. Inter-Actions: housing design in uncertain environments. JOVIS, Berlin (2018)

Learned in Italy. An Approach to Made in Italy Through the Lens of Fashion Education

Alessandra Vaccari(✉)

Università Iuav di Venezia, Santa Croce 191, 30135 Venice, Italy
avaccari@iuav.it

Abstract. This paper discusses the notion of "learned in Italy", in order to reflect on the meaning that fashion education may represent for critically rethinking the culture and history of Italian fashion in the 21st century. Learned in Italy is modelled on the label 'Made in Italy', but unlike the latter it has been the object of very little discussion and academic work. The paper suggests a vision of fashion culture not separated from its productive reality, which remains central to the Italian national context of the 21st century. "Learning in Italy" is an invitation to a research journey through the fashion manufacturing districts of Veneto Region. This approach stems from an ongoing research work carried out at the Iuav University of Venice and is an opportunity to reflect on the collaborative dimension of fashion and on the issues of generational turnover and offshoring-reshoring dynamics that are putting at risk the material and immaterial knowledges and skills of the Made in Italy.

Keywords: Made in Italy · Italian Fashion · Fashion Education

1 Introduction

This paper stems from the work of research I developed in the last decade on how to teach fashion in a university context in Italy. I shall explore the prospects of the concept "learned in Italy" in relation to the research possibilities it appears to open up. This concept is modelled on the label Made in Italy, but unlike the latter it has been the object of very little discussion and academic work. For Italy fashion has represented one of the greatest experiments of modernity since the end of the second world war, and a chance of redemption and regeneration after the fascism regime, or *The Italian Metamorphosis* to adopt Germano Celant's (1995) definition to highlights the achievements of the country in many artistic and creative sectors between 1943 and 1968. After the second world war, Italian fashion has grown in terms of identity, cultural impact, export volumes, and business, becoming in a few decades one of the country's most important industries and one of the most highly regarded aspects of Italian culture. The imbalance between the renown of Made in Italy and the niche concept of "learned in Italy" provided interesting insight for this research project, which has been developed since 2013. A body of historical and empirical work has been produced in the last decade with relevant insights into the figure of the *stilista* (fashion designer) (Volonté 2008; Segre Reinach 2014; Vaccari

2018), but critical reflection on the training of fashion designers at the university level has been lacking. In Italy, for example, we have no study fully comparable to Angela McRobbie's work on *British Fashion Design* (1998), dealing with the difficulties faced by graduates in fashion design from the prestigious London Art Schools when they begin to look for work. The research undertaken intends to make up for this lack.

2 The Genesis of "Learned in Italy"

The research has involved a small interdisciplinary group of scholars in the fields of fashion design and fashion history. I have developed the notion of "learned in Italy" together with the fashion designer and lecturer Maria Bonifacic. The first exploratory phase of our research date back to the autumn of 2013. At that time, we were looking for a concept that could help us understand the peculiarities of teaching and studying fashion in Italy and, more specifically, in our local academic context. If Italian fashion was globally recognised, was there also a specific way to teach and learn fashion in Italy? And what would be the historical, cultural, and social dimension of this concept? To answer these questions, we engaged in open and heated debate and brainstorming, in an exploratory investigation of what "learned in Italy" means. One result was the composition of long lists of words related to the concept. The first list included terms such as awareness; history, personality; humour; love; inheritance; gift; art; passion; originality; arts; territory; Veneto region; identity; international vision; local knowledge; craft, industry, small scale business and so on. This list reflects a special attention to the question of national identity and to global aspects of Italian fashion. In this regard, Bonifacic played an essential role, thanks to her transnational background and working experiences. On this regard, Bonifacic explained:

> When I think of Italy, I think of warmth, sensuality, slowness, family, and history. Everything in colours. Many of the things that come to my mind are associated with the processing and the pleasure of everyday life and material culture. It is the timeless refrain of 'Good food, good wine and good company' in a beautiful place. It is like an invitation open to all of us. When I think of Italian fashion, I think of dressmaking, design, textiles, handicrafts, relaxed elegance, and the tradition that is at home in today's world. These are generalizations, of course, but I think it is important for us to identify them and analyse them carefully" (Bonifacic 2014: 23).

3 The Research Hypothesis

This paper examines some keys aspects of the human, cultural, and environmental conditions of fashion design students enrolled in Italian universities. My aim is to provide some useful insights into the "learned in Italy" strategy. Such research could provide the following potential benefits: 1) fill a void in historical knowledge of Italian fashion through the study of the under-researched perspective of the fashion school; 2) contribute to a better understanding of the role played by fashion education, hitherto underestimated in Italy, both in academic contexts and in the fashion industry; 3) promote an enlarged vision of the knowledge and skills that Italy exports in terms of fashion graduates (considering both the Italians and the foreign students, who attend Italian universities and then return to their home countries to work as fashion designers and/or trainers expert in Italian fashion); 4) analyse the creative, cultural, and entrepreneurial energies that Italy is able to attract through the driving force of its fashion industry and try to understand the role of fashion schools in this process; 5) empower the learners to develop critical thinking as a means of questioning Italian Fashion and enable them to challenge dominant viewpoints on it.

There is a widespread belief that Italians have a natural bent for fashion, as if fashion were a hereditary characteristic that is genetically inheritable. This is obviously not true, however sometimes even Italian indulges in these visions. A vast literature demonstrates how behind the glamour, the *bella figura*, the relaxed elegance and the nonchalance – or *sprezzatura* in Baldassarre Castiglione's terms – attributed to Italians there are forms of the expression and the learning of fashion based on appearing visually and culturally suitable in complex, competitive, and geographically localised social environments, such as the Italian Renaissance courts (Greenblatt 1980; Collier Frick 2005; Paulicelli 2016).

Fashion in Italy is a cultural and environmental resource that one learns immersed in a space and in contact with others. Fashion is an embodied and spatially situated practice that relates the persons, things, and places in which these relations are performed. Such relation has been studied above all from the viewpoint of the formation of identity and of consumer practices, not from the viewpoint of the production and the teaching of fashion. In Italy the aspects of fashion production—in the sphere of production districts—are very closely related to specific locations, to human and creative density. It is a geographical context densely inhabited by people professionally involved in fashion industry and business. The system of production districts of Made in Italy was visually displayed in the form of a relational and interactive map in the exhibition *The Glamour of Italian Fashion 1945–2014* curated by Sonnet Stanfill (2014).

The research hypothesis states that Italian fashion schools cultivate strong geographical links with the districts where fashion is produced and can contribute to challenging the dominant perspectives on Italian fashion, disorienting old viewpoints and helping to create new ones. Furthermore, the perspective of "learned in Italy" makes it possible to explore immaterial aspects of fashion production, considering the human and relational factors of Made in Italy—embodied in the various professional sensibilities and production realities—apart from their minute local and regional scale. Furthermore, this perspective helps to identify, rationalise, and articulate the tacit knowledge's different components of Made in Italy. By saying tacit knowledge, I mean a set of not codified skills and experiences, according to Michael Polany (1967). Such a knowledge is crucial when it comes to Italian fashion production and culture, which became intelligible and sharable only within a community of practice, such as a fashion students' network.

4 Studying Fashion in the Veneto Region

The concept of "learned in Italy" has been analysed starting from the relationships between high education in the field of fashion on the one hand, and, on the other hand, culture and production of fashion. It has also considered the relationship between Iuav

University of Venice, where the research has been developed, and the territory in which it is located; that is Veneto region and its high concentration of fashion firms devoted to the production of high quality Made in Italy. The Veneto region hosts fashion firms of international renown, producing fabrics, accessories, footwear, eye-glasses, jewellery, knitwear and denim. Many firms do not possess their own brands but work for international fashion and luxury brands. There are also such firms as, for example, Bottega Veneta, Benetton, Luxottica and Diesel (OTB Group), whose history is deeply rooted in the Veneto. It is not by chance that Diesel chose Venice in 2014 for the show of the first collection by Nicola Formichetti, the brand's new artistic director, and that OTB Group funded the restoration of the iconic Venetian Rialto Bridge on the Gran Canal on 2015.

In the Veneto also host countless smaller firms connected with local sub-supplying and with the full Made in Italy supply chain. The undergraduate and graduate Fashion degree of the Iuav University of Venice are thus at the crossroads of one of Italy's most important fashion districts, offering an incredibly high level of fashion skills concentrated within the range of a few kilometres. Here, the students of fashion design come into closer contact with the wealth of knowledge around them, while progressively refining their designing skills. Many of the students experience the district around them as a great and widespread laboratory in which they find all kinds of workmanship, specialised techniques of skilled workers, and chances to work with them to develop the prototypes of the outfits of their own collections in industrial contexts of the highest level. The firms, for their part, are increasingly open to embrace this kind of collaboration—and here I do not refer to the many cases that come about directly through the university but rather to collaboration that stems from the autonomous initiatives of the students. This is the wealth that is the basis of a teaching of fashion design "learned in Italy", whose possibilities have not yet been full expressed. Taking this to the extreme, we could perhaps say that "learned in Italy" represents for the 21st century that which Made in Italy did in the 20th, considering both the on-going process of the "academisation" of fashion practice and design and, parallel to it, the shifting of fashion production far from the shores of Italy and Europe.

Summing up, the factors that have favoured the rise and can contribute to the growth of a fashion teaching "learned in Italy" are the following: 1) the coexistence on the same territory of the universities where fashion design is studied and of the firms where fashion is produced; 2) the current increased willingness of firms to experiment, also as a form of resistance to the economic crisis; 3) the presence in the Veneto Region, as in the other Italian fashion districts, of firms based on craftsmanship and/or on the family model, which makes the relations between firms and students more immediate and informal. To better understand how "learned in Italy" works, I would like to relate what Beatrice Zannini, a 2014 Iuav graduate, wrote in her Masters thesis. For the projected thesis—her subject was upcycling—she contacted a number of firms, some local and others not, chosen on the basis of her design needs, with a request for a free supply of discarded fabrics of the types she needed for her project.

In addressing the firms, she explained: "without any sort of preliminary contact or 'recommendation'; without a business card or website; or any previous experience with related projects, which could guarantee a correct use of the material. My only guarantee is the name Iuav, as the point of reference for my project" (Zannini 2014). Most of the

Fig. 1. Iuav students experimenting with a manual loom at "La Colombina", textile company founded in 1895 in Badoere, Treviso, Italy, 2021. Photo Veronica Spano.

positive replies she received were from local firms. Some firms went beyond her highest expectations. In another part of her thesis, she wrote:

> I had a positive reply from an important firm that produces clothing and accessories for motorcyclists, the Dainese firm. [...] located in a small town in the Vicenza hinterland, hard to get to, but I solved this problem by having them send me the material directly. This firm, unlike the traditional fabric cutters I had contacted, gave me material already selected—quality material in line with the project for the collection (Ibidem).

In this case Dainese took great pains to understand and to respond as precisely as possible to the student's requests, selecting the material ad hoc and even sending it to her home, as if she were one of their clients.

Let me make it clear that the relations with firms I am speaking about are relations managed autonomously by students seeking to take fullest advantage of what their local district could offer. At Iuav "learned in Italy" is not, at present, an official teaching method and is not part of the official curriculum. Rather, it is a form of learning that sets out from an individual research experience and is based on a detailed exploration of the territory. It is something it would be hard to imagine—in this form—outside of Italy and far from its production districts.

Another example, among the many I could give in the university context in which I teach, is that of Anna Fregolent, who graduated in Fashion at Iuav in 2014 with a thesis on the relationship between fashion design, identity, and local history. In her Master thesis she described the local territory as: "a basin for knowledge, professionalism, creativity, and the right place for the development of design skills" (Fregolent 2014). Her thesis involved the creation of a territorial collective of cultural and craft institutions and firms in the fashion sector of Trieste. To this end she developed a capsule collection, referred to the history and professional experience of the knitwear designer Anita Pittoni "as an element of sharing" with the other members of the collective.

The concept of "learned in Italy" has also been at the core of the research project FabbriCrafter, carried out by the author of this paper at the Iuav University of Venice from September 2020 and September 2021, and funded by the Veneto Region under the scheme of the European Social Fund (ESF). The project intended to question the usefulness of engaging the figure of the crafter in the context of the industrial fashion manufacturers, to outline the possible futures of a collaborative network system between education, territory, and manufacturing districts. FabbriCrafter has experimented socially and environmentally sustainable practices within the manufacturing of fashion through key themes: upcycling, bio-materials and zero-waste. The interventions and the researches were carried out in collaboration with three companies in the Veneto area, partners of the project: Ileana Pasin, Womsh, and Tessitura La Colombina (Fig. 1). The figure of the crafter was an element of reflection in relation to industrial processes and design according to a participatory approach from many points of view, which however underline the need to recover "the value of thinking through the intelligence of hands" (Dal Buono, 2014). The proposed approach pointed out the human factor as an integral part of the design process, in the parallel attempt to establish a collective link with the local community and with the dynamics of industrial innovation.

5 Conclusions

Italy has a geographical context densely inhabited by people professionally involved in fashion industry and business. As this paper demonstrated, "learned in Italy" is a useful approach for better understanding and improve local resources that firms over the territory can offer in terms of knowledge and research capabilities. It is also a useful educational tool for fashion design students who wish to refine their communication skills and to formulate their requests for projects in the Italian production context, be it on craft or industrial scale. Students can fit their requests for projects into the periods of lesser intensity for the local fashion enterprises—for example, taking advantage of the gaps between production peaks. This permits the students to have access not only to refined workmanship, but also to the wealth of knowledge that skilled masters can pass on to them. Moreover, the "learned in Italy" approach is an attempt to establish a collective link with the local community and with the dynamics of industrial innovation in this challenging era that urges to reflect critically on the future of design education.

In conclusion, this approach can help in making visible the tacit knowledge of the Made in Italy. It is an invitation to a research journey through the production districts of Italy and an opportunity to reflect on the collaborative dimension of fashion and on its interdependence with respect to environmental elements. It is also a form of cultural, social and environmental sustainability of Made in Italy, facing nowadays some relevant issues of generational turnover and offshoring-reshoring dynamics, putting at risk of extinctions material and immaterial knowledges and skills of Made in Italy.

References

Bonifacic, M.: Learned in Italia. In: Lupano, M., Vaccari, A. (eds.) Insegnare il design della moda, pp. 20–25. dCP-Aracne, Roma (2014)

Celant, G. (ed.): The Italian Metamorphosis, 1943–1968. Guggenheim Museum Publications, New York (1995)

Collier Frick, C.: Dressing Renaissance Florence: Families, Fortunes, and Fine Clothing. Johns Hopkins University Press, Baltimore (2005)

Dal Buono, V.: Crafters, Makers e Re-Makers. Futuro è condividere una cassetta per gli attrezzi. DiiD Disegno Industr. **57**, 90–96 (2014)

Fregolent, A.: Progettare Anita Pittoni: esperienze nella moda di un collettivo territoriale. Master thesis, Università Iuav di Venezia (2014)

Greenblatt, S.J.: Renaissance Self-fashioning: From More to Shakespeare. University of Chicago Press, Chicago (1980)

Paulicelli, E.: Writing Fashion in Early Modern Italy: From Sprezzatura to Satire. Routledge, London-New York (2016)

Polanyi, M.: The Tacit Dimension. Routledge & Kegan Paul, London (1967)

Segre Reinach, S.: The Italian fashion revolution in Milan. In: Stanfill, S. (ed.) The Glamour of Italian Fashion Since 1945, pp. 58–71. V&A, London (2014)

Stanfill, S. (ed.): The Glamour of Italian Fashion Since 1945. V&A, London (2014)

Vaccari, A.: Gli stilisti nel tempo della moda in Italia: 1966, 1977. In: Borgherini, M., et al. (eds.) Laboratorio Italia. Canoni e contraddizioni del Made in Italy, pp. 224–241. dCP/Mimesis, Milano-Udine (2018)

Volonté, P.: Vita da stilista: il ruolo sociale del fashion designer. Bruno Mondadori, Milano (2008)

Zannini, B.: Sfridi. Design dallo scarto. Master thesis, Università Iuav di Venezia (2014)

Sustainable Fashion Product Development: The Importance of Skills in New Materials and Processes in the Academy and Industry Context

Marta Bicho[1,3(✉)], Madalena Pereira[1,2,3(✉)], Teresa Raquel Barata[1,3], Nuno Belino[1,3], and Rui Miguel[1,3,4]

[1] UBI, Covilhã, Portugal
{marta.bicho.fernandes,madalena.pereira,traquel,belino, rmiguel}@ubi.pt
[2] UNIDCOM/IADE, Research Unit in Design and Comunication, Lisbon, Portugal
[3] FibEnTech, Research Unit, Covilhã, Portugal
[4] CiAUD Research Unit– FA-UL, Lisbon, Portugal

Abstract. Materials and processes play an increasingly significant role in the development of new fashion products and can significantly affect product sustainability and longevity. These aspects, like many others, are part of the numerous issues associated with sustainability [1, 2]. The Fashion ecosystem has a high impact on society and the environment. Sustainability is an important issue for the textile and clothing industry associated with the different stages of the product life cycle and their impacts. The skills of specialists in fashion product development (FPD) and respective training will have to evolve towards increased knowledge of materials and processes; however, the academy is not keeping up with these changes that go along with the rapid transformation of the industry. There is a need for new competence acquisition models that are tailored to the present-day requirements. The objective of this investigation concerns the use and validation of a set of methodologies for the acquisition of skills by students, which allows for the development of sustainable fashion products for the market. Active methodologies that are being used, include workshops, competitions, industry partnerships, and PD in the context of companies and academy, and using trainers with experience in the industry. It concludes a progress in the knowledge of sustainable materials and processes combined with motivation.

Keywords: Sustainable Fashion Product · Academy · Fashion Ecosystem · Skills · ITV Industry

1 Introduction

The Brundtland report defines sustainability [3] as "…development that meets the needs of the present without compromising the ability of future generations to meet their own needs". The sustainable strategy is a plan of action that seeks to balance economic development with environmental protection and social development, ensuring the use of resources in a responsible and balanced way to meet the needs of the present without compromising future generations. Elkingtnon [4], also includes the governmental/institutional dimension and Capra [5], includes not only the institutional but also the cultural one. The Ellen MacArthur Foundation [6], also promotes the transition to the circular economy - a model that aims to make the product life cycle longer – resorting, for example, to recycling, upcycling, etc., contrary to the linear economy practiced until now. Finally, McDonough [7], reinforces the strategic component for regenerative sustainability in terms of product and service innovation. The concept of a "sustainable product" is tricky because the impact that any product has on the social and ecological environment depends not only on its manufacture, but also on its use and the life of the product until it becomes waste [8]. Furthermore, Marques et al. [9], refers to the need to classify pre- and post-consumption waste in a hierarchical way so that sustainable strategies can be subsequently developed together with companies and society.

According to Niinimäki and Hassi [10], the fashion industry primarily emphasizes technical and cost-related factors, such as the affordability of clothing and improved production efficiency, at the expense of sustainability concerns.

The development of sustainable fashion products is a crucial issue nowadays, as the fashion industry is known to be one of the most polluting in the world. According to the European Commission, European consumption of textiles and clothing is the fourth largest impact on the environment and climate change, after food, housing and mobility. Additionally, it ranks among the top three contributors to water and land use stresses, as well as being one of the five key factors influencing raw material consumption and greenhouse gas emissions. In this sense, it is important that the academy and industry work together to develop new skills, materials and processes that can contribute to more environmentally sustainable fashion. The Textile and Clothing sector employs more than 1.5 million Europeans, making it necessary to reformulate business models, encourage local versus global production, innovate in fashion product development (FPD) and invest in more sustainable processes and materials, as well as strategies for a circular economy. While, simultaneously, aiming to preserve employment opportunities, enhance working conditions, and improve wages, among other factors, thereby contributing to the realms of social and economic sustainability.

At the European level, it is defined in the Pact for Skills [11], embodied in a commitment to the development of skills, along with social responsibility, in the Green Pact and Digital Transformation. Regarding the European Ecological Pact - Green Deal [12], it is established that all emissions of greenhouse gases are eliminated by 2050.

At the same time, it seeks not only to promote economic growth disaggregated from the excessive use of resources, but also to ensure that, socially and environmentally, no person or place is left behind. These objectives are obviously reflected in the need to foster new skills and knowledge in terms of the entire school and training path, with the requirement for teaching adapted to this new reality in the document skills anticipation in Portugal [13].

2 Sustainable Fashion Product Development

2.1 Product Life Cycle

The fashion business has undergone several transformations over time, one of the most significant being the change from mass production of products of the same model, to the production of a greater number of models in smaller quantities. These changes in production, driven by changes in consumption, have changed the product's life cycle, as fashion brands seek to win over the consumer, with the permanent development of new products. The trend towards seasonal demand means that the life cycles of fashion products are even shorter. When compared with the life cycle of other products, it appears that the fashion product cycle reaches the peak of maturity quickly and quickly entering the decline phase. Consumers, manufacturers, and designers all play a decisive role in the product lifecycle. The new challenges of sustainability in the field of fashion have a direct impact on the product's life cycle, as there is a concern for valuing the pieces and consequently extending their useful life. Another interpretation of the life cycle of a garment is based not on the sales of a particular item, but on the chain of processes involved in the life of the product, from the extraction of the raw material used in its manufacture to its disposal of their waste [14]. In this perspective, the impact of the ecological footprint must be evaluated, as products are currently tending to have an increasingly shorter useful life, due to changes in consumption. The big challenge in the fashion business and for specialists in fashion product development is to create the product based on sustainable development.

The life cycle assessment is based on the ISO 14040 and ISO14044 standards, in which it is possible to quantify the environmental impacts associated with a system throughout its entire life cycle, be it a product or a process. This methodology makes it possible to assess the environmental impact through calculation parameters (with real data on raw materials), industrial processes, transport routes and waste generated, among other data, and are very important for improving products and systems in terms of sustainability.

There are companies that, based on life cycle assessment (LCA) reports, have eliminated certain raw materials, increased the consumption of recycled raw materials, readjusted production processes and adopted policies that are more in line with sustainability guidelines worldwide. There are already some examples of LCA in the textile and clothing industry (for example in Patagonia, Levi Strauss & Co or Adidas), as a strategy to assess and mitigate the environmental impacts of their products, but also as a means of communication and transparency of organization's values.

It is imperative that the Fashion Designer has clear knowledge of the life cycle of the product he intends to place on the market, from design, raw material, manufacturing, to the end of the product's life. New knowledge is required in a plural sector such as fashion.

Currently, the focus on the selection of resources with low environmental impact for the design of the product life cycle or eco-design of the products, expands to the design of an eco-efficient product-service system and to the design of equity and social cohesion [15].

2.2 Materials, Manufacturing, and Industry

According to the European Commission report, EU Strategy for the Sustainability and Circularity of Textiles [16], the production and consumption of textile products continues to increase, even an increase of around 63% is expected (textile and footwear) by 2030. In this sense, the European Union has developed several strategies to promote the sustainability and circularity of textile materials. Among which, the circular economy strategy, which aims to reduce waste and promote the reuse, recovery, and recycling of materials. This strategy has ambitious targets for the textile sector, in which it is expected that by 2030 textile products will be produced mostly from recycled fibers, along with a commitment to social and environmental responsibility. Regulation in the textile sector has been equally crucial, namely in terms of chemical products, such as REACH, with a view to protecting human health, as well as reducing/eliminating risks to the environment from exposure to hazardous chemical substances.

Equally important are certifications aligned with sustainability, such as the Global Recycle Standard (GRS), which validate and encourage companies to adopt more responsible practices throughout the supply chain and consumers to make more sustainable choices.

Another strategy promoted by the European Union is public-private partnerships, as a way to promote the sustainability and circularity of textiles. Along with financial incentives, measures and funding programs have also been put in place to promote innovation, the development of sustainable technologies in the Textile and Clothing Industry (ITV) and the adoption of BREFs [17]. Likewise, research and creation of high added value products have been encouraged, aiming at a sustainable bioeconomy, though, for example, the BE@T Project – Textile Bioeconomy in Portugal [18]. Being an example of a project with the participation of industry, technological centers, and universities. This project assumes a relevant role both in the research phase and in a later phase, in the transfer of knowledge, and although the number of scientific investigations published internationally in sustainability in ITV is small compared to other countries [19].

Portugal is a country that in the last 10 years has seen growth in the production textile and clothing sector, but also in the training of fashion designers. In 2022 alone, Portugal exported 5.7 billion euros in textile products (from January to November), an increase of 17.4% compared to 2019 [20] highlighting, among the type of products, knitted clothing and accessories, and mainly in the European market. The importance and need for knowledge of materials is currently one of the important skills for the industry and referred to in scientific studies on the matter [21]. In terms of human resources, ITV is estimated to increase by 138,000 in 2018 [19, 22]. The ATP Roadmap study refers to the urgent need for qualified labor for the ITV sector with new skills, some of which are essential in terms of sustainability, circular economy, and bioeconomy. But according to the study: Prospective Vision and Strategy ITV 2030, ITV could lose from 11 thousand to 56 thousand jobs by 2030 [23]. Are the companies' sustainability, circular economy, and bioeconomy strategies sufficient to minimize this estimate in terms of added value in the product? Or will we have specific needs in terms of human resources that we are not preparing in time?

3 Sustainable Fashion Product Development

3.1 Problem

Currently, fashion designers and fashion product development specialists will have to align themselves with the new European directives associated with sustainability and the circular economy, whose objectives are defined for 2025, 2030 and 2050. But the existing skills and training in Portugal are recent, limiting the supply of human resources with this profile, which consequently constitutes one of the limitations for the evolution of the industry and market in the coming years [24].

It should also be kept in mind that we live in a time where "there is an increasing need to retrain human resources throughout their working life, and universities have to better tackle this task, finding new ways to collaborate with companies in co-creation of educational paths and new training in work experiences" [25]. The industry is looking for human resources that have knowledge and skills associated with new sustainable materials and processes.

Conti [26], states that the transformation to a digital economy is redefining the ways in which people work, so it is necessary to think of "new" ways of working in academy that reflect the types of skills needed and in which innovation becomes the key to success.

In view of the above, the following question arises: What methodologies can be employed and implemented in sustainable fashion product development to enhance the skills of future students and specialists and meet the labor market demands in Portugal?

3.2 Methodology

The research work was developed based on active methodologies based on interviews, questioning, experimentation, project development in the classroom and industry. In a first phase, interviews with ITV entrepreneurs are used to understand what skills are necessary for the development of a sustainable fashion product, namely the importance of knowledge of materials (principals and secondaries) and processes. In a second phase, the Project Design Methodology was applied for the development of a sustainable fashion product in a pedagogy of acquiring skills "learning to do" and "learning to learn". Within Design, several authors have developed active design methodologies for solving society's problems.

Bonsiepe [27], the methodology serves as "(…) the modalities of action in a given field of problem solving". Bonsiepe [27], methodology aims to provide guidance for the product design process and consists of several stages. Table 1 presents this methodology adapted for the Project developed by the students in this investigation.

Table 1. Design Methodology adapted from Bonsiepe (Source: own)

Problem	General: Excessive consumption of resources in the textile and clothing industry and the use of unnatural resources that harm the environment Specific: The lack of knowledge and skills by fashion design students and future specialists in sustainability at ITV in the use of main and complementary sustainable materials in garments, selection of processes, techniques and technologies among others to contribute for the transformation the industry more sustainable and circular in FPD
Analysis	An analysis is made of the main and complementary materials, existing in the Portuguese textile industry, through workshops and visits to fairs and their respective aesthetic and functional characteristics associated with sustainability and with reflection and discussion between teachers and students
Problem definition	Identified natural resources and processes for reduce the excessive consumption and harm the environment. How can create a FPD more sustainable in terms of selection materials? What skills we need for that? The skills necessary for the development project of a sustainable fashion product are analyzed together with the industry. Active methodologies are identified and analyzed in partnership with the industry for an effective transmission of knowledge and acquisition of necessary skills. The definition of the product more sustainable, definition of materials more sustainable and definition of requirements and the form and space in which to transmit them are carried out
Draft/Generation of alternatives	Ideas and alternatives are generated for create a Fashion Product, not only in the pedagogical methodologies "learning to do" and "learning to learn" (business and academic environment), but also in the practical development of the first prototype solutions, followed by analysis and testing of prototype changes with discussion and reflection about sustainability
Project	The final solution is produced and presented to an expert external jury for validation and evaluation. The final solution is valid against a set of requirements: sustainability, aesthetic quality, selection of materials/fabrics and production techniques. Students present orally to the jury and justify their choices. This is followed by a discussion between students and the jury in a classroom – workshop environment. A prize is awarded in recognition of merit Finally, the students were invited to present improvement solutions and to transmit feedback on their learning Fashion Show and award for students and social media for explain the sustainable concept

Phase 1 – Problem Identification and Analysis

The profile of fashion designers will have to adapt to the new paradigm of sustainability, which is a growing concern in academy alongside industry. The skills of fashion designers are facing a paradigm shift due to society's relationship with design and the demand for a sustainable future for existence [28]. To respond to new requirements for industry, "new knowledge and skills are needed, for example in relation to new materials and processes. In this context, it is crucial that the academy adjusts quickly and promotes the transfer of knowledge in new sustainable materials and processes in academy" says CEO's of companies of clothing in interviews. This includes providing students with access to information and resources on biomaterials, sustainable production processes, design and the circular economy, and the opportunity to experiment and work with these new materials and processes in academic projects [16]. In the fashion industry, the importance of skills in new materials and processes is increasingly evident. The new requirements for the design of sustainable products include the imposition of textile products having minimum values of recycled materials, along with a digital product passport, providing the user/buyer with information about circularity and other environmental data.

The industry must gain knowledge and accelerate this paradigm transition, in the development of new sustainable materials or in the use/optimization of available materials and know how to use them in their products. These materials include organic, recycled, and biodegradable fabrics, as well as materials with a low environmental impact. To this end, the adoption of more sustainable production practices, the use of renewable energy, more effective waste management and a significant reduction in the carbon footprint are some of the steps in this sustainability cycle.

It is also imperative that at the pedagogical level, the academy adopts new teaching models, through active methodologies, such as workshops or competitions with industry and FPD, in the context of companies and academy, to provide students with the necessary skills to develop products of sustainable fashion involving local socio-economic actors is fundamental nowadays [29].

By working together with teachers/trainers with experience in the ITV industry, students can acquire more in-depth and knowledge-transmission-oriented, problem-solving and soft skills-focused knowledge, exploring multidisciplinary knowledge and thus improving their skills and motivation in relation to sustainability.

In this sense, one of the good practices in higher education are partnerships with companies and the promotion of visits to specialty fairs, classes in a company context and workshops, which are dynamic in the context of the academic classroom.

The company Tintex was an example of this pedagogical practice in this work, having gone to the academy, promoting a workshop, in which it shared the company's experience, the productive processes and innovation in terms of products and processes, always with the lever of sustainability. Being an opportunity for students to hear directly from the company's interlocutors, this workshop served to discuss issues such as: the relevance of materials in the life cycle of products; creative processes and sustainability; use of waste in textile substrate; innovation in terms of fibers, construction of fabric or mesh, or in the process of ennobling. Likewise, the importance of communicating the company's values associated with the product, such as natural, sustainable, recyclable, etc., was highlighted. One of the aspects also presented in this workshop was the importance of certifications as a process of recognition and transparency in the value chain. Finally, in the practical component, students observed, touched, and felt different samples of the company's sustainable knits, and observed garments made (see Fig. 1). Furthermore, Tintex has established partnerships with other companies to promote sustainability, innovative solutions for the industry and circularity in the textile sector.

Other examples, from the Workshop, of the company in the academy, were the European companies of good practices in terms of sustainability such as Freudenberg and Lenzing, the latter with a presentation of textile fibers that are so well known and relevant regarding sustainable materials, as an example, the fibers of Tencel™, Ecovero™

e Veocel™. Through this workshop, an active approach was promoted with the discussion of the role of the Fashion Designer in the conception of the entire product, from the choice of materials, such as fabrics, accessories, trimmings, and labels, to the packaging material and product presentation. At the same time, it also presented questions such as: chemical and mechanical recycling, and respective fields of application and limitations given the structure of the fibers; the company's social and environmental responsibility in its local and global operations; the different productive processes versus type of fibers; and, finally, the minimization of the environmental impact, through the implementation of innovative technologies in its production process and the use of closed production processes, reducing the consumption of water and energy, significantly reducing the emission of pollutants. Another aspect addressed was the relevance of partnerships between companies as a strategy to drive innovation and development (see Fig. 1).

Fig. 1. Workshop Tintex, Freudenberg Lenzing (Source: own)

These workshops allowed students to come into contact with innovative and differentiating textile products, developed and manufactured under the prism of sustainability and circularity. The arrival of industry to university, as a learning space, not only represents the sharing of the state of the art, but also contributes to awakening students' curiosity for themes that are so crucial at the current time. In this sense, the physical contact with the materials that are on the market was fundamental and differentiating. Classes were also taught in the context of a local clothing company to learn about the processes in a real context - Twintex. The Fashion Designer while being creative, but also under the seal of circularity, is also a promoter of new materials and, for that, must know how to find partners and establish synergies with the different players in the value chain. This interaction between industry and academy resulted in a very beneficial pedagogical methodology for the acquisition and sharing of knowledge, in a factory and informal environment, as opposed to other classical, classroom and expository methods. At the end of this action, the students were asked to understand at what point this action had influenced their knowledge and motivation for the development of a sustainable fashion product for the next phase.

Phase 2 – Development of a Sustainable Fashion Product Project
Faced with the new challenges around sustainability, the fashion industry has been reinventing itself and transforming itself, not only in terms of adopting good sustainable production practices, with strong investment in innovation, but also in terms of communicating these same environmental concerns to the client/consumer. The work that has been carried out by the industry, across the entire value chain, to keep up with new market demands and build the necessary bridges in a fundamental contribution to the future of sustainable fashion is remarkable. That said, the industry has been promoting partnerships through various types of activities for young students of Fashion and Design Schools, based on the pillar of sustainability. The objective is to raise awareness of its products and their application in projects developed in the academic field, which allow responding to the production of sustainable fashion products. At the same time, they enable the involvement and commitment of new designers with the theme of sustainability.

Thus, Freudenberg Performance Materials Apparel, world leader in the production of various materials for the clothing industry, such as quality and sustainable interlinings, horsehair, and stuffing pastes, currently has a wide range of articles that are produced with recycled polyester, biodegradable materials, and natural fibers, to which certifications for the aforementioned purpose are additionally associated. For Fletcher and Grose [29] most sustainability has as its starting point the innovation of materials and processes. Innovation regarding new sustainable materials has been encouraging fashion brands to adopt more responsible production, focusing on the development of sustainable fashion products.

To leverage cooperation between students and industry, it was proposed to develop a sustainable garment, using one or more products from Freudenberg Performance Materials, a project whose design methodology is presented in the Table 1 considering the life cycle, conservation, and extension of the life of the product through Design solutions, selected materials and processes and with the generation of ideas and alternatives. Ten projects were developed whose sustainable materials were one of the essential aspects and selected based on the knowledge acquired in the workshops, which are presented in the Table 2.

The execution of the projects was carried in and out at the academy in partnership with the local industry. The validation of the projects before a group of experts was presented, justified, and discussed by the students in multiple dimensions: sustainability requirements, aesthetic quality, material/fabric selection, materialization processes and techniques, and end-of-life solutions. This action allowed the Designer to reflect on his options in the development of a sustainable fashion product and to develop his communication, synthesis, and argumentation skills.

3.3 Analysis of Results and Discussion

The first phase of the learning process, in which active methodologies were resorted to, in an informal, industrial environment and with elements from the industry and materials from the market, it is verified in this case applied and analyzed an increased motivation by the students in fashion design and improvement in the acquisition of knowledge of sustainable materials with the methodologies described above. This motivation, curiosity

Table 2: Characteristics of the Materials selected for the projects (Source: own)

	Design*	Main Material	Complementary Material *Freudenberg Performance Materials*
Project 1	Fragility Jacket	100% Silk - pre-consumer dead stock from industry	wadding HO 606 10 - 100% Tencel (Lyocell) – Biodegradable in 57 days
Project 2	About Temperature Contrast	100% Wool - pre-consumer dead stock from industry	Interlining RCY50 100% recycled polyester
Project 3	Wear Tobacco	Upcycling ounces of tobacco	Wadding HO 606 10 - 100% Tencel (Lyocell) – Biodegradable in 57 days
Project 4	2IN1	100% Organic Cotton	Wadding Fiberball thermal insulation HO 29R 70% Polyester fibers from post consumer recycled PET bottles
Project 5	Pro to Type of Love	Upcycling	Interlining RCY50 100% recycled polyester
Project 6	Pro to Type - Vest	Upcycling	Hair linning RCY272 Eco 100% Recycled Polyester
Project 7	Wool Armor	98% Organic Cotton 2%EA	Interlining RCY50 100% recycled polyester
Project 8	Hoody	100% Wool - pre-consumer dead stock from industry	Interlining RCY50 100% recycled polyester + Wadding HO 606 10 - 100% Tencel(Lyocell) Biodegradable in 57 days
Project 9	CobWeb	100% Organic Cotton	Interlining RCY50 100% recycled polyester
Project 10	Percibir	100% Silk (skirt)	Wadding HO 606 10 - 100% Tencel (Lyocell) – Biodegradable in 57 days

* The Designs and processes of the products were studied to extend the life of the product and with prior analysis for a 2nd life through easier reconstruction and Upcycling techniques

and participation in the discussion could be an indicator of the acquisition of additional skills. One of the problems that the academy is experiencing, in this post COVID'19 phase – lack of skills "learn by doing".

In the second phase: the challenge of developing a project within the scope of a curricular unit centred on a real problem of a sustainable fashion ecosystem for public validation by a specialist jury (academy and industry) with a final award. It appears that students, in addition to motivation, had a greater commitment in all phases of the design methodology, innovating in the selection of main and complementary materials with sustainable characteristics, in design solutions, applying in some cases multifunctionality and modularity, in the selection processes and techniques, and in complementary services, in the technologies used in the materialization of the project and in the communication of the project.

Fig. 2. Examples of industry classes with students. (Source: own)

In the Fig. 2, 3 and 4 we can observe examples of the classes in industry, presentation and results discussions with jury obtained from the projects developed in terms of communication, presentation, and type of product. The commitment and results obtained demonstrate that the reinterpretation of learning spaces, active methodologies, involvement with the industry, rewarding merit, professors with experience in the industry, partnership with the national and international industry are fundamental in obtaining better results that consequently are the reflection of the knowledge and skills acquired.

Added to these factors is planning and organization of different actions in time and responsibility transferred to students for a free way of thinking throughout the entire learning process. These results are in line with that reported by Ceschin and Gaziulusoy [29], Agarwal [28], and the European Commission 2030 strategy for sustainability at ITV.

Fig. 3. Presentation and discussion with a Jury (Source: own)

Fig. 4. Fashion Show (Source: own)

4 Discussion and Conclusions

Sustainable fashion is an emerging trend in the fashion industry, driven by growing environmental and social concerns around the world. However, the development of sustainable fashion products requires specific skills and knowledge about new materials and processes, both in academy and in industry. At the university, it is important that fashion courses include solid training in sustainable materials and production processes, allowing future fashion professionals to have a clear understanding of the environmental and social impacts of conventional fashion and to develop skills to design and produce fashion products sustainable. It is also up to the academy to introduce new teaching practices, more attractive for students and effective in terms of meeting objectives, knowledge and learning skills, given the needs of the industry. In this sense, the academy must ensure proximity to the industry, developing strategies so that the skills of Fashion Designers are aligned with the expectations and requirements of the industrial sector, on a global market scale. Agarwal [28], refers that today the curriculum in fashion design must develop practical skills and creative sensibilities that can satisfy the needs of the industry. The different types of partnerships and practices in the academic-industry context are necessary and fundamental. As far as the industry is concerned, it should continue to have, along with the university, a relevant role in the research and development of new materials and production processes to create more sustainable fashion products, but not only. It is also up to the industry, together with the academy, to promote discussions on the new skills and knowledge needed for future professions, to design activities such as: new pedagogical methodologies, or reformulation of existing ones that contribute to valuing and increasing the component of "learn to do". Attracting human resources at ITV with skills in sustainability is a crucial element for the management and implementation of sustainable practices in the value chain and consequently with an impact on the life cycle of the fashion product. If this type of professional profile is not produced, the industry will not be able to evolve in the face of the challenges imposed on it by the European Union's ITV 2030 Sustainability Strategy [16, 24]. In short, it is necessary to resort to active and non-passive pedagogical methodologies in collaboration with the industry, in the training of new Fashion Designers, through the reinterpretation of learning environments, promotion of visits to specialty fairs, workshops, classes in an industry context and other challenges proposed by the industry in academia, promoting freedom of thought. These projects, which use materials and technologies from the productive industry and are validated by an external and internal jury of the academy, must be associated with a curricular unit program, or in intersection with several curricular units, which contribute to the acquisition of new knowledge and skills by students, but also increases motivation in acquiring these same skills in the case study of this work. There are certain limitations to this investigation as it is currently being implemented exclusively in the field of public education, specifically targeting Fashion Design students (textiles and clothing) from a public higher education institution in Portugal during its initial phase. In the future, it is intended to develop more case studies within the scope of other curricular units, and in the connection between them with other public and private schools, to carry out a national comparative analysis. The contribution of this research from a practical point of view is fundamental for the development of reformulations and new formations based on case studies at the university, which, in view of the above, is lagging in terms of sustainability, in relation to the international context, and it is necessary to promote the sharing of practical knowledge of the ITV industry, which is extremely relevant in Portugal. These conclusions are in line with the results obtained by Seixas et al. [30], in which he states that "the textile designer needs to be prepared to respond to the challenges of the textile industry and strengthen the dialogue between university and industry".

Acknowledgments. The authors acknowledge the financial support from integrated project be@t – Textile Bioeconomy (TC-C12-i01, Sustainable Bioeconomy No. 02/C12-i01/202), promoted by the Recovery and Resilience Plan (RRP), Next Generation EU, for the period 2021–2026.

The authors acknowledge the financial support by national funds financed this work through FCT – Foundation for Science and Technology, I.P., within the scope project UIDB/00711/2020 from UNIDCOM/IADE - research unit in Design and Communication.

The authors are very grateful for the support given by the research unit Fiber Materials and Environmental Technologies (FibEnTech-UBI), on the extent of the project reference UIDB/00195/2020, funded by the Fundação para a Ciência e a Tecnologia, IP/MCTES through national funds (PIDDAC).

References

1. Fletcher, K., Grose, L.: Fashion & Sustainability: Design for Change. Hachette, UK (2012)
2. Cooper, R.: Design research–Its 50-year transformation. Des. Stud. **65**, 6–17 (2019)
3. Brundtland, G.: Report of the world commission on environment and development: our common future. United Nations General Assembly Document A/42/427 (1987)
4. Elkington, J.: Towards the sustainable corporation: win-win-win business strategies for sustainable development. Calif. Manage. Rev. **36**(2), 90–100 (1994)
5. Capra, F.: The Web of Life: A New Scientific Understanding of Living Systems. Anchor Books (1996)
6. Ellen MacArthur Foundation: Towards the circular economy: economic and business rationale for an accelerated transition (2013). https://ellenmacarthurfoundation.org/towards-the-circular-economy-vol-1-an-economic-and-business-rationale-for-an. Accessed 28 Apr 2023
7. McDonough, W., Braungart, M.: The Upcycle: Beyond Sustainability–Designing for Abundance. North Point Press (2013)
8. Stegall, N.: 'Designing for sustainability: a philosophy for ecologically intentional design. Des. Issues **22**(2), 56–63 (2006)
9. Marques, A.D., Marques, A., Ferreira, F.: Homo sustainability: circular economy and new business models in fashion industry. SN Appl. Sci. **2**, 1–5 (2020)
10. Niinimäk, K., Hassi, L.: Emerging design strategies in sustainable production and consumption of textiles and clothing. J. Clean. Prod. **19**(16), 1876–1883 (2011)
11. Pact for Skills (2020). https://ec.europa.eu/social/main.jsp?catId=1517&langId=en. Accessed 28 Apr 2023
12. European Green Deal (2019). https://commission.europa.eu/strategy-and-policy/priorities-2019-2024/european-green-deal_en. Accessed 28 Apr 2023
13. Skills anticipation in Portugal (2022). https://www.cedefop.europa.eu/en/data-insights/skills-anticipation-portugal. Accessed 28 Apr 2023
14. Salcedo, E.: Moda ética para um futuro sustentável. Editorial Gustavo Gili (2014)
15. Vezzoli, C., et al.: Design for Sustainability: an introduction. In: Vezzoli, C., et al. (eds.) Designing Sustainable Energy for All. GET, pp. 103–124. Springer, Cham (2018). https://doi.org/10.1007/978-3-319-70223-0_5
16. Estratégia da UE em prol da Sustentabilidade e Circularidade dos Têxteis (2022). https://eur-lex.europa.eu/legalcontent/PT/TXT/HTML/?uri=CELEX:52022DC0141&from=EN Accessed 28 Apr 2023
17. Roth, J., Zerger, B., De Geeter, D. et al.: Best available techniques (BAT) reference document for the Textiles Industry. EUR 31392 EN, Publications Office of the European Union, Luxembourg (2023)
18. Be@t Project: Bioeconomia na Indústria Têxtil (2023). https://bioeconomy-at-textiles.com/. Accessed 28 Apr 2023
19. Abbate, S., Centobelli, P., Cerchione, R., et al.: Sustainability trends and gaps in the textile, apparel, and fashion industries. Environ. Dev. Sustain. 1–28 (2023)

20. Jornal, T.: Exportações têxteis disparam 14,6% até novembro (2023). https://jornal-t.pt/noticia/exportacoes-texteis-disparam-146-ate-novembro/. Accessed 28 Apr 2023
21. Haug, A.: Acquiring materials knowledge in design education. Int. J. Technol. Des. Educ. **29**, 405–420 (2019)
22. Pamésa, C.: A Fileira Têxtil e Vestuário No Horizonte 2025, Promotor Associação Têxtil e Vestuário de Portugal - ATP (2019)
23. Pamésa, C., Agis, D., Vaz, P.: Visão prospetiva e estratégias ITV 2030, Edição Associação Têxtil e Vestuário de Portugal - ATP (2021)
24. Nabais, J.: (2023). https://eco.sapo.pt/2023/03/16/empregos-verdes-dao-salario-20-acima-mas-escassez-de-talento-trava-progresso/ Accessed 18 Mar 2023
25. Bertola, P., Mortati, M., Vandi, A.: Future capabilities for creativity and design: how digital transformation is reshaping fashion careers and higher education. Mandragora, Florence (2020)
26. Conti, G.M., Motta, M.: Bringing responsible fashion approaches to the fashion industry: a new educational model. In: Proceedings-GFC 2021 Warsaw, pp. 1–15 (2022)
27. Bonsiepe, G.: Teoria e prática do design industrial. Centro Português de Design, Lisbon (1992)
28. Agarwal, V.: Sustainable fashion education in changing world scenario. Eur. J. Sustain. Dev. **7**(1), 365 (2018)
29. Ceschin, F., Gaziulusoy, I.: Evolution of design for sustainability: from product design to design for system innovations and transitions. Des. Stud. **47**, 118–163 (2016)
30. Seixas, S., Montagna, G., Félix, M.: Weaving the skills of the textile designer as a contribution to the construction of a textile ecosystem of the future. In: Broega, A.C., Cunha, J., Carvalho, H., Providência, B. (eds.) CIMODE 2022, pp. 505–517. Springer, Cham (2022). https://doi.org/10.1007/978-3-031-16773-7_43
31. Murzyn-Kupisz, M., Hołuj, D.: Fashion design education and sustainability: towards an equilibrium between craftsmanship and artistic and business skills? Educ. Sci. **11**(9), 531 (2021)

Women, Fashion Design and Ancestrality: Reflections on the Past and Future Possibilities

Fernanda Enéia Schulz(✉) and Joana Cunha

Centro de Ciência e Tecnologia Têxtil, Universidade do Minho, Campus de Azurém, Guimarães, Portugal
ferschulz.design@gmail.com, jcunha@det.uminho.pt

Abstract. The design profession is not immune to the systematic oppression of patriarchy, although it has its own gender-specific histories. From the history of design, it is possible to understand the relationship and participation of women in its development and understand the need for other approaches, especially those associated with fashion design and the relationship with female textile ancestry. Currently, there is a considerable number of studies that address female relationships, the role of gender and female textile ancestral knowledge both in relation to the area and design processes. These add data and values to the question of how gender is constructed through design processes, which can directly contribute to more sustainable results, as well as to the configuration of meaningful and culturally relevant products. This article aims to address the marginalized relationships in the history of design: women - fashion design - female textile ancestry.

Keywords: Fashion Design · Women in Design · Feminism · Gender inequality · Craft education · Textile ancestry

1 Contextualization

Historically, what is verified are data that reflect the bases of the design of the 19th century with a masculine and separatist character, where feminine knowledge is interpreted as domestic and masculine knowledge is associated with production [1]. For Moura [1], such gender and class hierarchization is explicit in the language used by design to characterize the workspaces and the way in which design itself is practiced, where it is necessary to emphasize that spaces such as the atelier maintain a connotation distant from the ateliers of domestic spaces, since the design profession is not seen as an activity connoted with domestic and female activities. With regard to the segmentation of areas in design, Moura & Ramos [2] refer to the Bauhaus, a school that had a great impact on industrial design in the 20th century, shaping and inspiring many academic courses, and training a good part of the world-renowned designers. According to the authors, the school promoted the discourse of equality among students, but in practice what happened was a sexist separation, where women were mostly enrolled in the textile area, an occupation for which they were considered "vocational", connoted to domestic spaces and to feminine characteristics, such as calm, patience, obedience, delicacy and submission. Corroborating with the discourse, historical and statistical, Lupton [3] and Hall [4] point to the great absence of women in the male-dominated profession. A survey carried out by the Design Museum UK in 2018 showed that only one in five professional designers are women, despite these being seven out of ten design students [5].

The lack of female representation in the academic environment, through lectures, publications, knowledge from the female universe and the erasure of these references in the history of design itself, contributes to the propagation of the male model of design, making it difficult for female students to pass from the classroom to the labor market [5]. All these shortcomings not only make access to women and minority genders difficult, but also generate uncertainty, lack of confidence and insecurity to assume creative or leadership positions. For Ellen Lupton [3], it is not enough for women to create support networks, it is important that the faculty of design institutions explore contexts in which anyone can develop intellectually. According to the aforementioned author, currently, designers are looking for tools and practices that develop design in an inclusive and sustainable way, since the insertion of cultural, collective and collaborative processes is part of design projects [6].

It is in this context, between the historical and academic space of design, that the need arises for didactic implementation capable of recognizing the intellectual and creative value of women, as well as a sense of historical reparation, through the inclusion of activities, methods and methodologies that encourage the development of female ancestral techniques. Almeida [6] defends the point of view of gender discussion in design relations, as a sociocultural construction, pointing to the development of semiotic values and the need to identify erasures and invisibilities in the process. Through an inclusive gender perspective, it is possible to understand feminine design as an approach and invitation to contemporary explorations, falling short of the uniform discourse of the design field [7], since design is also the result of cultural and social experiences.

With regard to fashion design processes, an area marginalized by design itself, due to its direct association with the feminine, which for a long time was conditioned to domestic and family spaces [8], it is important to remember that the area is directly associated with cultural expression and interacts directly with social values and canons. For Schulz and Cunha [9] it is through understanding the relationship between clothing, design and handicraft that values of cultural and symbolic identity are re-inserted into society through fashion production. In this way, it is possible to observe the absence of female ancestral craft processes (crochet, knitting, embroidery, weaving, sewing, etc.), classified as distinct from design, and consequently minimized or excluded from the historical context by historians and academics [10], since that these are part of the female domestic context and do not emphasize the relationship with modernism and industrial concepts, which Buckley [8] determines as selective and exclusionary design. It is important to remember that the craft techniques associated with textiles have always been part of the global female context, both in individual contexts of production and use, and in social contexts through communities and groups of artisans [9]. For the feminist designer Middleton [11], when design denies the feminine, as much as the femininity expressed in projects, it is affirming that in its space, of design, there is no appreciation of female professionals. Designer and writer Anja Neidhardt [5] adds that it is through hierarchization that men are seen as creators and women are seen as muses, which strengthens disbelief in women's creative potential.

It is from a reactive and avant-garde perspective, which causes a rupture with the historical and traditional concepts related between design and fashion design, transcending the masculine barriers imposed on the discipline of design, that we seek to explore the feminine universe, in the field of fashion design education, as a way of exposing realities in the creative, ethical, innovative and culturally sustainable fields.

2 Female Ancestral Knowledge, Fashion Design and Academia

We understand the importance of female creative and intellectual approaches in academic spaces as a way to change the traditional rhetoric of design history. In this way, the approaches that follow, start from a critical analysis of the fashion design education in Portugal and its relationship with the erasure of female textile ancestry, as well as pointing out evidence of a male scenario in fashion design at global and Portuguese levels.

2.1 Feminine Ancestral Knowledge and Fashion Design Education

Reflection on women's participation in historical, cultural and economic development represents a cross-cutting need in several areas, including design, an area whose capacity is to make possible the reproduction of artisanal textile typologies that are part of the various cultural lexicons. Understanding the relationships and transmission of ancestral and craft knowledge, associated with female cultural learning, provides the rescue of identity, cultural and symbolic values [12] of textile craft techniques and the appreciation of female intellectual work. In this process, the designer plays an important role in the designer-craftsman interaction process, as he must have creativity, ability and sensitivity to unite traditions, cultural and emotional values, innovation and vision to create products that tell stories [13]. The project "Voces De La Artesanía" [14] develops artisan-designer relationships, where artisans feel listened to and valued, and designers recognize artisans as living sources of wisdom. It is understood that sustainability in design must overcome the ecological barriers associated with the term, meeting values such as quality and social and cultural integrity [9]. Transcending barriers imposed on handicraft techniques and seeking to explore the universe of possibilities through the designer's eye, it is fundamental to insert women's textile cultural knowledge in design teaching, using methodologies that explore memories, cultural ties, identities, traditional and geographic knowledge [1], as it is in the academic space that the role of the designer is shaped, being a mediator between local and technical knowledge, connecting historical human factors and technologies [15]. It is essential to develop new methodological strategies for design education, encompassing factors and experiences beyond those already contained in traditional curricula [16]. Crafts are part of everyday life and it is necessary to propose reflections on the many ways in which design is produced, especially in the women-design axis [8].

2.2 Exclusion of Crafts – The Portuguese Case

The fashion design education in Portugal has its bases in product design itself, and in the 1990s it underwent curricular adaptations to better meet the global needs of the fashion industry [16], which contributed to the rupture with regard to cultural and ancestral heritage. In this way, it is based on the traditional models of design education, which, among other issues, are supported by the idea that "making" involves giving shape to the material through the designer's project, with the final objective being the created product [17]. Thus, it is common for education to be orientated towards solving problems, largely supported by drawing skills (manual and digital), knowledge of materials (as raw material) and creation of prototypes of clothing.

In this view, the materials are the textile surfaces that conform to a given design, which is why, in general, education in fashion design only values the practice of sewing as a technique for assembling this material, leaving aside other millennial textile practices that have enormous potential for the development of the designer's tacit knowledge.

Recent works have shown the importance of giving greater emphasis to craft processes in designer training, Ingold [18] argues that in the design process the primacy should be in the process of shaping the material and not in the final product, which is why it becomes more important how the material is transformed, that is, the techniques used. On the other hand, Niedderer et al. [19] argue that craft activities are activities that, when well executed, give individuals a sense of pride and confidence, which motivates them to share their knowledge and promote experimentation. Kokko & Räisänen [20] stress the importance of craft disciplines as a way of sustaining the local and global changes necessary for sustainability.

Although several studies conducted in the last decade have made it clear that craft disciplines are important for the training of designers, in the case of fashion design, it is noted that the curricula of courses in Portugal does not include an approach to these practices [21]. The reasons for excluding artisanal textile practices in the academic approach to fashion courses are related either to their direct association with the feminine, as previously mentioned, or because they are practices perceived as having a lower status due to either their low economic value or lack of intellectual rigor as referred by Niedderer [19]. In any case, these are indisputably ancestral female practices associated, in many parts of the world, with female learning, in which girls learned to become women by spinning, weaving, embroidering and actively participating in the production of the trousseau that marks their entry into married life [22].

The recognition of the association between fashion design and handicrafts corroborates the way designers understand human ancestry, converting these into creative, sustainable materials that can be linked to innovations and the development of differentiated products, endowed with these ancestral cultural values [9]. It is in the handicraft-design interaction that dialogues can be developed that allow the insertion of aesthetic and symbolic values into products [23], imbuing them with their own poetics [24]. In this process, the union between design and handicraft gives rise to new possibilities for creation and makes room for the creation of practices capable of making traditions relevant in the contemporary and future scenario of fashion design. On the other hand, according to Kokko [25], the inclusion of craft practices as subjects in higher education gives value and visibility to crafts while at the same time reinforcing their cultural identity.

The result of this discourse, between craftsmanship and design, leads to an appreciation that is felt by the difference, authenticity and uniqueness of production [23].

Despite being a timid and slow process, in Portugal initiatives are already emerging that aim to promote the various forms of cultural expression associated with handicrafts, such as the example of the "TASA" project, born in 2010, which promotes dialogue between Portuguese artisans and designers, with the aim of combining tradition and innovation, in a set of products of interest to the consumer, associated with social and cultural sustainability [26]. Another example worth mentioning is the "Portugal Manual" project, born in 2018, and which disseminates manual and cultural knowledge through products designed by artisans and designers. Associated with fashion design, it is possible to find names of women who work with traditional Portuguese handicrafts, revisited by design and made through the hands of expert artisans, as is the case of designer Helena Cardoso [27]. In a more recent example, we mention Béhen, created in 2020 by designer Joana Duarte, who, through collective work methodologies, associated with design practices and methodologies, recreates traditional Portuguese embroidery, not only with the aim of reviving the technique, but also to work crafts from a cultural and feminine point of view, valuing the identities and individual creative value of each craftswoman [28].

2.3 Gender in Fashion Design: Global Vision and Portugal Vision

The idealization and conception of fashion design, as education or as a creative area, has always been associated with the feminine, due to the nature of the activity itself, endowed with femininity, delicacy and knowledge associated with the domestic, and therefore it is not surprising that the female gender thus constitute the majority among students in design courses. Ironically, what is observed is the male occupation of the profession of fashion designer [29]. Although in recent decades we have been witnessing a slow feminization of the fashion designer (or fashion creator) profession, the fact is that the vast majority of renowned fashion designers/creators continue to be male [30]. Contrasting this scenario, what is recorded is, in many cases, the use of female knowledge and craftsmanship, which, far from the spotlight of the fashion scene, remain hidden, even if essential to creators. Duran [30] exposes the case of Karl Lagerfeld's interview for the New York Times in 2006, where the designer, now deceased, mentioned the term "*petites mains*" [little hands] when referring to the countless women behind his work.

In 2017, the BoF[1] website carried out a survey, cited by Marques [29], referring to the main world brands present at international fashion weeks, that revealed male superiority in the creative direction of these brands. In 2016, Pike [31] reports that only 14% of the top 50 world fashion brands are intellectually directed by women.

Given the scarcity of scientific literature on this subject, we sought to find out if these data were maintained in the present. For this we conducted an internet search, using a new private browser window, using the search term "Best Fashion Designers"[2]. The analysis focused on the first 5 search results (Fig. 1) and on the images of the first

[1] The Business Of Fashion - https://www.businessoffashion.com/

[2] Search conducted in 23, April 2023.

50 designers obtained (Fig. 2). Given the scope of the search term, the results refer to the universe of fashion designers from the 20th century to the present.

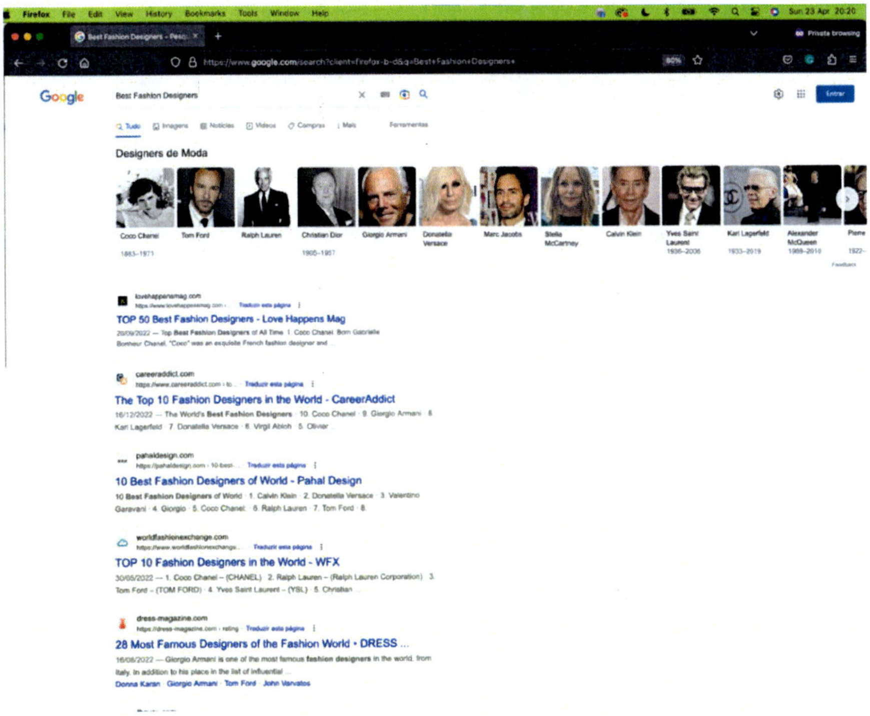

Fig. 1. Search results for "Best Fashion Designers".

Among the 50 designers in Fig. 2, only 16 are women. A similar relationship is obtained from the analysis of the first 5 articles found. The article "TOP 50 + Best Fashion Designers"[3] names 57 designers of which only 18 are women; The CareerAdict article, a publication by Joanna Zambas entitled "The World's Best Fashion Designers"[4] features only 2 women in the Top 10, Coco Chanel and Donatella Versace; The pahaldesign.com article entitled "10 Best Fashion Designers of the World"[5], presents 3 women in the Top 10, they are Coco Chanel, Donatella Versace and Kate Spade; "TOP 10 Fashion Designers in the World"[6], presents 3 women in the Top 10, they are Coco Chanel,

[3] TOP 50 + Best Fashion Designers. (2022). *LoveHappens*. [online] Available at: ttps://www.lovehappensmag.com/blog/2022/09/20/top-50-fashion-designers/ [Accessed 23 Apr. 2023].

[4] Zambas, J. (2022). The World's Best Fashion Designers. Career Adict. [online] Available at: https://www.careeraddict.com/top-10-fashion-designers-in-the-world [Accessed 23 Apr. 2023].

[5] Pahaldesign.com. (2012). *10 Best Fashion Designers of World | Pahal Design*. [online] Available at: https://pahaldesign.com/10-best-fashion-designer-of-world/ [Accessed 23 Apr. 2023].

[6] TOP 10 Fashion Designers in the World. (2022). *WFX*. Available at: https://www.worldfashionexchange.com/blog/top-10-fashion-designers-in-the-world/ [Accessed 23 Apr. 2023].

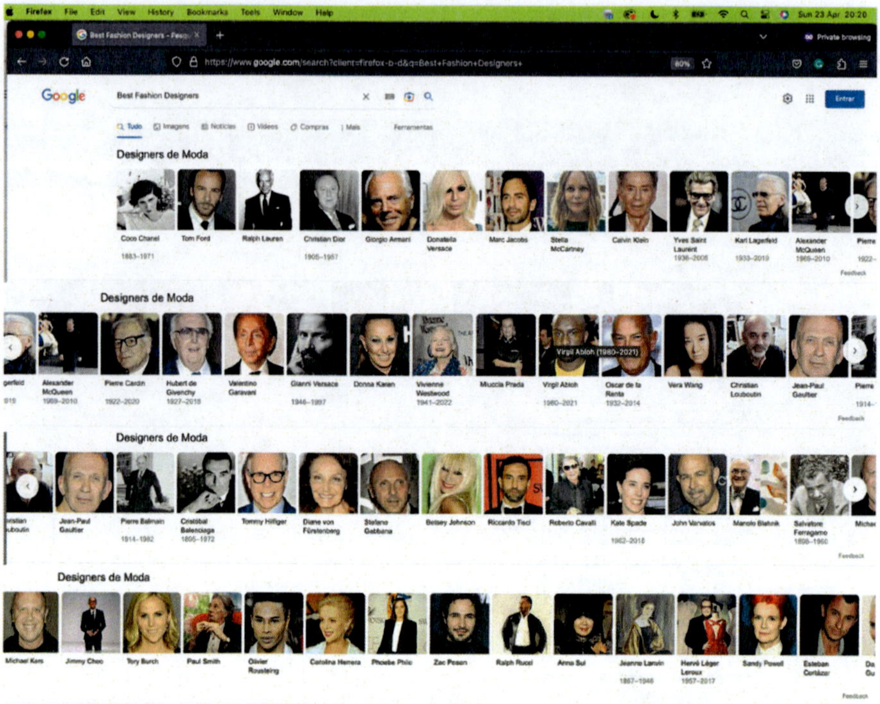

Fig. 2. Images obtained in the search with the term "Best Fashion Designers".

Donatella Versace and Stella MacCartney; Finally, the Dress Magazine article entitled "28 Most Famous Fashion Designers Who Changed the Fashion World"[7] presents 6 women out of 16 designers of the moment, but when referring to the 12 most famous designers of all time, only one woman is mentioned in the group, Coco Chanel.

In the Portuguese universe, reality differs from the world scenario regarding notoriety. Repeating the search now using the expression "*Os melhores designers de moda portugueses*"[8] [The best Portuguese fashion designers] the image search results only present 12 designers, of which half are women (Fig. 3). The first entry of the search points to an article by Vogue Portugal entitled "the 7 Portuguese designers who are shaping the future of National Fashion"[9] where of the 7 designers 5 are women; The second entrance leads to the Lisbon Lux website[10] where out of 9 designers/stylists that

[7] Berg, H. (2022). *28 Most Famous Designers of the Fashion World*. [online] Available at: https://dress-magazine.com/rating/most-famous-fashion-designers/ [Accessed 23 Apr. 2023].

[8] Search conducted in 23, April 2023.

[9] Silva, M. (2021). *Os 7 designers portugueses que estão a moldar o futuro da Moda nacional*. [online] Vogue.pt. Available at: https://www.vogue.pt/designers-portugueses-emergentes [Accessed 23 Apr. 2023].

[10] www.lisbonlux.com. (n.d.). *Moda em Lisboa - Os Melhores Estilistas*. [online] Available at: https://www.lisbonlux.com/lisboa/compras/estilistas-de-moda.html [Accessed 23 Apr. 2023].

the website presents, 4 are women. The third entry leads to the website Modalisboa[11] where of the 24 designers listed, only 10 are women; and finally, the fourth entry points to the article entitled "TOP 5: Portuguese stylists to keep an eye on!"[12], where only 1 is a woman.

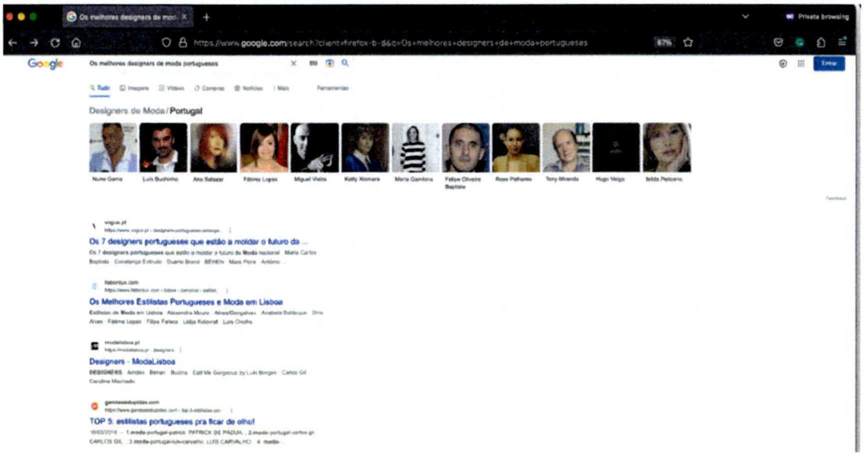

Fig. 3. Search results for the term "The best Portuguese fashion designers"

3 Discussing Female Ancestral Knowledge, Fashion Design and Academia

When analyzing the exposed facts, both from the historical point of view of design, as well as the professionalization of fashion design, the imbalance present in gender rhetoric is perceptible, as well as the lack of information. In a recent study carried out by Almeida [6], it was found that the lack of collections of women's work constitutes one of the great gaps in terms of the reconstitution of the trajectory of women in the areas of design and the arts, thus contributing to the absence and the erasure of the female gender. In part, this is also due to the fact that research is centered on men, even when the data are separated by sex, as well as tending to underestimate female work, classifying activities as domestic or "household", while characterizing male activities. as profitable [32]. In this context, it should be noted that craft activities undergo the same discriminatory process, as in addition to being associated with the aforementioned domestic activities, they are also considered amateur, in parallel with male activities, which are of an "extra domestic" character and understood as professionals [33]. In

[11] Associação Moda Lisboa (2022). *Designers - ModaLisboa.* [online] Modalisboa.pt. Available at: https://modalisboa.pt/designers/ [Accessed 23 Apr. 2023].

[12] Coutinho, C. (2016). *TOP 5: estilistas portugueses pra ficar de olho!* [online] Garotas Estúpidas. Available at: https://www.garotasestupidas.com/top-5-estilistas-portugueses-pra-ficar-de-olho/ [Accessed 23 Apr. 2023].

addition to this assumption, discussed throughout the article, it is also worth noting that female ancestral practices, which require a high level of skill, attention and creativity, are seen only as fun and hobby, since they demand little capital investment [33].

The cultural heritage of women's work, from a historical and creative point of view, is still unknown, not only to the general public, but also to academics and professionals in the various areas of design, even though women are the main producers of handicrafts, since the craftsmanship is seen as female and domestic work, to the point that teaching in arts and design is conventionally male [33]. When craft activities gain notoriety, they are always associated with the masculine, as in the case of male appropriations of female jobs and crafts associated with the profession of fashion design. These activities, known as domestic activities, commonly known as sewing, embroidery, tapestries and knitwear, when they gain prestige and economic value, are taken over by men, who not only learn them, but also start to occupy female spaces, excluding women from their own cultural space [30], as observed in the survey on the occupation of creative positions in world haute couture brands. Therefore, when we observe these traits of masculinization of female work, we understand that the activities associated with fashion design are directly associated with social, economic and political areas.

In the aforementioned surveys on the "best fashion designers" globally, we noticed that in addition to the low number of female names, there is also a constant repetition of the few female names, which opens us up to questions within the proposed rhetoric, such as female erasure and male domination in fashion design positions, but also to a possible lack of opportunity for new names to emerge in the fashion scene, especially if these names are associated with textile, craft and cultural traditions, as already mentioned, marginalized arts labeled as entertainment.

The Portuguese scenario, although slightly better in terms of professionalization of fashion design, does not differ from the historical rhetoric already discussed, with numerous paths being pointed out for the lack of female representation in design: cultural issues associated with the female domestic role, political issues related to the dictatorship period and finally, issues related to the lack of possibilities to remain connected to the labor market, which may be directly related to the factors mentioned above. Moura [1] describes the Portuguese design profession as selective from the gender point of view, with design being associated with male activity and thus assuming the exclusion of specific design disciplines, such as fashion design. At this point in the discussion, it makes sense to recall the role of design, both as a unifier of the various disciplines (product, graphics, fashion, jewelry, etc.), and its social role, that is, its ability to train designers capable of relate directly to multiple cultures, knowledge, traditions and languages [1, 15].

In addition to the aforementioned case of female creative, cultural and ancestral exclusion, fashion design demonstrates how the male gender issue is imposed on women in several other ways, one of which is standardization through unisex or "gender neutral" fashion. In both cases, they are approaches of masculine character and of exclusion of feminine traits, which Middleton [11] describes as androcentrism. That is, involuntarily and even believing that they are creating inclusive fashion, fashion designers are creating products for men [34].

We live in a world designed by men and for men, we need more women in design leadership to correct design errors (such as gender-inclusive proportions, solutions that are suitable for women's needs, etc.) as well as to propose more inclusive solutions at the gender level regarding work, and intellectual and creative value. Even though this article was composed of an almost absolute majority of female intellectual academic works, what is observed in the academic field is a male predominance. When analyzing the main European design conferences in 2019, it was found that the average number of women speaking at events was only 35.7%, as well as the average speaking time was less than the time given to male colleagues [34].

Another relevant data pointed out through the bibliographies of this research, is the significant number of feminine and gender approaches around graphic design and architecture. As an example, we mention the project "Designing women" [5] which reflects on historical factors, intellectual values and female participation in graphic design. Also, in the field of graphic design, we cite the publication "Extra Bold: A Feminist, Inclusive, Anti-Racist, Nonbinary Field Guide for Graphic Designers" [3], which synthesizes a practice guide for new approaches within graphic design. Another example, the project "MoMoWo" [35] aims at gender equality at the social and professional levels associated with design, architecture and engineering. For this, the project is supported by academic productions, book publications and an online platform. Finally, we mention the recently published "Woman Made" [4], which presents a collection of projects with female authorship in the areas of product design and architecture.

In all the cases mentioned above, even though they have made a great contribution to the proposed reflections and have directly contributed to breaking the old paradigms around design, none of them presents cases of female authorship in fashion design, neither at a creative level nor on an intellectual level. Even in broader cases on gender issues, such as the publication "Ivisible Women, exposing data bias in a world designed for men" [36], where cases of design, engineering, everyday life, etc. are cited, we found no references associated with fashion design.

In this sense, we believe that an academic approach capable of inserting feminine cultural and craft content into fashion design education, developing artisan-designer-university interactions, can provide a new perspective and significant possibilities for the future of sustainable fashion, contribute to the participation full and effective participation of women and equal leadership opportunities [37], following the example of what has been developed in the aforementioned project, "*Voces de La Artesanía*" [14]. Specifically in the Portuguese case, we mention here two initiatives:

1. "*Portugal Manual*" [38], which aims to promote and encourage Portuguese artisans, designers and brands. Even though it is not a specific example of fashion design, the project directly contributes to the presentation of traditional handicrafts and especially of those who make them. On the platform, in the fashion area, we find several examples of artisans and designers who work with new perspectives on traditional crafts.
2. "*Cooperativa Capuchinhas de Montemuro*", a group created by women in the 1980s that specifically works with traditional Portuguese handicrafts, clothing made

from *burel*[13], linen and wool fabrics, woven on manual looms, transcribed in a contemporary vision through the reinterpretation of fashion design [39].

In the second case, we particularly emphasize the importance of the connection between artisans and fashion design, as already observed in the aforementioned cases of the *Béhen* brand and the designer Helena Cardoso. We understand and visualize the existence of Portuguese and international fashion brands (such as the renowned London fashion house Vivienne Westwood) that, through the work carried out with groups of artisans, disseminate cultures, values and identities of groups, erased, discriminated or forgotten by the great scenarios.

The aforementioned practices, associated with feminine crafts and the incursion into design by social initiatives and economic projects, alert to the need for these approaches in academia, specifically in the curricula of fashion design courses. For Maffei and Sandino [33] such relationships constitute a solid body of actions and tools that correspond to identity formation, with a strong impact on gender discourses, as well as on social and domestic life, capable of suggesting changes in traditional design discourses.

4 Final Thoughts

Through the proposed article, we point out several studies related to the effective participation of women in design and the importance of the traditions, inheritances and cultures that are part of the female nucleus to be inserted in academic contexts, mainly in fashion design. In this context, we observed the existence of emphatic approaches in specific areas of design, which made us reflect and question about the approaches of fashion design: "where are the academic, philosophical and cultural discussions centered on the representation (or exclusion) of women in fashion design?", "Is design itself constantly repeating the exclusion of fashion design from intellectual and creative approaches?".

In our understanding, the feminine cultural erasure in fashion design still lacks in-depth discussions, research supports and the construction of its own historical-reflective research material. We observe a vast organization in the areas of graphic design, product design, architecture and engineering, which has been seeking to remake history, through the appreciation of women's work, both in the past and today.

We recognize the work carried out by the Craft Council and the London College (both in England) and the *Métiers d'Excellence* project by LVMH (France), for the teaching, dissemination and development of artisanal manual knowledge, both as recognition of the intrinsic values of humanity, and as tools for new approaches to sustainability for the future.

Even so, through this article, we point out the importance of dialogues around the theme proposed here, as the systematic factors presented contribute to the absence of intellectual discussions around fashion design and gender relations. Therefore, changes are needed to combat discriminatory practices and include women in creative and decision-making positions, as well as to encourage greater representation and dissemination of female practices and knowledge in academic circles, design competitions,

[13] *Burel* is an ancestral wool fabric once used by Portuguese mountain shepherds still produced in an artisanal way in some localities in the interior of the country.

conferences, textile and clothing companies, among others that we may lack memory at the moment.

Acknowledgements. This work is financed by Project UID/CTM/00264/2021 of 2C2T – Centro de Ciência e Tecnologia Têxtil, funded by National Funds through FCT/MCTES.

References

1. Mário, M.: O design que o design não vê. Orfeu Negro, Dl, Lisboa (2018)
2. Moura, M.C.D., Romano, R.B.: Design e Feminismo: a iniquidade de gêneros na 12o Bienal de Design Gráfico (2019). www.proceedings.blucher.com.br. https://www.proceedings.blucher.com.br/article-details/design-e-feminismo-a-iniquidade-de-neros-na-12-bienal-de-design-grfico-33713. Accessed 13 Mar 2023
3. Lupton, E.: Extra Bold: A Feminist Inclusive Anti-Racist Non-Binary Field Guide for Graphic Designers. Princeton Architectural Press, New York (2020)
4. Hall, J.: Phaidon Verlag Gmbh (2021). Woman Made Great Women Designers. Berlin Phaidon
5. Diana, K.: Designing women — a project by Readymag exploring the impact of women in design (n.d.). Designing women. https://designingwomen.readymag.com/. Accessed 18 Apr 2023
6. Almeida, A.J.M.: Mulheres e profissionalização no design: trajetórias e artefatos têxteis nos museus-escola MASP e MAM Rio (2022). www.teses.usp.br. https://www.teses.usp.br/teses/disponiveis/16/16140/tde-16012023-175956/pt-br.php. Accessed 18 Apr 2023
7. McEntee, K., Andersen, J.G., Brandalise, I.: Conversation: a feminine approach to design. In: DRS Biennial Conference Series (2018). https://dl.designresearchsociety.org/drs-conference-papers/drs2018/conversations/14/. Accessed 18 Apr 2023
8. Buckley, C.: Made in patriarchy ii: researching (or re-searching) women and design. Des. Issues **36**(1), 19–29 (2020). https://doi.org/10.1162/desi_a_00572
9. Schulz, F.E., da Cunha, J.L.F.L.: Diálogo entre crochê artesanato, design de moda e comunicação para a sustentabilidade. Modapalavra e-periódico **14**(34), 85–110 (2021). https://doi.org/10.5965/1982615x14342021085
10. Sellers, L.: WOMEN IN DESIGN: pioneers in architecture, industrial, graphic and digital design from the... twentieth century to the present day. S.L.: White Lion Pub (2021)
11. Middleton, M.: Feminine exhibition design. Exhibition **38**(2) (2019). https://www.academia.edu/40660101/Feminine_Exhibition_Design. Accessed 18 Apr 2023
12. Daicampi, M.: O encanto dos fios que entrelaçam a moda: o artesanato nas passarelas do São Paulo Fashion Week 2014–2018. semanaacademica.org.br (2018). https://semanaacademica.org.br/artigo/o-encanto-dos-fios-que-entrelacam-moda-o-artesanato-nas-passarelas-do-sao-paulo-fashion-0. Accessed 17 Mar 2023
13. Fletcher, K., Grose, L.: Moda & sustentabilidade. Editora Senac São Paulo (2019)
14. Argentina, L.: VOCES DE LA ARTESANÍA Diálogos para prácticas sustentables (2022). https://www.fashionrevolution.org/wp-content/uploads/2022/08/Voces-de-la-artesania_ESP.pdf
15. Manzini, E., Cipolla, C.: Design for social innovation and cities. DESIS Network Association (2019). https://www.researchgate.net/publication/339015123_Design_for_Social_Innovation_and_Cites/stats. Accessed 5 May 2020
16. Schulz, F.E., Freire, L.H., da Cunha, J.L.F.L.: The 7 complex lessons from Edgar Morin applied in fashion design education for sustainability. In: Raposo, D., Neves, J., Silva, R., Correia Castilho, L., Dias, R. (eds.) Advances in Design, Music and Arts II. EIMAD 2022. SSDI, vol. 25, pp. 193–208. Springer, Cham (2023). https://doi.org/10.1007/978-3-031-09659-4_15

17. Cross, N.: Designerly ways of knowing. Des. Stud. **3**(4), 221–227 (1982). https://doi.org/10.1016/0142-694x(82)90040-0
18. Ingold, T.: The Textility of Making. papers.ssrn.com (2010). https://papers.ssrn.com/sol3/papers.cfm?abstract_id=1540398. Accessed 28 Apr 2023
19. Niedderer, K., Imani, Y., Overton, M.G.: Craft as a Tool for Multidisciplinary Collaboration and Design Practice. researchprofiles.herts.ac.uk (2011). https://researchprofiles.herts.ac.uk/en/publications/craft-as-a-tool-for-multidisciplinary-collaboration-and-design-pr. Accessed 28 Apr 2023
20. Kokko, S., Räisänen, R.: Craft education in sustaining and developing craft traditions: Reflections from Finnish craft teacher education. Techne serien - Forskning i slöjdpedagogik och slöjdvetenskap **26**(1), 27–43 (2019). https://journals.oslomet.no/index.php/techneA/article/view/2911
21. Teodoro, J.D.O.: O ensino do Design de Moda em Portugal: contribuição para uma análise crítica da educação para a sustentabilidade. repositorio.ul.pt (2014). http://repositorio.ul.pt/handle/10451/12184. Accessed 28 Apr 2023
22. Auslander, L.: Culture matérielle, histoire du genre et des sexualités. Clio **40**, 171–195 (2014). https://doi.org/10.4000/clio.12163
23. Providência, B., Cunha, J.: Do artesão ao design de autor como metodologia. 6a Congresso Internacional de Pesquisa em Design (CIPED) (2011)
24. Krucken, L.: Design e território: valorização de identidades e produtos locais. Studio Nobel, São Paulo (2009)
25. Kokko, S., Almevik, G., Bentz Høgseth, H., Seitamaa-Hakkarainen, P.: Mapping the methodologies of the Craft Sciences in Finland, Sweden and Norway. helda.helsinki.fi (2020). https://helda.helsinki.fi/handle/10138/322242. Accessed 30 Apr 2023
26. Projecto TASA. (n.d.). Projecto TASA - Técnicas ancestrais · Soluções atuais. http://projectotasa.com/
27. Izabel, M., Bruno, C.: Helena Cardoso: impressões da paisagem e texturas entretecidas. Estúdio: artistas sobre outras obras **10**(26), 104–114 (2019)
28. behenstudio.com. (n.d.). B É H E N. https://behenstudio.com/. Accessed 30 Apr 2023
29. Marques, Â.M.D.C.: Sexismo na Indústria da Moda. ubibliorum.ubi.pt (2021). http://hdl.handle.net/10400.6/11636. Accessed 28 Apr 2023
30. Durand, J.-Y.: Bordar: masculino, feminino. repositorium.sdum.uminho.pt. Aliança Artesanal (2006). http://repositorium.sdum.uminho.pt/handle/1822/5480. Accessed 24 Apr 2023
31. Pike, H.: Female Fashion Designers Are Still in the Minority. The Business of Fashion (2016). https://www.businessoffashion.com/articles/fashion-week/less-female-fashion-designers-more-male-designers/
32. Caroline Criado Perez (2022). Mulheres invisíveis: o viés dos dados em um mundo projetado para homens. Editora Intrínseca
33. Lees-Maffei, G., Sandino, L.: Dangerous liaisons: relationships between design, craft and art. J. Des. Hist. **17**(3), 207–220 (2004). https://doi.org/10.1093/jdh/17.3.207
34. Kasay, D.: Designing women — a project by Readymag exploring the impact of women in design. Designing women (2018). https://designingwomen.readymag.com/
35. Dan, M.B., Lacrarui, R., Lacrarui, M., Dusoiu, E.-C.: Women's creativity since the Modern Movement. MoMoWo (2017). http://www.momowo.eu/. Accessed 28 Apr 2023
36. Perez, C.C.: Invisible Women. [S.I.]: ABRAMS (2019)
37. ONU (2022). ODS Objetivos Desenvolvimento Sustentável BCSD Portugal. https://ods.pt/. pahaldesign.com

38. Belo, F.: Início. Portugal Manual (2018). https://www.portugalmanual.com/. Accessed 24 Apr 2023
39. Hellena, V.: Contributos do Design para intervenção social na criação e desenvolvimento de produtos artesanais têxteis: estudos de caso: 'As Capuchinhas de Montemuro' e 'Mulheres de Bucos'. Handle.net. (2019). https://hdl.handle.net/1822/60317

Knowledge of Traditional Techniques in the University Education of Fashion Design

Alexandra Cruchinho(✉), Catarina Rito, and João Barata

CICANT, Lusófona University, Lisbon, Portugal
alexandra.cruchinho@ulusofona.com

Abstract. The importance of acquiring knowledge of traditional techniques, which validate the knowledge of the cultural construction of the identity of a people/society, require synergies between artisans and academia, in order to validate the skills of students, who choose the area of Fashion Design as a future profession. The richness, in Portugal, of traditional techniques resides throughout its territory, however, the gap of interconnection between them and future fashion designers is a concern that is growing on the part of the academy, with institutions that teach fashion design. Thus, it is central that students have the opportunity to learn from people who have the knowledge, which should be a consistent tool on the part of those who structure the syllabus of current courses and others that may arise in the near future.

One of the main objectives of this research is to think of strategies to develop activities that enable the acquisition of this ancestral knowledge. Also, to demonstrate the importance of this knowledge for the professional profile that Higher Education Institutions intend to train, in order to recognise that this knowledge is an important element to add value and differentiate these future professionals in a global market.

Keywords: Academia · Identity · traditional techniques

1 Introduction

The research work presented here seeks to validate the synergy between the teaching of fashion design and the knowledge of different traditional techniques, typical of various parts of the Portuguese national territory. The transmission of an ancestral knowledge, which resides in the craftsmen, composed of older generations, is fundamental with the younger generations, focusing on those who study areas such as fashion design. This transmission of techniques, typical in various regions of the country, validates both the local and cultural identity of a people, in this case the Portuguese.

The main objective of this research is based on the assumption that the training of young people propitiates the dissemination of knowledge and, consequently, perpetuates the knowledge of these techniques. Transmitting the knowledge to younger generations ensures its perpetuation and the continuity of the use of these techniques.

It is not too much to discuss this theme in these times when the return to traditions and the revival of traditional techniques, Arts and Crafts, is a strong trend. Remember the last collection presented by the brand Christian Dior Fall23, where Indian embroidery is the starting point and inspiration. Lace is also present in the latest Louis Vuitton Fall 23 and Channel Fall 23 collections.

This research aims to understand how the learning of traditional techniques can help to value the identity and culture of regions and how this knowledge can be transmitted to future fashion designers.

2 Cultural Identity

The question of cultural identity is essential for the construction of societies, having started to have a status of relevance more than 50 years ago, through anthropology and in postmodernist societies, which validated the growing importance both in research and in historical-historical interpretation. Geographic (Cabral, 2003). Fashion has the ability to unite the past with the present. It can bring together, in the same context, what time has separated, returning to a distant time, reusing and revitalizing what was declared dead (Agamben, 2009). In this article, the focus is on observing the Portuguese context and the way in which identity issues can influence the work of the future designer and the respective characteristics of the idealized product, not in a specific way, but throughout his career. Addressing the cultural value of Portuguese fashion design, reflecting on its meaning and the influence that factors such as history, society and culture have on the concept and development of contemporary products produced by fashion designers, apprehended throughout studies at the academy. Places of creativity optimize dispersed resources and generate new opportunities in the economy, however, their primary objective is the ability to encourage local development, promoting projects with the participation of various areas called creative activities: art and culture, knowledge of the economy and social initiatives (Gaddi, 2014; Bertola and Manzini, 2006). The study of culture, identity, sociology, anthropology and heritage, with a focus on fashion, reinforced by a global perspective, must go through a focus on local traditions, aiming at valuing the construction of the final product, with specificities that make it the same unique product, capturing the consumer's interest in obtaining differentiating creative ideas compared to that produced by mass consumption brands. As Fiorani (2006) reinforces, fashion represents the spirit of contemporary culture, due to its ability to bring together different dynamics, already existing, that reinforce the relationship between the individual and society. It is noteworthy that today's Fashion, recognized for its universal multidimensionality, is not limited only to change, trends and temporal changes, but to the combination of successive styles. Fashion is the most complete expression of industrial and cultural post-modernity, which found a way to express itself in the design process. The concerns introduced by this sector are leading a large number of designers, in various areas, to start taking a slow or more sustainable approach to design and the respective production of clothing. New solutions are emerging in the sense of redefining fashion design and the market, for example, the reference to local cultural practices, namely traditional techniques, is one of the contemporary manifestations in the search for authenticity and based on the slow fashion concept (Clark, 2008; Apadurai, 1996). Future generations of fashion designers, in order to validate their creative options, must have direct contact with those who have the knowledge, in this case the artisans who are still available to teach, thus achieving direct transmission to all those who can potentially the future of the local heritage with those who have creativity as a tool in their future work. Obviously, this knowledge can be reflected in the regular or occasional work of each designer/brand. It is important that the cultural tradition is not lost, motivated by a break in the transmission by those who know from those who can learn and recreate ideas from a past time combined with a contemporary vision, respecting the changes in concept/creation/style promoted by fashion.

2.1 Knowledge to Create

To create has much to do with the surrounding cultures. Designing is to create from and for the culture, this is why Manzini & Cullars (1992) state that designers are responsible for the artificial (forms of) manifestations in a certain community.

When one tries to define a certain culture, one highlights practices and values. Cultures are defined by symbols, heroes and rituals (Hofstede et al., 1990; Hofstede, Hofstede and Minkov, 2010); their designs have a cultural shape.

As important as the surrounding Cultures, Explicit Knowledge appears to be a topic of interest in the literature review regarding creativity. Several developments and approaches regarding the generation of creative outcomes highlighted cultural, implicit/tacit knowledge/information and explicit knowledge once they are fundamental variables.

2.1.1 Explicit Knowledge

Explicit Knowledge drives from training and it has to do with skills that can be shared using logic or instructions.

The cognitive styles rooted the first models to understand the process in which creative outcomes would occur. Guilford's Structure of Intellect (1956) identified the cognitive processes behind the generation of creative responses.

Later, the Component Model of T. Amabile (1988), later evolved into a dynamic model for organizational creativity (T. M. Amabile & Pratt, 2016), reinforced the need for domain relevant-skills, creativity-relevant skills and motivation for the task, influencing the creative stages in creative endeavors. Domain relevant - skills are related to specific technics and explicit knowledge, this component aims to prepare the task, select information and recourses and, in the end of the pipeline, relevant skills are crucial do evaluate and validate the outcomes. Creativity – relevant skills drive from cognitive styles, personality traits and implicit knowledge from past experiences; this component is useful in the preparation, generation of ideas and validation stages.

2.1.2 Implicit Knowledge

"No group can escape culture" (Hofstede, Hofstede and Minkov, 2010, p. 11) once the creative action shares rules and conditions from a given culture.

As defined by Csikszentmihalyi (2015) in Fig. 1 and Glăveanu (2010) in Fig. 2, creativity results from an interaction of shared information, rules, signs and practices within a defined community; this interaction is represented by open systems (with faded edges) or systems where novelty is constantly being introduced.

According to Csikszentmihalyi's (2015) Domain – People – Field system originally published in 1988, the creative outcome drives from all types of knowledge within a culture; different domains can provide information for the creator to generate a new outcome.

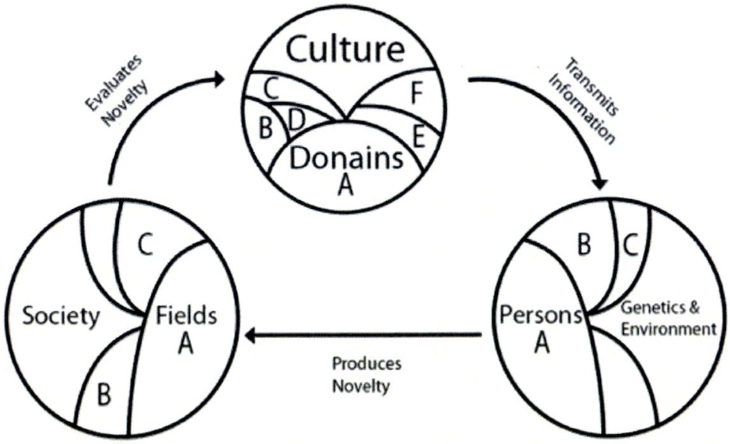

Fig. 1. Domain – People – Field system. Font: Csikszentmihalyi, 2015, p. 211

This principle is also present in Glăveanu's interactive framework of Cultural Interaction (2010), where the intersubjective signs and artifacts of a certain culture (belongings to a given community) are directly linked to the creative/creation action.

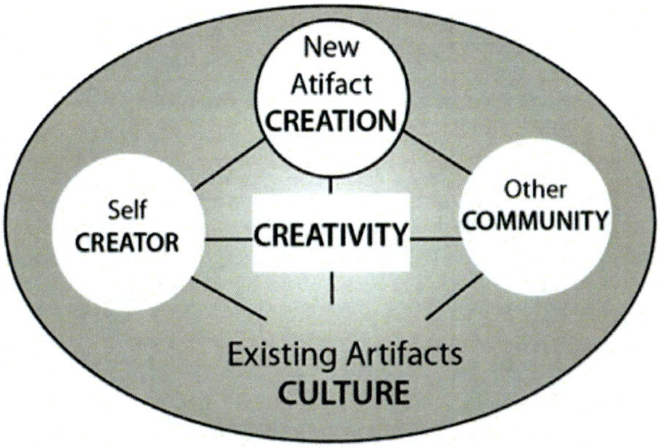

Fig. 2. Glăveanu's interactive framework. Font: Glăveanu, 2010, p. 87

In the Domain – People – Field system from Csikszentmihalyi (2015), novel productions are scrutinized by the field/society where the gatekeepers determine if the novelty is valid or accepted; only then, the creation is considered good and a new part of a domain inside the (accepting) culture. According to Csikszentmihalyi (2015) this new fragment on the domain can later be used by someone else to ignite his/hers creative process.

Glăveanu (2010) also concludes that creativity generates new creativity and that is why implicit knowledge is generative of more knowledge and creativity once it augments the repertoire inside the cultural framework. Both cultural shared knowledge and explicit knowledge play a big role when it comes to creativity.

3 Formation and Dissemination of Lérias Lace

Lérias Lace is a traditional element that marks the cultural heritage of the villages Póvoa de Atalaia and Atalaia do Campo. Made by elderly women who, while young, prepared their trousseau, it is a lace with its own characteristics associated with the region by adapting the stitch to a new aesthetic of the final product and by the symbolic and emotional meaning. To continue this tradition, a project is underway which, among other components, aims at training and promoting social sustainability by valuing the knowledge of the women who participate in it, involving services and companies in the region in synesthesia for the development of new products.

In general terms, the project seeks to maintain tradition, on the one hand, through the research that has been carried out and which, being recorded and published in various publications, perpetuates and perpetuates this knowledge and this traditional technology. On the other hand, the training guarantees the dissemination of this knowledge, augmenting the creative portfolio, as well as the continuity in the execution of the Lérias lace.

Thus, the development of an artistic residency with the theme – Lérias a Arte das Linhas, presents itself as an important activity for the dissemination of this knowledge.

25 higher education students of Fashion Design from a Portuguese institution got together with elderly ladies from the "land" who make lérias lace from an early age, for a week, to learn and teach how to make lace (photos 1, 2 and 3). The chosen location was the Eugénio de Andrade House of Poetry located in Póvoa de Atalaia, Fundão.

Fig. 3. Ladies from Póvoa de Atalaia teaching the students. Author's photo.

For a week, the students made contact with the people of the land, understanding their habits, customs and traditions. They listened to traditional songs while learning how to make lace (Fig. 3). The ladies did not just pass on knowledge about lace, but also left examples of life wisdom, humility and resilience, and patiently passed on what they also learned at a younger age than these students.

Knowledge of this traditional technology was transmitted and enthusiasm was evident as knowledge was consolidated.

After mastering some of the points, the students began to idealize possible uses for the income and creativity was unleashed throughout this exchange of knowledge.

The ladies were integrated and played an active and decisive role in this activity. Students determine how the continuity of this ancestral knowledge will be ensured.

After the Residency week, the students were confronted with a new proposal – to think of clothing products, accessories in Lérias lace, with Lérias lace and inspired by Lérias Lace.

The results will be presented in a Parade and, subsequently, the pieces will be presented at an international fashion fair.

4 Conclusions

In conclusion, it is important to highlight the importance of the connection between such different generations. On the one hand, the recognition of knowledge dominated by women of the land, on the other hand, the established proximity and the exchange of learning in an activity that touches on social sustainability.

Lérias lace is a very aesthetically rich traditional technology with the potential to be used in multiple applications in the field of Fashion and also in other areas of design, arts, architecture, etc.

Versatile technique that is still at the beginning of its creative exploration but reveals great potential for recognition in a national and international context.

The continuity of Renda das Lérias is assuredly guaranteed by training actions and transmission of knowledge that allow its dissemination and it is through activities such as the Residence of Lérias that this knowledge will reach a greater number of people.

Interest in Lace has been shown by the younger generations and, in the style of revivalism, these have been applied in multiple proposals by international brands, presented at the main fashion weeks in the world.

References

Agamben, G.: What is an Apparatus? And Other Essays, Stanford University Press (2009)

Amabile, T.: A model of creativity and innovation in organizations. Res. Org. Behav. **10**, 123–167 (1988)

Apadurai, A.: Modernity at Large Cultural Dimensions of Globalization. University of Minnesota Press, Minneapolis, MN, and London (1996)

Amabile, T.M., Pratt, M.G.: The dynamic componential model of creativity and innovation in organizations: making progress, making meaning. Res. Org. Behav. **36**, 157–183 (2016). https://doi.org/10.1016/j.riob.2016.10.001

Bertola, P. Manzini, E.: Design Multiverso. Edizioni POLI. Design Milano, Milão (2006)
Cabral, M.V.: A Identidade Nacional Portuguesa: Conteúdo e Relevância. Revista de Ciências Sociais, Rio de Janeiro, Brasil (2003)
Clark, H.: Slow + fashion an oxymoron or a promise for the future...?. Fashion Theory **12**(4), 427–446 (2008)
Csikszentmihalyi, M.: The Systems Model of Creativity - The Collected Works of Mihaly Csikszentmihalyi. Springer, Netherlands, Dordrecht (2015). https://doi.org/10.1007/978-94-017-9085-7
Fiorani, E.: Moda, Corpo, Immaginario. Edições Poli. Design. Milão, Itália (2006)
Gaddi, R.: Design for territorial enhancement: the cultural value of fashion design. In: Proceddings of 2o CIMODE, Politécnico de Milão, Milão (2014)
Glăveanu, V.: New ideas in psychology paradigms in the study of creativity: introducing the perspective of cultural psychology veanu. New Ideas Psychol. **28**(1), 79–93 (2010).https://doi.org/10.1016/j.newideapsych.2009.07.007
Guilford, J.P.: The structure of intelect. Psychol. Bull. **53**(4), 267–293 (1956)
Hofstede, G., et al.: Measuring organizational cultures: a qualitative and quantitative study across twenty cases. Adm. Sci. Q. **35**(2), 286–316 (1990)
Hofstede, G., Hofstede, G.J., Minkov, M.: Cultures and Organizations: Software of the Mind. 3rd edn, Cultures and Organizations. 3rd edn. McGraw-Hill Education – Europe, New York (2010). https://doi.org/10.1007/s11569-007-0005-8
Manzini, E., Cullars, J.: Prometheus of the everyday: the ecology of the artificial and the designer's responsibility. Des. Issues **9**(1), 5 (1992). https://doi.org/10.2307/1511595
Mendeley, n.d., Harvard Format Citation Guide. Available through. https://www.mendeley.com/guides/harvard-citation-guide. Accessed 06 Mar 2023

Designing for People Inside and Outside the Classroom: Service Design Proposals for Positive Social Inclusion of Refugees

João Sampaio[1,2](✉) and Bernardo Providência[1]

[1] University of Minho and Member of Lab2PT, Guimarães, Portugal
joao.sampaio@eaad.uminho.pt

[2] Member of ID+ Research Institute for Design, Aveiro, Portugal

Abstract. The scourge of migration associated with refugee status is a societal challenge in contemporary societies. This encourages the development of adequate responses for people welcomed by responsible organizations, to promote full and effective social integration in the society in which they are integrated. In a problem-based learning approach, a partnership was developed with the Municipality of Guimarães, where students of the master's degree in product and services design develop service design proposals for and with refugees integrated into the Guimarães Acolhe program. Based on participatory approaches, the students assessed the main anxieties and ambitions of the refugees, and developed, evaluated and proposed two services to respond to the questions of language learning and women's empowerment. This process of learning and interaction with people and real contexts demonstrated the importance of promoting these initiatives both through the learning process and the potential impact of academia on society.

Keywords: service design · refugees · problem-based learning · education

1 Framework

1.1 Proposal Framework

Over the past ten years, there has been an enormous and unprecedented increase in the number of refugees and asylum seekers. Last year, in Europe the population of migrants reached 12 million due to being forcibly displaced from their homes as a result of conflicts, persecution, violence or human rights violations. Of these, 6.7 million are refugees and 1.1 million are asylum seekers [1]. One of the biggest waves of forced immigration in recent memory was brought on by the civil war in Syria since 201Portugal is part of the group of countries included in the so-called Cluster 2, Emergent Destination Countries that includes the Czech Republic, Estonia, Spain, Italy, Latvia, Lithuania, Poland, Portugal, and Slovenia. The countries in this cluster are characterized by medium levels of GDP per capita, Gini Index, corruption perception index, and business freedom [2].

Portugal, currently has three different procedures for the reception of refugees, according to whether they are spontaneous refugees, resettled refugees or relocated refugees. Due to the refugee crisis in Europe, which compelled European governments to address this social issue, the classification of relocated refugees was originated in 2015 [3]. With respect to this status, the relocation and resettlement programs last for 18 months, and a variety of local organizations from 99 distinct municipalities are in charge of assisting in the reception of refugees and overseeing the process of integrating them alongside governmental agencies and intermediary organizations [4]. According to ACM [4] report until the 29th of November 2017, and after the formal end of the Program on the 26th of September, 1520 relocated citizens arrived in Portugal, 1190 from Greece and 330 from Italy. Of these, 261 are households and 533 are minors. As far as origin is concerned, Syrians (833), Iraqis (338) and Eritreans (311) are relocated. The largest percentage of people who make up these flows are male, with a strong presence of young and isolated men, mostly from Eritrea.

The Masters in Product and Service Design at the University of Minho promotes a training that seeks to place students before real problems and real partners. In the scope of the curricular unit of Project 2 (service design) of the 1st year, 2nd semester, in January 2022, and resulting from the partnership with the social department of Guimarães City Hall, a challenge was identified where students could emerge, intervene and have a potential impact through the development of service design proposals, along with the training process and development of skills and soft skills. The issue of positive integration of refugees was considered relevant and with greater potential impact, both in terms of solutions and in terms of framing the course content. In Guimarães city, "Guimarães Acolhe" (GA) program welcomes and provides guidance up to 60 citizens from several countries, including Eritrea, Sudan and Syria. The GA program began in 2016 and is composed of a consortium that seeks to promote positive integration in the Guimarães community by supporting and promoting the socio-cultural inclusion and autonomy of refugees.

1.2 Conceptual Framework

Relating to the contextualization previously defined, a conceptual framework was created for the proposed project presented (see Fig. 1). Higher education's main objective is to prepare students, providing them with the necessary material for the development of useful professional skills for the future, promoting their specific knowledge, complementing the acquisition of useful soft skills to solve problems, working in multidisciplinary teams and particular contexts. In line with the evolution of the training offered, the role and relationship of the Academy with society in general – inherent to contemporary societal and technological challenges – requires new teaching strategies. In this sense, we seek to safeguard a set of characteristics and capabilities to be performed by future designers, which become complex or even impossible to replicate in the classroom. In consideration of the above, that highlights the importance of project-based Learning strategies, that refers to an inquiry-based instructional method that engages learners in knowledge construction by having them accomplish meaningful projects and develop real-world solutions [5]. On that basis, the relationship with the refugees and local administration promotes this extended experience. This promotes student's development of strategies and soft skills promoting direct channels of access to information, which highlight the importance of convivial tools [6] in these processes. These do not invalidate the analysis of the existing relationship between the local administration and the refugees, where the strategies of data collection and analysis make it possible to identify various actors and the ecosystem to be intervened [7].

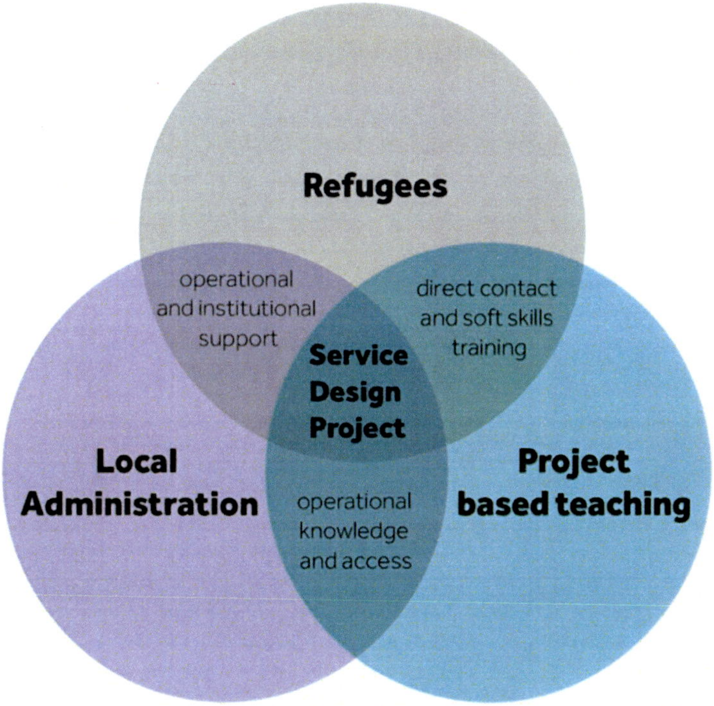

Fig. 1. Project conceptual framework

2 Project Based Teaching Developing Stages

For the didactic and operative process, exposure of contents, and consequent practical application of tools and strategies in the different project phases, a set of stages was defined that allowed to endow students with specific knowledge applicable to the different moments of the project. For the immersion process, a gradual unfolding method was planned that included a first contact with the topic presented by the lecturer, promoting an initial phase of autonomous discovery and consolidation of knowledge and ecosystem. With the involvement of the consortium's pivot, another phase began, which introduced the operating features of refugee reception in Portugal, the particularities and challenges of this program. This moment not only allowed the students to confront the information previously collected but to complement it through the perception of the operators and real ambitions and challenges related to the program. After these phases, the contacts with the refugees took place, which were supported by the contacts, information collected, and translated into the communication and participation strategies and tools defined and/or created. For the immersion process, a gradual unfolding process was planned (see Fig. 2), that reflects the process, stage, activities and references.

Pre-Workshop	**Stage** - Context framework - Stakeholders ID	**Activity** - Pivot's Consorcium meeting (in clasroom) - Ecosystem/context definition - Personas - Planning session - Designing/adapt Tools - Workshop Invitation (video)	**Referencies** Kumar (2013) Polaine (2013) Stickdorn & Schneider (2011, 2018)
1st Workshop	**Stage** - Pre-project - Framing needs and opportunities	**Activity** - Presentation of team and session goals - Ice-braker - How is my day? - Map of ambicions and problem	**Referencies** Sanders & Stapers (2014) Stickdorn & Schneider (2011, 2018)
Proposal development	**Stage** - Generative design	**Activity** - Insights managing - Brainstorming; - Service map - Jouney map - Model-making - Stakeholder ID	**Referencies** Polaine (2013) IDEO (2015) Stickdorn & Schneider (2011, 2018) Manzini (2012)
2nd/3th Workshops	**Stage** - Evaluative design	**- Activity** - Prototyping and testing - Planning (process map-ping, assigning roles, materials and skills)	**Referencies** IDEO (2015) Stickdorn & Schneider (2011, 2018)
Final Proposal	**Stage** - Presentation	**- Activity** - Design Review - Planning (task, strategies and communication, assigning roles, materials and skills - Construction	**Referencies** Stickdorn & Schneider (2011, 2018)

Fig. 2. Overall process, activities and references

2.1 Pre-Workshop Stage

After the presentation of the challenge, shared research was carried out class to enhance the amount of gather information to better define the context. As a complement to the autonomous research, the operational management of GA who has regular direct contact with refugees was also present in the classroom. This presence allowed the students to gain a clear profile of the refugees, and their situation and communication limitations. On the other hand, it was possible to validate some research previously carried out, as well as to inquire him about the structure needs and challenges to obtain a more holistic perception. This contact also impacted the design of the presentation strategy, tools and dynamics to be implemented in the participatory session, more specifically:

- Produce a presentation video of the students' team and the session, in Portuguese and subtitled in English, to encourage adhesion and initial proximity;

- Choose the GA headquarters as the location to hold the session, as it was a space already known to refugees;
- Create artifacts that facilitate communication using icons related to states of mind, work, culture, religion, transport, food, household, etc.;
- Create a simplified journey map that did not reflect so much on the time but on the concept of day and night and states of mind;
- Using a hierarchical target to organize what the refugees' problems or ambitions;
- Decrease the session time as it would take place during the Ramadan period, which made it impossible to organize a coffee break;
- Definition of a strategy for working groups, but without imposing their constitution, as this could create situations of discomfort.

2.2 First Workshop

The entire process of managing the invitation and sharing the video was done by the main pivot of the GA, which for the first time tried to bring together all refugees regardless of their nationality. The adhesion of the group and the success of the call were notable, although, after the initial presentation, several women (mostly accompanied by children and babies) choose to leave. This abandonment was justified by the pivot since they were in a place in the presence of men without their husbands. Something that in their cultural frameworks would not be acceptable. Apart from this issue, the pivot himself was quite satisfied with the result of the permanence of different nationalities in the same space. As identified (see Fig. 2), after the initial presentation (of the session, objective and team), GA pivot introduced everyone present to facilitate the first contact. Therefore, at the end of this moment, people were asked to spread out among several work tables (4 in total). Desk 1 consisted of 3 women (Colombian); at desk 2 a group of 6 men (4 Syrian and 2 Sudanic); at desk 3, there was a mother and two children (Ethiopian) who already spoke Portuguese quite well because they were integrated into the school; at table 4 which hosted two groups, as they did not all arrive at the same time. The first group was with a woman, her husband and baby (Eritreans) and finally 3 more men (Syrian). It should be noted that most men did not have any family members in Portugal.

At all desks, in the beginning the participants were asked if they had any questions and were informed that they could leave the session if they did not identify themselves, if they did not feel comfortable or if they had to perform some practice taking into account the Ramadan period. The first activity, the adapted journey map, was easily understanded and along with wooden pieces they started to communicate. At this stage, it was possible to perceive various levels of Portuguese language skills, and despite that in general everyone was very receptive and participating. At desk 1 and 4, respectively, communication was carried out in Spanish (since two students mastered the language) and in English, since the first group of this table did not have a basic level of Portuguese. The activity, on the one hand, allowed to break the ice and start the process of sharing the daily routines of refugees, as well as for the students to begin to gain this perception and start conversations about topics or deepen the relationship between the routine and the different states of mind in the course of the day. After this activity, the main themes associated with the ambitions and challenges of refugees were recorded, either by the participants, individually or with the support of the students (see Fig. 3). At desk 3,

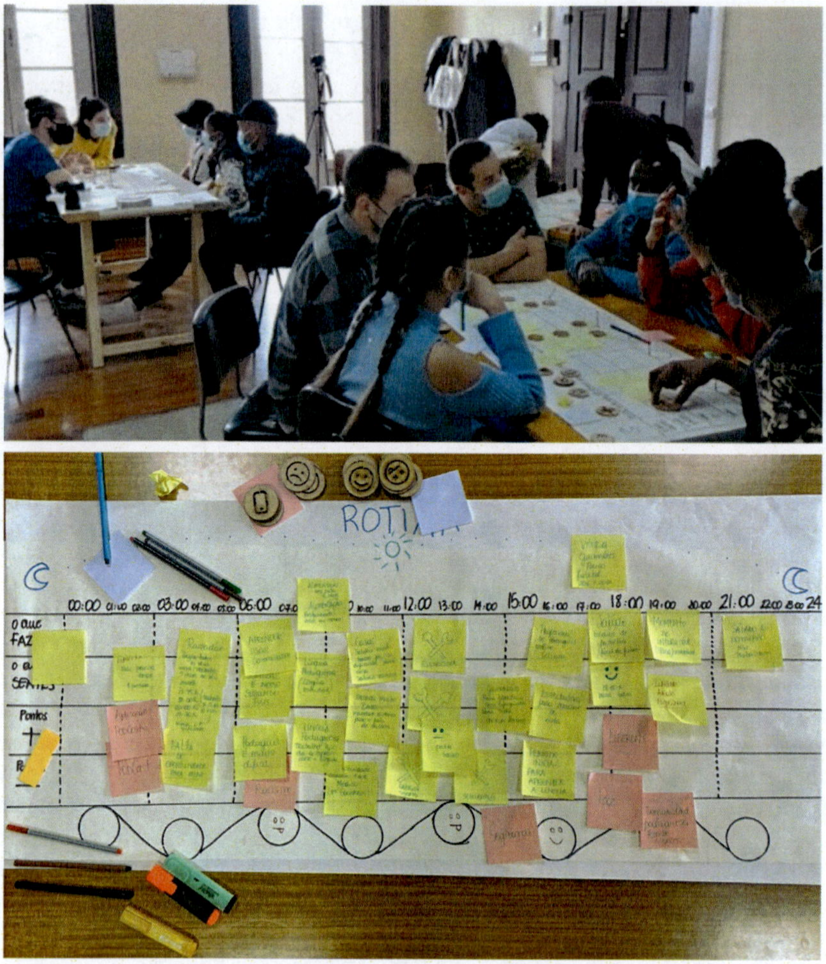

Fig. 3. Participatory workshop

in which the interaction was made only with the children as the mother did not speak Portuguese, it was quite interesting to understand the degree of success and the challenges of the children's integration, as well as the relevance or the added degree of responsibility that these may have in the process of integrating the parents.

1st Workshop	**Participants** Women, men and children from: Eritrea, Sudan Syria, Ethiopia and Colombia	**Gender** M - 10 F - 5 + 2 children, 1 baby	**Duration** 2h30	**Goal** Insights Aims and challenges
2nd Workshop Prosperas (project 1)	**Participants** Women from: Eritrea (3), Sudan (1)	**Gender** F - 4 + 2 children	**Duration** 1h30	**Goal** idea validation/feedback 1st activity testing
2nd Workshop Poder da língua (project 2)	**Participants** Men from Syria (2)	**Gender** M - 2	**Duration** 1h	**Goal** Usability test lo-fi mockup
3rd Workshop Prosperas (project 1)	**Participants** Eritrea (1), Syria (1) and Ethiopia (1)	**Gender** F - 3	**Duration** 2h30	**Goal** Material testing

Fig. 4. Characterization of the participants of the various participatory sessions, duration of the sessions and main objective

2.3 Proposal Development and Debriefing

After the 1st contact with the refugees, the gathered information was treated in the classroom. In this process, it was identified the most relevant ambitions and challenges for this possible group of reals, presented by the possible in Fig. 5 that were organized around the challenges/current reality, professional and personal ambitions. As showed in Fig. 5, it is possible to perceive the interdependence between the various subjects, which gives the notion of the vicious cycle on which refugees are hardly able to get out and evolve.

This analysis process was followed by a brainstorming session in order to create possible responses to the perceived relationships that could contribute to the evolution and improvement of the conditions and life of the refugees. The projects were developed in the classroom, in order to outline a proposal that would respond to the identified challenges, but also using a survey that made it possible to identify and contact possible partners with the aim of filling the structure provided by the Municipalities and their support networks. In this sense, in addition to identifying and contacting stakeholders, students consolidated ideas through the design of system maps, customer journey maps and models of various solutions. An interactive process that, as it was tested, validated, was consolidated at the strategic and operational levels. After reaching that stage, the contact was made in order to validate and optimize the solutions with the main potential users, the refugees. These sessions were rehearsed and analyzed in the classroom and with the teacher and colleagues. The overall process had several sessions indentifyed on Fig. 4.

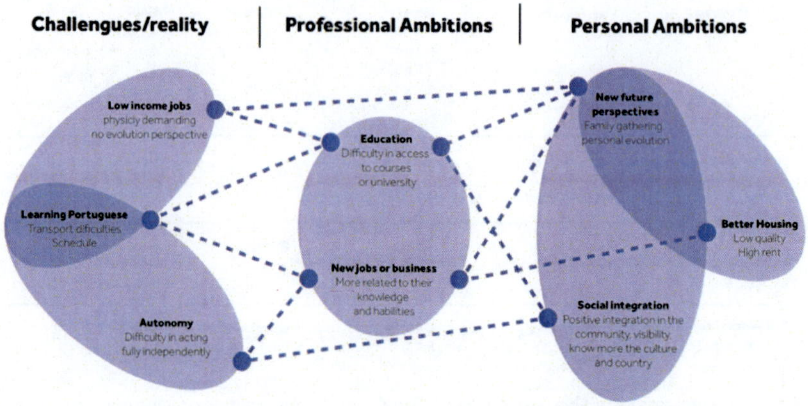

Fig. 5. Interdependence between topics regarding refugees.

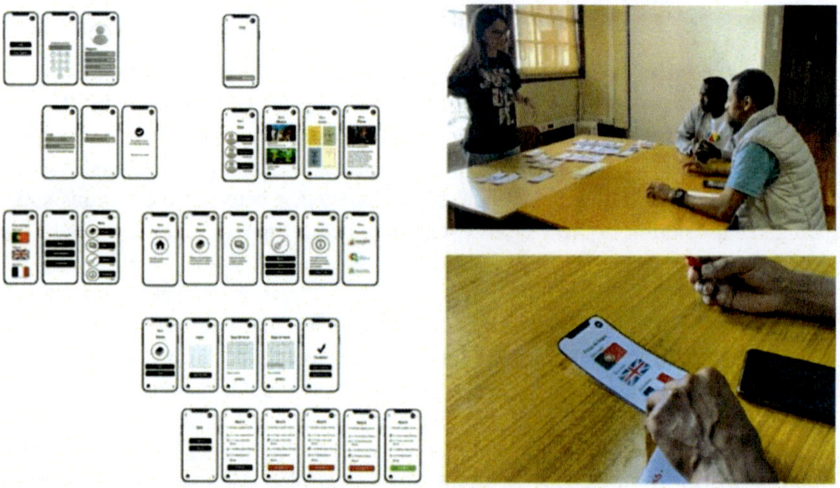

Fig. 6. Usability test session

2.4 Preliminary Proposals, Consolidation and Validation Stages (Workshops 2 and 3)

Group 1, focused on the empowerment of refugee women, held two extra sessions. These were organized autonomously, outside class hours and adapted to the availability of women. The first session aimed to validate the interest in the solution, as well as the communication strategies and supports created for the 1st phase. The second session already focused on rehearsing the following phases of the outlined program as well as new communication strategies. Although the sessions were brief and with few participants, they ended up being important for this process. Group 2, which organized only one session, carried out a usability test using a low-fi model, where the two participants were asked to perform tasks and to vocalize their actions and interpretations, as showed in

Fig. 6. This process also made it possible to validate the suitability or improvements to be included for the realization of the final solution.

3 Final Proposals Addressing Refugees Needs

Prósperas (Prosperous). is focused on the empowerment of refugee women, to promote their fulfilment and integration into society through their uniqueness and support and promotion of economic and social autonomy at the personal and family level. Through a diagnostic and training program/services of basic tools and supported guidance for the realization of a path that allows you to implement personal entrepreneurship projects. As part of this personal action, a digital platform is proposed to promote a network of contacts between the various refugee women to promote and give visibility to projects, sharing experiences (in direct contact between women), presenting success stories and implementation strategy.

A Força da Língua. (The Force of the Language). is a digital platform that aims to mitigate absenteeism in the Portuguese language learning process, being a complementary tool to the learning process. This add-on offers quizzes for the initial assessment of the Portuguese level, quizzes and games for autonomous development and training of Portuguese, as well as the promotion of the language through Portuguese culture, more specifically through access to Portuguese lyrics and music and books by Portuguese authors (Fig. 7).

Fig. 7. Prósperas proposal (left); A Força da Língua proposal (right).

In both proposals, they assigned the initial management and registration to the GA to allow the control of a secure network and protection of the refugee's location and identity. Posters created to present to GA consortium pivot. The students initially presented the final solution using media presentations and several supports/models of the produced solutions (both digital and analogue). After the presentation, the concept and content of the solutions, as well as the methods and forms used to materialize and communicate, were critically analyzed. This discussion was crucial not only for the teacher's feedback, but also for the participation of colleagues, who brought a positive viewpoint to the solutions and tactics that aided in the solutions' efficacy and communication. In addition to the main objective, the critical and analytical exercise, the tuning of the presentation made it possible to prepare the students for a second presentation of a broader nature, including the elements of the consortium. In this sense, suggestions were made for correction and/or adaptation of the presentation aiming the consortium. As a complement to the presentations, it was decided to produce two posters (Fig. 8) explaining the proposals that could be explored after the presentation to the guests and that included an abstract of the solution, the authors, visual references, QRcode to access the digital contents, and the service map. The central figure of the GA consortium was present for the second moment. His comments were highly insightful in terms of content and excellent from the standpoint of future implementation that was demonstrated regarding the future implementation perspective. This last possibility was relevant in terms of satisfaction and perception of the potential impact that students could have on the work carried out. In addition, an invitation was made to attend an extended meeting of the consortium to present the proposals to all elements. Session that took place and revalidated the relevance, suitability and interest in the solutions developed.

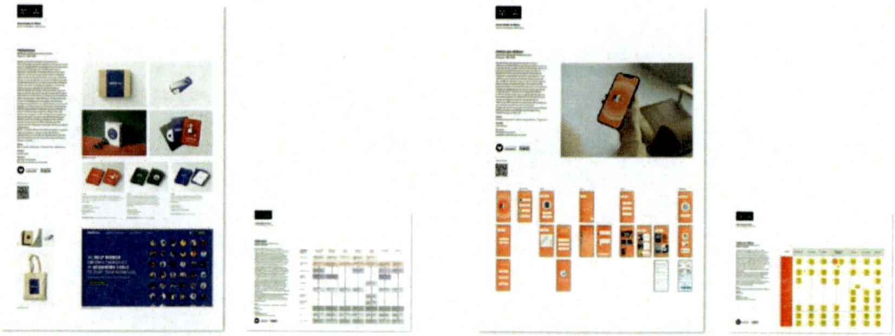

Fig. 8. Posters created to present to GA consortium pivot

4 Final Remarks

Through this process, it is possible to identify three levels of action and impact: (1) is related to the training of students; (2) is the demonstration of the potential impact of service design in the public sector, and finally, (3) is the actual potential to improve the conditions of refugees. In the first level, by basing the proposal on a problem-based learning approach and, the learner-centred approach encourages deeper learning about the context and topic, as well as working among groups to solve an open problem. Which enhances motivation and learning as well as the development of soft skills when students interact with various stakeholders, contexts and cultures. In this specific project, the empathic dimension gauged by the students - through contact with the GA pivots, as in the participatory workshops with refugees -, allowed the development of communication tools and the deepening of the designer's problematic and responsibility, impossible in projects with this degree of involvement. We should also highlight the promotion of initiative and autonomy developed by the students in the organization of initiatives outside school hours, as well as the management of the differences between the project's time and people's time. This last topic has provided the students with a degree of sensitivity to the needs of adapting the schedule to the people we are working with, as well as to managing the expectations of the participants and of themselves as leaders of this process. Regarding the 2nd level, although the area of service design is not extremely new, its presence and potential for impact is not registered or fully measured by organizations. In this sense, the partnership with the Municipality of Guimarães facilitated this change in perception, which highlights the need for strategic relationships between academia and the outside (whether with public or private organizations). As the process unfolded, it was possible to perceive a growing awareness of the possibilities and potentialities created by service design; which procedural and approach strategies to use in the project; and the degree of involvement needed from all those involved to enhance the achievement of positive results. The sharing of information and the involvement of GA employees in different academic moments and participatory design initiatives were also relevant factors for the consolidation of this partnership and for the perception of the designer's role facing these societal challenges. After the classroom presentation to the GA staff, they requested a second presentation to the other members of the consortium. The remaining members of the consortium showed equal interest in the developed proposals. Therefore, it was defined as the next step to facilitate the implementation of the developed solutions, through their integration into application processes that would support this action, that is, related to the 3rd level. This was satisfactorily accepted by the students, who showed interest and availability to join future implementation teams. With this last phase, the students will be able to support the implementation process of the project they have developed. Without neglecting the significance of the continued presence of the refugees as the main focus of the solutions and the fact that they will continue to play a significant role in validating and optimizing the solutions that are developed.

References

1. United Nations High Commissioner for Refugees (2020). https://reporting.unhcr.org/sites/default/files/gr2020/pdf/GR2020_English_Full_lowres.pdf
2. Manafi, I., Roman, M.: A way to Europe: new refugees migration patterns revealed. Sustainability **14**(2), 748 (2022). https://doi.org/10.3390/su14020748
3. Constantino, C., Minas, M.: My new home?: strategies and challenges of refugee integration programs in Portugal. J. Community Psychol. **50**, 3792–3808 (2022). https://doi.org/10.1002/jcop.22873
4. Alto Comissário para as Migrações (ACM) (2017) Relatório de avaliação da política portuguesa de acolhimento de pessoas refugiadas
5. Guo, P., et al.: A review of project-based learning in higher education: student outcomes and measures. Int. J. Educ. Res. **102**, 101586 (2020). ISSN 0883-0355.https://doi.org/10.1016/j.ijer.2020.101586

6. Sanders, E., Stappers, P.J.: Convivial Toolbox: Generative Research for the Front End of Design. BIS Publishers, Amsterdam (2013)
7. Polaine, A., Lavrans, L., Reason, B.: Service Design: From Insight to Implementation. Rosenfeld Media, New York (2013)
8. Kumar, V.: 101 Design Methods: A Structured Approach for Driving Innovation in Your Organization. Wiley, Hoboken (2013)
9. Stickdorn, M., Schneider, J., Andrews, K., Lawrence, A.: This Is Service Design Thinking: Basics, Tools, Cases. Wiley, Hoboken (2011)
10. Stickdorn, M., Schneider, J., Andrews, K., Lawrence, A.: This Is Service Design Doing, O'Reilly Media, Inc. IDEO (2015) The Field Guide to Human-Centered Design (2018)

Exploring the Body and Volumes Through Moldable Materials

Ileana Jalil Kentros(✉)

Universidad Iberoamericana, Prolongación Paseo de Reforma 880, Lomas de Santa Fe, 01219 Mexico City, México
ileana.jalil@ibero.mx

Abstract. This article presents a teaching-learning process to address the redesign of the body through clothing, using moldable materials as a didactic strategy prior to industrial pattern making; for the first semester of undergraduate studies in Fashion Design and Sustainable Textiles. The volume generated by the planes (fabrics) in relation to the body is explored, allowing to visualize and feel the space between the skin and the textile. The principles, the process, the results, and a reflection on this exploration with moldable materials are presented.

Keywords: Teaching · Apparel Design · Methodology · Malleability · Patternmaking · Three-dimensional sketching

1 Fashion and Body Volume

The relationship of textiles with the body is one that goes from two to three dimensions at various times, where two-dimensional planes -the fabrics- are superimposed on a three-dimensional solid -the body- (Ellinwood, 2011). The relationships that can be established between the two are multiple and it is in these linkages where most of the possibilities of clothing have been established throughout history, and where the changes sought by fashion will continue to be explored.

In this changing dynamic between the 2D-3D dimension, what remains constant is the body. Although it is always designed for the constant of one body-of different genders, ages, ethnicities, etc.-it is also under pressure to change according to the imperative dictated by trends in ideal figures. As Danielle Sinay (2022) comments, "Unlike clothing and accessories, body types cannot be 'tried on' and discarded, but that has not stopped the dominant culture". This practice that incentivizes body transformation to fit a trend has only led to problems of personal perception, eating disorders, and an unhealthy promotion of body interventions.

The chronology of ideal bodies throughout the 20th century is very evident. One can analyze the beauty icons of a historical moment, such as Camille Cliford in 1910 with the "Gibson girl": her pronounced hourglass silhouette and her 18-inch waist were very difficult to achieve for most, so the use of the corset became widespread, no longer to support the breasts and give stability as in other times, but to dangerously constrict the waist. One hundred years later, the hourglass figure with full curves (Greep, 2022) is now totally impossible to obtain without surgical interventions, to the point that the Brazilian butt lift - known as BBL - is the deadliest plastic surgery (Ellin, 2021), although one of the most popular.

No matter how much dedication a body is trained to transform itself, or as I have heard from faithful gym-goers, "to sculpt itself," and even considering increasingly sophisticated surgical interventions, the transformation of body volume has an anatomical and organic limit that can no longer be altered. On the contrary, recent ruptures defend a diversity of bodies beyond the canon of thin, young, and Caucasian beauty, which instead promote respect and recognition of other bodies without seeking to change them as Sonya Renee Taylor does in the series of books such as "My body is not an apology"; which expanded to become a cultural movement and activism.

2 Redesigning the Body Through Clothing

Considering the above, the greatest possibility of redesigning the body (continuously and radically) is in clothing. Any textile placed on the body is a redesign of it, and even if it only covers its contour with an adherent surface, it is already a change in its texture and color. The more it moves away from the skin to generate other silhouettes, such as insinuating or volumetric (Saltzman, 2004, p. 69), the more opportunities the designer must generate innovative proposals that reinterpret that base dimension.

Understanding this principle, which in textile and clothing design represents the possibilities of transforming the perception of the body through the fabrics that cover it, can have two possible effects:

1. Respect and work with bodies as they are. If the focus is on the textile and garments, as they are what the transformation is about and what the design is focused on, then the basic body can be accepted in its uniqueness.

   Part of the principles of professional modeling is that they are expected to be bodies adapted to the clothes, with measurements close to the size standard to facilitate the fitting and display of the garments. But people don't have to have this imperative; rather, the clothes should fit them, just like any other design that considers usability. It is not up to design and fashion to modify bodies, because it means imposing a deadly requirement as it has been accused for so long.

   To exercise greater freedom in the design of garments and the proposal of new silhouettes. If the body -any body- is only the base, then the space you must work with is everything after it. It can be a centimeter further away from the skin or it can be a cauda seven meters away from the foot, all that area is the designer's working possibility. A body can give a guideline for a type of design, and another for a different one, as it can be according to its changes in different stages of life or according to some activity that transforms it as sport; for this reason it will always remain essential to know well those bodies of foundation, and that is the work of a lifetime for an apparel designer; that is learned beyond the pattern books.

3 The Teaching-Learning of Clothing and Its Relationship with the Body

In disciplines such as design, where form is the central statement, it is important that students materialize and experience what they learn. Students acquire a specialized language and knowledge according to their discipline, but this is usually accompanied by an application. From what has been mentioned in the previous paragraphs, it is important for a student to know and reflect on the relationship of clothing to different body types. If you are learning that redesigning the body involves working on the space generated on its contour, it is important that you can understand, identify, and also experience it; both consciously and rationally as well as sensitively and intuitively.

The training of clothing designers, as well as textile designers, includes a great importance to tactile sensoriality. Textile is a material that accompanies us in every moment of our lives and that we always have close to our skin. We perceive its flexibility, roughness, softness, lightness, or weight at all times, and also how it moves with us no matter how small the movement.

This goes beyond knowing how to identify and choose a suitable fabric for each function and user. It means developing an aesthetic, creative sensibility, open to experimentation and search not only from research and reason: "On this path, the Pedagogy of Sensitive Knowledge enters the teaching space as an alternative approach to the Cartesian logic of knowledge construction, still prevalent in the educational context. It focuses on the construction of knowledge through aesthetic experimentation, whose relationship with perception and sensitivity is inseparable and originates, therefore, from an education guided by and for the sensitive" (Souza, B. P., & Pereira, A. C., 2020, p.3). In design students, knowledge is developed by sensitively experimenting with the relationship between skin and textiles, and how that relationship is constructed and structured through patternmaking.

3.1 Cutting and Sewing Systems and Manuals

The contemporary tradition of patternmaking - and its subsequent steps - cutting and sewing, are structured as a practice whose indications must be learned and followed.

The books for learning patternmaking have a technical didactics: they first establish the tools and materials -rules, fabrics, fittings- and then systematize the method by dividing it into sections and instructions.

For example, in Mexico, the emblematic CYC system -Corte y Confección- (Miranda et al., 2019) has the following thematic sequence: tools, hand sewing, use of the sewing machine, measurements and basic patterning for each type of garment and finally its transformations. Similar processes can be seen in systems such as Gloria Gomez in Buenos Aires (Gómez Correa, 2012), Antonio Donnanno in Italy (Donnanno, 2014), Dennic Chunman in Barcelona (Chunman, 2011), "Manual de Patronaje Básico e Interpretación de Diseños" of Colombia (Servicio Nacional de Aprendizaje, 2011), Pat Parish in England (Parish, 2015), to mention some outstanding systems.

This confirms that, "Although fashion trends come and go, the principles of pattern making do not change" (Datta, Seal, 2018, p.4). These methods have been proven effective for decades, both in schools in industrial technologists, as well as in universities where the fashion, apparel or textile design program is taught.

Learning the technique of flat or industrial patternmaking is essential in the training of these types of designers, and the process to learn it is usually followed as in the books mentioned above, in a methodical way as indicated by the technique. But since its focus has been mainly from the clothing industry, and not primarily considering a didactic for the training of designers, it does not usually offer an approach that facilitates the learning of this abstract, geometric and two-dimensional language of industrial patternmaking.

In this sense and considering that this analysis is related to university design students, the focus is on generating didactics that promote the understanding of the body and the exploration of its transformation through clothing, before learning industrial pattern making. At the University of Santa Catalina in Brazil, this has been done by introducing the student through modeling:

"(A) Perception first, then abstraction. It is argued here that the initiation of learning pattern making is through draping. [...] This technique is more connected with technical-creative aspects than flat patternmaking, which makes it difficult for the student to become aware of volumes, fabric and body movements. In this perspective, it is believed that the initial contact of pattern making through direct contact of the subject with the object, and not by theoretical means, will open a more fruitful path for the introduction to pattern making studies" (Souza, B. P., & Pereira, A. C., 2020, p.3).

This educational proposal distances itself from the idea that:

> "... Only by mastering the techniques of garment construction will it be possible to give free rein, but above all, viability to the most creative proposals" (SENA, 2011).

Likewise, in universities such as the UBA (University of Buenos Aires) in Argentina, patternmaking is not a prerequisite or condition for clothing design but is discovered and explored throughout the courses. In the words of Patricia Doria (2014, p.2):

> "... First one learns to design objects of clothing and then one learns in the very exercise of that doing [...] Projects in clothing design are simulations of experimentation, of the relationship between the textile material and the limits or body contours."

Based on the above, it is proposed to bring the student first to the experimentation of the relationship of the textile material with the body contours, in a sensitive and intuitive way that allows him to have a better understanding of the volume, only that, instead of a modeling on a mannequin with fabric, it is proposed a modeling on a mannequin with moldable materials.

3.2 Studying Volume from Volume: Moldable Materials

To build a garment in industrial pattern making, we go from the plane of the pattern to the plane of the fabric to then generate volumes on a body. In other words, it goes from two to three dimensions.

This process presents a challenge for the three-dimensional visualization of the garment on the volume of the body. Even if I start from a figurine, it is a two-dimensional representation where I am not able to dimension its three dimensions.

In this case, we propose a study of the volume generated on the body that allows a manual three-dimensional exploration (since programs such as Clo 3D allow a virtual 3D visualization), to explore the volume directly without going through the plane. That is to say, by means of a three-dimensional sketch. For this reason, moldable materials were chosen, i.e., plasticine, which is a common material in programs such as industrial or product design, but unusual in fashion or textile, because of our priority to flat materials.

The introduction of this unusual resource in apparel and fashion can be a creative trigger, as Mike Ashby and Kara Johnson comment: "…the use of familiar materials in an unfamiliar way is also a creative step. New materials, in particular, act as a trigger for inventive thinking, offering the potential for novel design" (2002, p.41).

Plasticine may be the rediscovery of a material associated with childhood in a totally new use: to generate a volumetric silhouette on a mannequin. These are a material invented in 1880 in Munich, Germany by Franz Kolb for his artist friends who complained that the clay dried too quickly (Quispe Yucra, E., 2021). Likewise, a student does not have to repeat an entire modeling while experimenting, he or she only must remove or add plasticine over the process gradually. In addition, it offers other unique advantages: "The modeling is performed on soft materials thanks to its plasticity, shapes are given and do not offer any resistance […]. The malleable, flexible and soft nature of plasticine fosters in the creator's ingenuity a level of flexibility that is not accessible in carving." (Quispe Yucra, E., 2021, p.18). Or, in our case, that it is not accessible in the same way in the fabric. In addition, for the training of novice first-semester students, it promotes other incentives for learning:

"Stimulates the sense of touch in the learner.

- Develops the three-dimensional aspect of objects (in this case, the clothes, and the body).
- Allows direct experimentation with different textures.
- The movement works" (Quispe Yucra, E., 2021, p.19).

The material with which a form or an idea is investigated will influence its type of results. If a student of fashion or clothing is presented with plasticine and a mannequin, his search will be very different than if he has fabric and a mannequin. The fabric folds, pleats, sews, falls; the plasticine molds, squashes, expands by accumulation. As Zdravko Radman comments on manual sensitivity and tools:

"Just as there is no such thing as a naked eye, there is no such thing as a naive body or an innocent hand. For example, the way a hairdresser sees hair depends on whether he holds a comb or scissors in his hand. Similarly, the way a painter sees a landscape depends on whether he is holding a pen or a brush. His modes of representation are defined by the type of tools he has at his disposal. In other words, what hair or landscape allows, or rather what shapes manual perception, is literally in the hands of a hairdresser and a painter and the tools they use. Similarly, our manual perception of food (and, indeed, taste) differs depending on whether we hold chopsticks or knife and fork in our hands" (Radman, 2013, p.376).

Thus, it is intended that the student can touch and model in flexible and soft materials the volume that is generated on the body (of the mannequin) and investigate the shapes generated without thinking about the pattern that would be required to solve it, that will be seen after reaching the desired volume. Can you reach different results? Can you promote a "manual thinking" (Radman, 2013, p.383) of the volume when working three-dimensional modeling without planes?

4 The Process of a Hybrid Methodology: Plasticine and Textiles

The objective of this process is that the student can explore the generation of volumetric silhouettes on a body (plastic mannequin to which the plasticine is adhered), with a material other than textile that allows him to visualize and sketch three-dimensionally; as a didactic strategy prior to learning the modeling of garments and industrial pattern making.

I call it a hybrid methodology because it consists of two different stages that later converge. The first stage is the modeling of the volume with plasticine, which is a process closer to the exploration of product design; the second is to extract a pattern on that plasticine through modeling and then refine it, trace it on fabric, cut it and make it on another body. It is a hybridization of product and clothing methods, to learn an essential element to both, the volume and, in our case, in the clothing in direct relation to the body.

4.1 Conditions and Materials

To explore these volumes that are generated on the body, it would be desirable for didactic purposes to meet three conditions:

a. Have a body to work on to maintain proportion and scale.

   Unlike laminates or textiles, moldable materials are heavy, so the first requirement is to scale down the mannequin to a portable and manageable one, hence from the supply of suppliers a 25 cm high plastic mannequin was chosen, with a plasticine adherent surface.

b. To have the possibility to explore the volume itself without the use of plans.

   For the reasons explained in the previous paragraph, moldable materials were chosen. After trying several options such as foam clay (which is not very precise and dries quickly), professional plasticine (it is harder, requires warming and tools to refine) and soft dough (it is too soft, imprecise and cracks when it dries), the most recommendable were the soft school plasticines; since they do not require heat to be malleable, do not harden with cold, do not dry out, can be reused and combined many times and are easily ductile with the hands, with sufficient precision and consistency.

c. Conclude an outfit with a basic pattern and made with textiles, even if this is not achieved through industrial pattern making. It is important that the project is linked to the fabrics and does not remain in the moldable materials, otherwise it would only be a visualization exercise and would not connect directly with the garment.

4.2 From Two-Dimensional to Three-Dimensional Sketching

The first step to detonate the exploration of volumetric silhouette attire is pencil sketching; this is done over a period of at least a week from a theme of inspiration. In the case of the images shown below, the theme was utopia-dystopia.

Subsequently, from these sketches, an alternative is selected to be worked in moldable materials on mannequin. The student must consider that the modeling work on mannequin is also an exploration, a three-dimensional sketching, so the initial modeling stage is flexible, promoting an iterative process.

In the end, the result is a solid silhouette on a body, in a way that does not lose human proportion or impede the exploration of volume projected by the designer, as seen in the images below (Figs. 1, 2 and 3).

Fig. 1. (From left to right) Project and photograph by Antonio Ángeles, student at the Universidad Iberoamericana Mexico City, prepared in 2021.

It is a type of modeling with moldable material, where the solid volume is formed and the void under the fabric and on the skin acquires solidity. It is very different to imagine a puffed sleeve and try to generate its volume by means of two-dimensional patterning or by modeling with fabrics on a mannequin, than by forming with plasticine the amplitude of the sleeve with the fingers on the spot and on a mannequin. To amplify the sleeve only asks to add more plasticine, to decrease it is to extract it or to mold it, and at that moment it is possible to observe if the desired point has been reached.

The process of working with plasticine requires patience, since, being a manual exercise, it takes time to become familiar with the material and to model it repeatedly until a purposeful, dynamic volumetric silhouette is achieved, in relation to the inspirational theme. The first tendency of the students is to want to use the plasticine as fabric, that is, to create planes that fall or adhere to the body. Once this first impulse is discarded,

Fig. 2. Project and photograph by Inti Hidalgo, student at the Universidad Iberoamericana Mexico City, prepared in 2022.

Fig. 3. Project and photograph by Karla Rivera, student at Universidad Iberoamericana Mexico City, prepared in 2022.

they begin to observe that the volume with this material is expanded masses, and that the fabric will be only its surface.

4.3 Pattern Extracted by Modeling

The next step -once the exploration and modeling in plasticine is finished- is to leave the didactic and sensorial experience that the clays provided and return to the two-dimensionality with which we textile and clothing designers work the most: the planes of paper and fabrics.

In this step, the volume of the plasticine is traced with a transparent plastic sheet or paper that can be fixed with pins (Figs. 4, 5 and 6). In this way, the surface of the volume that will later become the fabric is recorded. It is intended to function as an intuitive approach to pattern making, since in the end many of the pieces they generated resemble industrial patterns. For example, the pattern of a truncated cone in plasticine resembles that of a circular or semicircular skirt.

It also implies an independent learning by obtaining step by step the molds directly from the surface of the plasticine and not from a book with indications. I believe that it can give you a greater capacity for logic and autonomous resolution when you see that you can solve it on the model; like what happens when you solve a garment directly on the three-dimensional modeling with fabrics.

There are three elements that turn a sketch into a garment: interpretation, technique, and technology (Datta, Seal, 2018). In interpretation, a figurine, sketch, or flat sketch is understood in order to be resolved by patternmaking. This requires extensive 2D-3D visualization skills and sufficient technical knowledge, for which mainly experience is required (Datta, Seal, 2018).

A first semester student is still inexperienced and developing his visualization skills, so seeing and touching the volume modeled in solid and extracting the pattern is a didactic and formative support to solve the exercise of a volumetric silhouette.

Fig. 4. (From left to right) Project and photograph by Natalia Méndez, student at the Universidad Iberoamericana Ciudad de México, prepared in 2022.

Fig. 5. Project and photograph by Fernanda Cervantes, student at the Universidad Iberoamericana Mexico City, prepared in 2022.

Fig. 6. Project and photograph by Ainara Dacasa, student at the Universidad Iberoamericana Ciudad de México, prepared in 2022.

4.4 Tuning, Cutting, Sewing and Embroidery

The next stage of the exercise consists of refining all the molds that were extracted on the plasticine to then cut them into fabric. The fabric proposed in this case is one that does not fray, since being very small pieces and being their first approach to textile, it would complicate the process a lot if they had to deal with over-sharpening the edges or with persistent wrinkles.

With that in mind, the fabric of the final model was wool cloth or polyester felt, according to the student's choice. It was woven to be firm enough to hold the volumes proposed in the plasticine. To assemble the pieces, they could either add a sewing tab, an essential principle of pattern making, or sew to the edge with a scallop stitch. The decoration that can be seen in the photographs is a decorative complement that does not structurally alter their proposal and is based on the theme that inspired the exercise (Figs. 7, 8 and 9). This is an important closure since it consists of visually and tactilely enriching the surfaces and continues to be part of her aesthetic experimentation and sensitization.

Fig. 7. (From left to right) Project and photograph by Karla Rivera, student at the Universidad Iberoamericana Ciudad de México, prepared in 2022.

Fig. 8. Project and photograph by Inti Hidalgo, student at the Universidad Iberoamericana Ciudad de México, prepared in 2022.

Fig. 9. Project and photograph by Ainara Dacasa, student at the Universidad Iberoamericana Ciudad de México, prepared in 2022.

4.5 Presentation of Hybrid Method: Moldable Materials and Textiles

The final presentation consists of exhibiting the two models, the one made of moldable materials and the ready-made one, to compare the similarity between the two and the solution reached by the student through pattern making and sewing, since there is not only one possible interpretation (Fig. 10).

At this point it is possible to observe how the moldable materials made it possible to form solid and defined volumes, and how from these, also defined volumes were generated in the textile, although not solid, but with the void between the skin and the fabric.

Thus, students can experiment creatively and technically the possibilities of the form on the body, in this case with the didactic limitation that they are derivations of recognizable and defined volumes (primitive solids), since in case of opening it to any indeterminate volume, it would be difficult to extract the subsequent pattern and the complications would be diversified. This limitation did not prevent the students from making disruptive, dynamic, and creative explorations. In fact, it pushed them to keep looking for a solution that would satisfy them aesthetically.

Fig. 10. Project by Valeria Fierro, student at the Universidad Iberoamericana Mexico City, developed in 2022. Photograph by Dora Segrera.

5 Conclusions

I share this teaching experience for two reasons. The first is because I consider it important to take the time to document, analyze and share a classroom experience, especially when it is a new one; many times these processes take place during classes and at the end of the school period the chapter is closed, without giving room for individual reflection and dialogue with students and colleagues, when the most enriching knowledge is open and shared, the one that is water that flows and does not stagnate in the zeal of the private.

Despite the didactic intention of the exercise, it has proven to be a challenging project because it demands from the student a manual sensitivity and patience to model the same piece sometimes repeatedly, it offers the advantage of allowing correction without repetition, but being the same object tends to exhaust the patience of the student who wants novelty and speed. On the contrary, their greatest satisfaction comes when they start working with fabrics, threads and applications, since by their choice of career they have a predilection for those materials and in many cases, they have previous experience. In addition, it is at this point where they have greater aesthetic freedom, they can choose the color combinations and enrich the entire surface if they wish with embroidery and appliqué.

The exercise proves to be challenging for teachers as well, since moldable materials are not usually in the daily practice of textile and apparel designers, and this changes the teacher's evaluation and guidance with the students.

It can be observed during the project how students develop their ability to visualize volume as the exercise progresses, with the implicit difficulty of relating it to a body. It is not the same to imagine a cylinder as to imagine it inserted in a neck or waist, and even less so if it is turned to one side, or inflated, flattened, or fragmented, and this can be experienced with the eyes, with the mind and with the fingers.

Once this first difficulty has been overcome, the next one is how to extract the pattern from that volume, for example: how do you divide a sphere into sequential segments like a puffed sleeve? In sequential segments like a puffed sleeve. What do you do when there is fabric left over in the back, but not in the front? With darts. What do you do with the excess fabric in the armhole next to the bust? Again, with darts, and there are several ways to place them that the student discovers on the material. So, reflection, experimentation and autonomous learning are constantly encouraged, even with the constant support of their teachers.

The intention with this teaching-learning process is for students to develop a greater capacity for visualization, by considering the relationship of volume to a body, and to realize how it translates into patterns and flat textiles. All this considering that their creative focus is on garments and textiles, and moving away from the idea that bodies must change to the imperative of fashion. Finally, when students think of a puffed sleeve, to repeat the example, they should know that it implies a pattern composed of segments or intervened with a series of darts that give it that width, and this not because a manual says so, but because they know how to deduce it from their learning experiences.

The exercise described here can be replicated, although I would only recommend it for the first semesters, where complements or formative palliatives are required. Once the pattern making and volume generation is understood, it is a process that can be released and go directly to fabric modeling or to the transformation of the different methods of industrial pattern making. To conclude, this method reminds the student that the language of design is the form and that its possibilities are infinite, even that other atypical materials can be used for clothing, remembering that designers such as Hussein Chalayan and Paco Rabanne went down in history, among other innovations, for the use of materials other than textiles.

In closing, I join the conclusion made by Paganini and Italiano in their historical review on the teaching of patternmaking: "It was possible to identify that two-dimensional and three-dimensional techniques complement each other and if combined in their positive aspects can provide a more efficient teaching [...], integrating the mathematics of flat patterns to the visibility provided by the three-dimensional (Paganini, D. & Italiano, I. C., 2013, p.9)". As they recognized this 2D-3D relationship is essential to the teaching-learning of patternmaking; and just as there have been many methods so far, this hybrid approach may be yet another opportunity for further exploration to be considered by students and teachers of apparel design.

References

Chunman Lo, D.: Patternmaking. Blume, Barcelona (2011)

Datta, D.B., Seal, P.: Various approaches in pattern making for garment sector. J. Text. Eng. Fashion Technol. India (2018)

Donnanno, A.: Técnicas de patronaje de moda, vol. 1. Promopress Editions, Barcelona (2014)

Doria, P.: On Teaching Apparel Design. The creative challenge (teaching the method). Notebooks of the Center for Design and Communication Studies, pp. 37–47 (2014)

Ellin, A.: Brazilian butt lift on the rise, despite the risks (2021). https://www.nytimes.com/es/article/cirugia-bbl-riesgos.html. Accessed 9 Apr 2023

Ellinwood, J.: Fashion by Design. Fairchild Books, United Kingdom (2011)

Gómez Correa, G.: El lenguaje de los patrones de moda. Nobuko, Buenos Aires (2012)

Greep, M.: How the perfect body has changed throughout the decades: plastic surgeon reveals how beauty standards have evolved from the 1910s hourglass Gibson Girl to the 'bootylicious pop culture body' of today (2022). https://www.dailymail.co.uk/femail/article-10467643/How perfect-body-changed-decades.html. Accessed 11 Apr 2023

Miranda Ruiz, S., et al.: Aprenda corte de ropa para niña y adulta. Editorial CYC, Mexico (2019)

Paganini, D., Italiano, I.: Guidelines for patternmaking teaching and for didactic materials. Int. J. Arts Commer. **2**(9). São Paulo (2013)

Parish, P.: Fundamentals of creative patternmaking. The architecture of fashion. Promopress Editions, Barcelona (2015)

Quispe Yucra, E.: Modeling with plasticine developed artistic expression in early education students. University of Tumbes, Peru (2021)

Randman, Z.: The Hand, An Organ of the Mind: What the Manual Tells the Mental. The MIT Press, London (2014)

Saltzman, A.: El cuerpo diseñado. Paidós, Barcelona (2004)

Servicio Nacional de Aprendizaje SENA (2011) Manual de patronaje básico e interpretación de diseños. Textile and Leather Manufacturing Center, Bogota

Sinay, D.: For the Last Time: Body Types Are Not Trends, Glamour (2022). https://www.glamour.com/story/body-type-trend. Accessed 9 Apr 2023

Souza, B., Pereira, A.: Not all path is straight: limits and possibilities for the sensibilization of fashion design students through pattern making teaching. Revista De Ensino Em Artes, Moda e Design **4**(2), 237–242 (2020)

New Ways of Ageless Fashion: Project Development in Class Context

Alexandra Cruchinho[1(✉)], Benilde Reis[1], and Teresa Ramilo[2]

[1] CICANT, Lusófona University, Lisbon, Portugal
alexandra.cruchinho@ulusofona.com
[2] CPES, Lusófona University, Lisbon, Portugal

Abstract. Ageless Fashion is the name that best characterizes the research presented in this scientific article. Accepting aging and knowing how to deal with the various changes that the body undergoes is one of the great challenges that human beings, men or women, face with advancing age.

Feeling good about your body and the way you dress is one of the major concerns of older people and it is also one of the issues that arise as a starting point for the project that was developed – Ageless Fashion.

The main objectives are related to the understanding of the body and its natural changes for the best adaptation of the technical modeling of clothing and for the definition of an image where aesthetics and comfort are not forgotten.

Thus, the challenge was launched to a group of Fashion Design students who, for a week, intensively sought to find solutions and proposals for clothing suitable for seniors, where comfort and aesthetics were the focus of study.

Keywords: Ageless Fashion · Project Development · Age · Body Image · Teaching Methodologies

1 Introduction

Ageless Fashion is the term that served as a starting point for the development of this applied practical research and its main objective was to develop a project in Clothing Design considering new forms of ageless fashion.

The methodological approach chosen was to put into practice student-centered teaching through the application of design methodology, in short, the focus is on raising a problem so that students can find solutions in the classroom context.

In the proposed conceptual framework, it is about developing proposals for garments for a niche market where consumers have different requirements for clothing products based on age and, consequently, on the changes that the body undergoes.

With the projects developed, it is possible to demonstrate that the fashion industry needs to be more attentive to a Senior consumer to allow this public to feel good about their body and comfortable with what they wear.

2 Ageless Fashion: A Framework

Literacy and generational culture have a lot to do to progressively raise awareness in society in general and the elderly in particular, asserting their rights in order to counter the existing AGISM.

However, the media and the economy tend to increasingly value the image of the elderly due to the tendency to increase the representation of this group in Western society, the same happening in Portugal where they represent about a third of the total population (23.4%, INE, 2021), and there are 182 elderly people for every 100 young people (INE, 2021).

Certainly, for the Fashion industry, the 60+ represent an interesting niche market to invest in, in the near future. We proceed to develop some considerations resulting from the interdisciplinary reflection that we share in the Ageless Fashion project.

Therefore, special attention should be paid to clothing intended for the senior population, whose design must meet parameters that correspond to the changes that arise in the body of men and women associated with age, contextualizing the minimization of possible skeletal deformations and areas of pressure that must remain free, as in men, for example, the prostate region and in women, the legs to allow good circulation. The garments may be discreet (do not stand out) in relation to the areas of the body that the elderly person prefers or intends to keep intimate or protected from social gaze, as is the case with three-quarter sleeves, wide and/or long, which in addition to protecting from the sun or the cold allow you to play with the illusion of making the length of the arms more proportionate to the figure, since the anthropometric ratio of the arms to the trunk (whose height decreases with age) makes them appear longer.

Another value to be added to Ageless Fashion clothing will be the comfort of the materials, the associated design lines, the simplicity of the finishes as well as the handling of accessories such as buttons and other closures used in the clothing and their location, since the fine hand movements become, they are often limited in gripping, pinching and strength. Vision can also be a limiting factor.

The garments can be versatile and complementary to each other, producing a variety of ready-to-use "Kits", simultaneously valuing the practical, aesthetic, and economic point of view. We would even say that the "Kits" could be the object of training for professionals and aimed at the literacy of those targeted.

The lightness and sobriety, the pattern and color of the pieces can help shape and influence choices, so that the results can meet real challenges aligned with the "Ageless Fashion" idea, such as the introduction of sportswear lines or Freestyle as an achievement and diversity adapted to different occasions, going beyond the "old fashioned dress code" that leads them to opt for monochromatic black and gray clothing.

The influence of the seasons is no less important and adds, whether by color or pattern, not only aesthetics, but the transmission of the appropriate thermal effect. Light colors for hot weather. The introduction of cheerful, light, versatile clothing is certainly the subject of literacy that needs to be worked on and cared for. With age, the body must remain protected and exposure to the sun conditioned. The skin becomes thin and fragile with age. Light skins are the most susceptible if exposed to weather conditions, whether cold, heat, wind, UV rays from the sun cause discomfort and even subsequent harm. The skin needs protection and clothing may be the most appropriate solution, as it is not intrusive. However, sun exposure is also recommended in the early hours of the day and in the late afternoon. The choice of light, natural and protective fabrics allows you to filter out the sun's rays. In any season of the year, hats and scarves are essential protective and aesthetic complements.

With age and especially under the effect of a sedentary lifestyle, the body shows signs of alteration in body constitution, weight and necessarily posture. In the physical sense, posture concerns the positions of the joints and the correlation between the limbs (arms and legs) and the trunk. The posture results from the synergy and complexity of the skeletal and neuromotor system, whose components are – bones, muscles, joints and the nervous system that respond for the performance, which moment by moment, allows the balance of the body in space and its trajectory – the movement. Posture is related to the environment – the physical and social space. But posture is also the "attitude" that is adopted at a given moment with a given meaning. So, for example, I can run to meet someone, to escape, to catch transport, to shorten a distance or to help or compete with someone. Naturally, each motivation recreates its own posture. Also with age, the posture becomes characteristic, and, by the image, we recognize that a person is simply old. Elderly people tend to walk with pronounced leg spacing in search of balance (thus increasing their base of support). They are less fast and sometimes less efficient in their motricity, and some movements become unnecessary, but existing, as is the case of tremors that hinder, for example, the precision of the gesture.

Phrases like, you're not old enough to... or you're not old enough to... Are often heard in an intimately critical tone in the family or social nucleus, more broadly, recriminating or ridiculing the intentions or actions of the elderly. Clothing is undoubtedly one of the most susceptible to criticism if it deviates from the procedures considered and socially accepted as "appropriate for age".

Current generations aged 60 and over are increasingly sensitive to the challenges society poses. The active lifestyle (EVA) is recognized as a tool that has many supporters and scientific evidence proves its positive influence on the aging process. EVA reinforces the organism and provides more quality of life. PLACE of choice to LIVE does not always match the will of the elderly, but the probability of having continuity from adult life is frequent.

Reduced mobility is the situation in which the individual has limited movements. With disabled age, it becomes more complicated. If possible and appropriately, appropriate behavior can positively condition mobility. The physical condition can be assessed by the performance of the elderly person in the Activity of Daily Living (ADL). Regular and adapted training, when it accompanies the aging process, provides relevant protection for maintaining mobility and positively influences an active lifestyle.

Posture always provides a communicative and non-verbal amount of information. For example, mimes use body posture to communicate actions, emotions, and expressions. Socially, each culture has a code of communication that facilitates the relationship through the expression contained in its body language and that each one adjusts and personalizes, creating "their own style", which the Ageless Fashion project intends to emphasize.

In recent years, several brands have timeless fashion on their agenda.

Like Hermès, the prêt-à-porter designer of the luxury brand Nadège Vanhee-Cybulski designs collections with a unique mastery of luxurious minimalism, working with a timeless and timeless agenda, a winning strategy of the French brand (Flaccavento, 2015).

According to Matos (2018), in recent years we have seen an increasing number of older women taking over the fashion industry. They starred in campaigns, magazine covers and fashion editorials, as was the case with Anna Von Rueden in the editorial "First day of the rest of our lives" for Vogue Portugal in February. They are it girls, influencers, or bloggers; wear Chanel, Vetements, Balenciaga, Saint Laurent, Dolce & Gabbana, Valentino, or J.W. Anderson, and are the epitome of elegance that is a true inspiration for generations to come.

In this context, it is possible to gradually see a change that style has no expiration date, and models of all ages are gracing the catwalks, from Sydney to Milan (Venkataraman, 2023).

3 Methodologies in Class Context

The Ageless fashion project is the result of a set of questions that were raised in different moments of discussion in multidisciplinary work groups. Different areas of investigation crossed in a project where fashion, sport and social action come together.

The questions arise from the problem that stands out in the group: There is no suitable clothing in terms of aesthetics and comfort for seniors. Comfort is questioned by the materials and fittings used, by the choice of colors and elements that make up the aesthetics and silhouette of the pieces, but also, and fundamentally, by the technical component of clothing construction. There is no doubt that modeling is an important component for attributing comfort to garments. This modeling necessarily must be adapted to the changes that the body undergoes with advancing age.

The challenge was launched to a group of higher education students in Fashion Design and the problem posed was the starting point so that, during 1 intensive week, students could look for and think about the best solutions.

The students worked in groups following a design methodology based on Bruno Munari's design methodology, where the Problem arises from a need and it is through the resolution of a set of work steps that a solution will be reached (Munari, 2014).

Teaching is centered on the student, with the teacher taking on the role of facilitator of learning and the acquisition of skills, being present to help the student find the best solutions. Project Development in Class Context.

In the first phase, the Ageless Fashion Project was presented to our students, with a briefing to inspire engagement with this project actively in class Context. In this way, we demonstrate that ageing is a natural, biological, social and psychological process which develops throughout life (Fig. 1).

Both women and men age in distinct pathways, and gender is present in clothing choices. Body ageing assumes changes and transformations of silhouettes, which must be considered for clothing development.

In this context, it was proposed guidelines to our students to develop the creative process of two looks (for men or women, or both), based on: research of the theme, target

Fig. 1. Ageless Fashion Project Presentation. Font: Fashion Lusófona Archives.

audience constructed on a conception of a persona, the development of a mood board, the technical drawing of the clothes and respective materials and final illustrations.

After the project guidelines presentation, we gave an opportunity to our students for them to choose to form work groups or work individually, four groups were formed, and two students decided to work individually.

3.1 Fashion Research Process Phase

Deeply investigative research is necessary for a fashion designer to be innovative. By finding inspirational images, for fit in the proposed project: ageless fashion.

Fashion research is essential and serves as the foundation for each collection (Mbeledogu, 2022). Fashion designers' students may undertake research at any moment by studying the world around them, gathering and documenting the things, pictures, and ideas that inspire them. This type of research can be part of an ongoing process (Fig. 2).

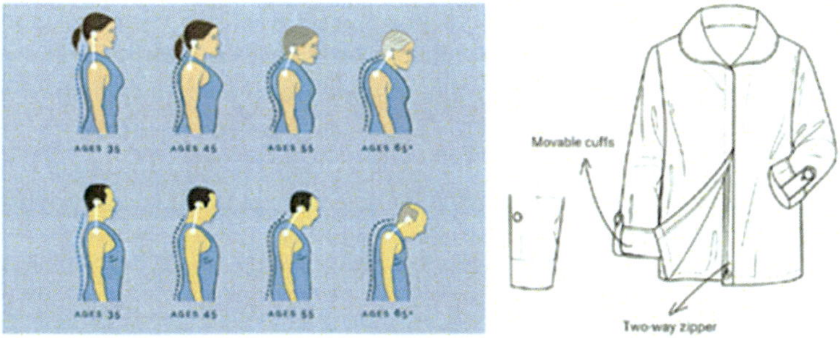

Fig. 2. Research details of ergonomics in project by Gonçalo Valentim and Kristina Hristova, Fashion Lusofona student.

The research phase, in fashion process development is crucial, to find inspiration, made cross referencing between collected data, take information to define silhouettes and proportions, materials (Fig. 3).

According to (Mbeledogu, 2022), in fashion design process the silhouettes are the overall outline of a garment or outfit, which is the first thing the eye sees before focusing on the details. Considering the proportion in a garment or outfit, is based on the silhouette. Proportion relates to how the body is divided into parts, and how the parts then relate to each other.

Fig. 3. Silhouettes and Proportions in "Old is the new young" project by Inês Diogo, Fashion Lusófona student.

3.2 Concept Development Phase

At a later stage in project development, finding a concept for a fashion collection is crucial. This phase is also based on the fashion research process phase, where the students could find inspiration images for their ageless projects (Fig. 4).

Once the designers have begun editing the research, it is ready to create a mood board. Within fashion design, concept boards, mood boards and storyboards are all essentially the same thing – they summarise a collection in which the themes, inspirations, concepts, colours and fabrics are communicated to the viewer (Mbeledogu, 2022).

Making a mood board is considered all research materials since the group of collected images, fabrics, and other materials selected to select the most interesting, dynamic, and informative ones that should reflect the mood of the collection (Fig. 5).

Deeply investigative research is necessary for a fashion designer to be innovative. By finding inspirational images, for fit in the proposed project: ageless fashion.

Fashion research is essential and serves as the foundation for each collection (Mbeledogu, 2022).

The six developed projects in ageless fashion were: "Old is the new young", "Old but Bold", "Timeless", "Back to my roots", "New wave, new age" and "Jazz and Free".

Fig. 4. Example of the concept development phase in "Old is the new young" project by Inês Diogo, Fashion Lusófona student.

Fig. 5. Moodboard of "Old but Bold" project by Maria Beatriz Lopes and Mariana Moura, Fashion Lusófona students.

3.3 Concept Development Phase

The design development process occurs when more consideration is given to these ideas. This is when you will focus on developing design details, fabrics, colours, silhouettes and proportions (Mbeledogu, 2022).

This phase can be developed in a sketchbook manually. However, majorly our students prefer to develop their designs digitally; nevertheless, some students made some initial sketching manually (Fig. 6 and 7).

Fig. 6. Moodboard of "Old but Bold" project by Maria Beatriz Lopes and Mariana Moura, Fashion Lusófona students.

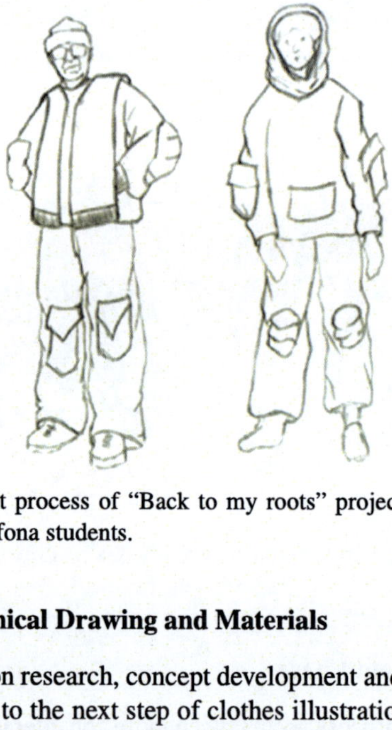

Fig. 7. Design development process of "Back to my roots" project by Maria Inês Caeiro and Beatriz Silva, Fashion Lusófona students.

3.4 Illustrations, Technical Drawing and Materials

After the phases of fashion research, concept development and design development, the fashion students proceed to the next step of clothes illustration development as well as the technical drawings and the association of materials to the developed garments (Fig. 8 and 9).

Fig. 8. Ageless Fashion Project: Technical drawing of a dress and the materials in "Old is the new young" project by Inês Diogo, Fashion Lusófona student.

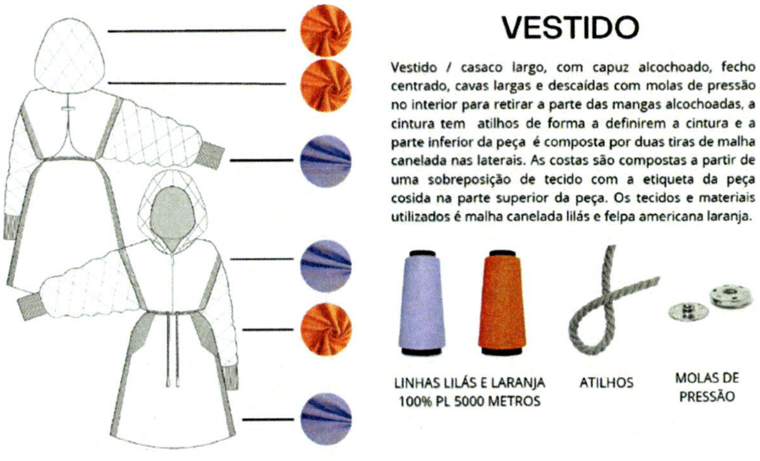

Fig. 9. Ageless Fashion Project: Technical drawing of a dress and the materials in "Old is the new young" project by Inês Diogo, Fashion Lusófona student.

4 Discussion and Final Considerations

We can conclude that clothing items can be versatile and complementary to each other by producing a variety of ready-to-use "Kits", simultaneously valuing the practical, aesthetic and economic point of view. These "Kits" could be the object of training for professionals and aimed at the literacy of those targeted.

The lightness and sobriety, the pattern and color of the pieces can help, form and influence the choices, so that the results can meet real challenges in line with the "Ageless Fashion" idea.

The introduction of sportswear or Freestyle lines as an achievement and diversity adapted to different occasions, going beyond the "old fashioned dress code" that leads to opting for monochromatic black and gray clothing.

Considering ageless fashion and the projects that were developed on a total of six projects, all of them accomplished cohesive looks from strong research applications where silhouette, proportion, fabric and color were effectively explored. The projects developed in ageless fashion were: "Old is the new young", "Old but Bold", "Timeless", "Back to my roots", "New wave, new age" and "Jazz and Free".

The proposals that result from the project are not just the representation of a work that demonstrates that it is possible to develop garments suitable for Senior bodies, but it is proof that our clothing industry still needs to evolve a lot in terms of satisfaction and comfort for clothing suitable for Seniors. It is also important to highlight the importance of developing and applying these methodological practices for students to acquire more structured skills.

The students carry out a project with an intergenerational dimension and the impact on both participating parties is highly relevant. On the one hand, students recognise and value a fashion intervention with a social and inclusive nature, on the other hand, the elderly consumers feel more integrated by the fashion system, increasing their self-esteem. This experience also improves consumer literacy and understanding of the way of dressing and its suitability for different everyday situations.

For students, raising awareness of the topic represents a specific recognition of the value of age in the community and society. There are more and more elderly people in our society. The trend will be to adapt modelling, the creative process and industry in the processes to create proposals for the recipients' well-being. This senior group no longer represents a minority or a niche market considering the gradual ageing of the Portuguese population.

In short, it is important to consider age-related comorbidities in the performance of daily life. Students' awareness of the physical limitations of this group enables an aesthetic and creative approach to proposals and the final image of the consumer.

Acknowledgments. Authors can in this section thank all individuals or entities that contributed directly or indirectly to the work, and that are not included in the list of authors.

References

Flaccavento, A.: Hermès' Ageless Agenda, Business of Fashion (2015). https://www.businessoffashion.com/reviews/fashion-week/hermes-ageless-agenda/. Accessed 24 Apr 2023

Matos, R.: Ageless: estilo em qualquer idade, Vogue Portugal (2018). https://www.vogue.pt/ageless-estilo-em-qualquer-idade. Accessed 24 Apr 2023

Mbeledogu, E.: Fashion Design Research, Second edn. Laurence King Publishing, Great Britain (2022)

Munari, B.: Das coisas nascem as coisas. Lisboa: Edições 70 (2014)

Venkataraman, D.: 41 older women who prove that real style is ageless, Vogue Australia (2023). https://www.vogue.com.au/fashion/news/stylish-women-over-50/image-gallery/4ba61f83517f6c830409171b3a8fb22f. Accessed 24 Apr 2023
www.clese.org. Accessed 22 Apr 2023
www.academyart.edu. Accessed 15 Apr 2023
www.mymodernmet.com. Accessed 27 Apr 2023

Shape as Experience: Analysis of the Application of the Expressive Categories Map as a Facilitator Way in the Configuration of Objects and Spaces

Maria Celeste Sanches(✉)

Nebrija University, Calle Sta. Cruz de Marcenado, 27, 28015 Madrid, Spain
tetisanches@hotmail.com

Abstract. Configuration of a design solution combine aspects of surface, volume, contour, space, and point of view, promoting a sensorial experience that determines the perception of the shape with a visual utterance that stimulates the production of meaning. Following this, this article addresses the educational field of design, analyzing facilitating strategies to synthesize expressive concepts and the configuration process during the project practice in class. With a cross-sectional exploratory methodological focus, that includes participant observation and documental analysis, it examines the outcomes of applying the Map of Expressive Categories Map (MEC) and the concrete experimentation with didactic tools in the scope of Fashion Design, Interior Design, and Production Design formation. The analysis noted that the students integrated aesthetic, ergonomic, and technical requirements in the articulation of the shape syntax of objects and spaces, fostering the perception of the shape as an objective and subjective experience. Furthermore, the evidence confirmed the effectiveness of the MEC as a way to organize project thinking and communication in collaborative environments.

Keywords: Design teaching · Project methodology · Expressive Categories Map · Shape

1 Introduction

The scope of design focuses especially on helping configure and reconfigure human context. Throughout the design process, means of interaction materialize with its surroundings, to the extent that an idea becomes tangible by articulating usage, production, and communication aspects. As a result, the group material/esthetic/informational arises, something which we denominate as shape.

Assuming Cardoso's [4] definition, it is considered that the concept of shape, in the scope of objects and space design, merges three components: the appearance, indicating what is perceived by looking; the configuration, which implicates the arrangement of the parts; and the structure, that refers to the constructive dimension and guides the configurative arrangement. In this line of reasoning, the perception of the shape derives from the convergence of surface, volume, contour, space, and point of view aspects, constituting an entity that expresses and communicates, especially through visuality.

Considering that the focus here is to analyze facilitating ways to articulate the relationship between configurative components, the terms configuration and shape assume the same connotation when used as attributes of an object or designed space.

From this premise, the research that bases this work is directed to the educational field of design, analyzing a framing that is focused on the synthesis of expressive concepts and the configuration process during the project practice. To do so, a cross-sectional exploratory approach has been used, with the main objective of analyzing the effects of applying the tool Expressive Categories Map (MEC), proposed by Sanches [14], in distinct areas of design formation.

With a pedagogical basis, the principles adopted have been the Meaningful Leaning proposed by Ausubel [2], the guidelines of *Ensinagem*, by Anastaciou [1], as well as Reflective Practices, supported by Shön [17], which indicated the importance of praising the student's prior knowledge and the experimentation as a way of reflection to an autonomous knowledge building. Under this instruction, we have analyzed the activities carried out in fashion design, product design, and interior design undergraduate programs, that are supported by three guidelines: to seize personal surroundings as ways to creative research; to apply the MEC as a facilitator of expressive concept synthesis; to stimulate idea generation through materiality.

Using Sanches' [15] recommendation as the basis, who focuses on the ability of expression and representation as a primordial aspect in the designer's action, either to create visual utterances that promote the semantic load of the shape, either to manage the knowledge derived in the passing of the design process, the results confirm the potential of the MEC and concrete experimentation as conception facilitators and in the development of the syntax of the shape. In this sense, visual-graphic thinking has helped the intrinsic processes of the project-thinking organization, and the materiality of rapid prototyping facilitated the perception and integration of the multiple dimensions that characterize a design solution, being a piece of furniture, the interior of an architectural space, or a garment.

2 Theoretical Bases

By bearing in mind that design implicates a creative action that configures a matter according to a defined purpose, in this article we have assumed that the essence of a process is to give shape to an idea, from the possible objective and subjective relations that are established between the projected object and the subject using it later. Thus, the shape materializes through the designer's decisions about the arrangement constituted among color, structure, texture, surface, etc., however, attributing meaning to such arrangement assumes the existence of a subject who is going to experience it physically and psychologically.

For this reason, project education demands the development of two essential skills: flexibility to observe the project context from distinct points of view, by identifying relationships; and the focus to synthesize such perceptions in an expressive intention. These skills directly affect the definition of an expressive concept to articulate the visual syntax of the shape and, consequently, the experience with the projected artifact.

2.1 About Shape and Syntax

The word shape englobes multiple aspects when used in the design field. Thus, it is important to clarify the connotation approached in this study, highlighting it as an expressive and perceptive space.

Costa [5, p. 20] focuses on the physical dimensions of shape, describing it as "the external appearance of things, the set of lines, surfaces, and volumes that determine its figure in the space; something we can see, touch, and manipulate". It is a visual stimulus is apprehended by perception, which can be a phenomenon (a cloud, the stars, or the rain), a thing (a stone or the mountains' outline in the sky), or an object (a shoe, a clock). However, the author also highlights an abstract dimension, locating the ideas as mental shapes. Ideas are internal effects that, before being transformed into visual shape and stimulating perception, are mental forms that emerge, disappear, and reappear until they stabilize, become understandable, and, lastly, are prone to expression.

Bomfim [3] uses the word figure, that results from the process of configuration, to name the set of aspects of an object (or space) that can be perceived, imagined, and represented sensorially. A figure is fulfilled through elements such as material, geometrical structure, texture, color, etc. Likewise, Löbach [11] points out the shape, the material, the surface, and the color as elements that compose a figure. Complementary, Wong [18] indicates that a shape is a set of interrelated elements according to a structure of organization.

Following this, as already mentioned, Cardoso [4] considers that all shape integrates three components: the appearance, the configuration, referring to the arrangement of the parts, and the structure, that guides the configurative arrangement.

From this, it is possible to highlight relevant aspects of shape in this reflection, considering that we analyzed the scope of objects and space design: materiality, articulation of related elements, sensoriality, and visual-aesthetic perception. At the same time, by marking the perceptive aspect, we inevitably enter the field of expression, assuming that a shape constitutes a visual utterance with the ability to produce feeling, that is to say, it carries a semantic dimension.

Hence, we recur to Niemeyer [13] to explain the four active dimensions in the configuration of a design solution: a) material dimension, constituted by the artifact's material properties, that is not considered separately, once in order to communicate something it depends on its articulation with other dimensions; b) syntactic dimension, that consists in the artifact's parts and how they are connected, including technical composition, covering compositional visual elements and relational procedures; c) pragmatic dimension, that implicates the practical use of the object, considering a set of relationships that it establishes with the user in the sociological or ergonomic field (who uses it and in which context is it being used); d) semantic dimensions, that gathers the expressive and representational qualities, anchoring references to the syntactic and material dimensions, its descriptors. It discusses how a product suggests, through shape composition, qualities of usage and cultural codes. So, the shape is only semantically accomplished in the interaction of material, pragmatic, and syntactic dimensions.

From this perspective, it is evident that project practices in the classroom need strategies that develop the competency for studying the sociocultural context, to identify that expressive contents will be led to the project, and, especially, to define that elements will be articulated to elaborate a coherent visual utterance that transmits a message. Following this, Munari [12] explains that a visual message includes two components: information and visual support. Information is the content to be communicated, it indicates the expressive intention, the concept that the configuration should transmit or stimulate at a subjective level. In turn, visual support is the set of elements that makes this information visible. It refers to the way the expressive intention is achieved through colors, textures, surfaces, volumes, etc. Thus, there is an interdependent relationship between the visual support and the information interpretation.

Shapes carry nonverbal messages, that are guided by an expressive intention and built through an ordered combination of elements, constituting a set of signs. This demands syntactic thinking, to relate these elements and to reach a visual narrative that transmits information in a recognizable code by the user [15]. It is worth highlighting that the adoption of the word syntax and language in the field of visuality addresses the characterization of compositional units from the shape and its interaction principles, however, this use is made with the full awareness that verbal and nonverbal messages are from very distinct places.

The visual substance of the shape is constituted by the integration of some basic elements. Thus, the visual utterance is made of selective choices and combinations from the referred elements. To facilitate such choices, there are principles and techniques that can be learned and articulated in class. Dondis [6] calls "syntactic foundations" the set of knowledge that covers visual elements and basic principles to match them. These foundations help plan the units of configuration and facilitate the expression of the aesthetic-symbolic guideline of a project.

However, Lessa [10] warns that the parameters to configure an artifact depend on the integration of utilitarian aspects, technological conditions and human interactions (physical/symbolic), among others. So, it is needed to establish parameters to guide the units that integrate the shape, since such organization is going to be fundamental to connecting with the user.

2.2 About Configuring the Experience

The above arguments reinforce the idea that the elaboration of the shape depends on a convergence of factors, that are linked according to the objectives of interaction and usage that guide the project. Hence, if the expressive quality of a design solution is given by its configuration, that in turn derives from the equating that defines the experience of use that is intended, it is pertinent to infer, in accordance with Lessa [10], that the shape is the concrete evidence of the artifact's concept. To clarify, it is considered the concept of the artifact as a synthesis of the artifact's traces and characteristics that fundamentally determine it the most, its acting essence in the human daily context [10]. Yet, Lawson [11] names it as a central idea.

From this, during project practices in the classroom, it is indispensable that the student assimilates that, by giving shape to an object or space, they materialize its practical utility and, jointly, determine the aesthetic language that must be connected to the universe of the user. For this reason, to foster the process of configuration, it is fundamental that the student knows strategies to define the expressive concept and articulates the visual syntax from context searching. To exercise such competence in classroom, we adopted the following aspects, listed by Sanches [15]: a) each configurative element integrates a visual utterance and, at the same time, constitutes a way for physical adaptation in the material surrounding; b) the visual utterance of the shape is subordinated to the sociocultural context; c) the syntactic arrangement depends on the communicational guidelines chosen and the operationality requirements of the projected artifact. About these premisses, the author synthesized the main aspects that influenced the configuration process (see Fig. 1).

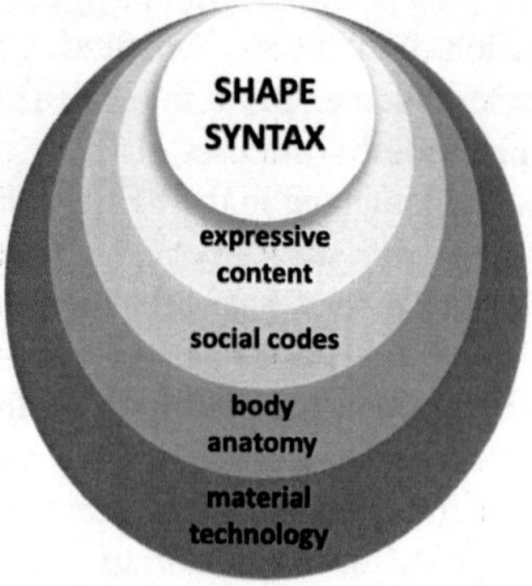

Fig. 1. Set of acting influences in shape/visual syntax. Source: Sanches [15]

The factors previously pointed out, even though they were extracted from a research directed to the educational context of fashion design, can be adapted to other educational design scopes, such as product and interior design. In this direction, it is possible inferring that the corporeal anatomy factor, that in the case of fashion is the underlying structure of the projected artifact shape, can be comprehended with greater amplitude, translating as physiological requirements of use and operationality.

Expanding the analysis further and transferring the focus from the shape itself to the experience it provides, once the study includes students of product, interior, and fashion design, three differentiated spaces of experience with the shape are distinguished (see Fig. 2), and they influence its perception.

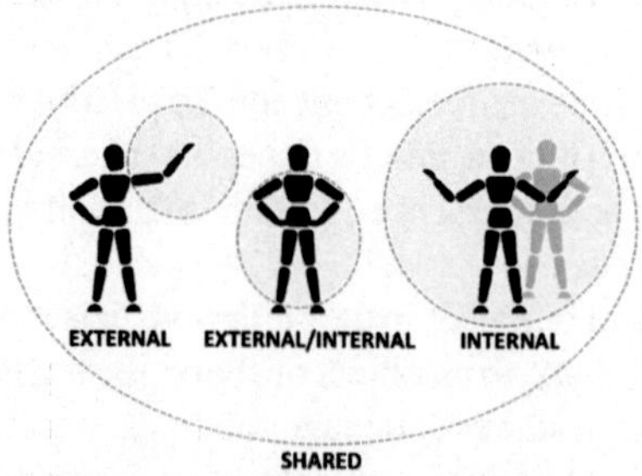

Fig. 2. Space of experience with the shape. Source: the author

In Fig. 2, three basic spaces of experience are schematized. They are part of the context of the general use of products, garments, and architectural interior space: a) external, when there is no, or when is very reduced, stimulus of perception of the shape's interiority, just as what happens with a stool or an iron; b) external-internal, referred to the partial insertion of the body in the shape, allowing the interiority and exteriority to provide simultaneous perceptions, as in the case of a garment; c) internal, that considers the whole insertion of the user in the interior of the configuration into an experience that can be shared.

In the case of garments, it is important to consider that the shape is an extension of the body and must be transportable. On the other hand, in interior design, it is relevant considering that the experience with the shape, most of the time, is shared with other people and can, or not, include the perception of external space, even though the user feels itself as part of the configuration interiority.

Beyond this, the presence of the surroundings and shared perceptions cannot be ignored, once the experience is always subjected to the context influences (physical, social, economic, etc.) in which the interaction occurs, as confirmed by Desmet and Hekkert [7].

Under this line of thought, it is attributed to the term experience all the possible subjective perceptions that derive from the interaction with the artifact (object of space) through its shape, including the instrumental and non-instrumental interaction. Desmet and Hekkert [7] clarify that instrumental interaction englobes practical usage, the operationality, or the management of products, while the non-instrumental refers to the interactions that do not serve a direct role in the operation of the artifact and that also cause affective responses. Hekkert [8] distinguishes three levels of experience: a) aesthetic pleasure, that derives from sensorial perception; b) experience of meaning, which implicates cognitive processes, such as interpretation, memory, and association, that allow recognizing visual metaphors, attributing personality or other expressive characteristics, evaluating the personal or symbolic meaning of an artifact; c) the emotional experience, referring to the evaluation of the importance of a stimulus to personal well-being, making an artifact beneficial or harmful to a user's concern, according to the emotions it provokes.

In the case of this study, it is ascertained that the tools used are especially centered on facilitating the synthesis of the communicational guideline and the definition of plastic components that will provide the sensorial and perceptive experience of the shape. To do so, this is underpinned by the research of the surroundings, the use of visual metaphors, and the concrete experimentation so that the student associates the instrumental and non-instrumental aspects during the configuration process.

2.3 Expressive Categories Map (MEC): How Does It Work?

Proposed and validated by Sanches [15], the Expressive Categories Map is a synthesis tool that works to collect and filter expressive codes, through a graphic structure that facilitates the connection of aesthetic references and the definition of elements of visual syntax. The tool was generated from a combination of four techniques: moodbard, scale of semantic differential, verbs of action, and mind map.

The application of the MEC is summarized into the following procedures: a) from the investigation of the user's sociocultural context, by defining keywords that synthesize the expressive concept that the artifact (product or space) must transmit; b) a subjective search of representative images of the concept is conducted; c) a map of connection of the images collected is elaborated, linking similar representations and distinguishing one or more possibilities of expression to the concept (expressive categories); d) with the help of a specific table (see Fig. 3), composed of a differential semantic scale and column to organize the configurative elements, the map of connections is associated to a set of sensorial perceptions (light, temperature, touch, and gesture/movement) organized in opposite polarities. When there is more than one expressive category on the map, each one will have its own line in the differential semantic scale with different colors.

The interpretation of the differential semantic graphics is the last phase of the MEC, converting data about temperature and light in a range of colors, the tactile indications in texture/material possibilities, and the gesture/movement sensations into schematic drawings of shape structures. Afterward, such elements foster the experimental manipulation of the shape that, through the rapid prototyping, studies them in different combinations.

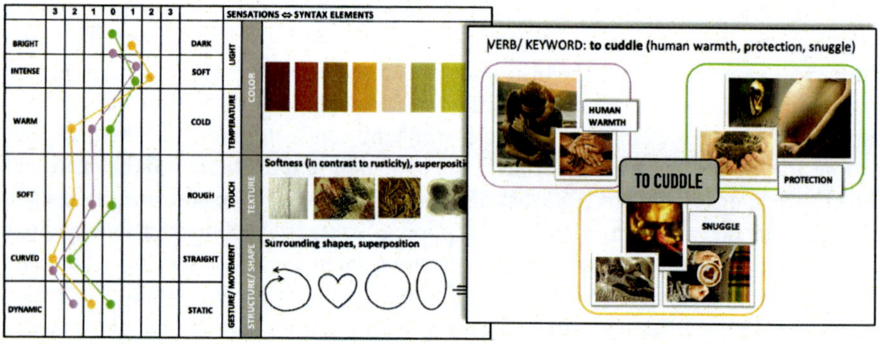

Fig. 3. Expressive Categories Map. Source: adapted from Sanches [14, 15]

In the Fig. 3 example, a table of image interpretations and an example of the result with three expressive categories identified from the verb "Embrace" are shown. Configurative elements that are found in this panel, as shapes schemes and chromatic ranges, are registered on the right of the table and derived from the differential semantic graphic, that is located on the left.

3 Practical Application: Procedures and Analyses

This study derives from the doctoral investigation of this research's author, conducted through a partnership among the State University of Londrina (UEL-Brazil), University of São Paulo (USP-Brazil), and Valencia Polytechnic University (Spain). The research purpose was to help the elaboration of more systematic strategies to project learning, validating unprecedented tools through investigation-action, among which the MEC is emphasized, its application currently unfolds in the scope of product and interior design formation.

The original research elucidated that the project process is constituted of recurring cycles of analysis, synthesis, and evaluation, to articulate information transversally, promote ideas, and produce a pertinent answer to the design issue. Thus, a systematic approach to such a process is assumed, once the way to the solution requires constant adjustments, analysis feedback, and, frequently, the connection of parallel lines of thought.

For this specific study, the proposal by Sanches [14, 15] was adopted. She organizes four groups of procedures to project practices in the academic sphere: a) procedures to contextualize and synthesize the project guideline (concept); b) procedures to promote and experiment possibilities to explore the concept, stimulus of ideas; c) procedures to evaluate the effectiveness and outcomes of the possibilities promoted; d) procedures to enable the proposals at a technical-productive level. The author warns that this organization does not implicate a linear chain of isolated procedures, on the contrary, simultaneous actions and successive cycles of interaction and feedback usually occur. Still, it is confirmed the study on the sociocultural scenario as starting point and it is observed a correlation between transformative actions that articulate themselves in three project focuses: explore the context, by identifying relationships and guidelines (DELIMIT); experiment combinations from guidelines and multiply the possibilities (PROMOTE); select and channel the possibilities to a feasible proposal (EVALUATE/CONSOLIDATE).

As a pedagogical basis, the contributions of Anastaciou [1] were essential. To her, apprehending is supported on a knowledge built based on the possibilities shared between the student and the professor. The author names *Ensinagem* [from portuguese: *ensino* (teaching) and *aprendizagem* (learning)] the process in which the leading role of the professor and the self-activity of the student takes place both ways and enables the connection of the student to the object of study. From this collaborative focus, the active posture of the student is reinforced, following the guidelines of Ausubel [2] on Meaningful Learning that takes advantage of students' prior knowledge as a cognitive anchor to build new knowledge. From this, in all of the projects, regardless of the course, the action of delimiting was started through exploratory research of the personal surroundings and the identification of an interesting topic, that served as an anchor to understanding, synthesizing, and applying an expressive concept.

Based on the principles above, the pedagogical practices undertaken in the classroom were oriented by a hybrid methodology, including an exploratory documental investigation and a practical experimental application, by using the following strategies: a diary photo of the personal surroundings, a mind map of the own lifestyle, expressive categories map, and experimental manipulation of the shape through rapid prototyping.

To analyze the outcomes of this practice in the classroom, which is the focus of this study, participating observation, and documental analysis were used. That includes registries of observations in class, individual documentation of the process, and oral defense of the results. To confirm it, documentation collected during two academic years was examined, with groups from Interior Design and Product Design programs from the European Institute of Design (IED-Madrid), and the Fashion Design program, from Nebrija University.

Observations in class showed that, to project shape as sensory space, it is primordial clarifying the relationship between shape/configuration (visual support) and content (information). Having in mind that the projected shape presents, in an intentional way, visual information that orients the use and stimulates the affective experience in the user's daily life. In this sense, the use of MEC facilitated the comprehension of what is the concept of the artifact/space and how to synthesize it into configurative elements, as well as broadened the debate about the relationship between syntax and semantics.

The prototyped experimental process took place with the assistance of the moulage, and the confection of mockups, using the human body as reference in real scale (manikins and real bodies) or reduced scale (anthropometric dummies). In this phase, schematic drawings of the gesture/movement of MEC served as starting point to moulage in the fashion field. In the case of interiors and product design, they boosted free experimentation of volumetric combinations. In the analysis of the photos of process, undertaken by the students, as well as in-class observations, an evolutive path of the shape was seen. It indicates that there was a first moment of assimilation and plastic exploration of the elements extracted from MEC in which, to the extent that the volumes generated interacted with the body, configuration thinking integrated more details of usability that, in turn, boosted new perceptions of the original shape and the expansions of the artifact's function.

It is worth highlighting that the students were encouraged to directly move from MEC to tridimensional idea generation since some of them showed difficulty in experimenting innovative paths when they used drawings as a way of expression. Two essential questions were observed: on the one side, if, without body volumetry and 360° vision to support special perception, some students had trouble extrapolating labeled configurations (jacket, bag, chair, etc.); on the other, once a drawing does not express the material properties, the elements of touch defined in the MEC ended up in second place.

It is noted that, in fashion and interior design programs, the notion of shape as a compartment of the body and the awareness of an internal/external space of experience expressively helped the debates about interfaces and usability of spaces and garments.

Regarding concrete results, the students from the three programs integrated requirements of aesthetics, ergonomics, and technical in the syntax of the shape articulation, showing that the perception of the shape as experience was assimilated, as for the oral discourse as for materialized solutions.

What concerns the reflection on the practice, two moments for sharing individual growth were promoted in the classroom: a partial presentation, exhibiting the major points of the research on personal surroundings and the associations that led to MEC construction, and the final defense, in which a prototype (or mockup) was presented, exhibiting the final proposal and particularities of its development. These moments promoted new perceptions from each MEC and, mostly, opened spaces for reflections about their own project journey.

4 Final Considerations

According to Sanches and Silva [16], the use of graphic-visual tools and concrete experimentations benefits the cognitive and metacognitive processes involved in the projective action, once it enables a multiple livingness perceptive, something that facilitates the association of information and building new perceptions about the situation being studied. Whit this background, it is considered that the tools implemented promoted ways to the development of the capacities to organize and decode the compiled variables in the research of context.

The evidence confirmed the efficacy of the activity as a didactic strategy, once, more than the creative and innovative solutions (shown in prototypes and mockups), the effective integration between conceptual aspects, of productive usability, and viability involved in the process of configuration was observed. Moreover, once the MEC offers a synthesis of configurative parameters, in the implementation of the ideas phase, it was noted the recurrence of spontaneous responses to the document, its use to support arguments about symbolic aspects, and affective sensations that the shape should transmit. Consequently, the MEC also works as a facilitator of individual interactions between student and professor, as well as optimizes communication during interdisciplinary activities.

About autonomous thinking, it was ascertained the advance in self-evaluation ability, once, during oral defense, all the students, spontaneously, inserted some sort of analysis about their own journey, indicating obstacles, findings, and criteria in decision-making. Furthermore, this retrospective look awakened the attention toward the importance of methods to organize the creative process and generate original ideas. Thus, it is possible inferring that the methods used stimulated reflexive practice and self-evaluation during the *Ensinagem* process. Hence, it is expected that it contributes to the advance of didactic strategies that aim at building project thinking within the scope of educational design, with a greater basis on systematic approaches that allow the adoption of non-linear, holistic, and flexible methodologies.

References

1. Anastaciou, L.G.: Ensinar, aprender, apreender e processos de ensinagem. In: Anastaciou, L.G., Alves, L.P. (eds.). Processos de Ensinagem na Universidade, pp. 17–44. Univille, Joinville (2012)
2. Ausubel, D.P.: Aquisição e Retenção do Conhecimento: uma perspectiva cognitiva. Plátano Ed. Técnicas, Lisboa (2003)
3. Bomfim, G.A.: Morfologia dos Objetos de Uso: uma contribuição para o desenvolvimento de uma teoria do design. In: Couto, R., Farbiarz, J., Novaes, L. (eds.). Gustavo Amarante Bomfim: uma coletânea, pp. 23–32. Rio Books, Rio de Janeiro (2014)
4. Cardoso, R.: Design para um mundo complexo. Cosac Naify, São Paulo (2012)
5. Costa, J.: La forma de las ideas. Cómo piensa la mente. Estrategias de la imaginación creativa. Experimenta, Madrid (2008)
6. Dondis, D.A.: La sintaxis de la imagen, 2nd edn. Gustavo Gili, Barcelona (2017)
7. Desmet, P.M.A., Hekkert, P.: Framework of product experience. Int. J. Des. **1**(1), 13–23 (2007)
8. Hekkert, P.: Design aesthetics: principles of pleasure in product design. Psychol. Sci. **48**(2), 157–172 (2006)
9. Lawson, B.: The Design Student's Journey: Understanding How Designers Think. Routledge, New York (2019)
10. Lessa, W.D.: Linguagem da forma/linguagem visual no âmbito do ensino de design: balizamentos teóricos; tópicos de pesquisa. Arcos Des. **5**(2), 69–91 (2009)
11. Löbach, B.: Design industrial: bases para a configuração dos produtos industriais. Edgard Blücher, Rio de Janeiro (2001)
12. Munari, B.: Diseño y comunicación visual, 2nd edn. Gustavo Gili, Barcelona (2016)

13. Niemeyer, L.: Elementos de semiótica aplicados ao design. 2AB, Rio de Janeiro (2010)
14. Sanches, M.C.: O projeto do intangível na formação de designers de moda: repensando estratégias metodológicas para a sintaxe da forma na prática projetual. PhD thesis. FAU-USP, São Paulo, Universitat Politècnica de València, Valencia (2016)
15. Sanches, M.C.F.: Moda e Projeto: estratégias metodológicas em design. Estação das Letras e Cores, São Paulo (2017)
16. Sanches, M.C.F., Silva, M.A.R.: Project thinking in fashion design: strategies for facilitating cognitive and metacognitive processes. In: Broega, A.C., Cunha, J., Carvalho, H., Blanco, M., García-Badell, G., Goméz-Chacón, D. (eds.) Reverse Design, pp. 551–558. CR Press, London (2018)
17. Schön, D.: Educando o profissional reflexivo: um novo design para o ensino e aprendizagem. Artmed, Porto Alegre (2000)
18. Wong, W.: Princípios de Forma e Desenho, 2nd ed. Martins Fontes, São Paulo (2010)

Sustainability in Fashion and Design

Teaching Perspective: Teaching Fashion for Sustainability in Fashion Design Courses

Bruna Lummertz Lima[1(✉)] and Suzana Barreto Martins[2]

[1] Instituto Federal de Ciência e Tecnologia de Santa Catarina, Adriano Kormann. 510, Gaspar 89111, Brazil
bruna.lummertz@ifsc.edu.br
[2] Universidade Estadual de Londrina, Campus Universitário, Londrina 86051, Brazil

Abstract. In recent years, especially in Brazil, we have advanced in terms of discussions on sustainability in the fashion chain. In order to contribute to studies in the area, this research aims to discuss teaching for sustainability, in undergraduate courses in Fashion Design, from the teaching perspective. The research started in a bibliographic survey about publications on fashion teaching in periodicals in the last ten years culminated in the creation of a questionnaire, which was applied with twelve teachers. As a result, it is expected to contribute to the discussion on the teaching of sustainability and the training of fashion professionals for a context more aligned with sustainable development.

Keyword: Graduation Course · Teaching · Fashion · Sustainability

1 Introduction

Undergraduate courses in Fashion Design in Brazil date back to the end of the 1980s [1]. In their origin, most of these courses emerged as an adaptation of foreign courses, without considering the specificities, often, of the Brazilian identity [2]. On the other hand, slow fashion, a term conceived to designate sustainable fashion, has been approached in the last twelve years to name the processes of development and production of products that consider sustainability, in the environmental, social and economic dimensions [3].

According to the National Register of Higher Education Courses and Institutions, today Brazil has 142 higher education courses in Technology and 53 Bachelor's courses in Fashion Design [4]. In this sense, with advances in discussions about sustainability and its insertion in companies and fashion projects, it is important to reflect on the articulation of teaching fashion and sustainability, so that these courses train professionals who are better prepared for the challenges of the market.

In this context, Lima [2] studied in his doctoral thesis how Bachelor of Fashion Design courses in Brazil have included sustainability in their curriculum. According to the author, the way sustainability is studied and inserted in fashion courses needs to change, especially because "society and its contemporary challenges are constantly changing".

It is worth mentioning two more significant findings of this work. The first is the way in which sustainability is inserted in the curricular matrices. According to Lima [2], inserting sustainability disciplines in isolation is not effective and does not motivate students. The second is the absence of models or strategies for this inclusion in curricula.

From this, together with the need observed by the authors, during the exercise of teaching in undergraduate courses in Fashion Design, to approach fashion sustainability in a more integrated way to the content of the disciplines of fashion projects, the motivation for carrying out of this research. This aims to identify which are the tools and strategies that have been used by professors to teach Sustainability in Graduate courses in Fashion Design.

For the methodology, a questionnaire was applied through Google Forms with twelve professors of undergraduate courses in Fashion Design. The professors were previously selected, through the analysis of publications on the theme of "teaching for sustainability" and through the analysis of research groups registered with CNPq - National Council for Scientific and Technological Development. From the results, categories of analysis were created and the data discussed.

2 Methodological Procedures

To carry out the research, a questionnaire was applied with a sample of professors from undergraduate courses in Fashion Design (Table 1). Considering the need for discussions in the context of teaching sustainability in fashion, teachers were selected who work with the theme, through teaching, disciplines or research, along with research projects registered with CNPq, considering the submission of projects with the identification of authors and their institutions. In addition, the annals of the Fashion Colloquium, especially regarding the Working Group on Fashion, Sustainability and Inclusion, was also researched.

The selection of the participating sample took place in the following steps: a) analysis of publications on teaching sustainability in fashion at the Colóquio de moda; b) analysis of publications on teaching sustainability in fashion in journals, based on research in the journal Capes and Google academic; c) Group Survey or Research Lines registered with CNPq.

2.1 Data Analysis

The organization of the questions considered the Google forms platform that would be used and sought to survey the teacher's professional trajectory, the subjects taught, training and experiences in teaching sustainability, considering projects, partnerships with companies and suggested good practices.

To start the survey, the first question questioned the teacher's performance time. Most participating professors (50%) have been teaching undergraduate courses in Fashion for over 10 years. Another 33.3% have been teaching for more than five years (see Fig. 1).

It should be noted that this data reinforces the importance of the sample selection criteria, thus proving that professors have a considerable track record in teaching undergraduate courses in Fashion Design.

Table 1. Selected sample.

Author	Congressional Publication	Journal Publication	Line or Group Research
1	X	X	
2	X	X	
3	X		
4	X	X	
5	X	X	
6			X
7	X	X	
8	X		
9	X	X	
10	X		X
11	X		
12			
13	X	X	
14	X		
15		X	
16	X		
17		X	
18			X
19	X		
20			
21			X
22			X
23			X

Search source: Authors (2023).

Regarding the disciplines taught by the professors (Fig. 2), 71.4% teach Product Development disciplines (fashion design, product design, collection design) and 64.3% teach sustainability disciplines.

This data points to an important issue: the presence of sustainability teaching beyond the theoretical teaching of fundamentals and dimensions (social, environmental and economic).

As some authors have been recommending for some time [3, 5] it is necessary that design for sustainability in fashion is also inserted in the development of projects, as well as in the conception of systems-products-service (PSS). For designers to consider

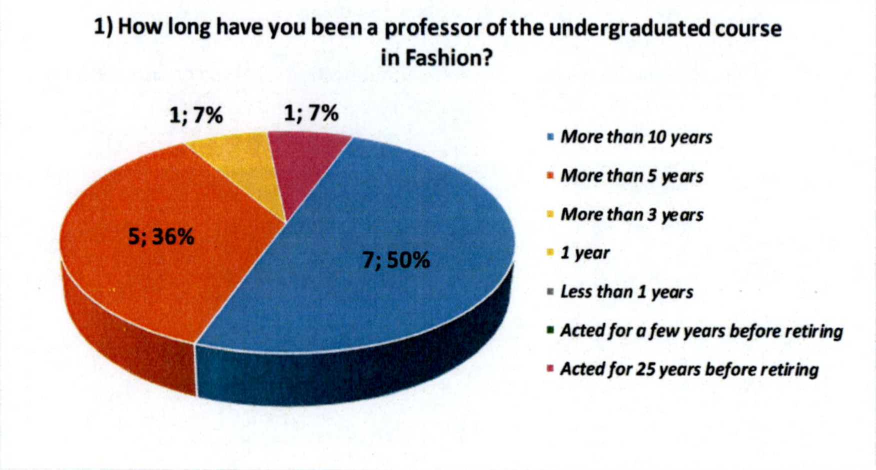

Fig. 1. Teaching Time.

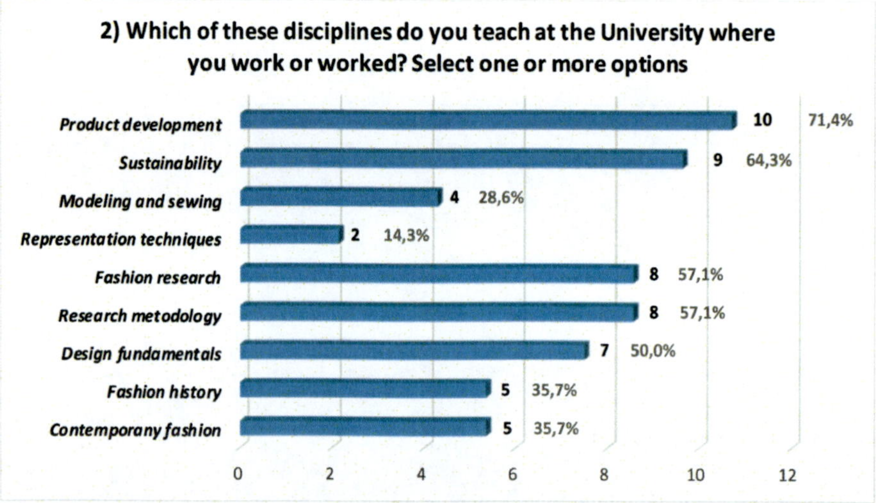

Fig. 2. Disciplines taught.

sustainable assumptions from the beginning of the design path, with the choice of references and research of materials and production processes, as well as analysis of the product life cycle.

The fact that the majority of the responding professors selected this option demonstrates that the teaching of sustainability in undergraduate fashion courses in Brazil has moved towards the natural insertion of sustainable concepts in academic projects.

Subsequently, the professors were asked about the approach to sustainability in the disciplines they teach (Fig. 3). For 78.6% of the respondents they informed that they teach through the insertion of Fundamentals of Sustainability (Dimensions of sustainability (environmental, social, economic), product life cycle analysis, system-product-service). Such information reinforces the importance of theoretical knowledge of sustainability for the effective and efficient application of concepts in fashion projects for sustainability.

In parallel, 64.3% reported that they approach sustainability through interdisciplinary work (academic projects in which several disciplines and professors participate, partnerships with companies and non-governmental organizations).

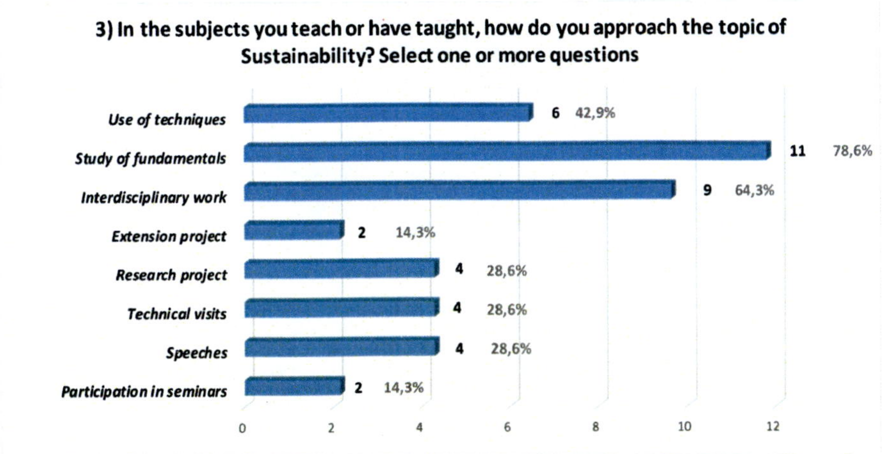

Fig. 3. Sustainability approach. Search source: Authors (2023)

In the fourth question, teachers were asked about their specific training for teaching sustainability (Fig. 4). 57.1% reported that they do not have specific training in sustainability (master's, doctoral or post-doctoral) and that they research the topic individually. Already 28.6% carried out Doctoral research on sustainability or topics related to it.

This data also reinforces an important bias: the teaching of sustainability has advanced in recent years, as a result of research and teaching practice in the classroom, which has evolved in terms of the results obtained in this research. On the other hand, it still shows gaps in terms of teacher training that is preparing new professionals to enter the market.

In question number 5, the professors answered about taking a course on Sustainability. 35.7% took short courses at Congresses and 21.4% took specialization courses in the area (Fig. 5).

Here it is worth emphasizing the importance of training, even if it is short, provided by Congresses in general and in particular those that address or focus on the theme of Sustainability. It is possible, for example, that from a short-term or extension course offered at a congress, a teacher starts to review his teaching practice or to incorporate other contents and formats into his teaching.

In this sense, specialization courses are also complemented, which are faster to be taken in relation to the formation of a Post-Graduate course (Master's and Doctorate).

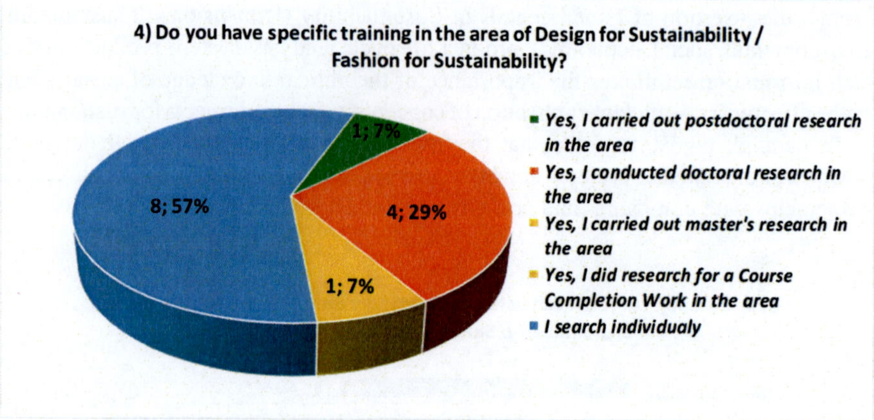

Fig. 4. Sustainability training.

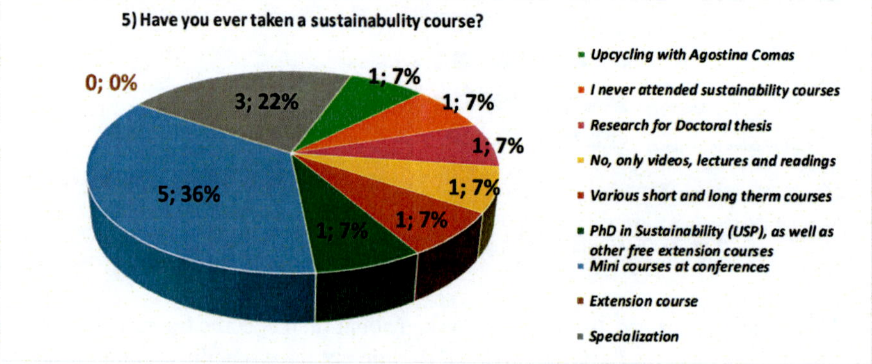

Fig. 5. Courses. Search source: Authors (2023)

As a sixth question (Fig. 6), teachers were asked about their own work in companies or partnerships with companies, seeking to highlight strategies used in the classroom for teaching sustainability.

Of the respondents, 35.7% have already entered into partnerships with companies, within the scope of the disciplines taught and 21.4% have already entered into partnerships with non-governmental organizations and the same percentage has already worked in companies.

This issue brings up an important speech: the one that underscores the purpose of teachers, to establish partnerships with companies and/or governmental organizations, bringing projects and real situations from the market so that students can work on solving problems.

And in contrast, it brings an important experience of the teachers, who have already worked in real companies in the market and can bring these speeches and professional situations to qualify debates and projects in the classroom.

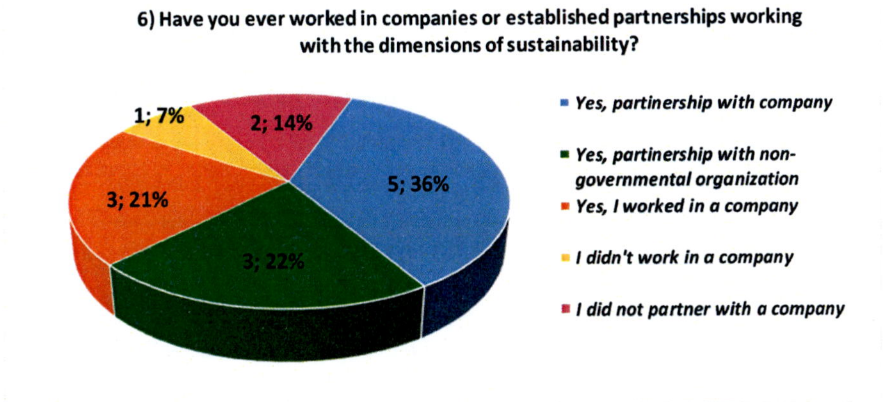

Fig. 6. Performance of teachers in companies. Search source: Authors (2023)

In the seventh question (Fig. 7), the professors were asked about their graduation, if they had any sustainability discipline and how the activities were developed. 71.4% of the respondents stated that they had not taken courses on the subject, which demonstrates that teaching sustainability in undergraduate courses in Fashion Design is something new, in the last ten years.

In contrast, 21.4% stated that they had sustainability disciplines and that activities were developed through interdisciplinary projects.

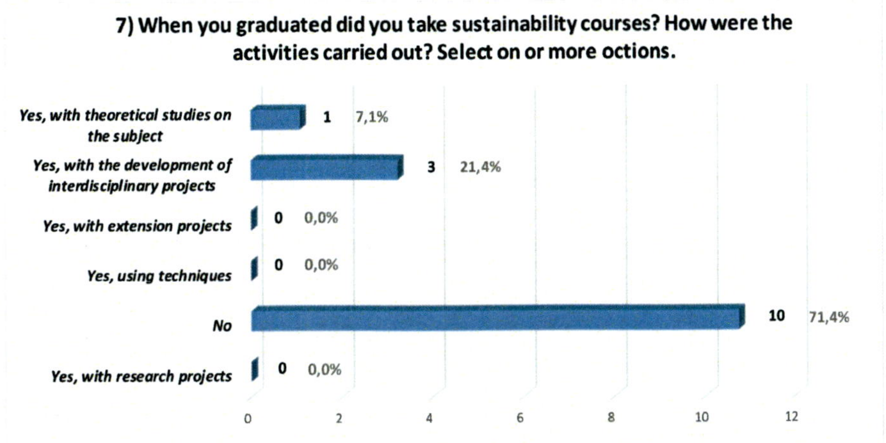

Fig. 7. Sustainability disciplines in the graduation itself. Search source: Authors (2023).

In the eighth question, respondents were asked about obstacles to teaching sustainability (Fig. 8). 57, 1% pointed out a lack of teacher training on the subject and 50% highlighted the lack of specific sustainability disciplines in Pedagogical Political Projects.

It should be noted that the two notes go hand in hand, since for the proposition of curricular units of sustainability, it is necessary teachers have proper training.

On the other hand, without teacher training and disciplines, Fashion Design courses will not be able to advance in teaching and training fashion designers who are more committed to the precepts of sustainability.

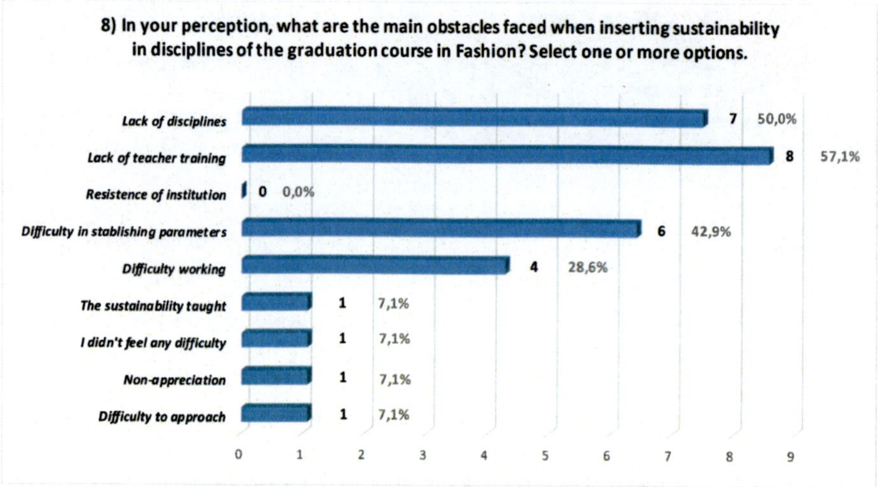

Fig. 8. Difficulties faced. Search source: Authors (2023).

After this selection, the questionnaire with ten questions (eight multiple-choice and two open questions) was sent individually to the selected professors, via e-mail. Of the twenty-three professors contacted, twelve responded to the survey. In the next section, data analysis will be presented.

Questions number nine and ten were descriptive and open. Therefore, to perform data analysis, we will use the content analysis technique [6].

From the analysis of the responses, six categories were created for question nine, considering the meaning of the respondents' statements (Table 2):

As a situation most explained by the respondents, partnership with a company appeared. In this aspect, the participation of companies occurs especially for the development of projects, linked to the development of new products, starting from a briefing or a problem, which may consider surplus materials, waste, among others.

Next, studies on sustainability appear. These are conducted by professors, normally in sustainability subjects, of a theoretical nature. Here, professors address the fundamentals of the subject, the dimensions of sustainability, the product life cycle and the product-service system.

As third and fourth components, the use of sustainable design strategies and the development of a product collection appear tied. It is believed that here both complement each other and that they can be developed concomitantly.

Table 2. Distribution of situations explained by respondents for the example of good practices in teaching fashion sustainability.

Explicit situation	Number of respondents
Partnership with company	7
Studies of sustainability concepts	6
Use of sustainable design strategy	5
Collection / Product development	5
Lecture / Conversation circle with professionals	3

Search source: Authors (2023).

Finally, there is a conversation or lecture with professionals from the market, who work in areas related to sustainability, in fashion companies, in different stages of the product development and/or production process.

In the next question, the categories created for implementation possibilities in sustainability teaching will be presented (Table 3).

Table 3. Distribution of situations explained by respondents for possible implementations in sustainability teaching.

Explicit situation	Number of respondents
Sustainability works across	8
Specific sustainability discipline	3
Product / Collection development project	2
Search group creation	1

Search source: Authors (2023).

For the last question, respondents were asked about the possibilities of implementing sustainability teaching in courses. Most teachers questioned suggested that sustainability could be implemented across the Fashion Design courses. In addition, it was stated that the theme could be worked on in any curricular unit, since it is a theme that covers a series of contents within the fashion production chain.

In a second moment, the professors mentioned the importance of undergraduate courses in Fashion Design to have specific sustainability disciplines in their Political Pedagogical Projects, especially in the first year of the course, in order to form the student in a critical way right from the beginning. of your graduation. It is noteworthy that this suggestion came from respondents who did not identify this alternative in the courses they teach.

Finally, it was suggested that the courses develop collections or products in partnership with companies or even create research groups, which encourage the in-depth study of students. In both cases, suggestions come from teachers who do not have these practices.

3 Discussion of Results

The data collected through the questionnaire made available to professors of undergraduate courses in Fashion Design brought important findings that contribute to a diagnostic understanding of this study. In this section, we will discuss them, separating the questions into points.

Teaching time and performance in disciplines: As shown in the data analysis section, the respondent professors had experience teaching fashion (50% more than ten years). On the other hand, 75% of respondents work in curricular units that address product development, especially in the creative field.

a) Teacher training: In the understanding of the authors, worrying data appeared here, especially in relation to the training of teachers during their graduation. Many professors reported that they did not have specific disciplines on sustainability in their courses. On the other hand, during the path of continuing education of these professors, most report training through individual research or mini courses at conferences.

At this point, there would be an important gap for the advancement of sustainability teaching in fashion courses: expanding teacher training.

b) Partnerships with companies: With regard to partnerships with companies, it is possible to highlight as a positive point of the teaching strategies for sustainability, since several institutions are already articulated with organizations, especially in their local productive arrangement, for the development of products or resolution of problems.

However, it is worth mentioning that these projects are, for the most part, related to the development of tangible products from the partner companies' own inputs. As a gap, in this regard, it is possible to highlight the lack of projects related to the creation of fashion services or life cycle analysis of existing products, for example.

c) Obstacles and possibilities for implementation: As main findings in this regard, we can mention: the need for all undergraduate courses in fashion to have at least one sustainability discipline in their curriculum and sustainability to be worked on in a transversal way, encompassing all the disciplines offered. However, for the two points to be put into practice, by educational institutions, it is necessary that another mentioned point be improved: the training of teachers in the theme of sustainability.

As a whole, in order to move forward on this path, it is important to reinforce teacher training, providing access to courses, allocating workload to specializations, among others.

As a possibility of expanding this discussion, we could consider the importance of teaching sustainability at a systemic level, where all curricular units of an undergraduate course in Fashion Design are important for this teaching and are part of it.

On the other hand, it is understood that the teaching of fashion, in a general way, is still linked to the teaching of projects that originate products, such as consumer goods offered to a chosen target audience. Here it is possible to mention the need to work on concepts and results that remove the obligation of products and, consequently, consumption, and that can generate other results, such as fashion-related services (systems-product-service, PSS), changes in consumption behavior fashion, aiming for new scenarios and more sustainable lifestyles, among others.

Such questions are in line with the higher levels of performance in design for sustainability, proposed by Manzini; Vezzoli [7, 8]: projects for new product-services and new sustainable scenarios and depart from a more strategic narrative for design.

4 Final Considerations

This study aimed to identify which are the tools and strategies that have been used by professors to teach Sustainability in Graduate courses in Fashion Design. Therefore, it is believed that it was possible to achieve it satisfactorily, since from the answers, tools and strategies were raised that are already being implemented.

On the other hand, it should be noted that this research allowed the authors to raise some gaps, especially as fashion teaching is still linked to product development, contrary to what the literature in the area of sustainable fashion and design for sustainability has been recommending for a long time. Some time (PSS, product life cycle analysis, dematerialization, review of the consumption process, among others).

To meet these gaps, there is another data raised: the lack of training of teachers who teach projects or disciplines related to sustainability. As mentioned, the respondents to the applied questionnaire were selected for their publications and/or participation in research groups and even so, a significant sample responded that they did not have a degree in sustainability for fashion.

During the course of this research, other data were collected, including a study to survey the state of the art of fashion sustainability teaching in Brazil, published at the Fashion & Sustainability congress held in November 2022, in Portugal.

On that occasion, the authors already showed that the teaching of sustainability was linked to the use of techniques and the development of products, which is a real problem and which needs to be changed.

In both studies, authored by the author, the results reveal that it is necessary to review the process of teaching sustainability in undergraduate fashion courses. Especially with regard to guiding principles so that fashion teachers can follow and qualify students' learning, focusing on the insertion of these new professionals in the market, which is increasingly concerned with sustainability.

As a suggestion for future studies, it would be a didactic proposition that would guide teachers regarding tools and strategies to qualify the teaching of fashion sustainability in undergraduate courses.

References

1. Pires, D.B., et al.: Design de moda: olhares diversos. 1st ed. Estação das Letras e Cores, São Paulo (2008)
2. Lima, V.F.T.: Superior em Design de Moda no Brasil: Práxis e (In)Sustentabilidade (Thesis). Faculdade de Arquitetura e Urbanismo da Universidade de São Paulo, São Paulo (2018). Homepage, https://teses.usp.br/teses/disponiveis/16/16134/tde-19122018-154908/pt-br.php. Last accessed 25 April 2022
3. Fletcher, K., Groose, L.: Moda e sustentabilidade: design para mudança. Senac, São Paulo (2011)

4. Ministério da Educação Homepage: Cadastro Nacional de Cursos e Instituições de Educação Superior Cadastro e-MEC, https://emec.mec.gov.br/. Last accessed 13 December 2022
5. Gwilt, A.: Moda sustentável: um guia prático, 1st ed. Gustavo Gili, São Paulo (2014)
6. Bardin, L.: Análise de conteúdo. Edições 70, São Paulo (2011)
7. Manzini, E., Vezzoli, C.: O desenvolvimento de produtos sustentáveis. Os requisitos ambientais dos produtos industriais. 1st ed. Editora da Universidade de São Paulo, São Paulo (2002)
8. Manzini, E., Vezzoli, C.: O desenvolvimento de produtos sustentáveis: os requisitos ambien-tais dos produtos industriais, 2nd edn. Editora da Universidade de São Paulo, São Paulo (2008)

Dyes and Identity as Objectives of Experimentation and Sustainability for Fashion Design

Angela Liliana Dotor Robayo(✉), Alis Pataquiva Mateus, and María Ximena Betancourt Ruiz

Universidad Jorge Tadeo Lozano, Bogotá, Colombia
{angelal.dotorr,alisy.pataquivam, mariax.betancourtr}@utadeo.edu.co

Abstract. This study shows the exploration and partial results of the use of ancestral and identity fiber dyeing techniques as current and sustainable processes to dye cellulosic textile materials such as fique and cotton.

For this purpose, elder fruits, turmeric and achiote (annatto) were used, with which the natural dyes were achieved, and different tones were developed through the use of different mordants such as alum or aluminum sulfate, ash solution, urea and others.

The application was made in the laboratory in small samples of fique and organic cotton fibers, and then it was done in larger quantities with a group of artisan weavers in Charalá (municipality of Santander, Colombia), sharing knowledge between the parties.

The study shows how dyeing with natural dyes with a simple technique from the artisans of the territory can be considered as a sustainable alternative, respectful of culture and the environment.

Keywords: Design · clothing · natural dyes · sustainability · identity

1 Introduction

The problems of the fashion industry and its environmental impact regarding the management of water resources in dyeing processes, and the dumping of waste in various water sources, have been a great concern in recent times. For this reason, the sources of colors and dyes, as well as the textile materials, are today an important object of study for researchers, academics, and foundations, who seek to establish the impacts and to improve or replace production processes and practices with friendly production systems, in accordance with the "system-product" framework of Manzini and Vezzolli [1], who reflect about design and resources for sustainable product development.

In this perspective, ancestral, artisanal and natural practices are very important due to the involved ethical, cultural and systemic views of the life of people and the planet. These practices are directly related to a subtle comprehension and care of the territory [2], making the fabric and its generating processes practices of social welfare.

The present study is part of an ongoing research that seeks sustainability of fashion and textiles in local Colombian contexts. Considering the intimate and cultural nature of some local fibers, work was previously done with fibers such as sheep wool [3] and silk [4].

The research carried out presents how these practices re-signify and promote textile design techniques such as natural dyeing, taking advantage of the resources provided by nature in the territories, but with a manifest sense of comprehensive care.

Taking into account the history of sustainable and circular productive processes for the production and design of products, it was investigated how the production and design of sustainable objects focused on the systemic and territorial vision allows to build processes and results with cultural roots and with a vision of care of the natural resources; in doing so, we wish to highlight the importance of knowledge and ancestral worldviews concerning the correct use of resources, conscious consumption, and sustainable development of textiles.

Results of the research of dyes and color with a territorial approach are presented. The selection of local fibers such as fique and cotton is relevant due to the tradition and cultural heritage of the area. Fique and cotton are "endemic" crops with broad cultivation and working tradition for fiber extraction, dyeing, and manual and loom weaving by artisans from the department of Santander, and previously by their ancestors, the Muiscas and Guanes indigenous people (Fig. 1).

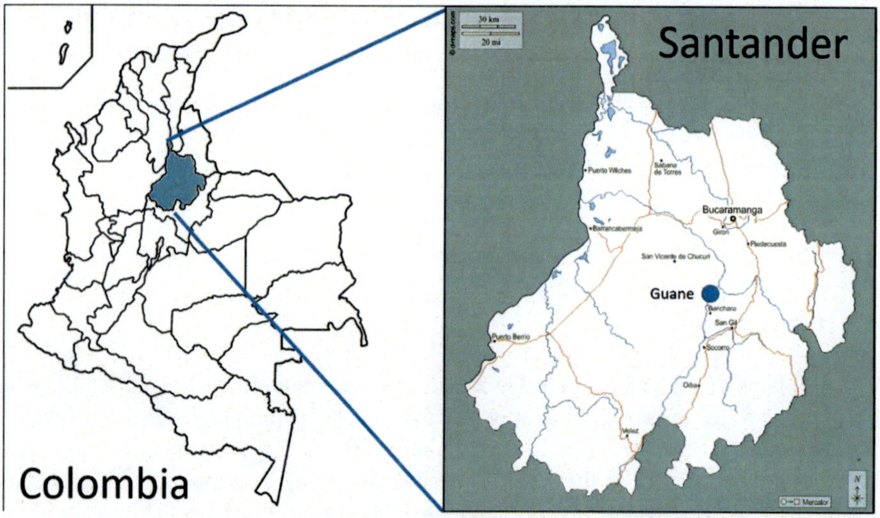

Fig. 1. Location of the Guane territory in Colombia. Modified from [5].

At the same time, the artisan dyeing process allowed us to recover methods and memories that today have cultural relevance not only in Santander, but also an important and current application to sustainable dyeing. The use of degradable materials, without synthetic chemicals that have high environmental impacts, contributes to the dyeing processes of local fibers; as a gain, we take care of the territory's environment.

2 Tradition and Identity

The historical and ancestral tradition of the cultivation of fique and cotton fibers is well known [6]. Also, weaving and dyeing techniques in the Santander region have been historically linked to the pre-Columbian Guane and Muisca cultures, indigenous people settled in the central highlands of the country, in the departments of Boyaca and Cundinamarca.

The Guane culture still has descendants. About 412 people who call themselves Guanes live in the province of Barichara and Guane in Santander, according to a study by the National Indigenous Organization of Colombia (ONIC)[1].

According to Raymond [6], these people practiced the association of crops; cotton and/or fique plants were usually associated with sugarcane and other crops. The associations facilitated rotative production; present peasant communities in the area still maintain those practices, even if there are also separated crops of fique, jute and cotton.

The tradition of weaving among the Muiscas and the Guanes was related to the production of blankets as an element of both ritual evocation and clothing [7]. They also woven girdles, backpacks (mochilas), and cotton hats with soft and noble fibers; often, these cotton fibers were differentiated according to their age, condition or trade, with a hierarchical, so to speak, use. Instead, fique and jute were used as resistant hard fibers in the development of stronger products such as bags to carry objects, baskets, and ties.

Although color was not a protagonist in pre-Columbian textiles of Colombia, there are remnants of a glove blanket dyed with achiote (annatto) in the National Museum of Bogota (Fig. 2), which evidences color application techniques and motifs with the use of stamps and fabrics.

At present, artisan weavers from Charalá tell you about and show you woven motifs discovered in cave paintings of the region; even if these have not been academically studied in depth, they are well known in popular culture and speech.

Regarding the dyeing process, vestiges are found in the Bolivar house museum in Bucaramanga, showing a cotton blanket dyed with brazilwood (*Haematoxylum brasiletto*), which dyes pink and fuchsia. Studies of pre-Columbian dyes have also revealed the use of chica (*Arrabidaea chica*), which dyes orange and brown; coralito (*Galium hypocarpium*), which dyes pink; turmeric (*Curcuma longa*), which dyes yellow; and the dividivi tree (*Caesalpinia coriaria*), which dyes from black to purple depending on the fermentation of the plant.

Following traditional practices, nowadays artisans dye with bighead onion, almond tree bark, guava tree bark, turmeric, achiote (annatto), dividivi, and brazilwood, keeping alive the ancestral practices. However, some of them have replaced traditional dyes with chemical synthetic dyes, or do washing with abrasive materials. It is for this that the recovery of memory and verification is important in order to recover and spread an ancestral knowledge that may contribute to today's sustainable development goals.

The craft of weaving is kept alive by artisans (mainly women) of Charalá. They weave on a loom with horizontal pedal and frame (Fig. 3); they also practice macramé with two needles. The cultivation of cotton and fique is part of their organic rural orchards;

[1] https://www.onic.org.co/pueblos/1101-guane#:~:text=Los%20Guane%20fue%20un%20grupo,20%20minutos%20de%20Barichara%20Santander.

Fig. 2. Guane dyed blanket (source: http://elguane.blogspot.com/2011/10/la-tintoreria-guane.html)

and the dyeing processes have been preserved by many of them, who are of peasant or indigenous origin. Oral transmission has preserved their cultural heritage.

Fig. 3. Horizontal rustic loom, made by artisans (photo by the authors)

The dyeing processes in use today normally use washing agents such as caustic soda. This is a problem since we know that "synthetic colorants have been cited as causing skin complaints, and illnesses and cancer. Close skin contact with colored textiles (e.g. leggings heavily dyed with azo and anthraquinone disperse dyes) has been a concern, and some cases of dermatitis have been reported" [8]. And that "synthetic dyes, which are widely available at an economical price and produce a wide variety of colors, sometimes causes skin allergy and other harmfulness to human body, produces toxicity/chemical hazards during its synthesis, releases undesirable/hazardous/toxic chemicals etc." [9].

The purpose of this study is then to recover healthy dyeing processes with the environment and the people. According to Ashis Kumar and Konar (*op. Cit.*), natural dyes are rapidly spreading in the industry and amog designers, due to their biocompatibility and benefits.

3 Materials and Methods

We intend to show progress on the dyeing of local fibers with territorial resources, as part of the construction of a sustainable or friendly textile system. For this purpose, four experimental stages were developed. In the first stage, the preparation of the dyeing solution was achieved; in the second, the conditioning or scouring of fibers; in the third, the etching process; in the last stage, the application of the dye with color modification and fixation.

To collect and verify the dyeing process presented here, three *immersions* were carried out, in which interviews were made with artisans who weave and dye fique and cotton. The ancestral knowledge was collected and recovered, then verified in the laboratory. The recovered dyeing recipe was presented to the artisans with some improvements to optimize resources such as water and material amount.

The following materials were used for the development of the natural dyes:

100 g of fruits, elder

100 g of turmeric

100 g of achiote (annatto)

1000 ml of water

5 g of salt

The fibers to be dyed were weighed to use a sample of 33 g of fique, and 33 g of cotton.

Stage 1: Preparation of the dye solution. For every 100 g of dye material, 33 g of fiber (ratio 3:1) to be dyed was considered. The ratio of the amount of dye to solution in water was 1:10 (for every 100 g of elderberry, achiote (annatto) or turmeric, 1000 ml of water were used).

In this stage we began by grating turmeric and macerating the elder fruit. Each one was placed in water and at medium heat so that it reached the boiling point, and left for 10 min; 5 g of salt were then added, and left to cook for 5 additional minutes. Finally, removed from heat and allowed to stand, the solution was filtered/strained (Fig. 4).

Stage 2: Scouring or preparation of the fibers to remove impurities. Based on the ancestral and artisanal knowledge, the fibers were washed with "ash bleach" (a solution

Fig. 4. Dye solution (photos by the authors)

Fig. 5. Scouring of the fibers (photo by the authors)

of water with charcoal ash) and left to rest for two days; then they were cooked for 45 min (Fig. 5).

Stage 3: In the preparation of the three types of mordants, alum was applied to the first at 10% of the dye solution; 20% lemon to the second; and 20% concentrated ash solution to the third. The fibers were then immersed in the alum mordent bath, cooked for 20 min, and left to rest for half a day until the dyeing process began. Three samples of each fiber were separated to be dyed later and to modify their color with the three mordant solutions in a subsequent wash.

Stage 4. Application and color fixation. Dyeing by cooking was done, in which the dye solution was put in an aluminum pot to boil. Jute, fique and cotton fibers were submerged and left to boil for 15 min; a branch or plant of tannin was added. The fibers were removed and immersed on the different mordants for 15 additional minutes. Then they were washed and allowed to dry. Color changes due to the fixant effect of the mordant appeared. Finally, the fibers were washed and allowed to dry (Figs. 6 and 7).

Fig. 6. Application and color fixation in laboratory

Fig. 7. Application and color fixation by artisans

3.1 Color Fastness Evaluation Using UV-C Light (100–280 nm)

The fibers of each sample were arranged on a Petri dish without lid, in a circular area of 50 mm in diameter, with direct access to UV-C rays at a focal distance of 40 cm, and for a 3-h period. Colorimetry measurements at the beginning (t = 0 h) and at the end of the tests (t = 3 h) were carried out with a Konica Minolta Chroma CR 410 colorimeter, which allowed obtaining the L*, a* and b* values necessary for the determination of ΔE according to Eq. (1):

$$\Delta E = \sqrt{\Delta L^{*2} + \Delta a^{*2} + \Delta b^{*2}} \quad (1)$$

where the ΔE values represent the difference between the values obtained from the blanks (samples at t = 0) and the values at the end of the test (t = 3 h).

3.2 Color Evaluation by Washing Tests

The washing tests were carried out manually according to Arroyo [10] and Palacios [11] using two test tubes with 1% neutral soap solutions (pH 7) and another with 1% alkaline soap. (pH 9), and the mixed in running water (pH 6.5). For this purpose, the fibers were immersed and manually shaken in the solution for 30 min, simulating 5 manual washes according to ISO 105-C06:2010 [12].

4 Results

The use of different mordants in the etching stage, or in the subsequent washing of the dyeing, causes the color tones in the fiber to change (Fig. 8). Alum, for example, apart from preparing the fiber for dyeing, maintains the color closer to the initial dyeing; urea, instead, darkens a little the tone and ash so does much more, allowing the achievement of different scales in the fibers.

Fig. 8. Natural color chart on cotton and fique

The observation of the dyed fibers through scanning electron microscopy allows seeing that the fibers retain their thickness, and the composition remains organic and degradable. Degradability of the fibers is not altered, nor their life cycle affected. Through all the process, we were happy to comply with the cradle to cradle circular theory [13], who proposes a biological cycle in design. In this way, an environmentally friendly result can be inferred.

Ultraviolet (UV) radiation is a form of invisible radiant energy that covers a wavelength range between 100 and 400 nm and is classified into three categories: UV-C (100–280 nm), UV-B (280–320 nm), and UV-A (320–400 nm): the shorter the wavelength, the more negative the impact on the fibers.

The results of color fastness and ΔE data are shown in Table 1. Comparing all the samples evaluated, the turmeric on fique sample showed the greatest color change after irradiation with UV-C rays. However, among the samples that had cotton as a substrate fiber, the achiote on cotton sample revealed the greatest sensitivity to UV-C rays.

Regarding color fastness and washing, when comparing the evaluated samples, the achiote on cotton sample presented the greatest color change after washing with alkaline soap. According to the washing tests of all the dyed fibers, greater color changes were observed with the use of alkaline soap instead of neutral soap. Even in the case of turmeric, there was a strong color change during the washing process; however, after rinsing and drying, the fiber reacquired a color very close to the initial one.

It can also be concluded that fique presents a greater color change than cotton.

Table 1. Color fastness data considering UV-C irradiation and resistance to different types of soaps/detergents.

	UV lamp 15W		Neutral soap effect		Alkaline soap effect	
Fiber	Exhibithion time	DE	Number of washes	DE	Number of washes	DE
Elder on cotton	3 h	1,0237	5	1,7	5	2,1
Annatto on cotton	3 h	10,3003	5	6,2	5	6,5
Turmeric on cotton	3 h	1,8199	5	2,8	5	3,9
Turmeric on fique	3 h	31,3785	5	6,2	5	5,4

5 Conclusions

We have given evidence of how an artisanal and ancestral knowledge of dyeing made by artisans descendants of the Muisca and Guane cultures is a favorable technique for dyeing. The procedure is simple, repeatable and environmentally friendly.

The results show how the mordants work as good fiber conditioners that open pores to receive the dye. But also, in the dyeing stage they can be employed as natural color modifiers, and to favor color fixation.

Industrial fixatives can be avoided to work instead with tannin plants, vigua or virgin salt. These can be added to the dye or to an independent solution after dyeing.

The color fastness tests carried out on the cotton and fique fibers dyed with turmeric and elderberry showed good stability to washing and to exposure to UV rays. This confirms their scalable potential for clothing design and production.

This study allowed us to understand and give new meaning to the knowledge of dyeing and weaving of cotton and fique from the artisan community of Charalá in Santander. It also strengthened the importance of collaborative and systemic design with local artisan communities to build design from its nature and with a territorial approach, contributing to the reconstruction of identity and autonomy from below, from the south, from and for the "pluriverse" [14], and to the sharing with others' views.

As aspects to improve future research processes we propose: to control variable temperature during cooking; to control the concentration of natural dye and mordant to determine the influence of different concentrations on the final results; and to explore the possibilities of reusing solutions and water for washing materials in order to establish their effect on the calculation of the life cycle of the possible products to be developed.

References

1. Manzini, E., Vezzoli, C.: O desenvolvimento de productos sustentáveis. Os requisitos ambientais dos produtos industriais. Edusp. São Paulo, p. 20 (2002)

2. Escobar, A.: Territories of difference. Place, movement, life, redes. Duke University Press, Durham, NC (2008). https://doi.org/10.1215/9780822389439
3. Castellanos Perilla, M.V., Sánchez Díaz, D.C., Pataquiva Mateus, A.Y., Echeverry Ocampo, M.A., Dotor Robayo, A.L., Ramos Kalmar, E.: Nanoestructuración y experimentación de fibras proteicas para superficies textiles y moda sostenible. Ingeniare (24), 53–75 (2018). https://doi.org/10.18041/1909-2458/ingeniare.24.5171
4. Dotor Robayo, A.L.: Thinking textile materials from their nature: ethical materials for fashion design with technological, social, and aesthetic sense. In: Muthu, S.S., Gardetti, M.A. (eds.) Sustainability in the Textile and apparel industries. sustainable textiles: production, processing, manufacturing & chemistry. Springer, Cham (2020). https://doi.org/10.1007/978-3-030-38013-7_2
5. Casas-Vargas, A., Gómez, A., Briceño, I., Díaz-Matallana, M., Bernal, J.E., Rodríguez, J.V.: High genetic diversity on a sample of pre-Columbian bone remains from Guane territories in northwestern Colombia. American J. Biolog. Anthrop. **146**(4), 637–649 (2011). https://doi.org/10.1002/ajpa.21626
6. Raymond, P.: El algodón y los tejidos en Santander. Origen, decadencia y renacer de una tradición textil. Ecoe Ediciones (2013). http://www.ecoeediciones.mx/wp-content/uploads/2016/08/El-algodon-y-los-tejidos.pdf
7. Fernández, M.: La manta Muisca como objeto de evocación. Kepes **10**(9), 285–296 (2013). https://revistasojs.ucaldas.edu.co/index.php/kepes/article/view/515
8. Mirjalili, M., Nazarpoor, K., Karimi, L.: Eco-friendly dyeing of wool using natural dye from weld as co-partner with synthetic dye. J. Clean. Prod. **19**(9–10), 1045–1051 (2011). https://doi.org/10.1016/j.jclepro.2011.02.001
9. Ashis Kumar, S., Konar, A.: Dyeing of textiles with natural dyes. In: Kumbasar, E.A. (ed.) Natural dyes, pp. 29–56. Intech (2011). http://www.intechopen.com/books/natural-dyes/dyeing-of-textiles-with-natural-dyes
10. Arroyo Figueroa, G., Dzul-Cauih, J., Vargas Rodriguez, L., Medina Saavedra, T.: Evaluation of the color fastness in the dyeing of wool and cotton with cocoa shell extract. Revista de Simulación y Laboratorio **16**(5), 10–13 (2018)
11. Palacios, C., Guillen, M., Siddons, D.: Evaluation of color fastness in wool and alpaca fabric dyed with biocolorants extracted from plants and animals. Revista SIEMBRA **8**(2), e2917 (2021). https://doi.org/10.29166/siembra.v8i2.2917
12. Norma ISO 105-C06:2010. Textiles. Ensayos de solidez del color. Parte C06: Solidez del color al lavado do-méstico y comercial
13. Braungart, M., McDonough, W.: Cradle to cradle: Remaking the way we make things. North Point Press (2005). https://mcdonough.com/writings/cradle-cradle-remaking-way-make-things/
14. Escobar, A.: Designs for the pluriverse: Radical interdependence, autonomy, and the making of worlds. Duke University Press, NC. (2018). https://doi.org/10.1215/9780822371816-001
15. Carvalho, C., Santos, G.: Global communities, biotechnology and sustainable design–natural/bio dyes in textiles. Procedia Manufacturing **3**, 6557–6564 (2015). https://doi.org/10.1016/j.promfg.2015.07.956
16. Rungruangkitkrai, N., Mongkholrattanasit, R.: Eco-friendly of textiles dyeing and printing with natural dyes. RMUTP International Conference: Textiles & Fashion 2012. Bangkok, Thailand (2012). https://www.academia.edu/67777728/Eco_Friendly_of_Textiles_Dyeing_and_Printing_with_Natural_Dyes

Fashion Alive, from the Unique to the Multiple. Design Experiments for Sustainable Fashion

Roberto Liberti, Ornella Cirillo(✉), Vincenzo Cirillo, and Chiara Scarpitti

Department of Architecture and Industrial Design, University of Campania Luigi Vanvitelli, 81031 Aversa, CE, Italy
{roberto.liberti,ornella.cirillo,vincenzo.cirillo, chiara.scarpitti}@unicampania.it

Abstract. The paper summarizes the objectives, methods, and results of an experimentation project on sustainability in the fashion sector, funded by the *European Community* in 2022, shared between three partners of different geographical origins. Among these, the Italian team, differentiating itself from the others who have chosen to work on *zero waste* or *recycling*, has explored *upcycling* as a technique which, going beyond the simple material reuse and recycling of trousseau and fabrics, in this case referring to traditional kits, triggers an upgrade process capable of raising its value compared to the original. The singularity of the experimentation is inherent in its multi- and trans-disciplinary character and in t in the complexity of the subjects involved, in the manipulation techniques adopted and in the construction of innovative and disruptive fashion concepts. The extension of the project brief to about 450 students and numerous teachers enhances the value of the experience, enhances the diversity of the garments in their being singular products but tuned to a common theme and amplifies the diffusion of the ethical message which is the basis of the objectives of the project. The result is a multiform repertoire, in which every expression contributes to the definition of a polyphonic choir, unique and multiple at the same time.

Keywords: Cultural Heritage · Local identity · Trousseau · Upcycling · Fashion conscious · Slow manufacture

1 Fashion Alive: A Slow Conscious Approach to Fashion Education

Many of the social and environmental concerns included in the term 'sustainability' are directly or indirectly related to design. Sustainability is a central issue for contemporary society as a whole [1], yet, despite this awareness of the critical relationships that exist between design strategies and production methods, energy use, environ-mental impact, the nature and politics of work, design for sustainability is not yet widely understood or practiced.

Designers and fashion designers can help define the man-made environment by paying attention to how the project becomes a product, how it is used and how long it lasts. This is the opportunity that presents itself to the young fashion designer today: that of learning to formulate a thought that is creative, but aware of the impact of one's fashion making on environmental and social ecosystems. A fashion designer aware of his role needs complete training, in which creativity, cultivated and enabled by digital knowledge, technical skills and cultural insights are able to deal with the themes of sustainable development. Fashion represents a powerful cultural lever, capable of influencing the consumerist economic model based on mass production that characterizes current markets, so it seems essential for educational purposes to contextualize this activity in the broader scenario of sustainable practice, creating an experimental methodological path that with Fashion Alive we have retraced through a multi-disciplinary approach.

A cultural paradigm shift as an alternative to unlimited growth, where reducing waste and limiting consumption is a way to respect environmental resources and regain possession of one's time, in a collaborative and non-competitive logic. In the paradigm of speed, words such as slow down, decrease often carry with them a concept of impoverishment, of involution, almost a return to the past; on the contrary, slow manufacturing wants to regain possession of those traditional knowledge that are being lost in the speed of our hyper-consumerist and globalized age. It is no coincidence that these issues are making a comeback in a particular moment like the one we are experiencing: perhaps, in all this global socio-political chaos and in the confusion of a 'world' dominated by uncertainty, the most desirable solution is precisely to change course and overturn the thought system.

According to these visions, the *Fashion Alive* project was framed, winner in 2022 of the Creative Europe tender (European Cooperation Project, Crea Cult 2021), together with *Cremodite* (*Asociación Cultural emprendidora para empresas de moda y Tecnologia*, Madrid) and the University do Minho, Departamento de Engenharia Textil (Portugal), where the Vanvitelli Research Unit wanted to study methods of sustainable fashion, raising awareness in a broader way on "environmental awareness". To this end, a multi and transdisciplinary project was structured through the involvement of 11 courses from the Bachelor's Degree Course in Fashion Design and the Master's Degree in Design for Innovation, Fashion Eco Design curriculum, and the support of 19 teachers and about 450 students. Through this experimental teaching path, the students analyzed some slow manufactures that in the time of digital fashion production they had often never studied in depth, even by listening to the stories of the generations who had created, preserved and guarded them for decades.

From the historical analysis of a piece of clothing that started the project in the history of fashion courses, we moved on in the Technical Drawing and 2D and 3D modeling laboratories to the representation of the various parts, analyzing their DNA, the embroidery stitch, the wearability, the quality of the fabrics. In the Laboratory of Technologies and Materials for Design, for example, natural dyes on fabric were analyzed which can create a new capsule collection from a set, with colors deriving from fruits, plants, minerals, or other elements of natural origin, to then channel the research carried out in the 5 fashion design laboratories that saw the birth of the various collections.

The final event of the *Fashion Alive* project is imagined in the cloister of the Abbey of San Lorenzo ad Septimum in Aversa, headquarters of the Department of Architecture and Industrial Design of the University of Campania *Luigi Vanvitelli* (Fig. 1). Even the conclusion of the project in a historic space in the Campania region wants to be the junction point of the different stories told so far. The results of the research will not be limited to a traditional fashion show but will participate in an event that amplifies and multiplies the results, translating the know-how acquired along the way, in the sense of the methodology defined by Joanne B. Eicher as an interdisciplinary cultural geography [2].

Fig. 1. The *Fashion Alive* event in the cloister of the Department of Architecture and Industrial Design of University of Campania *Luigi Vanvitelli.*

2 The Reasons for a Choice

In a research and didactic project centred on the broad and multifaceted theme of sustainability in fashion, the thematic declinations around which activities can revolve are many, and each one motivated by specific reasons. Starting from the assumption that sustainability, as is well known, not only has to do with the material production of goods, but also with the cultural dimension of products and their immaterial content, the chosen direction in Fashion Alive concerned the ethical sphere of the fashion project, with a focus that favours slow fashion and the revaluation of traditional working techniques, overshadowed by the more advanced processes of industrialisation. Hazel Clark [3] argues that slow fashion brings sustainability closer to dressing, as it brings into dialogue the valorisation of local resources, the transparency of production systems and the sensoriality of products, thus achieving "an actualisation of the ancient Greek concept *kalòs kai agathos*, beautiful and good" [4].

In this direction, studies have shown that this declination combines the eco-sustainability of production, solidarity fashion and second-hand fashion, which places recycling in its various forms at the centre of conscious consumption. The reuse hypothesis is a practice based on the use of materials that are recognised as having both an in-trinsic value and an aptitude for integration with innovative forms and ways. In this sense, considering that at present a large part of underwear and household linen produced in the 20th century is disused, because it is considered impractical or obsolete - often ending up in attics or in markets - we wanted to focus our activity on this ancient heritage, to create new garments characterised by an implicit respect for their materiality. Favouring *upcycling* practices, which go beyond the simple material reuse of the good, an upgrade process was undertaken that can raise its value compared to the original one.

In the multiple scales of intervention that this practice allows, space has been given both to simpler reuses of garments, slightly modified and improved, and, above all, to radical creative transformations that have foreseen the use of garments, or parts of them, in the realisation of completely different objects, ennobled by a contemporary translation of their provenance, even very distant from their initial destination. In this way, the potential of this kind of process was emphasised in the eyes of the young people involved, emphasising the aesthetic quality of the basic products, their uniqueness, but also the expressive capacities allowed by this practice. Beyond the ethical principle implemented, in fact, the possibility of expression offered to each student in the actualisation of the garments and the concrete participation in an action that goes from the idea to the realisation, in a circular vision with multiple meanings, was significant.

In this sense, before developing the modernisation and deconstruction phases of the reused garments, attention was paid to the study and evaluation of the qualities possessed and preserved over time by the fabric or its components: in addition to rarity, irreproducibility, the testimony of a past age, the craftsmanship aspects of production were taken into account, as well as the values and sentiments handed down by these artefacts, identified as strong points on which to base the subsequent creative process. Considering each garment as an asset worthy first of analysis and then of a preventive evaluation, capable of determining its suitability for conversion or vice versa for conservation, the ICCD's VeAC cards were compiled, dedicated to the cataloguing of Ancient and Contemporary Clothes, aimed at the students' acquisition of skills related to this type of activity. The analytical data (chronology, author, destination, material, technique, colour, decoration, measurements, etc.), together with the iconographic and oral sources and the state of preservation, constituted the first step in understanding the peculiarities of the materials, without which the subsequent project would have lacked a prerequisite.

As part of the international Fashion Alive team, the UdR of the Vanvitelli University has from the very beginning approached the project focus as an opportunity to express its sensitivity towards the memory of its own material culture - an inescapable critical precondition, which strongly distinguishes Italy from most of the European context. In this sense, in fact, the intention was to follow up on a pioneering experience carried out in 2022 on the occasion of Procida's nomination as the Italian Capital of Culture, during which it was possible to recognise a significant duality [5] in the trousseau: for many a special family treasure, for others waste material, a bulky surplus to be disposed of.

In southern Italy, the trousseau tradition has been perpetrated since ancient times, as a preliminary obligation prior to marriage and then as a ritual that has united generations, giving identity and recognition to women's work that would otherwise have been undervalued; since the 1970s, the results of consumerism and industrial production have halted this custom, making what is left in family heritages 'relics', precious evidence of an extraordinary and unrepeatable local history (Fig. 2). Their uniqueness lies not only in their non-reproducibility, but in the accumulation of signs and values concealed within them: each object is handcrafted and as such represents a form of expression connected as much to the collective dimension from which it springs, as to the individual one that conceived it; and in its physical components it bears all the 'signs' of this. Therefore, having recognised the exceptional nature of some artefacts, any use of them was excluded; otherwise, in the presence of common, worn-out, or not very valuable items of linen, their conversion into a second life could be admitted. This premise, by discouraging the use of virgin materials in favour of already available resources, has favoured a methodological orientation sensitive to the values of local communities, human labour and the conservation of resources, included in the broad framework of sustainability.

Fig. 2. Photographic images of the collection, selection, and recovery of ancient artefacts from Campania.

This choice, based on the wide availability of this resource, has ensured the homogeneity of the result, which by virtue of a common assumption has made it possible to demonstrate the versatility of raw materials and to emphasise the variety of design declinations to which the garments of a trousseau can lead. The non-ordinary nature of the recovered textile heritage by its very nature triggers heterogeneous creative directions, ranging from obsequiousness to desecration, from minimal intervention to radical manipulation, and initiates the students into concrete experiences with multiple ethical and educational purposes.

The result is a work that is by its very nature handcrafted, which amplifies the suggestion of the unique object made; it updates and converts the meaning of an ancient practice, used over time by people who were careful to rework garments with manual exercises that ensured their longevity [6]. On a cultural level, without losing an aesthetic and formal focus on the result, this approach helped raise participants' awareness towards respecting resources and controlling waste, and to feel their singularity as an integral part of a plural community. Overlaying the present with the past is a complex process that makes the identification of final solutions equally ar-ticult, which only a deep and serious design culture can teach one to evaluate. Deceleration, customisation, individual and collective memory, quality, intangible values, craftsmanship, were the main components of the process and their transfer into the experience of the many actors involved the first goal achieved.

3 The Drawing of the Trousseau as an Anticipated Vision of Inhabiting Body

Within the *Fashion Alive* project and as part of didactic experimentation conducted in the *Drawing Laboratory* (CdL in Fashion Design), it was proposed to the young designers to understand the cultural reasons related to the making and adoption of the traditional wedding trousseau as well as to study, through the methodological action of drawing, the geometric configuration of the textile surface and the habitability with the body of the same. The aim was to try to convey, on the one hand, how from the anatomy of the body and the textile motif arise the form and the language of the designed suit (which will be translated through the *silhouette*) and, on the other hand, how designing a suit means intervening on the topography of the body wearing it and reasoning in terms of the three-dimensionality of the textile surface that is projected onto the body, placing itself in contact or not with it [7, 8].

In the territories of southern Italy, the trousseau or 'nuptial trousseau' has always been traditionally understood as the set of clothes, linens, and other accessories that the bride brought to the marital home as a patrimonial and hereditary share arranged in relation to her economic possibilities [9]. This traditional culture of 'preparing' and 'pre-disposing' a trousseau of clothing well in advance (intended to last as long as possible) meant, in terms of accomplishment and permanence, taking into consideration all the gradual and progressive physical changes that accompanied the female body over time (youth, pregnancy, old age, etc.). In this sense, the trousseau (in the more specific adjectivization of a suit in general), understood both as the textile surface associated with the body and the surface boundary of the same, became an 'anticipated' means and project for establishing and constructing spatial and temporal relations with the surrounding reality at various stages of life. This multifaceted cultural meaning contains within itself what we now commonly refer to as principles of sustainability, reuse, durability, and recovery of local material-material traditions that, nowadays, affect numerous scientific fields, including fashion-related disciplines.

In the specifics of the laboratory teaching, the first phase consisted of the realization of photographs of a trousseau on a plane (owned by the students' family) and taken with the optical axis in the direction orthogonal to the planes of the lens of the one housing the garment itself. In this sense, the photographic shot represents the first record of documentation of the head (each one unique), then supplemented by morphological data. Then, through the scientific rigor of technical drawing, the second phase consisted of transposing the garment -a complex three-dimensional reality- into plane graphic representations (*plats*).

Fig. 3. The drawing of trousseau as an anticipated vision of inhabiting body. Three-year degree in Fashion Design, Drawing Laboratory. Prof. Vincenzo Cirillo.

The anticipated design of the trousseau meant that they intentionally did not correspond to a specific size, contrasting, until a few decades ago, with *prêt-à-porter* and industrial production on the fit of so-called 'tailored suit' that considered only the present body shape. It was therefore essential to refer the representation of the suit under investigation to a codified size plan system. Following the development of eidotypes [10], which were functional for the measurements to be taken, the *plats* of the investigated garments were then proportioned on the canon eight (by Lysippus) through specific proportional conversions and placed in correspondence with the reference anatomical parts.

The *plats* were accompanied by specific technical information such as the size immediately next to the size of the garment's dowry (deduced by taking the body circumferences of the garment itself), the fabric used, the color, and the surface modification techniques (fringing, cutting, folding, pleating, etc.). Next, however, was the depiction of *silhouettes*, with front and back views and in a standing position, with upper limbs adducted along the body, forearms in supination, and hands with extended fingers. Finally, based on the assumption that the form of dress "is related to motility, i.e., the body's ability to move its limbs and change position [as well as] the gestural expectations provided by the body" [11], posed representations and illustration were introduced to communicate with a narrative language the suit to a wide audience [12] (Fig. 3). This investigative operation, in the objective on the one hand to train educated designers who are attentive to the design aspects related to the understanding and recovery of local traditions, in which the discipline of drawing is here understood as a vehicle of knowledge, documentation, and acquisition of geometric methods and graphic codes to aid knowledge aimed at design, has within *Fashion Alive* the objective to forming a base of knowledge and documentation of the rich tradition related to the theme of trousseau. A theme that accommodates social, technical, and participatory design aspects for the development of new, unique garments.

4 Singularity and Plurality of the Project, Between Desecration of Trousseau and Textile Experimentation

The *Fashion Alive* project is positioned within a complex cultural and socio-political context, shaken by waves of populism and xenophobia, threatened by an impending climate apocalypse and the swirling acceleration of new technologies. We live in strange, to paraphrase Slavoji Zizek, particularly interesting times. Out of control, contemporary society, defined by some philosophers as fluid [13], of fatigue [14], or even of performance [15], escapes any attempt at definition. We live in a real and tangible crisis, where the continuous mutation of the contemporary places design culture on the border between human, nature, and technology, in a delicate balance between the dichotomies of the living and the artificial, the organic and the digital. As with every design discipline, fashion also confronts such questions, moving increasingly toward new horizons and sustainable visions.

The purpose-built brief for *Fashion Alive*, developed around ancient nuptial trousseaus, traditionally handed down in the Campania region, inevitably clashes with such a context. Hence, the need for a project intended as a critical interpreter of contemporaneity, in the deconstruction of obsolete systems and interpretation of the new values of society, innovating tradition to change.

In this direction, the Fashion Design Laboratories of the Three-year BA and MA have worked and constructed new imagery and visual productions. The design approach has been twofold: on the one hand, the sustainability of upcycling, the reuse of textiles and processing waste; on the other hand, the disruptive approach towards classic codes and a way of looking at femininity, social hierarchies, marriage, and roles finally within society. With this double approach, many working groups constructed 80 capsule collections of dresses (Fig. 4).

Thus, starting from a critical interpretation that each team developed around the trousseau, the Laboratories conducted fashion-oriented experimental design by hybridizing particular craft techniques, textural manipulations [16], and advanced production processes to develop new creative, conscious, and resilient manufacturers.

For all the workshops conducted by the different teachers, the color palette strictly followed the theme of white and its color variations: beige, ivory, pink, sand, inserting only a few notes of black or red in order to highlight specific graphic or concept elements. Fine fabrics such as linen, cotton, silk, hemp, wool, almost always pure composition, were recovered through the combined action between a different way of observing and working with textiles.

Through the collection of trousseaus among local markets and students' families-from, the selection of textiles and material scraps related to this ancient textile heritage-each group subsequently manipulated and breathed new life into the different types of recovered textiles. Curtains, tablecloths, sheets, nightgowns, collars, dishcloths, doilies, and lace were deconstructed and sartorially reassembled through different techniques such as draping, moulage, patchwork, smocking, fraying, hand, and digital embroidery, digital or hand printing, dyeing, and biocouture.

Fig. 4. Capsule Collections created as part of the Bachelor and Master's Fashion Laboratories. Prof. Patrizia Ranzo, Maria Antonietta Sbordone, Roberto Liberti, Caterina Fiorentino, Simona Ottieri.

Among the various Fashion Laboratories, the two Laboratories in the first and second years of the Three-Year Program followed a particular design process (Fig. 5). Methodologically, these Laboratories were conducted through a teaching divided into steps of process and material transformation. The research started by drawing a mindmap of keywords that could identify some unresolved sociocultural issues related to the trousseau tradition on which to base the collection concept (e.g., gender, marriage, religion, freedom of expression, etc.). The selection of specific keywords concerning the chosen communicative concept was supported by in-depth image research, functional for constructing mood boards capable of inspiring the collections and enriching their visual imagery from a formal, chromatic, and content point of view. In combination, several case studies about the world of fashion, art, and contemporary jewelry were selected according to a period relating to the last two decades.

Regarding textile manipulations, however, these were associated with actions, real or metaphorical. The use of desecrating actions towards textiles and the construction of breaking concepts towards the stereotypes traditionally linked to the trousseau strengthened the link between making and thinking, and structured collections have a strong aesthetic and conceptual impact.

Fig. 5. Capsule Collections created as part of the Bachelor Laboratories. Prof. Chiara Scarpitti.

In this sense, a critical reflection regarding the material and its transformation arises as an inescapable element of the project to the extent that, through a capacity for critical investigation and artisanal reworking, it arrives at a result that surprises and innovates, beyond the most conventional methods. Intersections, inclusions, segmentations, overlaps, subtractions, hyperboles, metaphors: all these and more are just some of the possibilities of textile investigation. Participation in matter, in the sense of a deep understanding of its processual, aesthetic, and symbolic possibilities [17], is one way to grasp its nature and transformative potentialities. Analyzing the methodologies and dynamics capable of triggering divergent thinking about the matter is a complex issue that deserves reflection at a time when contemporary fashion [18] increasingly needs to explore this dual speculative-practical mode [19].

Categories of inquiry, such as sexuality, the sacred, and the family -often left on the margins of design culture- are here put back at the center of design thinking. The strength of such reinterpretations is inherent in freedom of research that does not exclude the most intimate aspects of human experience through a structural critique of society. Mixing the sacred and the profane, bordering on the forbidden, the illustrated dresses break down and desecrate the outfit and, through inventive freshness, restore a high potential for communication and social transformation to the discipline of fashion design.

5 Conclusions

As part of the *Fashion Alive* project, the goal of the Italian research team was to explore the technique of upcycling as sustainable experimentation and an expedient for critical reflection in the field of fashion. The actions undertaken attempted, first, to unhinge the idea that upcycling today represents a new trend to follow, highlighting that, even before the great consumerist explosion of the last century, this practice has always been adopted in private contexts and by recognized fashion designers. Secondly, because of the great documentary value and uniqueness of each outfit, the research proposed overcoming the simple reuse and recycling of garments and textiles to trigger an upgrade process capable of elevating their values in contemporary times.

The interdisciplinary didactic experimentation described in this contribution has thus emerged as an excellent opportunity to learn to formulate creative and critical thinking, aware of the impact of one's fashion-making on environmental and social ecosystems.

The great awareness acquired by the young designers was determined by their understanding of how the concepts of sustainability, reuse, durability, and recovery of local material traditions are already inherent within trousseau, understood here as an 'anticipated' project to establish spatial and temporal relationships during the various stages of life.

Innovation related to the fashion project involved a visual and conceptual exploration of a contemporaneity, on the edge of tradition and future, memory and innovation, social conventions, and the overturning of systems. In this perspective, the Design Labs constructed visual imagery and fashion-oriented productions to offer a new and emerging way of looking at femininity, social hierarchies, and new societal roles in line with the most advanced principles of sustainable and conscious fashion design.

Acknowledgements. *Fashion Alive: Development of new audiences via innovative cultural and digital strategies to promote experimental sustainable fashion methods, CREA-CULT-2021-COOP-1.* The Italian team thanks its European partners: Creamodite (Madrid); Università do Minho (Portogallo, Departamento de Engenharia Textil). Student's project selected in the essay by: Giamel Benyahlou, Marco Coppola, Cristian Sorrentino, Noemi Toscano, Luigi Villano, Matilde Pinfildi, Mariavittoria Esposito, Claudia Del Gaudio, Noemi Davide, Francesco Velotti, Sara Canzanella, Sabrina Campanile, Virginia Corona, Benedetta D'Amato, Maria Vera De Meo, Anna Fortunato, Miriam Fusco.

Author Contributions. Roberto Liberti is mainly responsible for Chapters 1. Ornella Cirillo is mainly responsible for Chapters 2. Vincenzo Cirillo is mainly responsible for Chapters 3. Chiara

Scarpitti is mainly responsible for Chapters 4. All authors wrote the Abstract and Conclusions collaboratively.

References

1. Fletcher, K.: Moda, Design e Sostenibilità. Postmedia, Milano (2018)
2. Jenss, H.: Fashion Studies. Research methods, sites, and practices. Bloomsbury (2016)
3. Clark, H.: Slow fashion an Oxymoron or a promise for the future? In Fashion Theory **12**(4), pp. 427–446 (2008)
4. Segre, S.: Un mondo di mode. Il vestire globalizzato. Laterza, Roma-Bari (2011)
5. Cirillo, O., Fiorentino, C.C., Liberti, R.: Stories of Linen at Procida. An Upcycling Project for Fashion Conscious Design. In: Broega, J.C., Carvalho, H., Providència, B. (eds.), Advances in Fashion and Design Research. Proceedings of the 5th International Fashion and Design Congress, pp. 120–139. Springer (2022)
6. Danese, E.: La dimensione sostenibile del fashion design. In: Ricchetti, M.L. (ed.) Il bello e il buono. Le ragioni della moda sostenibile, pp. 49–56. Marsilio, Venezia (2011)
7. Avella, A.: Disegno di Moda. Corpo/abito/illustrazione. Aracne, Roma (2018)
8. Ciammaichella, M.: Animazioni di corpi rappresentati tra scienza e progetto. diségno, 1/2017 (2017)
9. Donanno, D.: Le parole della moda. Ikon editrice, Milano (2001)
10. Docci, M., Gaiani, M., Maestri, D.: Scienza del disegno. Città studi Edizioni (2017)
11. Di Giuseppe, R., Maggiora, G.: Spazio/Costume/Moda. Alinea, Firenze (2000)
12. Giannone, A., Calefato, P.: Manuale di comunicazione, sociologia, e cultura della moda: Performance. Meltemi editore, Roma (2007)
13. Bauman, Z.: Vita liquida. Laterza (2014)
14. Ha, B.C.: La società della stanchezza. Edizioni Nottetempo (2020)
15. Colamedici, G.: La società della performance. Tlon (2020)
16. Wolff, C.: The Art of Manipulating Fabric. Krause (2003)
17. Bachelard, G.: La poesia della materia. Red Edizioni (1997)
18. Rocamora, A., Smelik, A.: Thinking Through Fashion: A Guide to Key Theorists. Tauris (2015)
19. Scarpitti, C.: Oggetti Pensiero. Storie di design, organismi e nature postdigitali. Letteraventidue (2020)

Fashion Waste as Vibrant Matter. How Luxury Brands Are Taking Care of It

Paolo Franzo(✉) and Maria Antonia Salomè

Università degli Studi di Firenze, Florence, Italy
{paolo.franzo,mariaantonia.salome}@unifi.it

Abstract. The contribution investigates how the two leading multinational luxury fashion goods holdings – LVMH and Kering – have introduced circular design methodologies and waste valorisation in their creative, production and distribution processes. Starting from the hypothesis that the imaginary of fashion waste is transforming from a negative element, to be hidden and eliminated, to a vibrant matter and resource to be valorised, the websites of the holding companies and fashion brands that are part of them were analysed in order to map current circular practices. Examples of the valorisation of pre- or post-consumer waste and its reintroduction into the supply chain are not rare, but until a few years ago they came exclusively from the action of emerging designers and independent brands, alternative to the fashion system, and the existing bibliography focused on these. This contribution opens up a new line of research involving global luxury brands, starting from the hypothesis that circular practices have spread from below within the fashion system through a bubble-up effect.

Keywords: fashion design · fashion waste · luxury brands · care · vibrant matter

1 Introduction[1]

In its 2018 annual report, Burberry revealed that it had physically destroyed finished goods worth £28.6 million to protect its brand exclusivity[2]. This news was much discussed inside and outside the fashion system, prompting the British brand to develop alternative practices to solve the issue of unsold or defective garments such as, for example, recycling "52 tonnes of damaged garments into geotextile materials and 51 tonnes of pre-consumer textile waste into regenerated yarns, fabrics and automotive insulation materials"[3]. This is just one of many examples of how the widespread practices among

[1] The authors shared the approach of the text and jointly wrote *1. Introduction* and *5. Conclusions.*The paragraph 2. *Pre- and Post-Consumer Waste: From Deadstock to Vibrant Matter* was written by Paolo Franzo. The paragraphs *3. LVMH* and *4. Kering* were written by Maria Antonia Salomè.

[2] Burberry, Annual Report 2017/2018, p. 165. https://www.burberryplc.com/content/dam/burberry/corporate/Investors/Results_Reports/2018/Burberry_AnnualReport_FY17-18.pdf [Accessed 28 April 2023].

[3] Ivi, p. 46.

global luxury brands of destroying or disposing of unsold or defective materials and garments is one of the major issues related to the ecological transition and requires a new approach to what is considered waste[4]. While the theme of sustainability is declared and communicated by almost all fashion brands, often with actions that can be interpreted as greenwashing [1, 6, 9], the circularity practices implemented by global luxury brands are much more complex and little analysed. In the 21st century, several significant cases of valorisation of pre- or post-consumer textile waste, of re-introduction into the system, of development of innovative processes and methodologies capable of ensuring greater sustainability within the textile and fashion supply chain have emerged [5, 7]. However, these examples often coincide with small brands and emerging designers, who, also due to economic necessity and the limited quantities of garments produced, prefer to select materials from deadstock and leftovers. Stockists thus play a significant role in recovering part of the unsold textile production from larger companies, making it available to smaller ones. The rejection of big brands is thus translated into accessible raw material for all those emerging creative realities that self-produce and self-finance their own collections.

Until a few years ago, however, brands engaged in circular approaches were considered outside the fashion system, alternatives to the more trendy brands. It is only recently that some of these have made their way among the best known and most desirable brands, thanks in part to the attention of the press and dissemination through social and digital communication platforms. French designer Marine Serre, for example, works to demonstrate that circular approaches can play a significant role in redefining the concept of luxury in fashion. The 70 per cent of Serre's Autumn 2022 collection Hard Drive was made from regenerated – upcycled, deadstock and recycled – fabrics, an achievement reached through continuous investment in in-house processing for deadstock and vintage goods. She's developed her own supply chain in her Paris studio to disassemble garments and ready the upcycled materials for manufacturing[5]. Having won the 2017 LVMH Prize, organised by the French luxury holding company to support emerging fashion designers, and brought Marine Serre's work to the attention of the fashion system. A further connection between approaches to circularity and fashion luxury brands comes from a number of hacking actions that have helped change the imaginaries associated with textile waste. A significant example is the brand Avavav, founded in 2017 by Linda and Adam Friberg, which in 2021 created a collection by recovering textile waste from other brands, without asking permission and without hiding the identifying features of the materials, such as logos and exclusive patterns. The collection focuses on repurposing fabrics from luxury brands such as Fendi, Burberry and Jacquemus into wearable clothing and footwear deadstock patchworks which are embedded into the brand's silhouettes.

[4] The European Commission has also spoken out against the destruction of unsold or returned textiles. "EU strategy for sustainable and circular textiles", 30/03/2022, p. 4.

[5] https://www.voguebusiness.com/fashion/marine-serres-radical-reset-all-female-leadership-to-push-deadstock-as-luxury [Accessed 28 April 2023].

2 Pre- and Post-Consumer Waste: From Deadstock to Vibrant Matter

The paper analyses the approaches of global luxury fashion brands in relation to the reduction and reuse of textile waste within their creative and production processes. In particular, the two main global fashion holding companies LVMH and Kering are analysed. The choice to investigate this type of case study results from the hypothesis that the circular practices previously implemented by small independent brands and emerging designers have spread in recent years among the big luxury brands with a bubble-up approach, i.e. a fragmented and unstructured bottom-up contamination. According to the thesis proposed here, these approaches to textile waste have gone from being considered undesirable to being founding elements of a new idea of luxury within the fashion system. Starting from the consideration that sustainability today is no longer an option, but the only possible way to ensure a future [7], brands have been questioning themselves in recent years on how to modify their design, production and distribution processes.

From a theoretical point of view, the contribution interprets waste as "vibrant matter", according to the definition of the American philosopher Jane Bennett [2]. In this perspective, the objects and matter that surround us are not simply inanimate objects, but have their own life, their own agency and their own power of action. After gaining awareness that the waste and residues of fashion production constitute an environmental and social problem, also thanks to the intervention of public opinion and legislators, global luxury brands have been trying in recent years to redefine their relationship with what was habitually considered "dead matter" to be hidden, burnt, eliminated. Instead, the idea of considering waste as "vibrant matter" implies a new awareness of the power and agentivity of matter and its effects on the world and human life. This shift encourages greater responsibility in the production and consumption of fashion, developing new relationships between waste, people and the natural world. This vision is part of that return to a new materialism investigated by Anneke Smelik [10] in the field of fashion studies.

In recent years academic literature has begun to investigate approaches to circularity and the valorisation of waste by emerging designers and independent brands, highlighting the poetic and affective capacity of revalued waste [3]. Indeed, in many cases, the proposed initiatives arise not so much from ethical and responsible choices, but from aesthetic and creative ideas [11: 143]. These fragmentary but continuous operations have contributed to radically changing the imaginary of waste: from being an object without a future, it is now increasingly perceived as a material with a past, a history, a memory to be put back into circulation, altering that linear vision of time that has often characterised fashion and that is now increasingly called into question [8].

From a methodological point of view, the research was conducted through a desk analysis of the websites of the two holding companies – LVMH and Kering – and the fashion brands that are part of them. Specifically, online keyword searches were carried out, associating the names of the brands analysed with terms such as: "circularity", "upcycling", "waste", "leftovers", "deadstock", "re-use".

3 LVMH

The analysis of the French holding company, which brings together 75 companies from six different sectors including 14 fashion brands, reveals a first set of initiatives and methodologies to reuse textile waste. Some are transversal to the whole group, such as Nona Source and Weturn projects, others are specific to the individual brands within the group. These initiatives are part of LIFE 360 strategy that LVMH presented in April 2021 with the objective of defining priorities to reduce its impact on the environment, with three key deadlines: 2023, 2026 and 2030. In particular, Strategic Priority 1 is entitled Creative Circularity and is a commitment to use only eco-design methodologies by 2030 and to encourage repair, reuse and upcycling practices.

A significant initiative is Nona Source, a digital platform launched in 2021 by LVMH for the resale of deadstock fabrics and leathers from the group's companies, excluding materials with patented patterns or recognisable logos. The platform has an autonomous site, but it is clearly specified that it is a LVMH project and the logos of all the brands involved, indicated as partners, are shown. The principle behind the platform is to offer young designers and emerging brands high-end fabrics and leathers at competitive prices to encourage creative recycling. Nona Source consists of an online catalogue where you can choose from a variety of materials divided by composition, weight, colors, and patterns. For each item in the database, a photo is provided, with the cost per linear meter and the quantity available in stock, the geographical area of production and the current location, even if the brand from which it comes is not specified. It is also possible to order a sample, download detail sheets, washing and care instructions, and watch a video in which the fabric is folded, pulled, and laid on a dummy to fully understand its characteristics. Another project that sees LVMH's commitment to the reuse of waste is the collaboration with Weturn, a start-up that transforms unsold textiles into new second raw material. Founded in 2020, it aims to create the first recycling chain for unsold textiles, fabric rolls and offcuts from large fashion companies. The company processes fabrics made from natural yarns such as cotton, wool and cashmere and fabrics made from blends in which synthetic fibers make up no more than 20%, entrusting the recycling operations to partner companies distributed throughout Europe. The process, once concluded, allows the client company to obtain a report indicating: the rate of valorisation per channel, the fabrics and products developed, their impact, as well as all traceability and legal compliance documents. The start-up has an autonomous website in which the collaboration with LVMH is never mentioned, while the French holding company, on its own site, illustrates the partnership, indicating the recycling rates and emphasising the transparency of the supply chain managed by the French start-up.

After analysing the actions within the LVMH group, the research focused on individual brands with the aim of identifying approaches and methodologies for the valorisation of waste, which can be found on website pages. Louis Vuitton presents a series of references to projects that the company has implemented over the last few years, through a page on its website dedicated to its commitment to circular creativity[6]. The actions

[6] https://it.louisvuitton.com/ita-it/magazine/articoli/committing-to-circular-creativity [Accessed 28 April 2023].

implemented in the design phase are described, emphasising the importance of a process that is more sustainable and circular, such as the adoption of 3D prototyping, the reuse of waste materials, upcycling, and the repair service. A significant example is the LV Trainer Upcycling project, born under the creative direction of Virgil Abloh, which is part of this reorganisation of the creative process and proposes to use waste in the production of new shoes with the intention of giving new life to recovered materials. However, it should be pointed out that there are some misunderstandings in the description of the products related to this project. For example, it is stated that version #54 in white and green "is made of 90% sustainable materials" and that the upper is "made of recovered polyester", generating confusion between generically sustainable materials and recovered materials[7].

Fendi has developed strategies and initiatives along the entire product life cycle to reduce material waste in the pre-consumer and post-consumer phases. In particular, for the management of textile waste Fendi says it has partnered with Green Line and Reverso™, while for the circular management of metal accessory waste the brand collaborates with Safimet. The Italian brand has also developed collaborations with social associations and cooperatives, including Lai-Momo, Roma Best Practices Award Mamma Roma e i Suoi Figli Migliori, and Quid project, to which end-of-production textiles and leather have been donated. Another collaboration is with Cartiera ethical fashion workshop, to which, in addition to materials, machinery and equipment were donated for the production of new products, including cotton dustbags covering Fendi products. A further area of circular initiatives concerns fur, the brand's identity sector: Fendi Furever deals with giving new life to furs through a circular approach; Fur Restyling is a service available in selected boutiques, which offers the possibility of renewing and giving new shape to furs; Fur Upcycling is a project for recovering furs from previous collections which are used, through patchwork, in new collections under the name of Fendi Vintage. The Italian brand's approaches to circularity also extend to bags, with the Peekaboo Spa project, launched in 2021 with the aim of giving a second life to the iconic bag model. The project consists of a boutique event dedicated to a care and maintenance session for the bag, with the intervention of the brand's artisans who support customers in extending the life of the products.

The initiatives communicated by Dior on the subjects of this contribution are limited. One example is the use of discarded materials and other objects from the brand's ateliers for the window dressing of the Avenue Montaigne shop in Paris[8]. To the Spring-Summer 2022 season belongs the collaboration between Dior Homme and Parley for the Oceans, in which creative director Kim Jones worked with the environmental association to develop a series of eco-sustainable garments made from recycled plastic[9]. In this case,

---

7 https://it.louisvuitton.com/ita-it/prodotti/sneaker-lv-trainer-nvprod3710064v/1AAGXF [Accessed 28 April 2023].

8 https://www.dior.com/it_it/fashion/news-savoir-faire/folder-novita-ed-eventi/le-incantevoli-vetrine-di-30-montaigne [Accessed28 April 2023].

9 The association is responsible for recovering plastic from the oceans, sorting it, shredding it and then turning it into a Parley-patented yarn called Ocean Plastic.https://www.dior.com/it_it/fashion/news-savoir-faire/folder-novita-ed-eventi/la-nuova-capsule-collection-beachwear [Accessed 26 April 2023].

circular design methods do not act on the waste produced by the brand itself, but are used to introduce a new regenerated material.

The latest examples of circularity-oriented actions by LVMH Group brands are: Loewe, which in 2021 launched The Surplus Project with the intention of recovering waste and leftover materials from previous collections[10]; Berluti, which has a page on its website entitled Creative Circularity in which the brand's circularity actions are described with some data, including the recovery of unused fabrics through Nona Source platform, the optimisation of packaging and the reduction of plastic, the invitation to care and maintenance of products, including video tutorials, the customisation of products as a tool for reducing unsold and returned goods, the recycling of leather and processing waste transformed into fertiliser[11].

The analysis shows that the group's other fashion brands do not declare specific actions to reuse and valorise textile waste, but only more or less generic commitments to environmental or social sustainability issues, including tracking the production chain, reducing energy consumption, and protecting the rights of workers and suppliers.

4 Kering

The second case study analysed in this contribution is the holding company Kering, which groups 6 fashion and leather goods brands as well as jewellery and eyewear brands. The investigation carried out reveals various projects and collaborations, with public and private entities, to collect textile waste from production processes and fashion shows in order to recycle and reuse it, as stated in the Circularity Ambition report published on the group's website. Textile waste, upcycling and recycling projects include several partnerships with NGOs and cooperatives to optimise the management of textile and leather waste. Since 2018, Kering has been working with La Réserve des Arts, an NGO that collects waste materials and offcuts in Paris and Marseille by acting as a link between manufacturing companies and creative people interested in circularity methodologies. Tissons la Solidarité, a network of 70 companies whose main activity is the collection, sorting, resale and making-up of second-hand textiles, has also been working with Kering since 2019 within the framework of vocational reintegration projects. The partnership launched with Aquafil, a company that holds the patent for ECONYL, a regenerated nylon yarn obtained from waste, pre-consumer waste, fishing nets, carpets, and with Worm Again, a start-up that has developed a solution to separate polyester polymers from cellulose, allows the group's brands to reuse waste to produce new nylon and new fabrics from a fully traceable supply chain. Also worth mentioning is the participation of the entire Kering group in the Full Circle Textiles Project – Scaling Innovations in Cellulosic Recycling of the global movement Fashion for Good, to support and accelerate the development of industrial recycling infrastructures. The pilot project for chemical recycling brings together actors from the fashion industry and aims to convert cotton

10 https://www.loewe.com/eur/en/stories-collection/surplus-leather-project.html [Accessed 28 April 2023].

11 https://www.berluti.com/en-it/sustainable-development/creative-circularity/ [Accessed 28 April 2023].

and cotton-blend textile waste to produce new synthetic cellulose fibres using cutting-edge technologies. Finally, other initiatives of the French holding company involve collaborations with fashion universities, to which unused stock fabrics have been donated since 2015.

Analysing the brands of the Kering group, Gucci has developed several projects. One is entitled Gucci Vault, a digital project dedicated to the circular approach and upcycling of the archive. It consists of a virtual concept store, containing unique pieces created by independent designers using materials and models from the Florentine brand's archive. It contains three projects, two of which are strongly connected to the issues analysed in this contribution: Gucci Continuum is a set of capsule collections designed by designers and brands, including Collina Strada, Vans, Alpha Industries and Slam Jam, with the aim of enhancing materials left over from previous collections; Vault Vintage reworks a series of selected garments from previous Gucci collections through the intervention of the brand's artisans to approach new stylistic trajectories. Another Gucci project is the Circular HUB, which is presented on the Kering website but not on the Gucci website. The project, which is not yet active, was created with the intention of promoting the circularity model within the Italian fashion production system and envisages the creation of an open innovation platform to design and create circular products and solutions.

An analysis of the group's other brands reveals fewer initiatives and actions specifically related to circularity. In 2020, Bottega Veneta developed the Kraft Handbag Collection project, a capsule of bags made of 100% recycled FSC-certified cardboard paper and lined in leather, although there is no longer any trace of this project on the brand's website. On the brand's sustainability page, it is indicated – but without any dated and verifiable information – that 90% of leather waste is recycled and the various percentages of plastic, paper and packaging "certified or recycled", bringing together with these two terms different approaches and thus generating confusion. Another Bottega Veneta project moving in the direction of reducing pre- or post-consumer waste in fashion is the Certificate of Craft, lifetime warranty programme on bags announced in October 2022. It consists of a document on which the product code is stamped and which allows the customer to benefit from an unlimited number of repairs free of charge, with the aim of keeping products in use for longer, reducing the need to replace them and generating waste. It seems significant that this project is also not communicated within the brand's website, but was only released through a press release at the time of the launch.

Balenciaga, to date, is only working on waste reduction in the post-consumer phase. In particular, it has launched, in collaboration with the digital resell platform Reflaunt, the Re-Sell programme, a service for reselling used Balenciaga garments in a network of boutiques with the aim of encouraging the "circular fashion revolution", as stated on the page dedicated to the project on the brand's website.

The last brand of the Kering group whose initiatives are related to the reduction or reuse of waste is Saint Laurent, which has launched circularity programmes aimed at recovering leather processing waste, which is reintroduced into production through upcycling processes both in leather goods, in particular in the production of bags in the Monogramme and Sunset lines, and in womenswear. The brand also collaborates with fashion schools and the non-profit organisation La Réserve des Arts, based in Paris, which supports artists, art students, craftsmen and cultural workers through circular practices, and recovers waste materials, offcuts and surpluses by working with the social enterprise Le Relais, which transforms them into insulating panels that are also used in the brand's premises.

5 Conclusions

The analysis conducted on the two main global luxury holding companies – LVMH and Kering – and the fashion brands that are part of them, showed that the introduction of circular approaches in the design and production processes of global fashion brands is still at an early and unstructured stage. It emerged, in fact, that some brands are engaging in various projects and have decided to emphasise this on their websites, considering it significant for communicating a new idea of luxury; others, on the other hand, do not yet seem ready for this transition and do not consider it strategic to associate their image with an idea of valorising waste and circularity. However, it is clear that, also thanks to new EU regulations and an increasingly demanding public opinion sensitive to these issues, all brands will soon have to introduce circular fashion design and production methods in a stable and structured manner.

This contribution represents the start of a research project – Re-Waste –, financed by the NextGeneration EU programme, which aims to investigate the production of pre-consumer textile waste with the objective of developing solutions for the management, valorisation and redesign of waste. The identification and analysis of global case studies, therefore, will make it possible to identify methodological approaches that can be developed as models to be applied to the Made in Italy manufacturing supply chains.

Acknowledgements. Project funded under the National Recovery and Resilience Plan (NRRP), Mission 4 Component 2 Investment 1.3 – Call for tender No. 341 of 15/03/2022 of Italian Ministry of University and Research funded by the European Union – NextGenerationEU. Award Number: PE00000004, Concession Decree No. 1551 of 11/10/2022 adopted by the Italian Ministry of University and Research, CUP B83C22004890007. MICS (Made in Italy – Circular and Sustainable).

Spoke 2: Eco-Design strategies: from materials to Product Service Systems – PSS.

Spoke 7: New and consumer-driven business models for resilient and circular supply chains.

References

1. Adamkiewicz, J., Kochanska, E., Adamkiewicz, I., Łukasik, R.M.: Greenwashing and sustainable fashion industry. Curr. Opin. Green Sustain. Chem. **38**, 100710 (2022)
2. Bennett, J.: Vibrant Matter: A political Ecology of Things. Duke University Press, Durham (2010)
3. Binotto, C., Payne, A.: The poetics of waste: contemporary fashion practice in the context of wastefulness. Fash. Pract. **9**(1), 5–29 (2017)
4. Brydges, T.: Closing the loop on take, make, waste: investigating circular economy practices in the Swedish fashion industry. J. Clean. Prod. **293**, 126245 (2021)
5. de Freitas Netto, S.V., Sobral, M.F.F., Ribeiro, A.R.B., et al.: Concepts and forms of greenwashing: a systematic review. Environ Sci. Eur. **32**(19) (2020) https://doi.org/10.1186/s12302-020-0300-3

6. D'Itria, E.: Textile from the waste: an overview of the state of the art in the leather production process. Curr. Trends Fashion Technol. Text. Eng. **1** (2017) https://doi.org/10.19080/CTFTTE.2017.01.555565
7. Epifani, S.: Sostenibilità digitale: Perché la sostenibilità non può prescindere dalla trasformazione digitale. Digital Transformation Institute, Rome (2020)
8. Evans, C., Vaccari, A. (eds.): Time in Fashion: Industrial, Antilinear and Uchronic Temporalities. Bloomsbury Visual Arts, London (2020)
9. Kaplan, E.N., et al.: From take-make-waste to circular economy: tackling greenwashing and artificial demand creation in fast fashion. Int. J. Interdisc. Res. Methods **9**(3), 1–21 (2022)
10. Smelik, A.: New materialism: a theoretical framework for fashion in the age of technological innovation. Int. J. Fashion Stud. **5**(1), 33–54 (2018)
11. Vinken, B.: Fashion Zeitgeist: Trends and Cycles in the Fashion System, p. 2. Berg, Oxford (2005)

A Reflection on Sustainable Actions in a Plus Size Lingerie Manufacturing Company in São Paulo

Rachel Rios Scherrer[1](✉), Eduardo Romeiro Filho[2], and Heloisa Nazaré Dos Santos[1]

[1] Escola de Design – UEMG, Belo Horizonte, Brazil
rachel.scherrer@uemg.br
[2] UFMG, Belo Horizonte, Brazil

Abstract. This article aims to analyze the production processes of a lingerie brand, with a focus on sustainable production and eco-design practices. The fashion industry plays a significant role in social, political, psychological, and environmental aspects. The study utilizes a qualitative approach, based on a case study and an interview with the brand owner. The research is grounded in the Eco-design Strategies Diagram, which seeks to classify the brand's main product as sustainable. The analysis of production processes considers concepts such as sustainability, circular fashion, and the plus-size segment. The results highlight the eco-design practices adopted by the company, its commitment to sustainable production, and environmental responsibility. The study also addresses consumer awareness and the importance of transparency in production processes. It is expected that this work will contribute to the advancement of sustainable production in small and medium-sized companies in the fashion industry, promoting the integration of eco-design practices and environmental responsibility.

Keywords: Fat body · Circular fashion · Plus size · Sustainability

1 Introduction

Fashion is a broad term that, when used in various contexts, provides a shared framework and reflection for various aspects of social life. At its core, fashion encompasses a temporal dichotomy between the old and the new, the present and the past, mobility and immobility. It is through the experience of appearances that fashion manifests itself, serving as both a function and an aesthetic content. Specifically, the term fashion refers to the social phenomenon of cyclical change in customs, habits, choices, and tastes that is collectively validated and almost obligatory (Calanca, 2002).

Challenging the idea that fashion belongs to all epochs and civilizations, these analyses are based on the conception that fashion has a locatable beginning in history. In contrast to the notion that fashion is an inherent phenomenon in human-social life, it is presented as an exceptional process inseparable from the emergence and development of the Western modern world (Lipovetsky, 1989).

Thus, fashion, inspired by historical events, plays a fundamental role in promoting variations in clothing and in how people dress and behave. Additionally, political, social, and environmental contexts exert an increasingly significant influence on fashion trends and its own history (Carvalhal, 2016).

Over the centuries, a beauty standard has been solidified, exalting the thin body as the ideal and beautiful. This idealization of beauty becomes challenging for teenagers who do not fit into this established standard. As a result, the fat body has been marginalized, no longer being represented and valued as it was in Renaissance paintings. However, it is intriguing to observe the figure of the Venus of Willendorf, a sculpture dating back thousands of years, depicting a woman with a fat body. This figurine has become a symbol of resistance for the fat body, challenging established standards and emphasizing the importance of diversity and inclusion in contemporary fashion (Oelbaum, 2022).

In this context, Oelbaum (2022), in her project titled "The Visual Resistance of the Venus of Willendorf 2005–2021," recontextualizes the iconic prehistoric statue to challenge prevailing aesthetic standards and promote the inclusion of fat bodies. Using newspaper clippings and diets as symbolic elements, the project seeks to question the oppression and stereotypes imposed on these bodies. By highlighting the importance of diversity and inclusive representation, it aims to promote empowerment and raise awareness about the rights of fat bodies. Figure 1 illustrates a photomontage of photographs and drawings centered around the theme of the Venus of Willendorf.

According to Aires (2019), the contemporary consumer market redefines the fat body by associating it with the capacity for production and consumption. In this context, individualistic consumption practices are essential for the inclusion and redemption of the fat individual. Therefore, promoting respect, diversity, and inclusion in the fashion industry is crucial to value diverse bodies and promote a positive and healthy body image for everyone (Aires, 2019).

In addition to the need to promote respect, diversity, and inclusion in the fashion industry to value diverse bodies and promote a positive and healthy body image for everyone, it is important to highlight the connection between the fat body and the environmental issue. Contemporary consumption redefines the fat body by associating it with the capacity for production and consumption, which implies a higher consumption of materials to develop clothing that meets the needs of these bodies. During the prototype development, it is necessary for patternmakers and designers to keep a detailed record of consumption, which will serve as a reference for various calculations related to the garment, such as fabric consumption estimation, price determination, and cost evaluation (Fraga, 2020). Fat bodies may require more fabric and resources for the production of larger-sized clothing. This implies a higher expenditure of water and other raw materials in the manufacturing process of these garments.

According to Wise and Pitt (2021), in 2017, the clothing sector was responsible for generating 92 million tons/year of textile waste, ranking second among the industries that most harm the environment. Since the last century, environmental actions have been implemented by both developed and developing countries to mitigate the socio-environmental impacts caused by anthropogenic activities. The Conference of the Parties (COP) takes place annually to review the emission inventories presented by the member countries of the United Nations Framework Convention on Climate Change. Its goal is

Fig. 1. Photo panel of the Venus of Willendorf - created by the author.

to stabilize greenhouse gas concentrations in the atmosphere. The conference has been held annually since 1992, initiated by the UN, with the involvement of the private sector, bringing together representatives from 178 countries to deliberate on decisions to reduce environmental degradation and ensure the existence of future generations.

Berners-Lee (2020) emphasizes:

> About 100 years ago, during World War I, we couldn't have destroyed the entire planet even if we tried. But 50 years ago, especially with nuclear energy, it became clear that we could completely destroy everything if we made big enough mistakes. Today, we don't need to make any mistakes; if we don't make enough effort, we will destroy the entire environment. And in 50 years, if energy consumption trends continue, the world will become even more fragile compared to our ever-increasing power.

According to Aires (2019), the current consumer market redefines the fat body by associating it with the capacity for production and consumption. In this context, individualistic consumption practices are essential for the inclusion and redemption of the overweight individual. Therefore, promoting respect, diversity, and inclusion in the fashion industry is crucial to value diverse bodies and promote a positive and healthy body image for everyone (Aires, 2019).

In addition to the need to promote respect, diversity, and inclusion in the fashion industry to value diverse bodies and promote a positive and healthy body image for everyone, it is vital to highlight the connection between the fat body and the environmental issue. Current consumption redefines the fat body by associating it with the capacity for production and consumption, which implies a higher consumption of materials to develop clothing that meets the needs of these bodies. During the prototype development, patternmakers and designers must keep a detailed record of consumption, which will serve as a reference for various calculations related to the garment, such as fabric consumption estimation, price determination, and cost evaluation (Fraga, 2020). Fat bodies may require more fabric and resources for the production of larger-sized clothing. It implies a higher expenditure of water and other raw materials in the manufacturing process of these garments.

According to Wise and Pitt (2021), in 2017, the clothing sector generated 92 million tons/year of textile waste, ranking second among the industries that most harm the environment. Since the last century, environmental actions have been implemented by both developed and developing countries to mitigate the socio-environmental impacts caused by anthropogenic activities. The Conference of the Parties (COP) takes place annually to review the emission inventories presented by the member countries of the United Nations Framework Convention on Climate Change. Its goal is to stabilize greenhouse gas concentrations in the atmosphere. The conference has been held annually since 1992, initiated by the UN, with the involvement of the private sector, bringing together representatives from 178 countries to deliberate on decisions to reduce environmental degradation and ensure the existence of future generations.

Berners-Lee (2020) emphasizes:

> About 100 years ago, during World War I, we couldn't have destroyed the entire planet even if we tried. But 50 years ago, especially with nuclear energy, it became clear that we could completely destroy everything if we made big enough mistakes. Today, we don't need to make any mistakes; if we don't make enough effort, we will destroy the entire environment. And in 50 years, if energy consumption trends continue, the world will become even more fragile compared to our ever-increasing power.

In this context, a global awareness movement, "Who made my clothes?" has emerged, seeking transparency in the fashion industry's production processes. Vavolizza (2020) presents a global initiative:

In this scenario, it becomes crucial to rethink the fashion industry's business model, incorporating sustainability principles, ethics, and social responsibility. Maeda (2007) indicates the search for more uncomplicated, efficient actions to drive the economy. Currently (21st century), the most significant concern lies in the life cycles of fashion products, from the extraction of raw materials, encompassing the stages of production processes and fair and local labor, to the disposal of the product and its components, with proper waste management and reduced environmental impact. Circular fashion aims to minimize the ecological effects caused by such inadequate and toxic waste, mitigating those generated in Brazil and worldwide.

> A global fashion movement is in vogue with the call "Who made my clothes?" asking, "Who made ly clothes? We already kwow that our clothes are produced by a group of industries compesed of many people. If we delve into the motivation behind the activism, we understand its essence lies in demanding transparency in production processes.

The understanding of the motivation behind this activism reveals that its essence lies in the demand for transparency in the production processes of the fashion industry. As mentioned on the official website of the Fashion Revolution movement, the initial motivation arose from the collapse of the Rana Plaza garment factory in the capital of Bangladesh, which resulted in the death of approximately 1,138 people and left others injured in 2013 (Vavolizza, 2020). It is essential to highlight that our clothes are produced by numerous industries involving numerous individuals.

In recent years, a significant shift has been observed in consumer behavior, with a growing demand for sustainable fashion, conscious consumption, and practices such as eco-design, recycling, reusing, and upcycling. Fletcher and Grose (2011) emphasize that upcycling uses scraps of fabric and trimmings to remodel garments, adding value through careful repair. In turn, Berlim (2020) points out that upcycling is based on using materials whose useful lives are nearing the end due to actual or perceived obsolescence in form, function, or materiality. In this context, fashion companies must commit to sustainable production, demonstrating transparency and offering products aligned with these proposals.

Based on a case study, this study aims to contribute to sustainable production in small and medium-sized fashion companies. This study's general objective is to analyze a lingerie brand's production processes, identifying ecodesign practices and their relationship with sustainable production. To achieve this objective, the following purposes were established: describe the concepts of sustainability and circular fashion, highlight the importance of the environmental damage caused by the fashion industry; and explore the application of ecodesign strategies in the analyzed lingerie brand.

To support this research, a qualitative approach was adopted, with the primary data source being an interview with the owner of the lingerie brand. This dialogue aimed to understand the brand's perspective on sustainability, its production processes, environmental actions, and customer awareness. The Ecodesign Strategies Diagram was also used to guide the improvement of the product's eco-efficiency profile throughout its life cycle.

Addressing these topics is expected to advance sustainable production in the fashion industry, emphasizing the importance of integrating eco-design practices and environmental responsibility. It is believed that this research can serve as a foundation for small and medium-sized companies in the fashion sector and other stakeholders interested in promoting sustainability and ethics worldwide, aiming for a more conscious and environmentally responsible future.

2 Literature Review

Fashion has always been presented for slim bodies, which has bothered women who don't fit into this standard. In the last few decades, Brazil has experienced a significant increase in the number of people with obesity. According to IBGE (2020), the proportion of obese individuals aged 20 years or older more than doubled between 2003 and 2019, going from 12.2% to 26.8%. Female obesity rose from 14.5% to 30.2%, compared to 9.6% to 22.8% for males (Gurgel, 2021). Lieberman (2015) emphasizes that a body with excess fat is at risk of developing diseases and that being overweight doesn't always indicate an unhealthy body.

Lipovetsky (2016) highlights the narcissistic culture of the human body, in which the slim body is considered "unattainable." Aires (2019) adds that fashion, as a tool for contemporary identity, ignores the presence of the fat body. It is difficult to find fashionable products for the plus-size segment, starting from size 46, whose bodies are constantly stigmatized in current culture. The author further states:Observamos que há um silenciamento quanto aos corpos gordos no âmbito dos discursos sociais e mercadológicos, que o invisibilizam e excluem. Não há no mercado, por exemplo cadeiras projetadas para indivíduos que pesem acima de 200 kg, e se há, são especiais e devem ser sob encomendas.

The plus-size fashion segment is increasingly present in the market. Runway shows presented on important catwalks around the world reveal this, even though the showcased fat body is not as large as the ones seen on the streets. Gay (2017) asserts that a person with a chubby body is not blind and knows that:

> It's difficult for thin people to know how to talk to fat people about their bodies, regardless of whether their opinions have been solicited. I understand that, but it's offensive to pretend I'm not fat or deny my body and its reality. It's cruel to assume that I somehow lack awareness of my physical appearance. And it's an insult to presume that I am ashamed of myself for being fat, regardless of how close that may be to the truth.

Ferraz (2017) highlights the importance of understanding and embracing one's body. According to him, it is necessary to respect the fat body and incorporate it into the market by seeking to adapt and improve the ergonomics and fit of clothing for this natural body. Jimenez (2022) adds that accepting the body as it is can provoke changes in the concepts of beauty for the female body, making it possible to consider it an expression of resistance.

As stated by Gurgel (2021), fat people are often placed in the same box, as if they were abnormal beings who only think about eating and need to lose weight through diets, and that being thin is the correct and healthy state. The woman of the new millennium is independent, works, makes her own choices, and accepts her body as it is. The physical changes in the body also lead women to change how they dress and adapt to fashion trends.

Reinforcing this statement, Aires (2019) adds:

> A dynamic comes into play in which the market sphere appropriates difference, reorganizing it according to certain principles: on the one hand, it seeks to erase

otherness, but it also leads to the empowerment of fat women, who experience transformations in their identity construction, self-esteem, social belonging, and social interactions.

Sustainable development is based on social justice, economic viability, and environmental preservation. According to Sachs (Berlin, 2020), sustainability is established, in an interdisciplinary manner, on the following foundations: social, cultural, ecological, environmental, territorial, economic, national political, and international political. The concept of ecodesign is grounded in sustainable development through waste and emissions reduction and the analysis of a product's life cycle. As such, it must meet the demands of the present and future generations.

Vavolizza (2020) points out that many Brazilian brands consider themselves environmentally friendly by developing upcycled products (a term used to reintegrate materials that would otherwise be destined for waste into the production process to create new products). However, not just this action will make them truly sustainable. The circular economy encompasses the need for sustainable growth in the face of pressures on production and excessive consumption, which contribute to environmental destruction and the scarcity of non-renewable products. The linear economy, which relies on the extraction of natural resources, is no longer viable in today's market, which seeks a new sustainable positioning for preserving the planet.

Ohde (2018) adds:

> In the circular system, the product goes through several cycles involving product reuse, component reutilization, and material recycling. As the material is never discarded, it becomes a readily available raw material for new processes, making extraction less critical. It is a model inspired by nature, where the concept of waste or disposal does not exist

The exchange and sale of used clothing (secondhand clothing market) are less common in the plus-size fashion segment, as its users typically use them until the end of their lifecycle. At the BH Estilo Plus fashion fair held in Belo Horizonte, there is a digital influencer known as the "Fake Cinderella," who, due to having an extensive collection of clothes, always holds a bazaar at the end of the year. The BH Estilo Plus fashion fair is recognized as the leading plus-size fashion fair in Minas Gerais and one of the largest in Brazil. In addition to selling fashion products for plus-size bodies, the event offers opportunities for dialogue and artistic presentations. The featured plus-size lingerie brand in this article always participates in this fair. Carousel-style photos are showcased on Instagram (@use.uderline), as illustrated in Fig. 2.

Sustainable development has a deeper understanding of the environment, encompassing social and political aspects that influence the quality of life and environmental protection. Carvalhal (2016) acknowledges that society is moving towards a more conscious and sustainable fashion.

Fig. 2. Brand's Instagram (@use.undeline)

3 Methodology

For the development of this research, the theoretical framework was used as the basis, along with qualitative research through an interview with the brand owner, as shown in Table 1.

The Ecodesign Strategies Diagram, a tool that guides the improvement of the product's eco-efficiency profile within its life cycle context, was also created.

The Ecodesign Strategies Diagram (ESD) consists of eight items representing ecodesign strategies. Connections, interpositions, and relationships were established among the practices to classify whether the main product of the Underline brand is sustainable or not using this diagram.

With a focus on this new reality - conscious consumption, ecodesign, recycling, reutilization, and upcycling - the fashion market must incorporate sustainability, ethics, and social responsibility into its companies, demonstrating transparency and presenting products aligned with these principles.

Figure 3 shows the brand's website page with the new size.

Table 1. Form responded by the owner of the analyzed brand

1) How long has the brand been in the market?	**2)** What is the brand's target audience?	**3)** How is the creation and production process carried out?	**4)** What is the flagship product?
5) For how long do customers typically use your products?	**6)** What do they do with your products when they stop using them? Do they donate or dispose of them?	**7)** Is there any benefit for customers who return Underline products after the product's lifecycle?	**8)** What do you plan to do to address this issue?
9) Is your customer aware of your environmental concerns and actions?	**10)** What type of energy do you use in your manufacturing?	**11)** Do you plan to implement cleaner energy sources?	**12)** How is your stationery? Do you use recycled paper?
13) What were the reasons that influenced you to create a brand focusing on sustainability?	**14)** 13.Do you consider that your brand incorporates eco-design principles?	**15)** Can you mention the names of the fairs you participate in?	**16)** Could you provide some insight into your product delivery service via bike? Why did you implement this service?

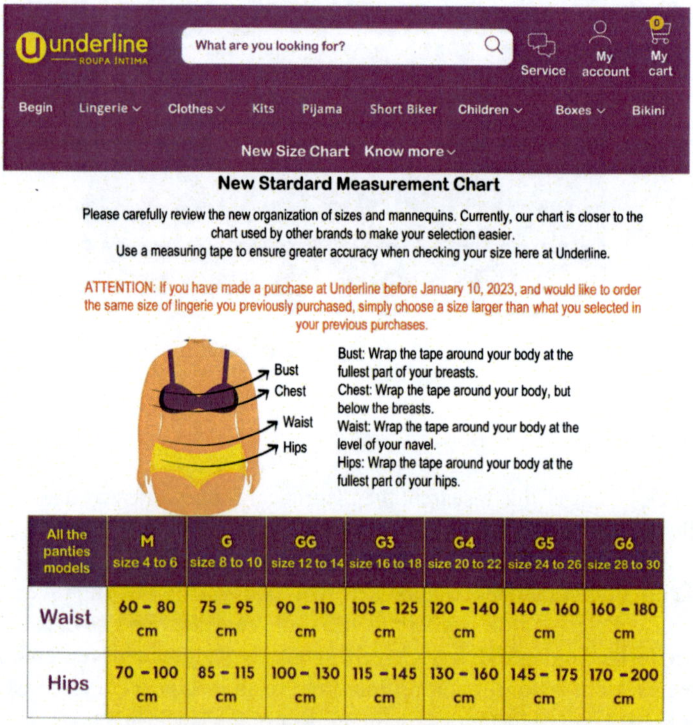

Fig. 3. Website page with the new size chart.

4 Results and Discussion

The plus-size lingerie brand analyzed in this study was established in 2017 in São Paulo, Brazil. Its sales are conducted through its website and at plus-size fashion fairs. The brand owner emphasizes the importance of reducing environmental impact, recognizing the significant contribution of the garment sector to global pollution. In 2020, amid the pandemic and increased sales, the brand became the owner's sole source of income.

As an entrepreneur, being a plus-size woman and from a family involved in the intimate apparel industry, the brand owner recognized the need to develop personalized products for plus-size women at an affordable price. She mentions experiencing difficulties finding properly fitting underwear in her childhood, which led her to wear cotton briefs instead. The market primarily offered lace, microfiber, and other synthetic fibers, without providing options for cotton underwear. Therefore, the brand's products feature natural fibers based on personal demand and comfort.

All the lingerie pieces from this company are based on basic lingerie patterns from the family business, adapted for plus-size bodies. The design process is straightforward. As the brand aims to sell more affordable pieces, the practices have fewer intricacies to reduce machine production time. The focus is on creating essential elements deeply ingrained in the brand's DNA.

The plus-size fashion fair market is booming. Even if sales are not always finalized, brand exposure is crucial. Purchasing lingerie online can be challenging: customers visit the website, get to know the brand at fairs, and then need to return to the website to make a purchase.

The lingerie manufacturing company participates in two virtual fairs: Pop Plus in São Paulo and BH Estilo Plus in Belo Horizonte. The brand has previously participated in a fair in Rio de Janeiro called @hashtagmodaplus, but the results fell well below expectations, making further participation unfeasible. These fairs occur three to four times yearly and have been in the market for over five years. They resemble the former Mercado Mundo Mix, an essential movement in the 1990s that aimed to promote youth culture and launch new fashion, music, and design artists. Prominent fashion figures like Alexandre Herchcovitch and Chilly Beans began their careers there and established their brands.

The analyzed brand offers a cotton hot-pant model, their best-selling item. It was created from scratch and adapted over time. It is a straightforward piece to manufacture, with three closures (two on the sides and one at the bottom) and overlock stitching at the hem. The company can produce 300 to 500 panties in two days of machine production, followed by two more days for finishing and quality control. They have a team of six external seamstresses (outsourced). This hot pant is the most profitable product, with an average monthly sales range of 300 to 380 units.

The manufacturing company plans to expand its team for better production control and logistics management. Due to the low average ticket price, selling a high product volume is necessary. The patterns are all manually made, the sewing is done in the family's workshop, and taxes and labor costs are shared between the two companies.

The analyzed brand has demonstrated compliance with ABNT NBR 16933 by making necessary adjustments to its size chart. This initiative aims to promote transparency for customers and eliminate any doubts that may arise regarding product measurements.

By adopting this approach, the brand demonstrates its commitment to providing clear and accurate information to ensure customer satisfaction and promote a more reliable shopping experience.

A spider web or radar chart represents the Diagram of Ecodesign Strategies (DEE). The eight items were evaluated using a rating scale of 0–5, indicating the ecodesign methods, as illustrated in Fig. 4.

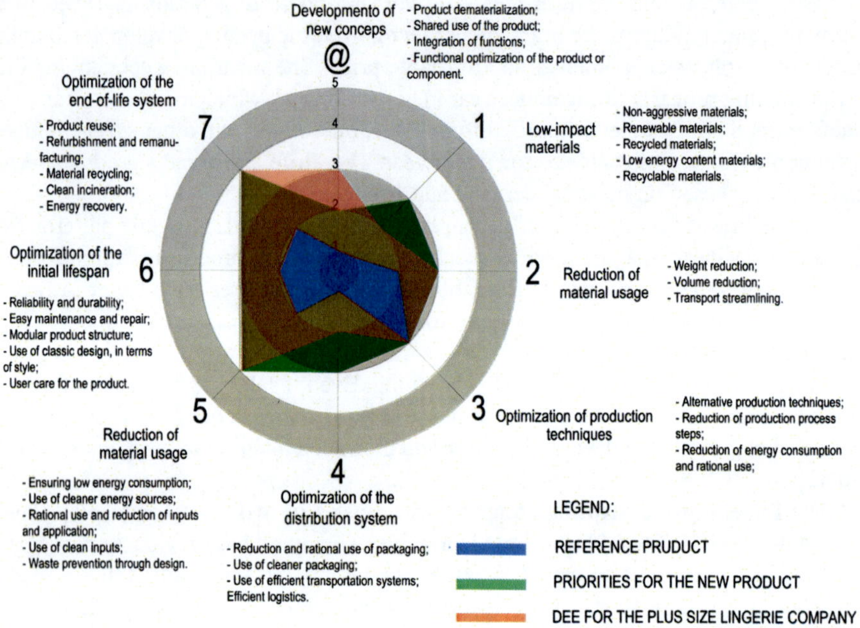

Fig. 4. Ecodesign Strategies Diagram of the Lingerie Manufacturing – created by the author.

The development of the DEE starts with the item Strategy @ - Development of New Concepts (dematerialization, shared product use, function integration, and functional product optimization). The analyzed company received a rating of 3.

Textile waste, which consists of fabric scraps generated from cutting fabrics, is also transformed into products to avoid environmental disposal. Some demolition is transformed into eco-pads (reusable facial cleansing pads), while others become hair ties. Larger fabric scraps are sent to other clothing manufacturers that create more affordable pieces in smaller sizes, as Underline lingerie tends to be significant. Specks are forwarded to an association for rug production. Textile waste has also been donated to companies that make punching bags and teddy bear stuffing. Some companies shred this waste and create popular blankets for donation to homeless individuals and people in vulnerable situations. The clothing manufacturer allocates a portion of the textile waste to these companies for separation.

Additionally, some bags of waste are kept for donation. In short, nothing is thrown away. The only debris discarded is the thread tail from the overlock machine, which is detrimental to the environment as it is made of synthetic fiber.

DEE Strategy 1 focuses on selecting low-impact materials (clean materials, renewable materials, materials with low energy content, recycled materials, and recyclable materials). This item received a rating of 2.

The creative process of Lingerie Manufacturing also includes purchasing rejected fabrics (with minor defects) that are returned and resold at a more affordable price, measuring over 10 to 20 m. Since it is intended for lingerie production, a smaller garment, it is possible to work around the defects. They also allow the production of pieces with minor imperfections, either donated or sold at lower prices. The company's production relies heavily on what is available in the surplus market. Reusing rather than discarding large fabric scraps measuring 5, 7, and 8 m is essential. This enables the brand to achieve a good profit margin due to lower fabric costs while considering environmental concerns. It represents a more organic production approach but one that is centered around the owner. Due to being a minimal brand, what can be produced will be within these parameters.

The brand's paper recycling can change its reuse of paper already used in sublimation. For example, this printing method transfers a design from paper to fabric using ink and heat. A fabric-cutting map needs to be created, but the company does not have a digitized CAD/CAM system, as its implementation cost is very high, and the timing is unsuitable for such an investment. All fabric-cutting maps are done manually by the owner. All patterns have their templates on cardboard-like paper to be copied onto this cutting map, where the production layout (cutting map) for the required product is made. Sublimation rolls are purchased in Brás, where several companies sell them in various weights. These are waste materials from printing that small clothing manufacturers can reuse. For recycling these papers, the fibers must not be crushed. Therefore, the owner separates them after use and deposits this waste in an eco-point provided by the municipality since there is no selective collection in the neighborhood.

DEE Strategy 2, which focuses on reducing material usage (weight reduction, volume reduction), received a rating of 2.

DEE Strategy 3, which relates to optimizing production techniques (alternative production techniques, fewer production steps, lower/cleaner energy consumption, reduced waste production, fewer/cleaner production inputs), received a rating of 3.

The distribution system, which received a rating of 2, is addressed in DEE Strategy 4, which includes packaging in lesser quantities/cleaner/reusable forms, efficient transportation, and logistics.

The brand owner avoids using plastic bags in the company as they are not environmentally friendly (it takes a long time to disintegrate). Products shipped by mail are packed in cardboard boxes and sealed with gummed tape, made of Kraft paper with a sticky substance on its surface. The owner brings her fabric bag to avoid plastic packaging when purchasing production fabrics. Products delivered to customers by air transportation use plastic for wrapping invoices and delivery slips. In manufacturing, waste materials such as thread spools and cardboard straws are donated to material recyclers. The company's website states that some deliveries can be made by bicycle. However, the entrepreneur mentioned the difficulty of implementing this service due to the high cost and longer delivery times, which can lead to customer dissatisfaction. Even though they encourage environmentally friendly practices, it is ultimately up to the customers to decide. The brand also plans to deliver with a small team of women, subject to their approval.

DEE Strategy 5 considers optimizing the distribution system (lower energy consumption, "cleaner" energy source, reduced need for inputs, cleaner inputs, rational use, and reduction of inputs during application). This item received a rating of 4. The company uses standard electricity from São Paulo. Recently, a quote was requested from a company regarding the cost of implementing solar panels for energy capture in manufacturing. Small businesses' price is very high, exacerbated by the lack of tax incentives. This investment will not bring significant returns because the family's other company has been experiencing financial difficulties for many years. Additionally, the cost is not budgeted, and there are other more urgent improvements to be made, such as replacing the sewing machinery, which has been used for over thirty years. Structural changes are also needed to improve the environmental and social comfort of the employees.

DEE Strategy 6 encompassed the optimization of the initial lifespan (reliability and durability, maintenance and repair, modular product structure, classic design, and strong product-user relationship) and was rated with a grade of three.

The average lifespan of the pieces created by Underline ranges from three to four years, as reported by customers who have been consuming the brand since its embryonic phase. Regarding the donation of items in the product's post-life stage, when it comes to panties, a gift is not encouraged because the Brazilian Health Regulatory Agency (Anvisa) even prohibits fitting intimate clothing due to the sufficient risks it poses. If trying on a panty already represents a biological risk, imagine donating a panty that has been used multiple times. Used panties still have no post-product use due to their contact with the intimate area - this part cannot be utilized for hygienic reasons. Based on the entrepreneur's research on the subject, the most common disposal method for panties ends up being organic waste disposal.

DEE Strategy 7 focuses on the end-of-life system (product reuse, remanufacturing, material recycling, and safe incineration) - this item was evaluated with a grade of four. This stage has the most significant impact on the environment.

Cotton can decompose in nature much faster than synthetic fiber. That's why using the latter is avoided, and there is a preference for natural inputs. The lingerie manufacturing process encounters many issues with threads since they are made of synthetic materials requiring more time to degrade. At Underline, only the sewing is done by a team, while one person carries out the other activities. The possibility of the company's products returning to complete the cycle is entirely unfeasible since no program can generate that within the company. This issue is prevalent for small businesses, as they lack fiscal incentives.

The lingerie manufacturing company is a small family business that seeks to fulfill its social and environmental responsibilities strictly. Therefore, it does not self-classify as a sustainable company.

5 Conclusions

The fashion industry has been widely recognized as one of the most environmentally impactful industries due to high consumption levels, large-scale production, and inadequate disposal practices. The fashion industry is responsible for approximately 10% of global greenhouse gas emissions, consumes significant amounts of water and energy, and generates substantial textile waste. In this context, adopting sustainable practices, such as circular fashion, becomes crucial to reducing the sector's environmental impact and promoting a more conscious and responsible approach to natural resources. Studies indicate that implementing circular strategies can result in waste reduction, resource savings, increased energy efficiency, and new sustainable business opportunities. Therefore, circular fashion benefits small businesses and contributes to the positive transformation of the fashion market as a whole, aiming for a more sustainable and balanced future.

The plus-size lingerie company analyzed in this study demonstrates a commitment to sustainability by seeking to minimize environmental impacts throughout its production process. The analysis of collected data revealed that the company adopts sustainable practices, such as collecting and sorting garments, donating those in good condition to institutions, and transforming others into new products. Fabric remnants are either donated or used as gifts for customers. Regarding the disposal of products at the end of their useful life, options are suggested for conscious customers, such as transforming used underwear into new products, provided that the fabric from the intimate area is discarded. In this regard, creating cleaning cloths, a car trash collector, and even storage organizers is suggested. Videos showcasing these transformations can be presented and encouraged on the brand's social media platforms. This way, customers could also receive some benefits on future purchases, as the product offered by the brand is not frivolous.

Acknowledgments. The author expresses her gratitude for the opportunity to pursue her doctoral degree in the Postgraduate Program at the School of Design of UEMG, especially to Aline Lima, the owner of the Underline brand, which was the subject of study in this research.

References

Aires, A.B.: De gorda à plus size: a produção biopolítica do corpo nas culturas do consumo – entre Brasil e EUA. Tese (Doutorado) – Escola Superior de Propaganda e Marketing, Programa de Pós-Graduação em Comunicação e Práticas de Consumo, São Paulo (2019)

Berlim, L.: Moda e Sustentabilidade [recurso eletrônico]: uma reflexão necessária. Barueri, SP: Estação das Letras e Cores (2020)

Berners-Lee, M.: Não há planeta B: um manual para evitar o fim do mundo/Mike Berners-Lee; tradução Marcelo Barbão. – São Paulo: Blucher (2020)

Calanca, D.: História Social da Moda. São Paulo: Editora, SENAC (2002)

Carvalhal, A.: Moda com propósito: manifesto pela grande virada./André Carvalhal. — 1a-ed. — São Paulo: Paralela (2016)

Gay, R.: Fome: uma autobiografia do (meu) corpo. Gay, R.: tradução Alice Klesck. 1 ed. São Paulo, Globo (2017)

Ferraz, D.: Moda sob medida: guia prático de moda para a vida real. Ferraz, D., Moraes, P.: São Paulo: Editora Senac São Paulo (2017)

Fletcher, K., Grose, L.: Moda & sustentabilidade: design para mudança. São Paulo: Editora Senac São Paulo (2011)

Wise, S., Pitt, K.: Moda circular: para una industria sostenible de La moda. Blume (2021)

Fraga, D.G.F.: O Zero Waiste frente à pragmática do consumo no setor de corte da confecção do vestuário: a falácia do aproveitamento na modelagem com foco na reducao de resíduo. Tese (Doutorado em Tecnologia Ambiental) – Universidade de Ribeirão Preto - UNAERP, Ribeirão Preto, 2020

Gurgel, A.: Comece a se amar. Bestseller (2021)
Jimenez, M.: Lute como uma gorda / Malu Jimezes. – São Paulo: Jandaira (2022)
Lieberman, D.E.: A história do corpo humano: evolução, saúde e doeca / Daniel E. Lieberman; tradução Maria Luiza X. De A. Borges – 1a Ed. – Rio de Janeiro: Zahar (2015)
Lipovetsky, G.: O Império do Efêmero: a moda e seu destino nas sociedades modernas. Tradução de Maria Lúcia Machado. São Paulo: Companhia das Letras (1989)
Lipovetsky, G.: Da leveza: rumo a uma civilização sem peso / Gilles Lipovesky; (tradução Idalina Lopes) – Barueri, SP: Manole (2016)
Maeda, J.: As leis da simplicidade: design, tecnologia, negócios, vida / John Maeda: tradução Fernando Lopes Dantas. – São Paulo: Novo Conceito Editora (2007)
Ohde, C.: Economia Circular: Um modelo que dá impulso à economia, gera empregos e protege o meio ambiente / Carlos Ohde, (organização); colaboração Helio Mattar... [et al.]. - São Paulo: Netpress Boooks (2018)
Oelbaum, B.: The visual resistance of the Venus of Willendorf Project 2005–2021. Fat Studies In Routledge Taylor and Francis Online – 2022. Available at https://doi.org/10.1080/21604851.2021.2014120© 2022 Taylor & Francis. Accessed on 4 July 2023
Vavolizza, R.: Design Sustentável para a Moda: uma abordagem sistêmica para a indústria têxtil e de confecção. – 1.ed. – Curitiba: Appris (2020)

Weaving Diversity with Critical Thinking Approach: Exploring Geopolitics of Fashion of Second-Hand Clothing and Sustainable Literacy

Bárbara Pino Ahumada[1,2](✉)

[1] IULM University, 20143 Milan, MI, Italy
barbara.pino@udp.cl
[2] Diego Portales University, 8320000 Santiago, Chile

Abstract. Over 4 trillion dollars and 4.1 trillion kilograms of second-hand clothes are exported worldwide (United Nations 2017). People use a garment an average of seven times, purchase five times more clothing than in 1980, and produce 100 trillion garments yearly, with 20% of the total remaining unsold (Thomas 2019). These figures depict an unsustainable system when considering its repercussions in terms of overproduction, environmental impact, and social consequences. Specifically, it is essential to observe and understand the Fashion System from a diverse perspective (Pierce 1903), which can shed new light on current problems. The exploration focuses on whether it is possible to construct a Fashion System that operates and establishes new relationships internally and among its actors, fostering equitable, ethical, sustainable, and horizontal relations. Analysis tools are provided to future creative talents so that, through comprehensive and systemic education, they can consider proposing solutions for an industry and ecosystem that owe a debt to the environment, ethics, labor laws, and social issues. To accomplish this, the circuit of second-hand clothing, specifically the importation of used clothing through the port of Iquique, the Free Trade Zone ZOFRI, is examined.

Keywords: Geopolitics of Fashion · Second-hand clothes · Actor Network Theory

1 Introduction

This research aims to observe and conceptualize the Fashion System as a whole, adopting a transdisciplinary, global systemic perspective that does not separate its operational and production modes from its narratives and influence on the masses as a cultural product. The Grounded Theory by Glaser and Strauss (1967) serves as a starting point, seeking a fresh perspective where "scientific truth" emerges from the act of observation. In this approach, individuals involved in a joint observation project produce an ongoing interpretation of meanings, viewing empirical reality as a result." (Suddaby 2006).

First of all, the Fashion System is analysed through the lens of popular geopolitics, examining concepts such as soft power and colonialism, which delineate the intricate network system through which the Fashion System has historically unfolded across the globe and disseminated narratives from the fashion hubs and creative centers to the rest of the world and its consumers. To observe the Fashion System as a complex assemblage, Latour's (2022) Actor-Network Theory (ANT) is adopted, recognizing the implication of an interwoven network of human and non-human actors and overlapping markets whose purpose is to sell fashion items (Entwistle in Rocamora and Smelik 2022.

In the case of second-hand clothing, the media discourse (vintage, sustainable, "*on-trend*") starkly contrasts the mechanisms, markets, and networks underpinning this pre-owned ecosystem. In other words, the positioning, establishment, and significance given to second-hand clothing differ paradoxically from the reality of what is actually happening "behind the scenes" along its production chain.

As an example and case study, the second-hand clothing system in Chile, specifically in Iquique – Alto Hospicio, is examined, along with the issue of excessive stock of such garments arriving from the Global North: more than 130,000 tons annually arrive in Chile in 2022 and 2021, 60% which end up in illegal landfills like Alto Hospicio, resulting in environmental, legal, and social repercussions. The methodology employed involves empirical observation and tracking of actors following the approach of Latour and Callon to understand and identify the links, connections, and networks formed among the actors by tracing the journey of the garments and observing "how products complete their journey, are selected, displayed, sold, and the qualities they assume throughout this process" (Entwistle in Rocamora and Smelik 2022: 393).

Finally, in the conclusions, the study explores ways to reimagine the fashion ecosystem, moving away from binary approaches such as North/South, center/periphery, wealthy/low-income countries, colonialism/decolonialism, which have dominated Fashion System analyses and supported the existing system. Instead, it advocates for alternative perspectives (systemic, interdisciplinary) capable of generating new solutions (O'Connor and McDermott 1998). By doing so, this investigation aims to articulate, investigate, and open doors to alternative forms of relationships and networks that embrace and validate the diversity and uniqueness of all actors, thus creating more equitable, horizontal, and inclusive fashion ecosystems.

2 Literature Review

The existing literature considers the Fashion System from various perspectives, with the most common ones being: fashion history (Yiping, Naldini, Riello, Soldati), sociology of fashion, or fashion as a social phenomenon (Simmel, Barthes, Ruocco, Baldini, Wilde), cultural perspective as a consumer object (Veneziani, Brooks), its representation and relationship with the media (Kawamura), identity and the appearance it gives to individuals (Bovone), gender (Butler), its connection with politics (Marquetti), or as a productive and industrial system (Thomas).

However, when thinking about the new generations of creative talents and considering Peirce's Pragmatism, if the goal is to explore new ways in which the Fashion system could unfold and adapt to the demands of sustainability, clean production, social justice, and ethics in the 21st century, it is necessary to try to understand fashion from new approaches capable of offering diverse perspectives beyond those previously executed, which have only perpetuated an unequal and dependent system (Brooks 2015).

To achieve this, it is necessary to think in new systemic and transdisciplinary terms to comprehend and uncover, beyond the visible facade of fashion and its meanings, all those connections, contradictory elements, and even paradoxes that intersect with its narrative.

Thinking of fashion as an ecosystem, a complex network of interconnected connections and subsystems within it, which deploys its networks globally, establishing a dense fabric of relationships, assemblages, and connections (Latour) that are not always considered or evident when observed separately from different angles of the social sciences and design. In this way, beyond its aesthetic visibility on runways and its narrative as a cultural product (Entwistle in Rocamora and Smelik 2022), its entire machinery and mechanism emerge, positioning it as a powerful industry with Soft Power, similar to other cultural industries. To achieve this, from a transdisciplinary perspective, geopolitics serves as a conceptual framework of reference.

2.1 Popular Geopolitics and Soft Power

Understanding geopolitics as "the study of power relationships and strategies established by different actors, primarily states, but also others, with spatial implications" (Méndez 2011: 11), it offers itself as an interpretative resource to explain how the Fashion System extends, influences, conquers, and embeds itself in diverse territories and identities, thousands of kilometers away from major fashion capitals like Milan, Paris, London, and New York.

The next item is about whether It is possible to understand the Fashion System in its geopolitical dimension as a tool of Soft Power (Nye 2021), which acts by colonizing through cultural and commercial characteristics (Aspers and Lise 2006) and the development of brands and trends (Popovic 2017). According to Joseph Nye, soft power is "the ability to influence the behavior of others and get the outcomes you want" (Nye 2021), a description that is relevant to apply to the oblique Fashion System. It influences collective behavior by dictating trends, and styles, assigning meanings and values to garments and collections, just like other cultural industries such as music or film, capable of seducing and co-opting audiences, managing to install and colonize certain discourses in territories and markets distant from dissemination centers (Atlas of Culture 2021).

Considering that information and culture are two strategies used by geopolitics since World War II (Mattelart 2002), conceptualized as "popular geopolitics," meaning "everyday geopolitical discourse in which citizens are immersed" (Dittmer and Boss 2019: 15), through mass media (Dittmer and Boss 2019), fashion, understood as the need for everyday clothing and identity construction, communicates global policies to its audiences (McFarlane and Hay 2003: 213 in Sullivan and Lee 2018), as well as other forms of popular culture. Through the discourses provided by the media, fashion manages to "generate meanings and realities for audience consumption, shaping their geopolitical imaginaries in both domestic and international realms" (Mawdsley 2008: 516 in Sullivan and Lee 2018). Everything we know about the world beyond our immediate experience comes to us through different media, which, in the words of Dittmer and Boss, even colonize our personal experiences.

Taking this information into account, one can understand that the narrative surrounding the use of second-hand clothing, which is portrayed as sustainable, fashionable, ecological, circular, "cool," and vintage for younger generations, is exported from developed centers along with bales of clothing to territories like Latin America and effectively colonizes them. However, this green ideological discourse of saving the planet and the world by using second-hand clothing completely opposes the local reality. In practice, these pre-owned garments predominantly turn into tons of waste, leading to pollution, the emergence of illegal parallel markets, a substantial carbon footprint, precarious labor conditions, and the destruction of local manufacturing companies. "Clothing production is only half the story. The other half is the enormous market for used clothing that reaches Africa, Asia, and South America, from the richest to the least developed countries, worth 4 trillion dollars and tens of billions of garments" (Brooks 2015).

The Observatory of Economic Complexity (OEC) highlights this through its figures where the largest exporters of used clothing are the United States (16.1%), China (13.7%), the United Kingdom (7.46%) - the global north – and the largest importers are Ghana (4.13%), Pakistan (4.48%), and the United Arab Emirates (3.34%) – the global south. Chile is the largest importer (3.01%) and exporter of used clothing in the region, ranking fourth globally.

2.2 Colonialism/Dependency

Colonialism was born from mercantilism, and its underlying principle was to conquer a place, preferably one with resources desired by the colonizer, and impose colonial administration, "it was possible to restrict to whom the colonized could sell their resources (thereby lowering the cost by lowering the demand for the resources) and simultaneously monopolize the supplying of markets in the colony". (Dittmer and Boss 2019: 75). This concept can equally be applied today when considering the outsourcing of textile manufacturing from centers to peripheral countries, where cheap labor constitutes the desired resources for the modern colonizer. As Andrew Brooks points out in the book "Clothing Poverty," to understand the relationship between clothing and poverty, one must first look at the manufacturing and sale of clothing and, secondly, examine how and why unequal geographic relations have emerged in the production and consumption of garments. To do this, he reviews the history of clothing and capitalism, highlighting how dynamics established during colonial times between producer centers/the Global North and peripheries/the Global South have laid the groundwork for an unequal and dependent system of relationships. "The textile and clothing sectors have played a leading role in making different regions of the world, some impoverished and others rich" (Brooks 2015: p. 104), with the Global South providing raw materials (such as cotton from Africa and India) and cheap labor for garment production, later compounded by protectionist practices perpetuated by European markets to this day, while simultaneously promoting a contradictory discourse of free trade to the rest of the world. All of this is made possible by an economic paradigm like capitalism that constantly seeks to expand and maximize profits, "a system that finds its main socially accepted goal in the pursuit of profit" (Brooks 2015: 63).

To this equation, the industrialization of emerging countries in the '50s, following decolonization in Africa and Asia, adds another factor. After the rise of mass consumption and improved living standards post-World War II in the USA and Europe, these areas presented themselves as good source of cheap labor. "Clothing was one of the first manufacturing industries to expand and increase wage levels in Europe and the USA," eventually becoming unable to compete with the cheap labor of countries like China or South Korea (Brooks 2015: 98).

3 Methodology

3.1 ANT. Actors and Networks

Given the preceding information and the complexity of the Fashion System, the approach will consider the systemic nature of the industry and utilize Bruno Latour's Actor-Network Theory (ANT) to gain a comprehensive understanding of its magnitude. In addition, from this perspective, as Joanne Entwistle points out, ANT can be applied to the field of fashion because it "presents itself as a hybrid of nature/culture" (Entwistle in Rocamora and Smelik 2022: 387), allowing for an encompassing understanding of the Fashion System in all its dimensions.

> *" Although fashion appears to us – in businesses, in our wardrobe, Etc. - already complete and finished, it represents the final realization of a complex intervention in the natural world and is entangled in a network of actors worldwide, both human and non-human (from water and chemicals used to cultivate cotton at one end of the process, to hangers and other devices for displaying clothing)." (Ibid 387).*

From this perspective, actors are not only people but also objects, artifacts, and even technologies, all of which are "involved in creating or assembling the world and shaping the way the world becomes one" (Entwistle in Rocamora and Smelik 2022: 388). The focus lies on generating relationships rather than interpreting causality. The inclusion encompasses the entire Fashion System, covering its production, consumption, design, and hybrid culture/nature practices, effectively treating it as a system. According to Latour, "Science does not observe a world of independent objects from a neutral and objective distance; instead, it assembles objects of a particular type, and these assemblages actively construct and stage the world they describe" (Entwistle in Rocamora and Smelik 2022: 385).

Therefore, the aim is to observe and follow the actors, tracking their actions, movements, and the objects they engage with in specific assemblages. "In this way of thinking, fashion is not a 'thing' but a complex ensemble of a heterogeneous range of actors. In fact, following Callon, we can think of it as a set of overlapping markets that bring together many diverse actors – human and non-human – to sell particular types of goods labeled as 'fashion' or 'trendy'" (Entwistle in Rocamora and Smelik 2022: 392); in this case study, mainly second-hand clothing associated with sustainability, a green agenda, using vintage items, and saving the planet.

4 Case

4.1 Second-Hand Clothes in Chile

In 2021, the global used clothing trade amounted to $5.17 billion. The United States was the leading exporter, with $834 million, while Ghana was the primary importer, with $214 million. These exports experienced a 31% growth between 2020 and 2021 (OEC 2021).

Considering the imported amounts of used clothing in Chile, we can observe a growing market. As indicated, and according to data from the OEC, Chile represents "the importer with the highest growth in import value." Furthermore, in 2021, Chile imported $156 million worth of used clothing, experiencing a growth of $78.7 million. Additionally, Guatemala ($56.4 million) and Malaysia ($55.3 million) followed in imports for the same year.

In the South American context, Chile is the largest importer of used clothing, accounting for 81.7% in 2021, with continuous import growth. In particular 1995, Chile imported 2.19% of the world's used clothing (which constituted 70.8% of the total in the region). More specifically, by 2021, this percentage had risen to 3.01% of the global total and 81.7% of the regional total.

Until the 1980s in Chile, used clothing consisted mainly of remnants donated by the wealthier classes to the less fortunate. This issue could occur directly to household service personnel, families with lower purchasing power, or through charitable institutions such as Rotary, Lions Club, Red Cross, Etc. This situation creates a disparity between the upper class, who followed current fashion trends, and the lower class, who wore outdated styles. The first used clothing store dates back to 1976, advertised as a means to alleviate the budget constraints of specific sectors unable to afford new garments (Montalva 2015 in Calvo and Williams 2022: 3).

Currently, the colonization of cultural and commercial characteristics is taking place (Aspers and Lise 2006), and through the development of brands and trends (Popovic 2017), Chile has embraced and adopted the discourses exported by the global north regarding the "value" of used clothing.

Wearing a second-hand branded garment from the global north advertised as "American clothing"-is perceived as a contribution to the protection of the planet (saving water, energy, fewer crops) and therefore becomes a trend among younger generations and Generation Z. Used clothing ceases to be a poorly regarded garment, someone else's discard, and becomes "circular fashion," "vintage," that is, an exclusive and hard-to-find piece rescued from the past. More specifically, a value attributed to a fashion product that, as will be seen, is far from being one.

In Chile, there are no official figures regarding the number of stores that sell this type of clothing. However, several stores adhering to this aesthetic and messaging have emerged in recent years, generating new economic circuits. These include chains of used clothing stores such as Orange Blue, Nostalgic, Bow!, as well as individual stores like Ecoropero, I love Vintage, or platforms like Ropera, which brings together 150 used clothing stores. Additionally, there is a parallel circuit of used clothing through charity shops like Coaniquem and Debra and informal markets such as street fairs, home sales, micro-enterprises, and websites (fardo.cl, mercadolibre.cl).

When analyzing how, why, and from where second-hand clothing enters Chile, it becomes evident that the network supporting this market reveals the paradox of wearing such garments concerning the attributed meaning, especially considering how this system perpetuates the logic of center/periphery, north/south divisions, wealth/poverty, clean production/pollution. Brooks emphasizes that understanding second-hand clothing necessitates considering fast fashion; the two go hand in hand. It is precisely the overproduction of low-quality fast fashion garments, with numerous seasons within a year and only five weeks for production, that generates tons of excess clothing that becomes obsolete within months, either due to its poor quality, lack of sales, and ends up in charities, resale markets, and landfills. Entwistle emphasizes that although fashion encompasses various markets such as high fashion, designer labels, fast fashion, and second-hand, which may appear "relatively disconnected," they interconnect in different ways within a circuit that can be summarized as follows:

1. High fashion from centers like Paris, London, and Milan (the global North), sold at exclusive stores at high prices, is then copied and disseminated in **2**. Fast fashion stores, and simultaneously, designers may produce collections for **3.** Large department stores at lower prices (both in the Global North and South). Moreover, the second-hand market follows fashion trends "with exclusive second-hand businesses selling vintage designer items" (Entwistle in Rocamora and Smelik 2022: 392,393).

The circuit continues as the clothing from department stores and fast fashion either deteriorates or goes out of style in less than a year and continues its journey towards **4**. Charities and NGOs in the Global North and **5.** What is not sold or donated are either exported directly to the Global South or via reselling countries like Pakistan to **6.** Countries in the Global South such as Chile. From there, it goes to **7**. Importers and distributors of used clothing in Chile, who **8**. in turn, trade with formal and informal resellers, who sell to **9.** End-users either discard the remaining clothing in illegal landfills, where it is burned or buried.

4.2 300 Hectares of Textile Waste in the Atacama Desert

In July 2022, a research trip was made to Iquique, which included visits and interviews to gather information about the circuit followed by used clothing that enters Chile and, as widely reported in the international press, ends up mainly in the illegal landfill at Paso La Mula, Alto Hospicio, with mountains of burned garments. The researcher collected information by visiting various places such as the customs office, the Free Trade Zone, used clothing importers and distributors (Zofri Importers Association), the largest illegal landfill, and speaking with people like Manuela Medina, who takes care of it, as well as regional authorities such as Sebastián Vergara, municipal councilor, Andrea León, environmental coordinator of the Municipality of Alto Hospicio, Jaime Soto, planning manager of Zofri, Zona Franca Iquique, and popular markets like La Quebradilla in Alto Hospicio. Every piece of information was recorded:

1. In 2021, Chile received 156,707 tons of used clothing (Central Bank 2022), three times more than in 2020, and it is estimated that 60%, which means about 93,000 tons end up in clandestine landfills in Atacama. In this territory, the **textile waste** occupies an estimated 300 ha, which, due to the lack of relevant legislation, ends up

being burned in the middle of the desert, causing damage to the local ecosystems. More specifically, in this respect, the Department of Environment of the Municipality of Alto Hospicio estimates that 60% of imported clothing is waste or disposable.

2. The imported used clothing enters Chile mainly through the port of Iquique (free trade zone), originating from the United States, Canada, and Europe as the primary sources (Customs 2023). In other words, the imported bales of clothing have a significant carbon footprint associated with their transportation, coming from distant origins, unlike, for example, used garments manufactured in Chile.
3. Considering the highly permissive and primarily administrative nature of **Chilean legislation**, in contrast to neighbouring countries such as Bolivia, Peru, and Argentina, which either prohibit or regulate this type of import, it becomes evident that Chile emerges as the dominant importer and exporter of used clothing in the region. Being Iquique a "free trade zone," it allows for "all kinds of goods, whether or not included in the list of prohibited imports, except for weapons or their parts and ammunition and other species that go against morals, good customs, health..." (art.7, DFL 341 of 1977, DFL 2 2001). Specifically regarding used clothing, it regulates the sanitization of these imports, but understanding it as "...any process of washing with dry or wet heat or fumigation that guarantees the hygiene of the product and the elimination of viable sanitary vectors" (Health Decree 2389, 1995). Next, they only require a sanitization certificate from the country of origin, which indicates the procedure to which it was subjected, specifying "fumigation with formaldehyde, methyl bromide, or other fumigants with similar effects" (Decree 2389). Notably, important to say that formaldehyde (NIH, 2023) associates formaldehyde with cancer, and methyl bromide (Ministry of the Environment, 2015) links methyl bromide to cancer, other illnesses, and damage to the ozone layer, which is why Chile ceased its use as a fumigant for agriculture.
4. **Importers** bring containers with an average capacity of 22 tons and an average value of $25,000. From the total received used clothing, bales of 45, 40, 23, and 20 kg are prepared, which are commercialized in different modalities, depending on the type of garment and quality. For example, it was observed in an interview with the Zofri Users Association bales shipped from Pakistan to Chile was classified under the category "articles of apparel," tariff code 63.09; however, they could have also been imported under code 63.10 (rags, cords, of textile material, waste, and unserviceable articles).
5. These bales are sold sealed to **distributor-resellers** or opened, and new bales are prepared according to the type of garment or quality (only jeans, only dresses; quality A, B, C) also for commercialization. The reseller, in turn, sells in their own business or resells to unregulated micro-entrepreneurs, such as open-air market vendors, through home sales, websites, Etc.
6. Once the garments distribute throughout this circuit, a significant number of bales remain unsold, estimated to be 300 hectares of clothing (Municipality of Alto Hospicio). Tons of garments find their way to the **De Paso La Mula landfill**, where they incinerate, and individuals with fewer resources collect some pieces for sale or personal use. The precise manner in which these tons of clothing reach the landfill is unknown, but it is believed that the same importers dispose of these excesses by transporting them in trucks.

5 Conclusions

Observing the Fashion System as a whole is necessary to assess the impact of second-hand clothing, the networks that sustain it, and its social, environmental, and commercial repercussions. The media narrative that propels it globally, with a discourse that is often far from reality, should not separate it, especially in countries in the Global South like Chile.

The second-hand clothing market and the surplus of used garments ending up in landfills are directly related to the overproduction of fast fashion. And following Brooks, one can only be understood by the other. Similarly, permissive legislations in countries like Chile, enabled through bilateral free trade agreements (e.g., with the United States), perpetuate unequal power models in which the periphery takes care of the textile waste from the Global North, damaging their ecosystems and undermining labor security through precarious and often informal market networks.

In the 21st century, commercial systems continue to reflect colonialist ideologies, where more powerful countries use smaller countries literally as textile dumping grounds, which is sustained in part thanks to the dominance of a global capitalist free market system in which profits take precedence. This system disregards the cost of cheap labor in third countries like China, with limited legal regulation to protect workers. Establishing other types of agreements, systems, and fashion circuits that are more ethical, sustainable, and equitable for all countries and stakeholders is possible. In that case, achieving it can only be done through research considering systemic perspectives, new viewpoints, and holistic global approaches. This research should provide space for participation by all stakeholders, acknowledging their local and specific realities rather than relying solely on the media narratives that drive the fashion industry.

Lastly, there are arguments for conducting research from an Actor-Network Theory (ANT) perspective, which emphasizes the inclusion and visibility of all actors in the fashion ecosystem as integral and essential parts rather than solely considering their commercial and economic weight in the global context. This knowledge will be crucial for nurturing creative talents who need to weave new networks and connections to generate an ethical and sustainable Fashion System for the entire globe. Hence, there is a need to continue researching tools and critical thinking to expand the scope of knowledge.

References

Aduanas: Bases de Datos Dinámicas de Importaciones (Dynamic Import Database) (2023). https://www.aduana.cl/base-de-datos-dinamicas-de-importaciones/aduana/2020-11-19/125330.html. Accessed 21 Jun 2023

Aspers, P., y Lise, S.: Encounters in the Global Fashion Business. Curr. Sociol. **54**(5), 802–813 (2006). https://doi.org/10.1177/0011392106066817

Pecqueur, A.: Atlante de Cultura (Atlantean Culture). ADD Editore (2021)

Banco Central: Indicadores de Comercio Exterior (2022). https://www.bcentral.cl/web/banco-central/buscador?categoria=Publicaciones/Estad%C3%ADsticas/Comercio%20exterior

Brooks, A.: Clothing Poverty: The Hidden World of Fast Fashion and Second-Hand Clothes. Zed Books (2015)

Calvo, S., Williams, G.: Ropa usada: mercado nacional y regulación en Chile y extranjero (Used clothing: domestic market and regulation in Chile and abroad). Biblioteca del Congreso Nacional de Chile/ Bcn, Asesoría Técnica Parlamentari (2022)

Cooperativa ciencia: EFE. Las Cifras Negras de la Basura Textil en Chile (2022). https://www.cooperativaciencia.cl/2023/06/29/las-cifras-negras-de-la-basura-textil-en-chile/

El Mostrador: BBC News. Disponible en (2022). https://www.elmostrador.cl/generacion-m/2022/01/26/hemos-transformado-nuestra-ciudad-en-el-basurero-del-mundo-el-inmenso-cementerio-de-ropa-usada-en-el-desierto-de-atacama/. Accessed 21 Jun 2023

Entwistle, J., Slater, D. Reassembling the cultural: fashion models, brands and the meaning of culture after ANT. J. Cult. Econ. **7**(2), 1611–1677 (2014)

Glaser, B., Strauss, A.: The Discovery of Grounded Theory: Strategies for Qualitative Research. Sociology Press, Mill Valley, CA (1967)

Latour, B. (2022). Riassemblare il Sociale. Actor – Network Theory. Meltemi Editore

Mattelart, A.: Geopolítica de la Cultura (The Geopolitics of Culture). Bogotá D.C, Colombia. Ediciones desde abajo. ISBN: 958-8093-27-9 (2002)

Méndez, R.: El Nuevo Mapa Político del Mundo (The new political map of the world). Tirant Lo Blanch, Valencia (2011)

Ministerio del Medio Ambiente: Noticias (2015). https://mma.gob.cl/ministro-del-medio-ambiente-anuncia-que-chile-pone-fin-al-uso-de-bromuro-de-metilo/. Accessed 21 Jun 2023

NIH: Formaldehyde, Report on Carcinogens. https://www.niehs.nih.gov/health/topics/agents/formaldehyde/index.cfm. Accessed 21 Jun 2023

Nye, J.: Soft power: the evolution of a concept. J. Polit. Power (2021). https://doi.org/10.1080/2158379X.2021.1879572

O'Connor, J.M.: Introducción al Pensamiento Sistémico (Introduction to Systemic Thinking). Urano press (1998)

Pierce, C.S.: The Essential Pierce: Selected Philosophical Writings, vol. 2. Indiana University Press, Bloomington (1903)

Popović, M.: The roles and practices of fashion in cultural diplomacy. In: Šešić, M.D. (ed.) En Cultural Diplomacy: Arts, Festivals and Geopolitics. Creative Europe Desk Serbia, Ministry of Culture and Media of Republic of Serbia (2017)

Rocamora, A., Smelik, A.: Pensare Attraverso la Moda: Una guida agli autori classici (Thinking Through Fashion: A guide to classical authors). Meltemi Editore (2022)

Sloan, G., Gray, C.: Why geopolitics? J. Strateg. Stud. **22**(2–3), 1–11 (1999). https://doi.org/10.1080/01402399908437751

Suddaby, R.: From the editors: what grounded theory is not. Acad. Manag. J. **49**(4), 633–642 (2006)

Sullivan, J., Lee, D.S.: Soft power runs into popular geopolitics: western media frames democratic Taiwan. Int. J. Taiwan Stud. **1**(2), 273–300 (2018). https://doi.org/10.1163/24688800-00102003

Thomas, D.: Fashionopolis, The Price of Fast Fashion and the Future of Clothes. Head of Zeus (2019)

Trajectory of the Fashion, Sustainability and Inclusion Workgroup: Advances and Achievements in Fashion for Sustainability

Suzana Barreto Martins[1](✉), Bruna Lummertz Lima[2], and Ana Mery de Carli[3]

[1] Universidade Estadual de Londrina, Campus Universitário, Londrina 86051, Brazil
barreto@uel.br

[2] Instituto Federal de Ciência e Tecnologia de Santa Catarina, Adriano Kormann. 510, Gaspar 89111, Brazil

[3] Universidade de Caxias do Sul, Francisco Getúlio Vargas, Caxias do Sul 1130, 95070, Brazil

Abstract. This article presents an analysis of the works published in the working group – GT Fashion, Sustainability and Inclusion, since its creation in 2009 until the last edition in 2022 at the Colóquio de Moda congress. For this, a Systematic Bibliographic Review (RBS) was carried out in the Annals of the Fashion Colloquium, where 99 works were surveyed, which were analyzed according to the themes presented, based on the analysis of their titles and abstracts. As a result, it was possible to show the evolution of sustainability research in Brazil.

Keywords: Fashion colloquium · Work group · Fashion · Sustainability · Inclusion

1 Introduction

Fashion and its system have undergone significant changes since the beginning of this 21st century. Not only because of the maturation of higher education courses in fashion, demonstrated by the qualification of professors, but also because of the recurrent insertion of good professionals in the market. The academy embraced fashion, books and magazines, congresses and colloquiums, masters and doctorates multiplied throughout Brazil. The recognition of Fashion as a science did not happen for free, but has its course traced in the perseverance of professors-researchers who, through the production of knowledge, promoted its emancipation.

In this context, the merit of the Fashion Colloquium in the Brazilian scenario is undeniable. The Colóquio inaugurated its activities in 2005, being created by professors and researchers Dr. Maria de Fatima Costa Mattos and Prof. Dr. Kathia Castilho Cunha in the city of Ribeirão Preto (SP). To this day, it maintains its annual periodicity, during the pandemic the events took place online. The Colloquium's policy of openness and welcoming different research groups, added to the itinerancy policy in the country, provided the involvement and commitment of professors, students and universities in an efficient and consolidated community.

Over the years, it has established itself as the largest Fashion Congress in Brazil and is among the largest in South America, and is also characterized as an international event. It has had the regular participation of institutions and researchers from several countries such as Portugal, Spain, Italy, United Kingdom, Chile, Argentina, Colombia, among others.

From 2009 onwards, a working group was created – GT Fashion, Sustainability and Inclusion. Since its conception, it has accompanied the growth of research in fashion design for sustainability in Brazil (participations in congresses, book launches, insertion of disciplines in universities) and increases in master's and doctoral studies [1].

In this sense, the creation and solidification of this WG accompanied the growth of fashion research on related topics in the country. Considering this scenario, this study proposes an analysis of publications from 2009 to 2022 to identify the GT's trajectory, the topics discussed there, published articles and the development and evolution of research in the environmental, social and economic dimensions, with emphasis on socio-ethical criteria – environmental.

2 Theoretical Foundation

2.1 The Fashion System

Fashion was organized as a system in the mid-nineteenth century, when Frederic Worth started to hold fashion shows, in seasonal releases, emphasizing novelties, favoring the aestheticization of clothing. Fashion developed throughout the 20th century, moving the economy and changing hierarchical domains. The Fashion system, as we know it today, started with the great couturiers in power and haute couture dictating the rules.

In the 1960s/1970s, the desires of urban consumers in western democratic capitalist countries began to show themselves to be more imposing in the face of the authority of the great couturier. The fashion system, as well as all other stable and authoritarian systems (school, church, family) begins to collapse with the shocks of the counterculture. There is a famous saying: "children no longer want to look like their parents". Open Fashion [2] represented by prêt-à-porter, imposes itself on haute couture and, democratically, expands consumer access. Fashion begins to segment itself, first catering to active minorities: young people, women (who become professionals), homosexuals and blacks. From then on, it enters the path of multiplicity that aims to serve markets that are increasingly differentiated socially and economically speaking.

The euphoric sense of economic progress that began in the post-war period sets up a frenetic match between young people in search of the latest fashion and fashion designers in the creation of the next. The values worshiped for fashion become novelty, democracy and multiplicity that, added to the culture of the American lifestyle, disseminated to the four corners of the cinema, promote the rise of consumption. For Lebow [3] the order to consume was so incisive that even today its echo is present. The surplus of goods produced by the effectiveness of industrial productivity needs to be consumed.

At the end of the 20th century, the individual assumes power in the fashion hierarchy. The consumer is the protagonist of the choices, he reorganizes the use of fashion. France is no longer the center of Fashion, there are many couturiers emerging in different parts of the world. And Lipovetsky [2] summarizes: the offers make up a rich patchwork of fashions in the plural. The consumer is the articulator of personal combinations, enjoying fashion like an à la carte menu. All classes are carried away by the intoxication of consumption and change, which goes beyond clothing fashion and gains the universe of objects.

However, in the beginning of the third millennium, the installed consumerism goes through sieves and other values beyond the aesthetics start to star in fashion. Ethics starts to control exaggerations, the excesses of environmental and social exploitation in the search for a more sustainable global economy. A warning signal is triggered. Many thinkers, sociologists, environmentalists, economists, philosophers, designers, politicians and even renowned fashion labels wield the flag of sustainability. Our role as research professors of fashion and design was to pay close attention to the zeitgeist, the spirit of the times, which moves the bowels of fashion.

Analyzing some critical works such as Metamorphoses of liberal culture [4] and Terra-Pátria [5] we dazzle a new phase for fashion, which we risk calling "Fashion of announced ethics." This phase is based on four evidences of the Metamorphosis in progress listed by Lipovetsky [4] plus the historical foundation of Western democracy, reviewed by Morin and Kern [5], who call for the necessary social inclusion in progress.

2.2 Announced Ethics Fashion: Five Pieces of Evidence

According to Lipovetsky [4] the first evidence refers to the preservation of the environment, where the unmeasured exploration of the planet's natural riches, typical of modern people, is replaced by awareness of environmental sustainability and preservation. Sustainable development aims to meet the needs of the present without compromising the ability of future generations to meet theirs. Morin and Kern [5] recall the manifestation of ECO 92, which took place in Rio de Janeiro, which recorded the vital need for humanity to safeguard the integrity of the Earth. According to the authors, despite the unilateral views of countries, considered in their sovereignty, there is a planetary entity to which we belong, there are properly global problems, which lead us to planetary consciousness. Thus intermittently but multiple times, the global mind develops.

The second evidence of metamorphosis is the crusade against corruption that reaches all levels. Corruption, possibly arising from the decline of basic institutions and the abandonment of absolute truths, generated a loosening of rules of conduct, allowing for an à la carte morality, which calls for limits. The control and denouncement of corruption has been taken over by the third power: the media, which in truly democratic countries, plays the primordial role for the preservation of respectability, for ethics in politics, business and education.

The third piece of evidence from the study by Lipovetsky [4] is the priority given to working conditions. The human factor brought to the center of the company. Collaborators are promoted, no longer cataloged as "human resources" or "labor", from the era of serial production and the precariousness of labor laws. Business transparency in neoliberal societies requires an administration that thinks about and improves human relations, and respect for employees.

The fourth evidence is **communication and marketing focused on morally** based values, such as solidarity, green markets, corporate citizenship, honesty, finally, institutional transparency in its acts and contracts. The concern of companies is the relationship of love or dislike that their brand can awaken in their consumers [4].

Based on this review of the ideas of contemporary thinkers about the ongoing changes in the neoliberal economy, we consider it important to base sustainability and inclusion in Fashion that companies in the textile and clothing sector fulfill their homework, respecting and practicing the evidence of Fashion from announced ethics: preservation of the environment; crusade against corruption, humane conditions at work; morally based institutional values; and fraternity as the willingness to include the disabled and excluded in fashion. Finally, consider the gospel of Morin's fraternity: "save the planet, civilize the earth, realize human unity and safeguard its diversity". This spirit of the times, zeitgeist, is certainly contained in the theory of sustainability and inclusion of our Fashion and Design researchers.

2.3 Design for Sustainability

The area of action of design for sustainability has been expanded, beyond the development of products that cause less impact, but to different ways of promoting innovations that are limited to the creation of new technologies. In this way, social innovation and systemic innovation become part of the repertoire of contemporary design. In this context, publications have built conceptual frameworks to represent this evolution of design for sustainability [6, 7]. Despite the many similarities between the different frameworks related to the first levels related to the development of products and services, what constitutes the last level is still being discussed, as well as the role of design in it. While Santos *et al.* [7] highlight the promotion of lifestyles compatible with sufficient consumption, Ceschin and Gaziulusoy [6] raise other perspectives related to design for social innovation and design for systemic innovations and transitions [8].

Although there is a diversity of dimensions of sustainability, here the most common division is adopted: environmental, social and economic, since this subdivision contributes to the direction of strategies for its effectiveness, and which are interconnected and inserted in the third and most comprehensive strategy in which you intend to achieve; the following stand out among them: 1) Upcycling, which seeks to return a product or material to the production cycle while preserving its technical qualities. It consists of giving a new and better purpose to a material or product that would otherwise be discarded without degrading its quality and composition, maintaining a quality equal to or greater than that of its original. It should be noted that upcycling differs from recycling, as the latter requires chemical processes to restore new raw materials, while upcycling maintains its original structure. It is a strategy for transforming by-products, waste, useless or unwanted products into new products, offering several benefits including environmental, social and economic benefits [9].

Here we highlight 2) the Circular Economy, which is a form of restorative and regenerative economy in principle, also encompasses the previous strategy, upcycling. Its aim is to maintain products, components and materials at their highest level of usefulness and value by distinguishing between technical and biological material cycles. This approach seeks to "decouple economic development from the consumption of finite resources and eliminate negative externalities from the economy". [10]. Also understood as a proposal for an economic model that integrates several schools and lines of thought, including Industrial Ecology, Life Cycle Engineering, Life Cycle Management and Performance Economics [12].

In turn, the Ellen MacArthur Foundation highlights the need for a systemic change in order to generate: long-term resilience; economic and business opportunities; environmental and social benefits [11].

Weetman [13] establishes an analogy between circular economy and nature, "where the waste of one species is the food of another, and the sum provides energy. The circular economy cycles through valuable materials and products, producing and transporting them using renewable energy.

Thus, when considering the life cycle of products, in the circular economy, resources are accounted for during the phases of their life cycle, from pre-production, use and post-use, when considering the useful life and optimization of their life cycle by acquiring new value chains and consequent productive efficiency and cost viability.

Finally, of extreme relevance, follows 3) Design for Sustainability, an expanded approach to Design, which also covers the strategies above; seeks to contemplate the main dimensions of sustainability (environmental, social and economic) in a balanced way, whether in the redesign of existing products, in the design of new products and services to replace current ones, in the design of intrinsically sustainable product-service systems, and even in the development of new scenarios of sustainable lifestyles. Here, design is supportive in the creation of effective solutions, contributing to the transformation of cultural profiles, industrial production and consumption in order to promote responses to real problems and effective changes [14].

Therefore, design for sustainability presupposes systemic thinking in addition to the creation of specific proposals and solutions that impact the various sectors of society.

3 Methodological Procedures

For this research, a survey (systematic bibliographic review) was carried out along with the annals of the Colóquio de Moda, between the years 2009 and 2022. For this purpose, the survey (presented in the annexes) considered the research published in the WG in the form of a complete article and expanded summary. It should be noted that scientific initiation researches that were also included in the annals and that had been presented in that period were excluded. As a database, the website of the Annals of the Fashion Colloquium [15] was selected. Table 1 shows the procedures used.

Table 1. Methodological Procedures of the Systematic Bibliographic Review.

Database	Electronic Annals of the Fashion Colloquium, accessed through a Google search
Search String	site:http://www.coloquiomoda.com.br/anais/
Keywords used	sustainability, sustainable fashion, inclusion
Inclusion criteria	Publication period: last thirteen years (between 2009 and 2022). Kind of publication: Article and expanded abstract
Exclusion criteria	Kind of publication: Scientific initiation article
Adopted filters	Filter: title and abstract analysis

Search source: Authors (2023).

From this previous survey, the works were organized in tables by year, where the titles were presented, as well as the name of the authors. Considering this organization, reading the titles and abstracts, the analysis process was carried out, as we will present below.

3.1 Data Analysis and Discussion

From the survey of 99 works (classified between WG articles and Oral Communication) and four round tables proposed during the thirteen years of WG on fashion, sustainability and inclusion in the Fashion Colloquium, the authors sought to identify the predominant themes in each year (Table 2) and draw a timeline (Fig. 1) that presents the evolution of research in Brazil during this period. Thus, based on the analysis of the titles and abstracts of the works, it was possible to identify, year by year, which were the themes most presented in the work group.

Table 2. Analysis of predominant themes.

Year	Authors	Quantity of works	Predominant themes
2009	Ross (2009) Serafim; Santos (2009) Manfredini Et. Al (2009) Schulte Et. Al (2009) Alexandre; Silveira (2009)	5	Identity, citizenship, social sustainability
2010	Schwambach (2010) Lima (2010) Martins; Daher; Pinheiro (2010)	3	Fashion, sustainability, inclusion, production

(continued)

Table 2. (*continued*)

Year	Authors	Quantity of works	Predominant themes
2011	Anicet; Bessa; Broega (2011) Carli (2011) Berlim (2011) Manfredini; Venzon; Rela(2011) Refosco Et. Al (2011) Silva; Broega (2011) Laschuk (2011) Almeida; Moura (2011) Lima; Garcia (2011) Borges Et. Al (2011) Moreli (2011)	11	Sustainable design, product sustainability, citizenship
2012	Queiroz (2012) Santos (2012)	2	Life cycle, sustainability
2013	Sampaio; Martins (2013) Carli; Peretti (2013) Brito (2013) Rodrigues (2013) Sóter (2013) Anicet; Rüthschilling (2013)	6	Life cycle, environmental impact, relationships between fashion and sustainability
2014	Rodrigues (2014) Manfredini; Venzon (2014) Navalon (2014) Ferreira; Martins (2014) Machado (2014)	5	Inclusive fashion, cultural identity
2015	Duarte (2015) Duderstadt; Santos (2015) Carli (2015) Abreu; Lima; Nobrega (2015) Maciel; Seibel (2015) Perini (2015) Manfredini; Venzon (2015) Lara; Carneiro; Fabri (2015) Martins; Paraguai (2015)	10	Sustainable actions, environmental impacts, upcycling, social innovation
2016	Perez; Santos (2016) Freire; Araujo (2016) Vallotto; Fabri (2016) Ferreira; Gomes (2016) Amaral; Vichy (2016) Nunes; Rocha (2016) Barbosa; Neira (2016) Silva; Nascimento; Ferreira (2016) Jordão; Broega; Martins (2016) Pessoa; Wanderley (2016) Chaves (2016) Aguiar; Wallace (2016) Perini (2016) Carli Et. Al (2016) Soares; Almeida (2016) Neiria (2016)	16	Teaching and learning, sustainable actions, consumption

(*continued*)

Initially, in 2009, the GT had the participation of five works, where it was possible to identify the presence of related themes, especially, sustainability in its social dimension,

Table 2. (*continued*)

Year	Authors	Quantity of works	Predominant themes
2017	Freire; Araujo (2017) Valente; Silveira; Dulci (2017) Carli (2017) Factum (2017) Camargo; Freire (2017) Freire; Araujo (2017) Perini (2017) Almeida (2017) Sampaio; Martins; Silva; Almendra (2017)	10	Crafts, waste, sustainability culture
2018	Dias (2018) Elman (2018) Pereira (2018) Aguiar (2018) Pimentel (2018) Hornburg (2018) Fernandes (2018) Massarotto (2018) Castro (2018) Castro (2018) Iamamura (2018) Anjos (2018) Puppim (2018) Franchini (2018) Ferreira (2018) Fernandes (2018) Martins (2018)	21	Handicrafts, experiments with materials and techniques, inclusion, extension
2019	Manfredini; Ross; Bressiani (2019) Freire; Pinheiro; Morais (2019) Silva; Teixeira (2019)	3	Brazilianness, culture, social innovation
2021	Costa; Broega (2021) Vieira (2021) Júnior (2021)	3	Circular fashion, inclusion
2022	Lima; Martins (2022) Faria; Barata (2022) Purpura; Mendes (2022)	3	Inclusion

Search source: Authors (2023).

extension projects of educational institutions that addressed the issue of citizenship and local identity. In the following year, in 2010, the three works addressed the culture of sustainability and the use of textile waste, linked to crafts and again to extension projects. In this sense, the works published in the first two years of the GT focused on generating income for a social group, especially for women.

In the year of 2011, the discussions proposed in the works brought Sustainable Design, addressing sustainability in products, through production techniques and product development processes that had less impact on the environment, considering the environmental dimension of sustainability and, again, citizenship, addressing, therefore, the social and environmental dimensions.

In 2012 and 2013, work began to study the life cycle of products, expanding discussions on the environmental impact of fashion product development and sustainability as a social issue in postmodern fashion design. It should be noted that here the focus of the works began to consider the fashion product as a tool to implement alternative techniques and processes, as well as, for the first time, the expansion of the life cycle of the products was considered.

Subsequently, between 2014 and 2015, the studies were able to bring the concept of inclusive fashion to the WG's discussions and again reinforce the possibilities of sustainable actions in fashion projects and the concepts of identity, especially linked to culture, social innovation and to the concept of upcycling. This year, inclusion appeared more prominently in the papers presented and, for the first time, the studies brought about the expansion of the product's life cycle, through a technique.

In the year 2016, studies began to address teaching and learning in sustainability, again bringing examples of sustainable actions in fashion projects and analyzing consumption. It is possible to point out that the works began to consider sustainability in fashion beyond the product, but contemplating consumption and teaching.

In the year of 2017 and 2018, most of the works discussed extension projects where crafts are used as an experience for using waste, experimenting with materials and techniques and enhancing inclusion. In this sense, they focused again on the social and economic dimensions, from the moment that crafts are used as a source of income and inclusion. The theme of inclusion was also reinforced by the round table, proposed by the WG coordinators that addressed the theme.

In the year of 2019, the works discussed about collaboration in projects, the Brazilianness, the appreciation of the local culture and the work of inclusion developed with immigrants from Haiti. This year, the social and cultural dimensions are now considered in the works. As a complement to the work of the WG, sustainable development was discussed in a round table.

In the year of 2020, the GT was not offered, as a result of the pandemic, but the coordinators proposed a round table entitled "Good Living" which discussed the socio-environmental problem based on the emergency issues brought about by the pandemic, which brought about the need to rethink our model of economy and civilization.

As of 2021, with the return of the WG after the covid-19 pandemic, work began to bring the concept of the circular economy, replacing the disposal processes, as well as the concept of sustainability linked to the use of more environmentally responsible techniques, but that did not consider extending the life of the product.

In the year of 2022, the works dealt with the teaching of sustainability in fashion design courses and experiences in modularity in clothing. This year, it was possible to observe that each work adopted a different theme, and that they ended up contemplating the three dimensions of sustainability. On the other hand, the WG proposed a round table that dealt with the "Bank of Textile Waste" initiative.

Therefore, it is observed that at the beginning of the WG, articles focused on social sustainability predominated, especially with an approach to handicrafts and also extension programs aimed at low-income or vulnerable women. Later, some also emerged with a focus on identity as a differential for sustainable fashion (slow fashion) and many cases were presented. As it matured, research on sustainability throughout the product's

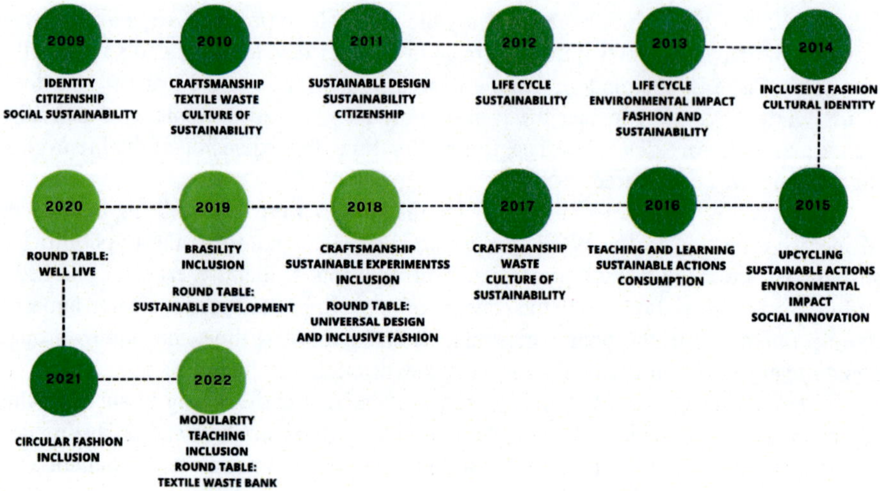

Fig. 1. Timeline. Search source: Authors (2023).

life cycle, sustainability in companies, cleaner and more recent production, replacing disposal with circular economy or reverse logistics proposals, emerged. Therefore, it is possible to affirm that there was a maturation of thinking about sustainability already in the product design, as well as signaling the emergence of the first works referring to inclusion.

4 Final Considerations

The review carried out along with the works published in the Fashion, Sustainability and Inclusion Working Group of the Colóquio de Moda allows an overview of the evolution of research on related topics in the last thirteen years, especially in Brazil.

This survey is an important instrument for mapping and categorization of research on sustainability and inclusion of fashion, especially as the research considers the aforementioned congress, which documents, through its annals, fashion research carried out in the last 18 years.

In the field of teaching, there were advances in the environmental dimension and in the third level of maturity of design for sustainability, which deals with the development of intrinsically more sustainable products, and some progress in relation to teaching approaches, with recommendations for sustainability to be treated as a cross theme. However, it is still necessary to implement advances in the social and economic dimensions of sustainability, in addition to the inclusion of new possibilities for professional performance, such as service design and design education for sustainability, with emphasis on product-service systems and the promotion of new scenarios of sufficient consumption, equity and social cohesion.

References

1. Martins, S.B., De Carli, A.M.: GT – 10 moda, sustentabilidade e inclusão. Revista de En-sino em Artes, Moda e Design **6**(3), 1–9 (2022)
2. Lipovetsky, G.: Império do efêmero. Companhia das Letras, São Paulo (1989)
3. Lebow, V.: Price competition in 1955. J. Retailing **31**(1), 1–7 (1955)
4. Lipovetsky, G.: Metamorfoses da cultura liberal. Sulina, Porto Alegre (2004)
5. Morin, E., Kern, A.B.: Terra-Pátria. Sulina, Porto Alegre (2003)
6. Ceschin, F., Gaziulusoy, I.: Evolution of design for sustainability: from product design to design for system innovations and transitions. Design Stud. **47**, 118–163 (2016)
7. Dos Santos, A., Ceschin, F., Martins, S.B., Vezzoli, C.: A design framework for enabling sustainability in the clothing sector. Latin Am. J. Manag. Sustain. Dev. **3**(1), 47–65 (2016)
8. Perez, I.U., Moura, M., Martins, S.B.: Inovação Social e decrescimento: desenvolvendo alternativas. In: Simpósio Design Sustentável 2019, Recife (2019)
9. McDonough, W., Braungart, M.: The Upcycle: Beyond Sustainability – Designing for Abundance. North Point Press, New York (2013)
10. Ellen Macarthur Foundation Homepage, Economia Circular. https://www.ellenmacarthurfoundation.org/pt/economia-circular/conceito. Last accessed 25 Jun 2021
11. Ellen Macarthur Foundation Homepage. O conceito de uma economia circular. https://archive.ellenmacarthurfoundation.org/pt/economia-circular/conceito. Last accessed 25 Apr 2023
12. Confederação Nacional da Indústria Homepage. Economia circular: oportunidades e desafios para a indústria brasileira. https://edisciplinas.usp.br/pluginfile.php/5532421/mod_resource/content/1/Economia%20Circular_CNI_2018.pdf. Last accessed 25 Apr 2023
13. Weetman, C.: Economia Circular: conceitos e estratégias para fazer negócios de forma mais inteligente, sustentável e lucrativa. Autêntica Business, São Paulo (2019)
14. Manzini, E., Vezzoli, C.: O desenvolvimento de produtos sustentáveis. Os requisitos ambientais dos produtos industriais, 1st ed. Editora da Universidade de São Paulo, São Paulo (2002)
15. Associação Brasileira de Estudos e Pesquisas em Moda Homepage. http://anais.abepem.org. Last accessed 10 Apr 2023

Sustainability in the Footwear Sector in Portugal Perceived by Consumer Habits

Gustavo A. de Morais and Ana C. Broega(✉)

Textile Engineering Department, Minho University, Guimarães, Portugal
cbroega@det.uminho.pt

Abstract. Today, the damages brought by the linear production system throughout history are apparent, and at this moment, the harm caused to the biosphere and the depletion of natural resources are perceived, as endangering the well-being and even the survival of humanity. In Portugal, the footwear sector becomes worrying due to the observed expansion of economic values and consequently the use of materials and resources, namely synthetic hides and leather. This research starts with a brief characterization of the footwear sector in Portugal, followed by a contextualization of the importance of the transition from the current production scenario to a Circular Economy. It ends with a study of consumption habits through a survey. It raises reflections on the number of products purchased per user, purchase motivations, forms of disposal, and familiarization with sustainability concepts. The collected information is analyzed under the principles of sustainability and especially the concept of the circular economy, which reveals impressive results and conclusions that can be extremely useful in developing new products and services for sustainability.

Keywords: Circular Economy · Footwear · Sustainability

1 Introduction

With a linear production system by tradition, in particular footwear and clothing companies need to adopt innovative processes to meet the demand for differentiated products and a transition to a more circular and, therefore, more sustainable system.

It is known that the rhythms of fashion lead to the manufacture of goods on an accelerated scale, which ends up causing the accumulation or premature disposal of products. This subject is accompanied by the wide use of natural resources and waste generation, contributing to the environment's degradation, which becomes even more worrying when the production numbers are analyzed.

According to the Ellen MacArthur Foundation, in 2015, the textile industry was responsible for the emission of 1.2 billion tons of CO_2 into the atmosphere. This production is based on non-renewable resources, including petroleum for the manufacture of synthetic fibers, fertilizers for cotton cultivation, and chemicals for dyeing the fibers. It is worth mentioning the use of water in the entire process, totalizing 93 billion cubic meters annually (Ellen MacArthur Foundation, 2017).

The global footwear market in 2018 was responsible for harm to the environment, accounting for 1.4% of global greenhouse gas emissions, equivalent to 700 million tons of CO2-eq (Quantis, 2018). This value is distributed within the production chain, being: extraction of raw material (20%), processing of raw material (14%), production of footwear components (43%), assembly (20%), packaging production (1%), transportation (2%). Within this process, the materials used for the manufacture of footwear are divided into synthetic (57%), leather (25%), and textiles (18%). For synthetics and textiles, the production phase represents the smaller level of impact, while for leather, the extraction and processing phases of the raw material (tanning) represent more than 50% of the climate impact (Quantis, 2018).

In 2018, 22.3 billion pairs of shoes were manufactured worldwide. Among the largest producers are China, India, and Vietnam, which represented 72.3% of the global production volume in 2017, followed by Brazil, Indonesia, Nigeria, Pakistan, Mexico, Thailand, and Italy (Abicalçados, 2019).

Furthermore, according to the Sector Report developed by the Brazilian Footwear Industry Association (Abicalçados) in 2019, the largest footwear producers are also the largest consumers. In China, domestic production is 3.5 times higher than domestic consumption, while in India, domestic production is practically equivalent to consumption. The report also highlights consumption in the United States, with 7.2 pairs of shoes per person, followed by Norway and Hong Kong, with 7.1 pairs per person. The study points out China as the largest exporter among the countries, with 8.3 billion pairs (2017), corresponding to 47.7 billion US dollars, followed by Vienna with 926 million US dollars, Indonesia with 407 million US dollars, and Germany with 281 million US dollars.

According to APICCAPS (Portuguese Association of Footwear, Components, Leather Goods and Substitute Manufacturers), by the end of 2017, the country (Portugal) had a total number of 1526 companies in operation, exporting in the same year about 83,257,409 pairs of shoes, and showing a growth of 2.9% over the previous year. This sector counts with the sale of more than 95% of its production to 152 countries in five continents. In 2018, the country exported around two billion Euros in shoes, representing around 3.4% of total national products exported. With these growing values of consumption and resources becoming more and more scarce, it becomes urgent to awaken to more conscious and responsible consumption.

The Circular Economy was born to oppose the current Linear System of production, then structured by extraction, production, and disposal. This Circular Economy has been developed over the last few years, having its bases in Regenerative Design, Cradle to Cradle, Industrial Ecology, Biomimetics, and Blue Economy. Its objectives are waste and pollution elimination from the beginning, maintenance of products and materials in use, and regeneration of natural systems. This new concept considers the existence of two cycles, the Biological, where natural raw materials are used to return to the environment through composting and anaerobic digestion, for example, and the Technical Cycle, where products and components are recovered or restored through reuse, repair, remanufacture or recycling (Ellen MacArthur Foundation).

According to the Footwear Technology Centre of Portugal, footwear companies will benefit if they choose to move away from the current linear production model and propose alternatives based on the circular model. It bets on ecodesign, an element that can guarantee the durability of the products, timeless styles, customization, reduction of chemical substances, absence of toxic products, comfort, lightness, minimization of materials, and repair, reuse, and recycling. The selection of materials is an essential factor, and it is suitable to use recycled, recyclable, biologically based, natural or renewable, ultra-light, durable, with a low environmental footprint or resulting from local production. All these processes must be covered in an ecological production, which suggests scanning, as the end of the prints on paper, for example, the production waste reduction, and use of renewable energy, among others. In this context, waste is seen as a great opportunity for raw materials and can allow different sectors' interconnection. Leather can be produced with biological origin materials, thermoplastic materials can be recycled and give rise to high-quality products, and post-consumption shoes can be recycled and transformed into new or raw materials for furniture and construction (CTCP, 2020).

Active in the transition process of this new economy, the Cradle to Cradle Institute of Product Innovation presents one of the most relevant certifications in recognition of products designed for the Circular Economy, guiding designers and manufacturers through a process of continuous improvement in the management of materials and resources. The certification is divided into five categories:

1- Material property, which seeks to ensure that products are manufactured with the safest possible chemical compounds for humans and the environment, through a process of inventory, evaluation, and optimization of the chemicals of the material;
2- Product circularity, which eliminates the concept of garbage and allows products to remain in continuous cycles of use;
3- Renewable energy and climate, which values the use of clean energy, promoting the reduction or vanishing of the climate changes caused by the greenhouse effect;
4- Water management, for the recognition of such resources, protection of springs and watersheds, and safe availability to people and other living beings; and finally
5- Social justice, which honors the environment and all those involved in the production process and product cycle, whether employees or the local community.

The certification can be granted on five levels: Basic, Bronze, Silver, Gold, and Platinum, to simulate continuous improvements. After a first analysis, the certification is renewed every two years, confirming if the improvement plans have been achieved and if leverage to another level is possible (Circular Idea, 2019). The circular economy of Planet Earth is only 9.1% (Circular Economy, 2018), i.e., numerous and rapid transformations are necessary for the search for environmental balance.

2 Methodology

As a tool for understanding the attitudes of the current footwear consumer, a survey was applied. Structured in three parts: questions of characterization of the sample, exploration of footwear consumption habits, and environmental concern of the surveyed sample. These were elaborated in digital format and applied online, and it was released by the University of Minho's contact database and the researchers' contact database. The digital format was available online for a month to answer. It is a convenience sample but directed to Portuguese citizens from the most diverse fields and occupations. The survey presented about 40 questions and required about 7 min to be answered.

3 Presentation and Discussion of Results

3.1 Sample Characterization

The results represent a sample of 167 individuals, distributed 56.3% female and 43.7% male, aged between 19 and 55 years, and an average of 30.4 years old in the female population and 34.7 years old in the male population. In terms of education, most have undergraduate or graduate degrees, 87.1% of women and 68.4% of men. Most of them are employees (52.6% of the men and 48.2% of the women) and students (18.8% of men and 10.5% of women). A significant part of the sample has professional actuation in the field of design and architecture (42,1% male and 40,0% female), in fashion (30,6% female), and in engineering (31,6% male).

3.2 Consumer Habits and Environmental Concerns

It was identified that more than half of the sample (62.4% female and 57.9% male) considers footwear an essential element in the composition of personal style, that 44.7% of the female public buys about 3 to 4 pairs of shoes per year, while 57.9% of men buy 1 to 2 pairs per year.

Regarding the shoes stored in the closet, on average, 40% of women have more than 16 pairs, and 31.6% of men have 7 to 9 pairs of shoes. The research pointed out that 38.8% of the female public almost "always plans to purchase footwear", but in turn 3.5% let themselves be carried away by "impulse". In turn 36.8% of the male sample "plans their shoe purchases", and 15.8% only buy when they "have to replace a worn out pair" of shoes.

Most of the respondents have a preference for purchasing in physical spaces, such as shopping malls.

Figure 1 presents the results related to the requirements considered indispensable in footwear. Both publics elected "comfort" as the most important element, followed by "design" and "durability" mainly for men. It is noted that flexibility in use is also a valued parameter for women, taking into account the use on several occasions.

When it comes to the footwear typology, the Sneaker appears first for both females (24.9%) and males (29.5%), followed by booties for women (15.6%) and sporty shoes for men (27.3%).

In terms of durability, most of the sample say that their footwear has a useful life of more than five years (51.8% of women and 42.1% of men). As for the minimum duration of the footwear, the sample pointed to more than one year (52.9% women, 57.9% men), with interesting answers in contrast, as footwear with a useful life of less than one month have 7.1% of women and 5.3% of men. The parts identified as those that are most rapidly damaged were the toecap and the soles.

Figures 2 and 3 show the values related to the disposal processes of footwear. The main reason why respondents discard footwear is when it is damaged and deformed.

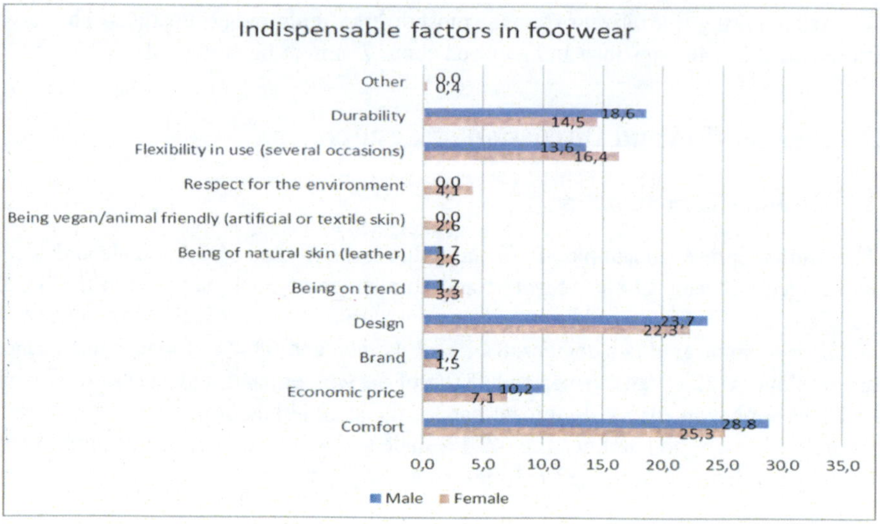

Fig. 1. Answer the question: "What are the indispensable factors in your footwear?"

Men throw disposed shoes in the trash (60%), while women not only throw them in the trash (30.6%), but also contemplate donating them to family and friends (30.6%), or even dispose it in clothing bin collectors (30.6%).

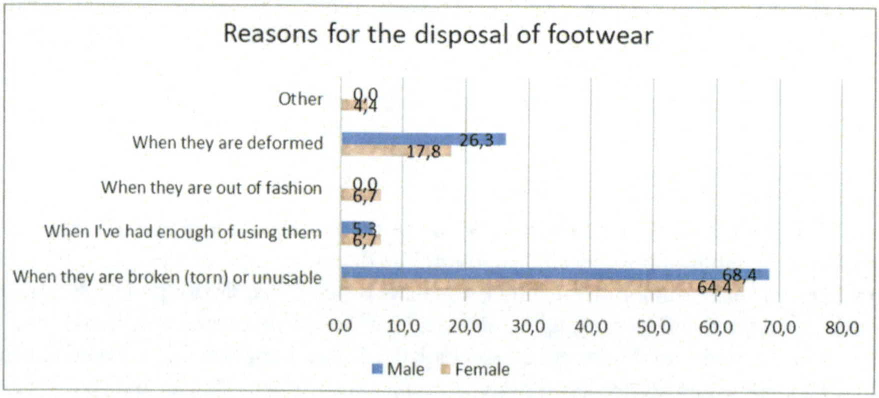

Fig. 2. Answer the question: "Usually why do you discard your shoes?"

A question about the use of second-hand shoes was presented, and the answer was divided into: 73.7% of the male sample did not accept such a condition, however, 54.1% of women accept to use of second-hand shoes.

When questioned about the materials considered more sustainable in the manufacture of shoes, textiles took the first position, followed by leather. Regarding the question: "Do you know Footwear Brands with ecological or sustainable models?" 60% of the women

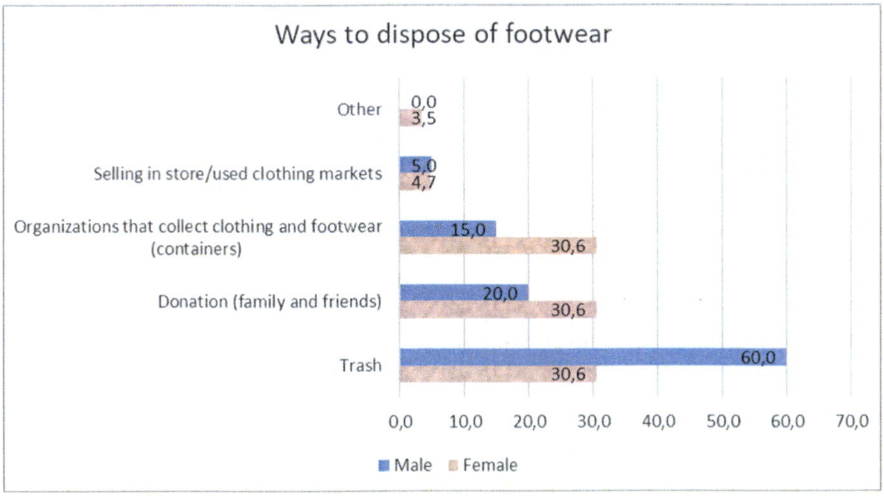

Fig. 3. Answer the question: "How do you dispose of the shoes you no longer want?"

said to know some footwear brand with these characteristics, while most men (57.9%) claim not to know ecological or sustainable brands.

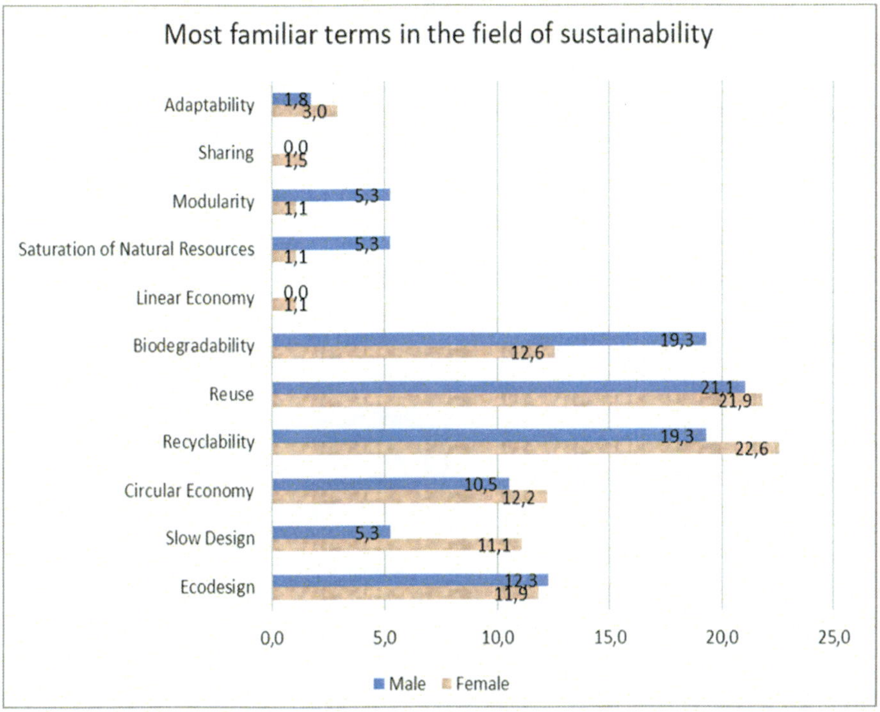

Fig. 4. Answer the question: "What terms are most familiar to you in the field of sustainability?"

To know the familiarity of the respondents with terms or concepts of sustainability, it was asked: "What are the terms which you are most familiar in the scope of sustainability? Among the most recognized answers are: "recycling", "reuse" and "biodegradability" (Fig. 4).

To allow a reflection on the respondents' perception of sustainable perspectives, statements were proposed to assess the agreement from 1 to 5, from total disagreement (1) to total agreement (5), passing through neutral value (neither agree nor disagree) with 3.

Thus in the statement of the phrase: "the footwear industry does not bring as much harm to the environment as the clothing industry", 34.1% of women agree (value 2), and 31.6% of men have no opinion (value 3).

As for the question "I need to know the origin of my product: where it is manufactured, how it is made, which raw material is used.", once again 36.5% of women agree with this need (value 2), and 31.6% of men answered neither agree nor disagree (value 3).

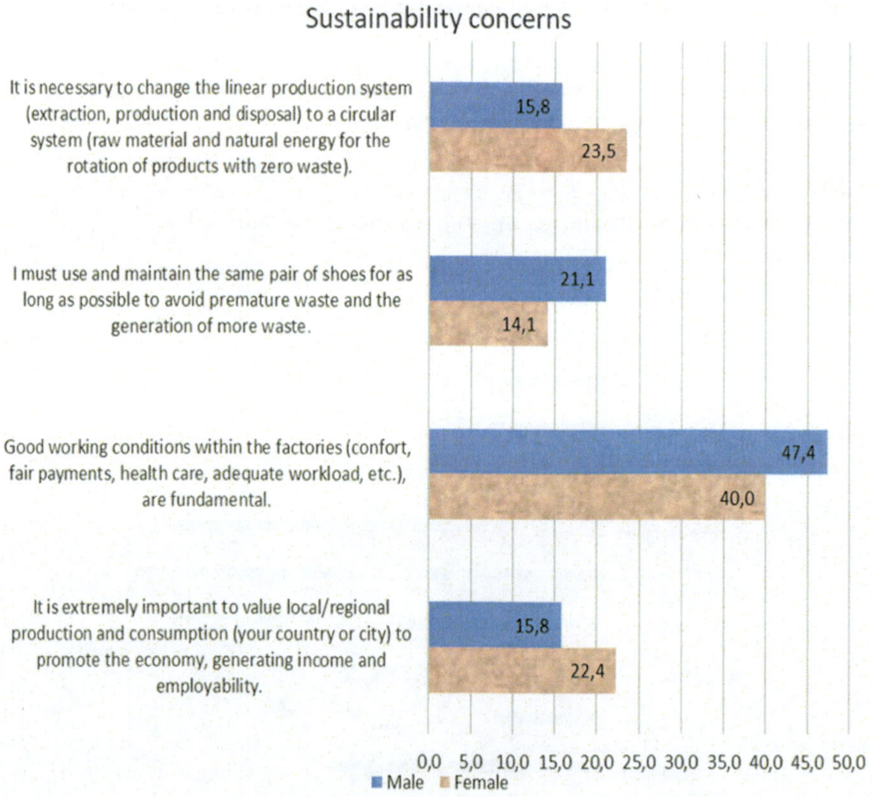

Fig. 5. Answer the question: "Which of the statements below best fits your concerns?"

Figure 5 presents the results regarding the environmental, social, and economic factors of sustainability. The main concern is the factories' "excellent working conditions" for 47.4% of men and 40.0% of women. However, 23.3% of women already show some concerns about changing the linear production system to a circular system and valuing local production and consumption to promote the local economy and employability (22.4%). Men for their part are more focused on maintaining their footwear for longer (21.1%), extending its Life Cycle, avoiding its premature disposal, and the generation of more waste.

4 Conclusions

Considering that Linear Economics is a process in transition, it is necessary to analyze sustainable perspectives since they are the ones that allow the direction toward a new circular model. This applies to any industry and especially the footwear sector, and this is reflected in the conclusions of this paper.

Footwear is considered an essential element in the composition of the personal style, both by women and men. The purchase of footwear in physical space is still privileged, perhaps due to the need for trying the product, since the "comfort" (adequacy of the feet) is still one of the most valued factors in this product, so the online purchase is not yet the preferred.

In general, besides valuing footwear design, women have a more fabulous taste for fashion issues, and maybe for that reason, despite planning their footwear purchases, they often let themselves be carried away by impulse purchases. They buy on average 2 to 4 pairs of shoes per year, and they accumulate more than 16 pairs of shoes in their closets. On the other hand, men accumulate fewer pairs (7 to 9), which can be explained by considering durability an essential factor, buying only when they are unusable.

Concerning footwear disposal, the figures are worrying because they show that more than 50% of men and 30.6% of women still dump them in the garbage. A good portion of the female public (30.6%), also usually leave their shoes in collectors (of used clothing), where organizations later collect it. It is worth remembering that the shoes deposited in the garbage end up in landfills or incinerated, bringing severe damage to the environment. Such attitudes occur because there are still no consistent policies regarding the correct disposal of clothing and footwear. In a new circular model, a possible solution to this problem would be to hold companies responsible for receiving their end-of-life products and recycling them for new products.

It appears that the majority of women surveyed know some brand of footwear with sustainable and ecological models, unlike men who are not yet awake to the cause. Such a scenario can be explained, considering the familiarization of the female public with fashion and an absolute lack of communication of brands to their values in terms of sustainability.

The terms Recycling, Reuse, and Biodegradability, about the concept of sustainability, are the most familiar to the respondents. Thus, in the context of the transition from an economic model, concepts may already have been perceived, such as Biodegradability recognition, which is an indispensable Circular Economy factor.

Regarding "concerns about the harm caused to the environment by the clothing industry and the footwear sector" and "I need to know the origin of the product: where it is manufactured, how it is made, what raw material is used," the subject seems to raise awareness among the female public, however, in terms of footwear consumption, men are the ones who buy the least.

We believe that because the female public has a more significant buying habit, and is thus directly linked to fashion, they have greater knowledge and positioning about such matters. Once again, we believe in the potential of fashion as a communication tool. Following this line of thought, there is perhaps an explanation for 54.1% of women accepting the use of second-hand shoes, and 73.7% of men not accepting these second-hand products.

In the search for the perception of sustainability concerns, both women and men have shown greater attention to fair working conditions, an essential factor of ethical fashion. It is believed that such a result arises from the fact that most of the inquirers are employees and can quickly put themselves in the position of employees of the footwear industry.

This study ends up recognizing the current footwear consumer, thus providing a vision of how it is necessary to pay greater attention to the values of circularity and its application in the face of current environmental demands.

References

Ellen MacArthur Foundation. A new textiles economy: redesigning fashion's future. https://ellenmacarthurfoundation.org/a-new-textiles-economy (2017)

Quantis. Measuring fashion: insights from the environmental impact of the global apparel and footwear industries. p. 18 (2018)

Abicalçados. Footwear Industry Report Brazil 2019. pp. 7–14 (2019)

APICCAPS. Footwear in numbers. Portugal, pp. 3–13. https://www.apiccaps.pt/

Ellen MacArthur Foundation. Circular Economy. https://www.ellenmacarthurfoundation.org/pt/economia-circular/conceito (2019)

Centro Tecnológico do Calçado de Portugal. How the footwear cluster can embrace sustainability and the circular economy. https://www.ctcp.pt/noticias/como-pode-o-cluster-do-calcado-abracar-a-sustentabilidade-e-a-economia-circular/4509.html?lang=1&newsletter=yes&utm_source=sendinblue&utm_campaign=CTCP_Notcias_Como_pode_o_Cluster_do_Calado_abraar_a_Sustentabilidade_e_a_Economia_Circular&utm_medium=email (2020)

Circular Idea. Cradle to cradle certification will win version 4.0 – know the main changes. https://www.ideiacircular.com/certificacao-cradle-to-cradle-versao-4-0/ (2019)

Circular Economy. The circularity gap report: our world is only 9% circular. https://www.circle-economy.com/news/the-circularity-gap-report-our-world-is-only-9-circular (2018)

Intimacy as a New Value. The Couture Practice of Shape for Slow Fashion

Maria Antonietta Sbordone[1](✉), Carmela Ilenia Amato[1], Martina Orlacchio[1], and Gianni Montagna[2]

[1] Faculty of Architecture, Dipartimento di Architettura e Disegno Industriale, Università degli Studi della Campania Luigi Vanvitelli, Via San Lorenzo - Abbazia di San Lorenzo, 81031 Aversa, CE, Italy
mariaantonietta.sbordone@unicampania.it
[2] Lisbon School of Architecture, CIAUD, Universidade de Lisboa, Lisbon, Portugal

Abstract. "Fashion Alive" is a winning project of Creative Europe – European Cooperation Project, Crea Cult 2021 call, whose partners are: the University of Campania Luigi Vanvitelli (Unicampania) – Department of Architecture and Industrial Design; Cremodite (Asociacion Cultural Entrepreneur for Fashion and Technology Enterprises); University do Minho (Portugal, Department of Engenharia Textil); promotes sustainable practices in the fashion and textile sectors, through the development of innovative fashion collections presented during ad hoc events. The program of exhibitions, fashion shows, conferences and workshops facilitates participation and sharing of new knowledge and discussion on issues related to sustainability in fashion and textiles. The initiative helps to raise awareness in the fashion industry and society towards circular, inclusive, and symbiotic (PNRR) production and consumption models (SDG's Goals). As a member of the Unicampania project team (of which: scientific director R. Liberti; members O. Cirillo, V. Cirillo, C. Scarpitti, M.A. Sbordone) during the courses collections of clothes were developed and prototyped, as well as critical-historical and narrative contents of the ancient trousseau tradition.

In particular, in the courses of the 3-B Fashion Laboratory, Fashion Ecodesign and in the Textile Design laboratories of the Master of Design for Innovation (prof. M.A. Sbordone, PhD students: C.I. Amato and M. Orlacchio) the Fashion Alive project was interpreted as the hub for the development of material contents (collections and textile concepts) and immaterial contents (enhancement of the ancient trousseau or trousseau) pertaining to the meaning of the creative and cultural industry, such as fashion represents and of the trousseau in the demo-ethno-anthropological sense.

Keywords: Intimacy for fashion · Ancient trousseau · Post-consumer fabric waste · Unique couture dress-concepts

1 Introduction

"Fashion Alive" is a winning project of the Creative Europe – European Cooperation Project, Crea Cult 2021 call –, whose partners are: University of Campania Luigi Vanvitelli (Unicampania) – Department of Architecture and Industrial Design; Cremodite

(Asociacion Cultural Entrepreneur for Fashion and Technology Enterprises); Università do Minho (Portugal, Departamento de Engenharia Textil), promotes sustainable practices in the fashion and textile sectors, through the development of innovative fashion collections presented during ad hoc events.

The program of exhibitions, fashion shows, conferences and workshops facilitate participation and sharing of new knowledge and discussion on issues related to sustainability in fashion and textiles.

The initiative helps to raise awareness in the fashion industry and society towards circular, inclusive, and symbiotic (PNRR) production and consumption models (SDG's Goals). As a member of the Unicampania project group (of which: scientific director R. Liberti; members O. Cirillo, V. Cirillo, C. Scarpitti, M.A. Sbordone) during the courses collections of clothes were developed and prototyped, as well as critical-historical and narrative content of the tradition of the ancient trousseau.

In particular, in the courses of the 3-B Fashion Laboratory, Fashion Ecodesign 2 and the Textile Design laboratories of the Master of Design for Innovation (prof. M.A. Sbordone, PhD students: C.I. Amato and M. Orlacchio) the Fashion Alive project is been interpreted as a hub for the development of material contents (collections and textile concepts) and immaterial ones (enhancement of the ancient trousseau or trousseau) pertaining to the meaning of the creative and cultural industry, such as fashion represents and of the trousseau in the sense of demo-ethnomethodological good[1]. Europe and individual European states have specific cultural resources and local traditions that need to be taken care of by introducing them into new circuits of meaning through creative practice. Over the years, the valorization of cultural contents has seen an increase in interest in the dimension of exchange and their economic valorization, through the "democratization" of access to many, posing the question of the differentiation between cultural and creative goods. The interest has shifted to the difference between the cultural and creative industries and for Sacco (2010) this means distinguishing: "The cultural sphere has a particularity: that of producing content that has no other purpose than to be experienced and appreciated in as such, with no further purpose: seeing a film, listening to a piece of music, reading a novel. The creative sphere, on the contrary, applies cultural contents to areas of experience in which there are other, important purposes: a design object can be very original but at the same time, if it is a chair or a laptop, it must be able to allow the user to sit comfortably or process data quickly, ergonomically, and effectively. (…) culture in general produces a relatively limited added value compared to creativity, but on the other hand much of the relevant added value produced by creativity is the result of contents and stimuli coming from the cultural sphere". Sacco proposes a classification of the industrial orientation and the relative creative content, in particular about the cultural and creative industries, the following difference is underlined: "The cultural industries, which have an industrial organization while maintaining a high density of creative, and are therefore based on the production of a potentially unlimited number of identical and wholly interchangeable copies: publishing, music, cinema, radio-television, video games; the creative industries, which maintain an industrial organization but present a

[1] https://www.soprintendenzabrescia.beniculturali.it/it/259/patrimonio-demoetnoantropologico.

relatively lower density of creative contents, in the sense that they also respond to non-cultural functional imperatives: architecture, design (including craftsmanship, fashion and, in perspective, food design), communication".

Fig. 1. Ancient trousseau from family of Campanian origin

It can be deduced that the contribution of creativity produces added value, in the case of the enhancement of common heritage goods of the populations, the creative activity, in addition to the recognition of the value, re-designs its use in the contemporary world. The knowledge and valorization of the ancient trousseau, as assets of the demo-ethnoanthropological heritage, are defined as follows: "(…) in general, material and immaterial assets are identified in this category which belongs to the traditions of European and no-European human groups and testifies the founding cultural fabric of the different communities. They are real estate, collections of movable assets and historical documents; but also, intangible assets such as ceremonies, music, legends and dialects or languages". The ancient trousseau is therefore identified as goods belonging to the populations of a given place, in this case of the Campania region, of Italy and of Europe more generally, which testify to the textile manufactures and typical craftsmanship. In this way, the courses with different teachings have oriented the students to the recovery and re-invention of the assets of their territory belonging to their families of origin and not (Eppinger, 2022). They have developed a perspective and an in-depth knowledge of textile specificities (Fig. 1), combined with ancient techniques and materials, exploring the still unexpressed possibilities that these offers in relation to the dimensions of sustainability and circularity that Fashion must face today.

2 The Project of Intimacy

The Design for Fashion project adopts a transdisciplinary approach that looks at the local dimension, people and real needs (Papanek, 2022), connecting different knowledge; from the sphere of anthropology, to that of material culture, to models of innovation, to advanced prototyping techniques, the model interprets current events and expands them. The goal of the project is to experiment with a hybrid system, using craft materials and techniques in a renewed context of use adapted to new needs, dialogue and confrontation. The goals of the project are multiple: the analysis of artisanal textile manufactures and techniques; experimentation through innovative hybrid approaches; the use of digital tools to support textile crafts; and the recovery of traditional trousseau elements for new uses. The analysis of goods belonging to the demo-ethno-anthropological category requires an in-depth study that goes beyond the sensitive data and proposes an anthropological reading of the issue of intimacy. According to, Lauren Berlant (2000, 2010) "intimacy constructs worlds, traces how people have come to identify life with intimacy and how the latter has been privatized (…)."

The economic model requires hetero-normed relationships that are divorced from conflict and are based on the concept of love and desires that are transparent, mutual and stable. The intimacy of the marital union in pre-modern societies is characterized by exclusivity; conversely, in modernity, the intertwining of the private and the public is affirmed since in lived experience it is not possible to distinguish between them; the concept of intimacy cuts across those of public and private. For Berlant, "the intimate is everywhere: you take it everywhere and it circulates everywhere. It registers as intensities of attachment and recognition, inferred and explicit, that cross people, groups and movements." Intimacy in Berlant's conception brings out the domestic and intimate world, transversal, in the era of neoliberal reorganization, to property and the public sphere, whereby events that happen daily (at work, in meetings, in travel,…) are manifestations of a real and lived intimacy. The social and economic dimension, then, of the production of the trousseau supported by the construction of the "patrimonial community", has its parallel in the anthropological dimension that definitively makes public the private substance of domestic intimacy.

3 Methodology. Slow Fashion for Active Participation

The three-year degree courses Fashion Design Laboratory III and the Fashion Eco Design II master's degree course participated in the Fashion Alive project by identifying educational and research contents aimed at experimenting with sustainable practices to be adopted in design paths in Design for Fashion.

In the Campania region, as in many other Italian and European regions, the kit has a great cultural and symbolic importance, representing a link with the tradition and the context of origin. The kit consists of a vast range of items made from weaving hemp, linen

Fig. 2. Fashion Laboraroty 3-B. First prototype detail made with craft and moulage techniques

and silk yarns, including: sheets, pillowcases, blankets, bath towels, tablecloths, decorated with great skill with hand-made lace, using traditional manufacturing techniques, such as embroidery, bobbin lace and crochet, according to the regions of origin.

The students were inspired by the sustainability objectives set by the UN SDG's (...), demonstrating through the awareness and skills of fashion designers to be able to accept challenges by responding with sustainable and innovative creative solutions to the green and digital transformation in the fashion sector. The courses are focused on the enhancement of the ancient heritage as a demo-ethnoanthropological heritage belonging to the local heritage community. The model followed is a hybrid, in which the ancient kits dialogue with the contemporary, defining advanced creative processes and textile manipulation techniques that go beyond the traditional ones.

The process consists of several steps: 1. Retrieval and acquisition of materials from typical local markets and/or from family trousseaux, with evidence collection; 2. Analysis and investigation of textile materials and techniques used, with emphasis on plant-based bio textile fibers, ancient textile techniques, hand weaving, loom work, embroidery techniques, and the use of natural dyes; 3. Design and prototyping of collections, through experimentation with hybrid textile concepts, flatwork and mannequin modeling techniques (Fig. 2 and 3); 4. Co-creative collaboration between experts, universities and students for innovative learning and practice paths; 5. Communication of demo-ethno-anthropological heritage through video storytelling, shooting and a final event capable of exciting and inspiring observers. During the creative process, workshops with experts are alternated to support and verify the challenges faced, in order to capture the intimate reflections and feelings that gradually define the concepts; and to safeguard textile memory by investigating the couturier's craftsmanship practice, which is a fundamental element for sustainable fashion. In a transdisciplinary vision, which connects different knowledge such as fashion, contemporary art, textile manipulation and advanced production processes, hybridization refers to local, endangered craft techniques with new visions of design. As Fiorani (2005) argued, "this is a plural type of design in which specializations, increasingly sophisticated and contextual, multiply, giving rise to a plurality of languages and specificity of methodologies (…)."

Fig. 3. Fashion Laboraroty 3-B. Second prototype detail made, craft and moulage techniques

4 Intimacy and the Practice of the Couturier for Unique Creations

'Talking Fashion Intimacy' is the title of the series of workshops and meetings with experts that substantiated the creative process developed by the students.

'Intimacy as a structure of feeling: a semiotic approach based on trend studies' gathers the testimony of William Cantú, professor of trend studies and contemporary art practices, at the School of Education and Social Sciences of the Polytechnic Institute of Leiria (ESECS). The workshop discusses the intersection between semiotics and contemporary practices, exploring the sociocultural aspect of intimacy. It is essential to understand sociocultural flows and relate them to fashion through the prism of semiotics on trends. Fashion, as an expression, power and reflection of a universal language using codes and symbols, plays a key role in the diffusion of each individual's identity, representing an ever-changing sociological symbol influenced by the surrounding environment (OKONKWO, 2007). A "Semiotics trends culture fashion" exercise on an interactive board followed during the workshop, thanks to the use of the Miro tool (miro.com/about). Participants worked collaboratively, creating concept maps, organizing ideas and sharing feedback in real time.

The first phase required students to divide into groups; in the second phase, they were required to identify themselves as a persona-type belonging to a specific subculture/tribe and to represent the identity through signs and images (e.g., surfer, skater, teddy boy, gen Z, cosplayer, cyberpunk, punk, hipster, emo); the third phase required each student to complete the persona-type identity, attributing clothing and accessories to them, and to justify their choices with post-it notes available on the board. In the final, concluding phase of the exercise, students exchanged reflections with the experts on the identified trends, making co-creative Fashion Design proposals.

The activity conducted had a positive impact on the students, defining the theoretical aspects related to the relationships with tangible and intangible signs that represent an identity, the intimate relationship with individual identity, and choices in relation to the sociocultural context.

'Intimacy for Slow Fashion' brings together the testimony of Gianni Montagna, professor of Fashion Design at the Faculty of Architecture, University of Lisbon (FAUL), head of the D_Tex Lab and the D_Tex: Textiles, Identity and Innovation conference. The workshop aims to enhance the common heritage of garment weaving, which represents the expression of craftsmanship in reproducing decorative codes handed down from generation to generation. The creative process of post-consumer upgrade revitalizes the elements of everyday life, which are then conveyed into the fashion system. The workshop focuses on the analysis, identification of materials and craft techniques to be used on specific fabrics, following the philosophy of slow fashion (Fig. 4).

This approach, based on respect for artisanal and artistic practices and the use of moulage, represents an initial phase of the Fashion Alive project, which aims to define the intimate and profound value generated by physical contact with materials, the memory of the fabric handed down by our ancestors, and the value of territory and cultural roots (Rissanen and Sørenssen, 2018). Handicrafts represent a form of resistance to the culture of fast consumption and disposability. Slow fashion promotes the use of natural techniques and materials in close contact with nature, creating the intimacy that requires slow rhythms and respect for the environment, thus fostering a culture of conscious consumption (Fletcher, 2007). During the workshop, students had the opportunity to immerse themselves in the experience of past wearers of the outfit, maximizing the tangible and intangible experience. The result is that despite Generation Z's predilection

Fig. 4. 2nd Workshop, TexLAB (Design LAB) "Intimacy for Slow Fashion, Moulding techniques" with Gianni Montagna

for modernity, there is a dutiful commitment to wanting to preserve those values. The dialogue with the expert was an opportunity to shape the fabric with moulage, with the challenge not to cut it. At the same time, it was necessary to enhance every single deteriorated detail and every important embroidery, highlighting them without hiding them, to create a narrative representing the processes of evolution of the material due to time. The approach taken is based on respect for craft and artistic practices, representing an important phase of the Fashion Alive project. In particular, the event aims to expand the intimate and profound dimension that develops through physical contact with materials, enhancing the memory of the fabric that represents an important cultural heritage passed down from generation to generation.

The result extends the life of the products, enhancing places, cultures and people, and allowing them to look at their territory and roots with a different and more conscious perspective (Fig. 5). Students engaged in both theoretical reflections and practical experimentation in the design and manipulation of textile materials. Students in the final year of the 3-B Fashion lab experimented with approaching the garment from flat geometries reported on the fabric, from which they proceed to moulage on a mannequin, adopting different techniques of fabric modeling. Students in the master's course Fashion Ecodesign 2, on the other hand, defined new silhouettes using ancient fabrics, creating new textile concepts through the insertion of ingenious structures to support the most fragile weaves, giving rise to reliefs, folds and so on.

Fig. 5. Master Degree Fashion Ecodesign 2 and the Textile Design Lab. Fashion Alive project final shooting

5 Conclusions

Craftsmanship knowledge constitutes a form of cultural heritage and territorial identity passed down from generation to generation. In the world of fashion, craftsmanship takes a central role in the creation of unique, high-quality garments. The use of techniques and materials that are often unavailable in mass industry makes it possible to produce clothing that lasts over time and possesses intrinsic value. Craftsmanship represents a form of resistance to the culture of fast consumption and disposability, promoting instead the use of natural techniques and materials in close contact with nature. Slow fashion thus promotes a culture of conscious consumption, based on the appreciation of unique products that require slow rhythms and respect for the environment (Fletcher, 2007; Togher & Steel, 2019; Broega & Teixeira, 2018).

The fashion project adopted focuses on the ethical sphere, with a focus on upcycling practices, in order to trigger an upgrade process that is able to elevate the value over the original, favoring the use of elements from traditional trousseaus. During the workshop, students had the opportunity to immerse themselves in the experience of someone who has worn a particular garment in the past, exploring the various aspects that characterize it. The dialogue with the expert was an opportunity to try to model the fabric with moulage, with the challenge of not cutting it (Aus, 2011). At the same time, it was necessary to enhance every single deteriorated detail and every important embroidery, highlighting them without hiding them, in order to create a narrative representing the processes of evolution of the material due to time (Williams, 2016).

The bachelor and master's courses, therefore, configure a creative ecosystem based on an approach characterized by respect for the artisan tradition and the manifestations of material culture with many qualitative aspects in artistic and productive terms. The Fashion Alive project as an innovation hub aims to define the intimate and profound value that is generated through physical contact with materials; the memory of the ancient quality textile productions of our territory and the European territories participating in the project; to the recovery of a fashion practice typical of the couturier which starts from the enhancement of post-consumer materials of first production to transform them into magnificent unique and unsurpassed fashion products.

References

Aus, R.: Trash to trend: using upcycling in fashion design [Unpublished doctoral dissertation]. Estonian Academy of Arts. https://www.digar.ee/arhiiv/en/books/15357 (2011)

Berlant, L.: Intimacy. University of Chicago, Chicago (2000)

Berlant, L.: Cruel optimism, Edizione Feltrinelli 2011. Duke University Press (2010)

Broega, A.C., Teixeira, F.: Slow fashion as a strategy for sustainability in fashion. Sustainability **10**(8), 1–15 (2018). https://doi.org/10.3390/su10082829

Eppinger, E.: Recycling technologies for enabling sustainability transitions of the fashion industry: status quo and avenues for increasing post-consumer waste recycling. Sustainability **18**(1), 114–128 (2022). https://doi.org/10.1080/15487733.2022.2027122

Fantin, V., et al.: Sustainable and circular design, production, distribution and consumption systems – Textile – Clothing – Fashion ChainPosition paper. 2020. (2021). Read in 10 May 2021

Fiorani, E.: I panorami del contemporaneo, p. 19. Lupetti, Milan (2005)

Fletcher, K.: Sustainable Fashion and Textiles, 2nd edn. Routledge, London (2014)

Fletcher, K.: Slow fashion: Estetica incontra etica. Teoria della Moda **1**, 11–22 (2007)

Global Fashion Agenda. Pulse of the Fashion Industry Report. https://www2.globalfashionagenda.com/publications/#pulseofthefashionindustry (2018). Read on 30 April 2021

Norman, R., Ramiréz, R.: From a value chain to value constellation: designing interactive strategy. Harv. Bus. Rev. **71**(4), 65–77 (1993)

Okonkwo, U.: Luxury Fashion Branding: Trends, Tactics, Techniques, Palgrave Macmillan, New York, NY (2007)

Oggetti_Correlati/Documenti/Cultura/2012/11/industrie-culturali-creative_sole24

Papanek, V.J.: Design per il mondo reale. Ecologia umana e cambiamento sociale. (a cura di) Clarke, J. A., Quinz, E. (eds.) Quodlibet (2022)

Ranzo, P., Di Roma, A., Sbordone, M.A.: Il design come mediatore dei processi di networking. MD Journal, vol. 4, Acocella Alfonso (2017)

Rissanen, T., Sørenssen, I.K.: The hand of the maker: educating for sustainability through making. Sustainability **10**(6), 1–14 (2018). https://doi.org/10.3390/su10061938

Sacco P.: Le industrie culturali e creative e l'Italia: una potenzialità inespressa su cui scommettere. Sole 24 Ore. Disponibile all'indirizzo: http://www.ilsole24ore.com/pdf2010/SoleOnLine5/_ (2010)

Sen, A.: Development as capacity expansion. In: Fukuda-Parr S., et al. Readings in Human Development. Oxford University Press, New Delhi and New York (2003)

Togher, K., Steel, C.: Crafting sustainability: L'importanza di competenze tradizionali in sustainable development. Sviluppo Mondiale **47**(3), 31–44 (2019). https://doi.org/10.1111/dech.12515

Williams, K.: Crafting sustainable fashion: a review of contemporary literature. Sustainability **8**(7), 1–19 (2016). https://doi.org/10.3390/su8070616

Zerubavel, E.: The language of time: toward a semiotics of temporality. The Sociological Quarterly': Vol. 28, No. 3. Conceptions of Temporality in Sociological Theory. Taylor and Francis (1987)

Design for Sustainability in the Footwear Sector: Survey on Strategies and Impacts Mitigation

Bruna Andrade[1(✉)], Aguinaldo dos Santos[1], and Ana C. Broega[2]

[1] Federal University of Paraná, General Carneiro St, Brazil 460, 80060-150, Brazil
{andrade.bruna,asantos}@ufpr.br
[2] Minho University, Azurém Campus, 4800-058 Guimarães, Portugal
cbroega@det.uminho.pt

Abstract. The environmental costs related to the production processes, the exposure of workers to inadequate conditions, the high competitiveness and the accelerated growth of the footwear sector, are results of the current production and consumption model, which also reverberates on the premature disposal of footwear and the lack of management of waste from its production process. Given these facts, this article seeks to understand which Design for Sustainability (DfS) approaches are being implemented in the sector. Through a literature review, the most used DfS approaches were identified and with a desktop research, a survey on examples was carried out, in order to mitigate the impacts through the concepts of Circular and Distributed Economy, models can enable greater resilience for small businesses in the sector, through local production and consumption. 16 examples of companies in the footwear sector were selected and analyzed, the study made it possible to carry out a critical analysis regarding the use of isolated strategies and a reflection on the incorporation of different DfS approaches. In addition, the study provides a wide repertoire of solutions and good practices for other designers in the sector.

Keywords: Shoes · Circular Economy · Distributed Economy · Design for Sustainability

1 Introduction

The footwear industry is responsible for several impacts on the environmental, social and economic dimensions of sustainability (UNECE 2018). The toxicity of materials and production processes; the difficulty in end-of-life solutions; the consumption of water and energy in the manufacture of raw materials and the exposure of workers to inadequate working conditions are the main impacts observed (Jacques and Guimarães 2011; Dominique Muller and Paluszek 2017; Ashton 2018; Guarienti et al. 2018).

The accelerated growth of the sector causes repercussions on the life cycle reduction of shoes, leading to the absence of solutions for managing the waste resulting from the production process and the premature disposal of products (Vezzoli et al. 2022). It is estimated that the sector annually generates 2.6 million tons of waste and that 91 million pairs of shoes that are not sold as a result of overproduction are incinerated annually worldwide (Kohan et al. 2020). The accelerated pace of this production model generates strong price competition between brands and manufacturers, resulting in significant impacts on the economic dimension (Boër et al. 2007; World Footwear 2022). In addition, there are violations of labor rights and a lack of transparency and traceability in the footwear value chain, compromising the sustainable development of the sector (UNECE 2018).

The footwear production usually occurs far from the places of sale, making it difficult for consumers to perceive the working conditions involved in the process and the exploitation of natural resources (CAVALCANTE 2020). In addition, the sector depends on a vast range of inputs and resources, which makes its value chain wide and complex (Silva et al. 2015).

Given these facts, Design for Sustainability (DfS) provides a series of strategies to think new configurations of the production and consumption model in the sector, implementing sustainability through systemic thinking (Rinaldi et al. 2022). In this way, this article seeks to contribute to the mitigation of the impacts caused by the sector through a literature review and a survey on examples, in order to identify design strategies oriented to Circular and Distributed Economy, concepts that can enable significant resilience of business through local production and consumption.

1.1 Main Impacts and Challenges for Sustainability in the Footwear Sector

Given the wide variety of models and styles of shoes on the market, the production processes are multiple (Rathinamoorthy and Kiruba 2020). A range of components made of different materials such as leather, rubber, plastics, fabrics and wood are required for its manufacture (Dominique Muller and Paluszek 2017). Footwear is a complex product and its structure comprises the junction of these materials, which are sewn or joined by adhesives so that the parts do not disconnect, this process makes reuse and recycling processes difficult (PASSOS 2014; Guarienti et al. 2018).

The life cycle of a product is made up of a set of activities and processes that consume resources and energy and generate emissions of various types throughout the pre-production, production, distribution, use and disposal phases (Vezzoli and Manzini 2008). The lifespan of a shoe is usually short and gradually shortens due to rapid changes in fashion trends and the consumer market (Morlet et al. 2019). Changing these practices is a complex challenge, which ranges from the formulation of problem-oriented public policies to consumer education (Vezzoli et al. 2014).

In the environmental dimension of sustainability, the consequences observed by the practices of footwear production chain are diverse: the Emission of Greenhouse Gases; Land use (habitat change, deforestation, urban development, agriculture, waste); Release of toxic chemicals; Depletion of water resources; Climate change (global warming); Depletion of the Ozone layer and effects on aquatic organisms due to toxicity and the formation of particles (microplastics) dispersed in rivers and oceans (Kim et al. 2022; Vezzoli et al. 2022).

In the social dimension, there are violations of human and labor rights caused by the lack of traceability and transparency in the production chain and social, racial and gender inequalities (Dominique Muller and Paluszek 2017; UNECE 2018; Fashion Revolution Foundation 2021; Vezzoli et al. 2022).

The economic dimension has the effect of barriers caused by patents and intellectual properties of large companies that affect small producers, and the great price competitiveness that leads to outsourcing (Boër et al. 2007; Santos et al. 2021; Scherer et al. 2021).

1.2 Design for Sustainability Contributions in the Footwear Sector

The most recent Design for Sustainability (DfS) approaches have more emphasis on socio-environmental issues, with strategies centered on people, collaborative, shared processes and communities (Gwilt 2020; Macêdo et al. 2022).

As described by Ceschin and Gaziulusoy (2020), the approaches are named: (i) **Emotionally Durable Design**: strengthens the user's bond and seeks to avoid premature disposal of the product; (ii) **Design for Sustainable Behavior**: makes people adopt a desired sustainable behavior; (iii) **Design for the Base of the Pyramid**: develops solutions to meet the needs of low-income people; (iv) **Design for Social Innovation**: design, development, promotion and expansion of social innovations; (v) **Systemic Design**: designs locally-based production systems where waste from a production process becomes inputs for other processes; and (vi) **Design for Sustainable Transitions**: seeks an expanded holistic view and focuses on transforming socio-technical systems through technological, social, organizational and institutional innovations.

The authors propose a theoretical model with these approaches distributed in 6 levels of innovation: (i) **Material**: interventions to gradually improve products, developing new materials or replacing them; (ii) **Product**: seeks to improve existing products or develop new ones, considering the entire life cycle; (iii) **Product-Service System**: these are integrated combinations of products, services, stakeholder value chains and business models; (vi) **Space-Social**: deals with the space-social conditions of communities, from neighborhoods to cities; (v) **Socio-Technical Systems**: design interventions that focus on promoting radical changes in the way society's needs are met; and (vi) **Socio-Technical-Ecological Systems**: focuses on systems in their entirety and envisions the field becoming Earth-centric, on all forms of life and on future generations.

According to Vezzoli et al. (2022), design intervention should not be limited only to the material level. To solve the complex problems of the product value chain, it is necessary to think about approaches and strategies located at the highest levels, as these are potentially more sustainable.

Gwilt (2020) presents a model to guide fashion designers in choosing the most appropriate strategies according to the phases of the product's life cycle, which can be multiple, represented by the term Design for X (DpX). In the pre-development phase, the model suggests **Design for empathy, low impact of materials and processes** and **for the use of mono-materials**. At the production stage, **Design for zero waste, longevity, efficient use of materials and resources, fair trade and production**. In the Distribution phase, the strategies are **Design for need, to minimize transport, reduce or reuse packaging** and to **engage local communities**. The use phase presents Design strategies for **multi-function, modularity, low-impact, customization, repair**, and **product-service systems**. In the last phase, at the end of the product's life cycle, the model recommends

Design strategies for **reuse, disassembly, recycling and upcycling, remanufacturing** and **closed-loop systems.**

According to Dwivedi et al. (2022); Fashion Revolution Foundation (2021); Koszewska (2018); Santos (2008) and Unece (2018) the most discussed strategies in the footwear sector today are: (i) **Efficient assembly and disassembly**; (ii) **Appropriate selection of materials and service providers**; (iii) **Development of DfS-oriented skills and competences of artisans**; (iv) **Consumer involvement in the processes of the footwear value chain, design and manufacturing**; (v) **Extension of the product's life cycle** and (vi) **Creation of product-service systems.**

1.3 Mitigation of Impacts on the Footwear Sector Through Circular and Distributed Economy

The Circular Economy (CE) concept is widely discussed in the footwear sector (Blume Vier et al. 2021). It is a regenerative industrial system that replaces product disposal with restoration and has as its main objective the elimination of waste through design (Ellen Macarthur Foundation 2013). Valtonen and Nikkinen (2022) believe that the industry is in transition to CE, which requires the skill of designers to develop products that can be used longer, **repaired** or **recycled**. However, the main challenges for implementing recycling processes in the footwear sector are technical barriers such as the lack of product separation technologies and mixed materials, logistics, and changes in consumer habits (MODEFICA, FGVces 2021).

Strategies such as **Design for the environment**; **modularity**; **recycling**; **reuse**; **disassembly**; **maintenance**; **product integrity**; and for the **end of life**, are examples that enable circularity in the footwear sector. Although they help to contain the negative environmental effects of the sector, they are influenced by the strong trend of downcycling, where there is a degradation in the quality of recovered materials that are normally reused in economic processes with low added value (Fletcher and Grose 2012; Christensen 2021). It is noteworthy that **reuse** and **recycling** do not prevent the production of waste and do not cause deeper changes in consumption habits. According to Fletcher and Grose (2012), it is important to consider that this model can encourage production and consumption, causing a rebound effect and increase the unsustainability of the footwear sector.

To overcome this challenge, it is necessary to balance measures that meet a holistic approach to sustainability to minimize any possible socio-environmental impact (Fletcher and Grose 2012). In this way, the **Distributed Economy** (DE) appears as an opportunity for a systemic change in the sector. This model promotes the **distribution of production** to regions where various activities are organized in small units that are synergistically and flexibly connected to each other (Johansson et al. 2005).

According to Vezzoli et al. (2018), the DE concept is locally based and can also be made open to non-local or global systems, associating **eco-efficiency** with equity and social cohesion. The proximity between end customers and manufacturing resources such as factories, workshops, **personal fabrication** labs (model where anyone can manufacture objects using accessible tools), FabLabs, Makerspaces, or mobile manufacturing units makes DE possible the use of **participatory design** and **co-design** strategies (Crul and Diehl 2006). This collaboration network can include amateurs, prosumers (consumers who become producers), producers, creative communities, and specialists from different areas (Santos et al. 2021).

The local-global potential of the distributed strategy can humanize production processes and provide an alternative to complex global supply chains. It is a possible solution for excessive consumption, characteristic of centralized mass production, through the use of **Open Design**, which promotes activities such as **Do-it-Yourself** (Armstrong et al. 2021). The model also generates new opportunities for use of digital technologies, such as **Digital Fabrication**. Enabling a fusion of digital and physical technologies, greater flexibility in manufacturing and service delivery, and knowledge sharing among local actors (Vezzoli et al. 2018).

2 Research Method

For a better understanding of the problem and how CE and DE have been explored in the footwear sector, a literature review was carried out. Publications peer reviewed from the last 5 years were selected on sustainability in the footwear sector, as well as the evolution of CE and DE concepts and strategies.

With the support of a desktop research, a survey on examples was carried out to identify and analyze cases related to the DfS approaches that were identified in the literature review. A total of 16 examples were selected and with the support of a table containing the name of the brand, project or company name, the place of origin and a brief description of the activities, it was possible to identify the most used strategies according to the literature review. The most used DfX (eg Design for multifunctional use) were identified and analyzed based on the model proposed by Gwilt (2020), the Ecodesign Guide for the Footwear Industry developed by the Life Green Shoes 4 All project (2020) and the theoretical model by Ceschin and Gaziulusoy (2020).

3 Results and Discussion

From the literature review, the DfS approaches and the DpX strategies most used in the sector were identified. From this data collection, and from the survey on examples, 16 companies that adopt CE and DE oriented strategies were selected and analyzed. The examples selected, their origin and the description of their activities and strategies, are detailed in Table 1:

Among the 16 examples detailed in Table 1, 11 adopted strategies predominantly oriented to the CE concept, while only 5 adopted the DE concept. The selected examples related to CE mostly adopt **Design for Reuse**; **Remanufacturing**; **Recycling**; **Upcycling; Emotional Connection**; **Low Impact of Materials and Processes**; **Longevity**; **Multifunction**; **Efficient use of Materials and Resources**; **Environment**; **Reduce or Reuse Packaging** and for **Sustainable Behavior**. In addition, they rely on local artisans and supplier's valorization. On the other hand, cases that are more DE-oriented adopt **Distributed Design** strategies; **Open Design**; **Co-design**; **Digital Manufacturing**; **Product-Service Systems**; **Do-it-Yourself** and **Personal Manufacturing**.

It is noted that the examples use the strategies and principles of the economic models discussed in this article in isolation, they do not adopt hybrid solutions of CE and DE.

Table 1. Examples of CE and DE strategies in the footwear sector.

Name	Origin	Description
Sujo	**Brazil**	Carry out **customization** and **personalization** of projects in **made-to-order** shoes, also customizes **second-hand** items **and reuses waste** in his creations. **Predominant model: EC**
RE49	**Italy**	The brand works with **waste reuse** from other companies and all its products carry a microship to enable **traceability** and **transparency** of its production chain and product life cycle. **Predominant model: CE**
WAO	**Italy**	The products are manufactured **locally**. The company collect the products at the end of their life and they are 97% **recycled**. **Predominant model: CE**
Marita Moreno	**Portugal**	The production is **local**, valuing the tradition in the manufacture of shoes that are only produced in **limited editions**. The products are made with **reuse** of leftovers from other productions and with **upcycling** techniques and all inputs are locally sourced. **Predominant model: CE**
Senaker Impact	**United States**	The project collects sneakers in different locations and they are sent to separate the parts that are sold to **partners** who do the **remanufacturing** process. It is a source of inputs for manufacturers looking to create products from post-consumer waste. **Predominant model: EC**
Blue View	**United States**	The brand uses **biodegradable** inputs to produce the shoes, the products are **durable** and at the end of their life cycle can be **composted** and works on design concepts for **multifunctional use**. **Predominant model: CE**
Undo for Tomorrow	**Brazil**	They carry out an **appropriate selection of materials**, and inputs are selected to obtain a higher **recycling** rate after the end of the life cycle. They are manufactured **locally** in a family-owned industry. **Predominant model: CE**
Helen Kirlum	**England**	Transforms shoes that would be discarded into new pairs. Each pair developed is unique, and can be **made to order**. Uses **reuse** of materials to manufacture parts. **Predominant model: CE**

(*continued*)

Table 1. (*continued*)

Name	Origin	Description
Pompeii Brand	**Spain**	100% of the materials used in production are **recycled** or **recyclable** and the products are **traceable**. They have full control and **transparency** over the supply and distribution chain, ensuring **good practices** in processes and respect for human rights. In addition, they **control toxic emissions** from products and develop solutions to **reduce** the impact caused by **packaging. Predominant model: CE**
ReyRey	**Denmark**	**Capsule collections** are launched only twice a year, so as not to promote excessive production and encourage **conscious consumption. Reuses** fabrics and also uses waste from the production to manufacture new products. Suppliers are selected with high criteria focused on environmental responsibility and shoes are designed for **longevity. Predominant model: CE**
Filling Pieces	**Netherlands**	Promotes **transparency** in the processes. Uses materials of natural, organic and **compostable** origin, seeks to **reduce the emission of toxic components** and publish a report with the assessment of the **life cycle** of its products, in order to **educate consumers** about conscious consumption. **Predominant model: CE**
Fctry Lab	**United States**	Promotes **distributed design** through the **sharing of resources** from an independent **technology** laboratory dedicated to the creation and prototyping of shoes. It focuses on helping young shoe designers and providing access to resources that are normally held by large corporations. **Predominant model: ED**
Brooklyn Shoe Space	**United States**	For designers and **makers**, it is a **shared space** for shoe manufacturing, offers courses to promote the traditional craft of shoemaking, and produces products for **small local** brands. **Predominant model: DE**

(*continued*)

Table 1. *(continued)*

Name	Origin	Description
Mercado Granaditas	**Mexico**	The Tepito neighborhood in Mexico City has a tradition of commercial activities, and many of its residents are dedicated to shoe manufacturing. Along the main avenue there are several stores supplying inputs for the manufacture of shoes, and ateliers offering production or **repair** services. Shoemakers share their resources for production and finished products are sold **locally** at the Granaditas Market, considered the largest shoe factory in the world due to the **collectiveness** and **sharing of resources**. **Predominant model: DE**
Sneaker Kit	**Switzerland**	The project provides **Do-it-Yourself** shoe kits for sale online with videos and tutorials for the customers. In addition, they offer several **workshops** for assembling shoes in person with the support of instructors, involving the **consumer in the process and design**. **Predominant model: DE**
The Parachute Collective	**UK**	It is a **shared space** for designers and makers to manufacture shoes, offers courses and enables **personal manufacturing**. **Predominant model: DE**

As pointed out by Vezzoli et al. (2022) in the literature review, designers should adopt an expanded holistic approach to solve the complex problems of the product value chain.

According to the Brazilian Footwear Industries Association (2022), the main practices carried out by companies in the footwear sector towards sustainability, are more centered on the environmental dimension, such as the proper disposal of waste, control of the use of restricted substances, use of ecodesign in product development and the use of renewable energy sources, which is corroborated by the examples analyzed in this article. Furthermore, Fletcher and Grose (2012) state that these strategies do not prevent the production of waste and do not cause deeper changes in consumption habits, an important DfS approach. In order to design a new configuration of the production and consumption model in the footwear sector, sustainability must be thought of from systemic thinking, using Design for Sustainability (DfS) approaches that prioritize socio-environmental issues (Gwilt, 2020; Macêdo et al., 2022 and Rinaldi et al., 2022).

According to Kohan et al. (2020), share resources is an efficient strategy for the development of sustainable footwear, thus, the life cycle concept that refers to exchanges (inputs and outputs) between the environment and the set of processes that accompany the phases of a product, might have more potential when designed for a locally based system. In this way, CE and DE can play a strategic role if thought of in a hybrid way, as both models are relevant to design new models of consumption and production in the sector. The growing demand from consumers for more sustainable fashion products and the increased discussion in the innovation agendas of several organizations (Grand View

Research, 2020; Footwear Distributors & Retailers of America, 2022), are also potential factors in this discussion.

4 Conclusion

This study contributes with reflections for mitigating the impacts caused by the footwear sector and makes it possible to think the combination of strategies that can carry out systemic changes in the production and consumption model. As pointed out in the theoretical model with the levels of the design intervention, strategies located at the sociotechnical level, where design interventions that focus on promoting radical changes in the way society's needs are met, are potentially more sustainable. Among the examples mapped for this study, it is noted that the use of DfS approaches in the footwear sector is being used in isolation and with greater emphasis on the selection of materials and reuse and recycling processes. Those strategies are located at the lowest levels of the design intervention, and has low impact. Thus, it is possible to conclude that the concept of CE has been much discussed and applied in the footwear sector, however, it can be enhanced if combined with the concept and strategies of DE. The strategies from both models, if thought of in a hybrid way and oriented to the practice of shoe design, combined with Systemic Design, can enable the promotion of greater local resilience for small businesses in the sector, and greater effectiveness in the implementation and realization of sustainable practices. Together, these models can guarantee the maintenance of production and consumption through the strength and union of communities, optimizing the flows and exchanges of resources due to their skills in local production and distribution.

The literature review allows inferring some strategies to implement such models, such as involving local suppliers; explore options for contracting local transport and distribution; form logistical consortiums with other companies in the community. Promoting locally-based sustainability and resilience in the footwear sector through a circular and distributed logic can bring socio-technical benefits to local actors, such as expanding access to infrastructure, enabling proper extraction, production, use and disposal. In addition, the data collected and categorized in the survey of examples, provide a wide repertoire of solutions and good practices for other footwear designers, and can contribute to mitigating the impacts caused by the sector if used as a reference for future projects.

From the perceptions obtained in this article, it is suggested for future studies to understand how the Systemic Design approach can promote resource flows between production units in the footwear sector in a practical way, as well as how Design for Sustainable Transitions can provide instruments in the transformation of the sector through technological, social, organizational and institutional innovations.

Acknowledgments. This work was financed in part by the Coordenação de Aperfeiçoamento de Pessoal de Nível Superior, Brasil (CAPES). Finance Code 001.

References

Armstrong, K., Bassi, E., Busnach, O.E.-B.Z, et al.: This is Distributed Design. Making a new local & Global Design Paradigm. Creative Commons, Barcelona (2021)

Ashton, E.G.: Analysis of footwear development from the design perspective: Reduction in solid waste generation. Strateg Des Res J **11**, 2–8 (2018). https://doi.org/10.4013/sdrj.2018.111.01

Blume Vier, M., Schreiber, D., Dalosto Jahno, V., Froehlich, C.: Análise Compreensiva Da Aplicação Dos Conceitos De Economia Circular Na Indústria Calçadista. Rev Gestão Sustentabilidade Ambient **10**, 159 (2021). https://doi.org/10.19177/rgsa.v10e42021159-181

Boër, C.R., Dulio, S.: Mass Customization and Footwear: Myth, Salvation or Reality? (2007)

Cavalcante, A.B.: Sustentabilidade na cadeia de suprimentos de calçados esportivos Estudo de Caso Veja Shoes. Fundação getulio vargas (2020)

Ceschin, F., Gaziulusoy, İ.: Design for Sustainability - A Multi- level Framework fromProducts to Socio- technical Systems (2020)

Christensen, T.B.: Towards a circular economy in cities: Exploring local modes of governance in the transition towards a circular economy in construction and textile recycling. J Clean Prod **305**, 127058 (2021). https://doi.org/10.1016/j.jclepro.2021.127058

Crul, M., Diehl, J.C.: Design for Sustainability. A Practical Approach for Developing Economies. UNEP Division of Technology, Industry and Environment, Paris (2006)

Muller, D., Paluszek, A.: How to do better: an exploration of better practices within the footwear industry (2017)

Dwivedi, A., Moktadir, M.A., Chiappetta Jabbour, C.J., de Carvalho, D.E.: Integrating the circular economy and industry 4.0 for sustainable development: Implications for responsible footwear production in a big data-driven world. Technol Forecast Soc Change 175 (2022). https://doi.org/10.1016/j.techfore.2021.121335

Ellen Macarthur Foundation: Circular economy introduction (2013). https://ellenmacarthurfoundation.org/publications. Accessed 26 Apr 2022

Fashion Revolution Foundation: Fashion Transparency Index (2021)

Fletcher, K., Grose, L.: Fashion & Sustainability. Laurence King Publishing, London, Design for Change (2012)

Guarienti, G.R., Pergher, B.S., Ximenes, F.S., et al.: Sustentabilidade no design de calçados: análise da comunicação de boas práticas. Des em Pesqui – Vol II Porto Alegre Marcavisual, pp. 117–138 and 163–190 (2018)

Gwilt, A.: A Practical Guide to Sustainable Fashion, 2ª. Bloomsbury, London (2020)

Jacques, J.J., Guimarães L.B.M.: Fim do Ciclo de Vida e Projeto para Desmontagem do Produto Calçado. 1–9 (2011)

Johansson, A., Kisch, P., Mirata, M.: Distributed economies – A new engine for innovation. J Clean Prod **13**, 971–979 (2005). https://doi.org/10.1016/J.JCLEPRO.2004.12.015

Kim, L., Kim, D., Kim, S.A., et al.: Are your shoes safe for the environment? – Toxicity screening of leachates from microplastic fragments of shoe soles using freshwater organisms. J Hazard Mater **421**, 126779 (2022). https://doi.org/10.1016/j.jhazmat.2021.126779

Kohan, L., Martins, C.R., Santos, H.N. dos, et al.: Brazilian Sustainability Outlook in Footwear Sector. In: Leather and Footwear Sustainability. Manufacturing, Supply Chain, and Product Level Issues. Springer, pp. 198–260 (2020)

Koszewska, M.: Circular economy - challenges for the textile and clothing industry. Autex Res J **18**, 337–347 (2018). https://doi.org/10.1515/aut-2018-0023

Life Green Shoes 4 All (2020) The Ecodesign Guide for the Footwear Industry

Macêdo, J. da S., Fernandes, L.M.A., Agra, A.C.M.G., et al.: Reaproveitamento de resíduos têxteis: aplicando upcycling na indústria calçadista. In: Ensino de design de moda: experiências, reflexões e desafios, 1ª. Bagai, pp. 60–71 (2022)

Modefica, FGVces R.: FIOS DA MODA. Perspectiva Sistêmica Para Circularidade. São Paulo (2021)

Morlet, A., Opsomer, R., Herrmann, S., et al.: A New Textiles Economy: Redesigning Fashion's Future (2019)

Passos, V.T.: Calçados artesanais e ferramentas digitais: proposta de um modelo híbrido de criação e desenvolvimento de produto para a prática do design de calçados no Brasil 153 (2014)
Rathinamoorthy, R., Kiruba, T.: Leather and Footwear Sustainability. Manufacturing, Supply Chain, and Product Level Issues. Springer (2020)
Rinaldi, F.R., Di Bernardino, C., Cram-Martos, V., Pisani, M.T.: Traceability and transparency: enhancing sustainability and circularity in garment and footwear. Sustain Sci Pract Policy **18**, 132–141 (2022). https://doi.org/10.1080/15487733.2022.2028454
Santos, A. dos, Vezzoli, C., Garcia Parra, B., et al.: Distributed Economies. Springer International Publishing (2021)
Santos, R.F. dos: Investigação do método de desenvolvimento de calçados no pólo calçadista do Vale do Rio dos Sinos e Paranhana no estado do Rio Grande do Sul (2008)
Scherer, K., Nicastro, M.L., Santos, A. dos: Product-Service System for Sustainability: Reflections From an Application on the Fashion Sector. In: VIII Sutainable Design Symposium. Curitiba, pp. 122–132 (2021)
Silva, A.H. da, Moraes, C.A.M., Modolo, R.C.E.: Avaliação Ambiental Do Setor Calçadista E a Aplicação Da Análise De Ciclo De Vida : Uma Abordagem Geral Environmental Assessment of Footwear Sector and Life Cycle Analysis Application : a General Approach **1**, 1–10 (2015)
UNECE: Traceability of Sustainable Value Chains. Enhancing transparency in the garment and footwear sector for informed and responsible choices (2018)
Valtonen, A., Nikkinen, P.: Designing Change. Aalto University, Joensuu, New Oportunities for Organisations (2022)
Vezzoli, C., Conti, G.M., Macrì, L., Motta, M.: Designing sustainable clothing systems The design for environmentally sustainable textile clothes and its Product-Service Systems. FrancoAngeli Open Access, Milano (2022)
Vezzoli, C., Kohtala, C., Srinivasan, A., et al.: Product-Service System Design for Sustainability. Greenleaf Publishing (2014)
Vezzoli, C., Kohtala, C., Srinivasan, A., et al.: Sistema Produto+Serviço Sustentável: Fundamentos. Insight, Curitiba (2018)
Vezzoli, C., Manzini, E.: Design for Environmental Sustainability. Springer, Milan (2008)
World Footwear: Business Conditions Survey (2022)

Hemp Fiber: The Textile Material as a Fashion Value

Graça Guedes[1(✉)], Rita Carvalhas[2], and Gonçalo Gomes[2]

[1] University of Minho, Guimarães, Portugal
mgg@det.uminho.pt
[2] University of Aveiro, Aveiro, Portugal

Abstract. The focus on sustainable textile materials opens a vital approach to industrial hemp. Hemp significantly contributes to the environment and local economies since it grows on an extensive range of latitudes, requires little water, is almost pest-free, crops benefit the soils, and requires low labor intensity. Fashion brands such as Patagonia, Lee, and Wrangler, among others, are adopting hemp to fulfill their quest for sustainability.

European Union incentivizes textile industries to use new materials and recommends hemp, of its sustainable characteristics, to support the new circular economy. In European textile industries, the Portuguese, among others, produce fabrics and knighted fabrics 100% hemp but mostly in mixtures of cotton/hemp for clothing and home textiles. The products can be labeled sustainable on all dimensions (ecological, economic, and social), durable, comfortable, and rated as high quality/high price products.

The research reviewed the related literature to refocus the problem on the current market conditions and presents how fashion brands, from global ones, such as Patagonia, and new brands, such as the Portuguese Sensihemp, reinforce their sustainability strategies through hemp. The foreseen perspective is that hemp will conquer a strong market share among natural textile materials at a very high pace, following the continuous augmentation of the amounts of this fiber that will arrive in the market from USA and China. Prices will tend to decrease progressively as the various countries, particularly in Europe, adhere to the comparative advantages that the use of hemp presents, and the textile production chains adapt to this fibre production.

Keywords: Industrial Hemp · Natural Sustainable Materials · Sustainable Fashion Brands

1 Introduction

Natural raw materials of plant and animal origin accompanied the evolution of human societies that, in clothing, used those that they had locally or to which they had access thanks to trade with more or less distant regions. The most refined fabrics were imported until the XV century by Italian, Flanders and English traders. Once the Portuguese opened the sea route to Asia, textile materials and sophisticated fabrics entered Europe abundantly. Linen fabrics and hemp were the least imported, as they had been produced throughout the territory since the pre-Roman period. In Portugal, the rural domestic economy was linked with linen, with bragal being the most produced fabric and burlap and scarf the finest fabrics. Due to its fibre length, hemp was mainly used to make sails and ropes (Sequeira and Melo, 2012).

Historical documentation of flax and hemp use dates back to six millennia B.C. (Gulati & Turner, 1929). Cotton is also known and used for over 7000 years, according to the National Cotton Council of America. In Pakistan and Egypt, it was spun and woven into clothing as far back as 3000 BC. Arab merchants would have brought cotton to Europe in 800 AD, and as Columbo discovered America, cotton plantations were established in the Bahamas. In the 16th century, cotton became known worldwide and started to be produced in the new North American colonies of Florida and Virginia. Cotton was used industrially in England from 1740 onwards, and England maintained world leadership as a cotton fabric producer until the 20th century, with the decline definitively occurring in the '70s (Tausif et al. 2019). Over the years, cotton became the most used textile fibre in industry and applied in a wide variety of products, from clothing, household and industrial products.

The production systems, both in the first and second sectors, were systematically oriented towards textile products based on cotton. Demand spread worldwide, with low wages countries and newly industrialised countries investing in cotton crops and, afterwards, in cotton-based textile industries.

The historical dependence on natural raw materials in general and cotton in particular in recent centuries is a factor that is central to how current consumers maintain their preference for natural fibres in clothing.

The negative impact on the environment of raw materials and industrial production processes became evident in the second half of the 20th century. In the last decades of this century, the use of organic cotton became a symbol of concern for the environment and human health, as the use of toxic pesticides affected not only the environment (and naturally those who lived in it) but also those who wore clothing produced with harmful chemicals. Organic cotton entered the market as a solution, and, at first, the products were aimed mainly at sensitive market niches, such as babies and susceptible skin people. An example of this movement is the Portuguese brand Natura Pura which finds an opportunity in this demand and launches organic cotton products without any dyes, in which the decorative elements motifs (reduced concerning the size of the pieces) were obtained using cotton threads in various natural tones of brown and green. In the 1990s, the rising interest in organic-coloured cotton led to the search to recover the plant's natural tones. Currently, six shades can be found: cream, beige, medium brown, reddish brown, chocolate brown and lilac. Between 1995 and 1996, organic cotton fibres reached values 65% higher than conventional ones, as prices reflected not only demand (higher than offer) but also the new consumers' values focused on the request for fair trade and environmentally clean materials. Consumers' movements begin to demand the reduction of the use of pesticides and fertilisers and the manual picking of cotton (Sigaard and Laitala, 2023).

The widespread use of cotton and consumer preference for natural fibres had led the market to make cotton a raw material used in products of all types. It made a high number of economies dependent on this fibre. Currently, this dependency becomes a significant barrier to change, even when it is evident that the overproduction of cotton, whether grown organically or not, has a high environmental footprint that makes it unsustainable.

On the other hand, industrial hemp had its' production forbidden for being of the same species as Cannabis, which produces THC and has become a controlled substance. As the sustainability paradigm raised in importance and consumers demanded environmentally friendly materials and responsible production systems, the search for new materials led to the rediscovery of the traditional natural raw material such as ramie, flax, jute, hemp, and abaca. Investments in new manmade fibres brought to market a new range of sustainable materials such as fibres of wood, bamboo, coconut, pineapple, seaweed, and other cellulosic and synthetic origin, such as Tencel, Lyocell, polyester, or Mylo, among many more (Harsanto et al., 2023).

2 Textile Fibers Market

Textile and fashion markets depend on materials, their quality, durability, and cost. In fashion and home textile products used in direct contact with the body, the comfort of natural fibers is a relevant value to consumers. The production of natural fibers is also an essential issue since millions of producers depend on them and their national economies (Sigaard, A.S. and Laitala, K., 2023).

The research in this field concluded that hemp presented interesting known characteristics, and several brands have begun using hemp in their products. Results showed this material has an excellent performance in cloth, shoes, bags, and textiles in general concerning durability, resistance, impermeability (using more dense fabrics in bags and shoes), presenting other rather interesting characteristics such as bactericidal, high levels of humidity absorption (keeping the body dry), easy to dye with organic dyes, and with mechanical and thermal high level of comfort in both forms, fabrics and knitted fabrics (Farag & Kayser, 2017).

Fashion brands introduced hemp in several ranges of products, mainly in cotton/hemp mixtures. From men's shirts to t-shirts and all kinds of cloth items and accessories, as well as home textiles, hemp is growing in importance. Jeans wear brands are particularly sensitive to the environmental impact of their industry, and hemp is the sustainable material soon for global brands such as Wrangler and Lee. Patagonia, the historic sustainable brand, also focuses on hemp but uses it in a much more extensive range of products.

Textile hemp fiber is gaining importance worldwide because the correct selection of textile materials opens a vital approach to environmental and social sustainability issues. On the one hand, sustainability requires fewer transport components, affecting from raw and intermediate materials to the final disposal of products at the end of life. Textile and fashion industries are particularly susceptible to costs. The impact on the environment of transport is also very high, with raw materials and final products flowing from low-cost wages to more developed countries and discarded textile products and garbage being sent from developed countries to poor economies by sea over thousands of miles (European Parliament, 2022).

Developing an industrial system that keeps most production factors close to final markets, from raw materials to consumer products, has considerable advantages. Manufacturing industries could obtain a more significant quantity of textile material if local production existed. Many European countries could produce hemp while reducing their dependency on raw materials. More important, research on hemp characteristics showed it presents some less-known advantages (Rehman et al. 2021). Considering the uses of hemp, it is possible to use it in the production of fibers for the textile industry, paper, and the lesser known, including production to construction (blocks and insulation panels are the main use of hemp in France), food industry where hemp seeds are a superfood, cosmetic industry and others such as polymers industry for hemp-filled polymers (Deshmukh, 2022).

Many materials, like animal beds, biomass, or energy, integrate hemp plants into their composition. Hemp acts positively and directly on the environment improving soil health, reducing soil contamination on heavy metals, and converting atmospheric CO2 into biomass (Donaldson, 2022).

Compared with cotton, hemp is wholly used from roots to seeds, does not require pesticides or chemical fertilizers, absorbs CO2 from the atmosphere, improves the quality of soils, and cleans them from polluting materials. Conversely, cotton offers fibers only from their capsules. Also, it is possible to obtain biomass from the plant when buried. However, cotton requires a high intensity of pesticides and fertilizers. Another indication against the use of cotton is its contribution to the significant water stress felt the in the regions where cotton production is intensive. Hemp requires ten times less water compared with cotton. Over more, while cotton grows on annual crops, hemp for industrial uses requires 70–120 days from seeding to harvest, according to the seeds variety (Fike, 2016).

Flax, with a presence among human societies as old as hemp, is closer to cotton in its characteristics since it attracts many pests. No bio flax is on the market because pesticides are necessary to produce them. It also requires fertilizers. The crops are annual, and the parts used are the fibers to make linen and the seeds used by the food industry for animal and human consumption. The advantage of flax over cotton is that it requires much less water. Cotton needs about 10000 L of water to produce one kilogram of fiber (Averink, 2015). Flax does not need much water (about one inch of water per week that can correspond to natural rainfall). It also grows in more extensive latitudes and is produced from India to Europe.

Until now, the most used textile plant fibers remain cotton, followed by flax. Cotton, flax, and hemp fibers present different compositions and sources of plant extraction. Table 1 exposes the characteristics of those fibers, considering their interest in clothing production.

Hemp stands out for some benefits, including long-lasting fiber resistance and its ability to resist insects, moths, mold, and rot. Given its characteristics, it is very suitable for clothes that need to be worn regularly, such as jeans, tracksuits, and professional attire, but also fashion items for consumers more concerned with the environment, that choose less but better fashion items in their closets.

Table 1. Main characteristics of cotton, flax and hemp for clothing production

Characteristics	Cotton	Linen	Hemp
Stregth	Fair to good	Very strong	Very strong
Elasticity	Very weak	Weak	Weak
Resistance	Low	High	Very high
Wrinkle	Very easily	Very easily	Easily
Comfort	High	High	High
Touch	Soft	Relatively soft, softer when washed	Soft and softer with the washes
Shine	When mercerized	Little	When finished
Absorption	Good	Very good	Good
Bleaching	Yes	Yes	Yes
Heat Conduction	Good	Good, feeling of freshness	Good
Damage by perspiration and bleach	No	Yes	No
Damage by insects, mold, rot and moths	Yes	Yes	No

Source: https://www.coats.com/en/information-hub/know-about-textile-fibres.

3 Evolution of Hemp Cultivation

The literature indicates that hemp was cultivated in Europe between the 7th-4th millennium B.C. in Romania. The first cultural and linguistic pieces of evidence of the use of hemp to produce textile fibers from the stem in Central Asia goes back to 2500 BC, although it is probable that hemp culture is older than 10000 years in that region (Mercuri, Accorsi and Mazzanti, 2002). The United Nations experts state that there are three botanical varieties of hemp Sativa, the regular hemp; Indica, the Indian hemp; and rude rails, the wild hemp (Allegret et al., 2013).

In Portugal, the hemp textile fiber appears closely linked to the flax fiber. The statistical data from ATP – Associação Têxtil e do Vestuário de Portugal (Textile and Clothing Association of Portugal) presents this close interrelationship because these two textile fibers share several characteristics: they are both natural fibers of plant origin, and the extraction can use similar processes (Sendin, 2015).

The literature review highlighted that the production of flax was spread all over Portugal and had a long tradition, observable since the pre-Roman period. It also states that the import needs of flax and hemp were minor during the 15th century (Garcia, s/d). Another reference to hemp textile fiber is present in Castro (1964, pp. 335–336) "(…) Hemp, a more resistant material, had applications mainly in naval articles, namely in the sails of ships (the so-called tréu cloth) and ropes".

Hemp was widely used around the globe until the 30s of the 20th century in an endless variety of textile products, such as ropes, rugs, and sails, among others. They point out the versatility of this plant, from which stand out the adaptation to different climatic conditions (temperate and tropical zones), given its compliance with low water levels; the plant benefits the soils where they grow, and fundamentally the resistance to pests was the reason for the rapid growth of its cultivation, spread all over the planet (Singh and Sardesai, 2016).

According to Batista and Santos (1999), hemp production gained prominence in Portugal, mainly in Torre de Moncorvo and Ribatejo. Until the 70s of the 20th century, the Spinning and Fabrics of Torres Novas' Factory spun the fibers. All the hemp waste was sent to the Matrena Paper Factory.

One of the reasons for the drastic decline in hemp cultivation throughout Europe and North America, and in 1985 in China, was its association with marijuana. The laws against cannabis consumption and regulating its medicinal production discourage industrial hemp research. Only at the end of the 20th century did hemp begin being largely studied (Meijer, 1995). The analysis allowed the economy and society to rediscover hemp as a promising solution for the world economy and the environment. Many different industries, like food, pharmaceutical, textile, paper, construction, energy, and others, as well as agriculture itself, can benefit from it. Karche and Singh (2019) predict that the global market for products made from hemp could increase several times in the coming years. If so, hemp will become the leading symbol for sustainable agriculture and a most promising raw material.

China remains the largest producer of hemp textile fiber, maintaining centuries of traditions and knowledge about cultivating and producing fabrics for clothing. Currently, hemp cultivation is growing across Europe and several other countries worldwide. However, a more significant effort is needed to create knowledge capable of generating investment in hemp textile fiber cultivation and other production processes (Farag & Kayser, 2017). The USA production is accelerating its investment in hemp crops and industry, and the United States National Agricultural Statistics Service (NASS) previews that national production will go from 115,000 hectares (ha) of industrial hemp planted in the United States in 2020 to 930,000 ha by 2023 (United Nations, 2022).

The European Commission shared a post on the official website titled "Hemp – Hemp production offers ample opportunities for the European Union's farmers, industrial sectors, and consumers." It analyses the extension of fields dedicated to hemp crops in Europe for five years. In 2015 hemp plantations used just 19.970 ha. Four years later, in 2019, the cultivated land grew by over 70%. France is the leading European producer (70%), followed by the Netherlands (10%) and Austria (4%).

However, in Europe, the textile industry still needs to start using hemp significantly. For example, some countries, including France, are exploring hemp as a multifunctional plant, using only 1% of it in the fashion value chain. Rehman et al. (2021) argue that hemp crops have a multipurpose industrial focus and are attractive to the increasing population in rural regions. In Italy, hemp is also growing in importance. However, if hemp is also multi-industry oriented, 83% of farms/enterprises have been created by young entrepreneurs in the last ten years, mainly under 35 years. Almost all farms produce products such as seeds, flour, decorticated seeds, hemp beer, animal seeds, seeds and inflorescences for cosmetics, and other products (Giupponi et al. 2020).

Currently, the quantities of hemp used in the textile and fashion industries are still low; Europe and the United States depend on China for hemp supply. As the U.S. recognize the importance of industrial hemp production and no longer wants to rely on imported raw materials, the investments in stimulating the interest in hemp crops among farmers and developing industrial equipment to defibrate hemp are growing. Hence the interest of the Secretary of Agriculture, Tom Vilsack, to announce in an interview that the USA will import hemp seeds from Ukraine (Fig. 1).

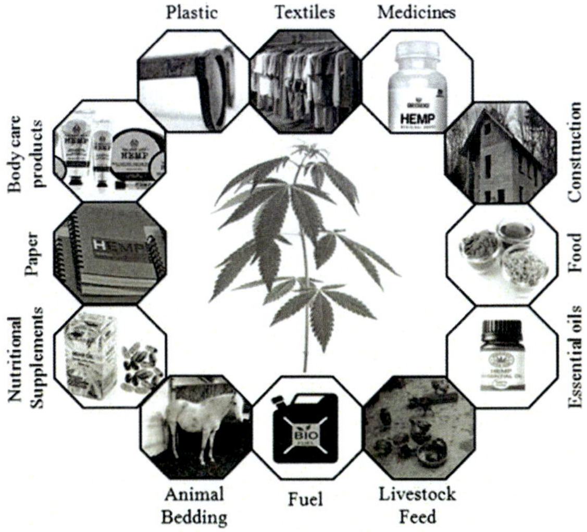

Fig. 1. Multiple products or uses of hemp. Multiple products or uses of hemp. Source: Rehman et al. (2012).

4 Fashion Brand's Role in Hemp Expansion

4.1 Global Sustainable Brands in the USA

Patagonia led in part the U.S. boom in the hemp industry. Yvon Chouinard, passionate about living outdoors in nature, founded Patagonia in 1957, supported by a genuine sustainable concept. This company goes far beyond the clothing it sells since Patagonia's most enduring legacy is its philosophy (Zint and Frederick, 2001). Of course, the brand has considerable weight in the market thanks to its environmental sustainability initiatives and the high quality and durability of its clothes that contributed to its continued expansion. Once again, Patagonia is leading the fashion industry to more sustainable approaches, focusing on hemp as a highly sustainable raw material. The brand supported organic cotton before, and the textile production chain accepted it more easily since cotton has a well-established production and market (Michel et al., 2019). The change from cotton to biological cotton did not require significant production changes. However, even if organic cotton is more sustainable than normal cotton, it still requires large amounts of water, takes longer to produce than hemp, and is more expensive.

Patagonia understands that they need to be able to guarantee a considerable volume of sales to justify growing hemp and develop the industrial production chain. The company adopted a proactive strategy, established agreements with farmers, even cosigned loans with them, and guaranteed them to buy the hemp production to make it happen. Patagonia recently launched its hemp collection, using mixtures of hemp, organic cotton, and recycled polyester in different compositions accordingly to the characteristics of clothing items.

Today, hemp processing is still unavailable in North America, but Patagonia is making its contribution and investment to bring hemp to the clothing industry and make it more sustainable. By building long-term relationships with their suppliers, Patagonia could get ahead of the competition in the organic cotton scene in the 90s and the hemp nowadays.

Patagonia is one of the leading brands. Once the technology and knowledge are there, it benefits the companies that initially invested in it and the whole industry. The others are already following as hemp becomes available in more quantities and at lower prices.

However, the strategic change from fashion industries to hemp as a sustainable raw material brings the need to introduce improvements in the production textile chain and, urgently, the availability of mechanic decorticators. This situation is being approached aggressively by U.S. industrial investors.

A significant example is the multibillionaire investment made by Panda Biotech, established in Dallas, Texas, which is developing industrial equipment for the hemp chain and projects to link industries and farmers around hemp. The Agriculture Improvement Act of 2018 (2018 Farm Bill, PL 115–334) brought a new context to industrial hemp production, and in May 2020, Panda Biotech offered Texas farmers 60 tons of hemp seeds to help them start with their crops.

The Panda Biothec mega project linked to USA fashion industries through collaboration agreements with Kontoor Brands Inc. That ensure the market for hemp crops. The fashion brands group is a global company with a portfolio led by Lee and Wrangler, with an annual production of over 175 million items and a sales volume of US$2.63 billion in 2022. Those brands already use cotton/hemp blends in jeans to increase strength and durability and promote water savings.

4.2 Local Fashion Brands Going Sustainable

Local fashion brands use style, materials, and inspiration sources as their main differentiation characteristics. As a sustainable textile material with a low environmental footprint, hemp offers new fashion brands opportunities to enter the market with new differentiation. Such is the case of Sensihemp, a new Portuguese sustainable fashion brand. Marta Vinhas founded it in September 2021. The brand concept is centered on sustainability and does not follow trends or launch collections; since it offers timeless design and long-lasting items 100% hemp, it is easy to mix them with other fashion items. The brand focuses on conscious fashion consumers, presenting them with low environmental impact, zero plastic, vegan, and ethically produced with high quality, guaranteeing high durability and comfort (Macedo, 2022).

The brand became a voice proclaiming the virtues of hemp in the Portuguese market, promoting its natural characteristics as core advantages for clothing and a contribution of fashion to the environment (T Jornal, 2019).

The brand's values are carbon-neutral, Fair Trade, Made in Portugal (locally produced), and zero waste. It also offers consumers other benefits linked with hemp characteristics, such as being bactericidal, retaining humidity, keeping the body dry and the user comfortable, being easy to keep, and being moth and mold-repellent. The brand also uses natural dyes. The waterproofness of hemp fabrics makes them adequate for producing fashion accessories such as backpacks, bags, and shoes.

However, the brand faces many problems maintaining its production in Portugal because this fiber still needs to be discovered by the textile industry since it specializes in cotton and mixtures. The brand buys yarn from North France and Romania at high prices because European production is mainly used in home industries, and small amounts arrive in the market.

Sensihemp presented its project to Worth Partnership Project and won Program support that included a financial prize, participation in international events, visibility, and professional links, among other advantages. The brand kept its focus on innovation and developed a t-shirt dyed with onion peel that produces a gold coloring on hemp knitted fabrics.

Fig. 2. Sensihemp onion peel dyed T-shirts. Source: Jornal Têxtil, 16 mars 2023.

The T-shirt was awarded Best Innovation Product at Barcelona, Spain's CBD Hemp Business Fair. The organization also invited the brand to be in the Milan Design Week, and this invitation was the opportunity to present the brand's new modular concept for fashion internationally.

Sensihemp was the first Portuguese fashion brand using; others are already following as Cura – Fashion Design Studio, which launched in 2019.

5 Conclusions

Industrial hemp presents exceptional characteristics that make it a genuinely sustainable plant that can offer raw materials for numerous products in different industries. In the fashion and textile industries, hemp is now used primarily in mixtures with organic cotton, mainly because of the small amounts of hemp in the market and the consequent high price of this fiber.

Investments in hemp production chains are growing all over the world. China and the USA are moving quickly in that direction, but Europe is also in motion, with France, Italy, and the Netherlands leading the hemp production. Global brands like Patagonia, Lee, and Wrangler assumed their interest in using hemp in their fashion products. They developed agreements with USA farmers in order to promote hemp production. China is also incentivizing their farmers to grow hemp and high-demanding crops such as wheat, soya, and tobacco to use hemp crops to recover their fields' fertility.

Many fashion and home textile brands already use hemp blends and even 100% hemp and knitted fabrics as a statement of their sustainability strategies. However, textile industrial systems still are focused on cotton, wool, artificial, and synthetic fibers. The change to hemp will require adaption, and competitive pressure and consumer demands for sustainable products are expected to accelerate the change process.

The deficit of industrial hemp raw material is growing daily even if the planted areas are growing fast worldwide. Many companies claim customers are eager to get high-end and premium price hemp products without suppliers having hemp fiber materials available to satisfy them. To spread industrial hemp farming know-how and adapt the textile industries to its products is an important market opportunity for European textiles and fashion brands. It is also important from the economic point of view since hemp is a natural fiber that can be produced almost in all European countries, conquering, this way, higher independence from imported fibers.

The Portuguese framework needs to catch up to the reality of other European countries, such as France, and international data worldwide. The Portuguese industry is producing many different fabrics using mixtures of hemp with other shorter fibers, like cotton, wool, modal, and polyester, among others, because the chain of long fibers, as is the case with hemp, is in deficit nowadays. There needs to be more cultivation, and the hemp production chain needs to be put in place since there are no decorticators or specialized yarn and fabric producers.

The European textile industries still need to invest in creating a hemp textile branch. Research on the subject and the consumer pressure to obtain more sustainable fashion products is expected to accelerate the process and allow Europe to compete with USA industries already focused on this old/new textile sustainable material.

Acknowledgments. This work was funded by the European Regional Development Fund (FEDER) through the Operational Competitiveness Program (POCI) and the National Foundation for Science and Technology of Portugal (FCT) under the projects UID/CTM/00264/2020 of Centre for Textile Science and Technology (2C2T).

References

Allegret, S., Bouloc, P., Arnaud, L.: The history of hemp. In: Bouloc, P. (ed.) Hemp: Industrial Production and Uses, pp. 4–26. CAB International, Boston (2013)

Averink, J.: Global water footprint of Industrial Hemp Textile. University of Twente, Netherlands. https://essay.utwente.nl/68219/1/Averink%2C%20J.%200198501%20openbaar.pdf (2015). Accessed 16 Nov 2022

Batista, C., Santos, N.F.: Cânhamo: aplicações papeleiras. Agroforum **17**, 37–40 (1999)

Castro, A.: A evolução económica de Portugal dos séculos XII a XV. Portugália, Lisboa (1964)

Deshmukh, G.: Advancement in hemp fibre polymer composites: a comprehensive review. J. Polym. Eng. **42**(7), 575–598 (2022). https://doi.org/10.1515/polyeng-2022-0033

Donaldson, L.: From illegal to ideal: the necessity of hemp production. Natural Sci. Educ. **51**(2), e20090 (2022). https://doi.org/10.1002/nse2.20090

European Commission: Hemp. Agriculture and development. https://agriculture.ec.europa.eu/farming/crop-productions-and-plant-based-products/hemp_en#committeesandexpertgroups (2023). Accessed 31 Jan 202

European Parliament: The impact of textile production and waste on the environment (infographic). https://www.europarl.europa.eu/news/en/headlines/society/20201208STO93327/the-impact-of-textile-production-and-waste-on-the-environment-infographic (2022). Accessed 20 Jan 2023

Farag, S., Kayser, O.: The cannabis plant: botanical aspects. In: Preedy, V.R. (ed.) Handbook of Cannabis and Related Pathologies. Academic Press, Amsterdam (2017)

Fike, J.: Industrial hemp: renewed opportunities for an ancient crop. Crit. Rev. Plant Sci. **35**(5–6), 406–424 (2016)

Garcia, J.: Os Têxteis em Portugal nos Séculos XV e XVI. Finisterra **21**, 327–344 (s/d.). https://repositorio-aberto.up.pt/bitstream/10216/28268/2/finesterrav21n421986000082969.pdf. Accessed 26 Jan 2023

Gomes, A.: Almalaguez – Tecer o Futuro com Fios do Passado. Tese de Doutoramento. Universidade do Porto e Universidade de Aveiro (2017). https://sigarra.up.pt/fbaup/pt/pub_geral.pub_view?pi_pub_base_id=260724. Accessed 28 Dec 2022

Giupponi, L., et al.: Overview on Italian hemp production chain, related productive and commercial activities and legislative framework. Italian J. Agron. **15**, 1552 (2020). https://www.agronomy.it/index.php/agro/article/view/1552/1197

Gulati, A., Turner, A.: A note on the early history of cotton. J. Text. Inst. Trans. **20**(1), T1–T9 (1929)

Harsanto, B., Primiana, I., Sarasi, V., Satyakti, Y.: Sustainability innovation in the textile industry: a systematic review. Sustainability **15**(1549), 1–21 (2023)

Kache, T., Singh, M.: The application of hemp (Cannabis sativa L.) for a green economy: a review. Turk. J. Bot. **43**, 710–723 (2019)

Macedo, C.: Sensihemp quer renascer tradição dos têxteis de cânhamo. Ambiente Magazine 14 Fev 2022. https://www.ambientemagazine.com/sensihemp-quer-renascer-a-tradicao-dos-texteis-de-canhamo/ (2022). Accessed 12 Sept 2022

Meijer, E.: Fiber hemp cultivers: a survey of origin, ancestry, availability and brief agronomic characteristics. J. Int. Hemp Assoc. **2**(2), 66–73 (1995)

Michel, G., Feori, M., Damhorst, M., Lee, Y., Niehm, L.: Stories we wear: promoting sustainability practices with the case of Patagónia. Family Cosum. Sci. Res. J. **48**(2), 115–211 (2019)

Rehman, M., et al.: Evaluation of hemp (Cannabis sativa L.) as an industrial crop: a review. Env. Sci. Pollution Res. **28**(38), 1–12 (2021). https://doi.org/10.1007/s11356-021-16264-5. Accessed: 21 Dec 2021

Sendin, B. (coord.): ATP 50 anos. ATP – Associação Têxtil e Vestuário de Portugal, Vila Nova de Famalicão (2015)

Sequeira, J., Melo, A.: A mulher na produção têxtil portuguesa tardo-medieval. Medievalista 11/2012. http://journals.openedition.org/medievalista/780 (2012). Accessed: 25 Feb 2023

Sigaard, A.S., Laitala, K.: Natural and Sustainable? Consumers' Textile Fiber Preferences. Fibers **11–12**, 1–23 (2023)

Singh, M., Sardesai, M.M.: Cannabis sativa (C annabaceae) in ancient clay plaster of Ellora Caves, India. Curr. Sci. **10**(5), 884–891 (2016)

Jornal, T.: Sensihemp won Innovation Award in Barcelona, 19 Jan 2019, p. 7. https://issuu.com/tjornal/docs/ti19_jan.23 (2919). Accessed: 20 Feb 2023

USDA – United States Department of Agriculture – Foreign Agriculture Service: 2019 Hemp Annual Report, Country: Peoples Republic of China. USDA Report Number: CH2020–0018 (2019). https://apps.fas.usda.gov/newgainapi/api/Report/DownloadReportByFileName?fileName=2019%20Hemp%20Annual%20Report_Beijing_China%20-%20Peoples%20Republic%20of_02-21-2020. Accessed 28 Jan 2023

United Nations: Commodities at a Glance. Special issue on industrial hemp 16. United Nations, Geneve (2022)

Zint, M., Frederick, R.: Marketing and advertising a deep green company. J. Corp. Citizensh. **2001**(1), 93–113 (2001). https://doi.org/10.9774/GLEAF.4700.2001.sp.00010

Emotional Design and Fashion

Assemblage + Waste + Memory = Jewel

Mónica Romãozinho(✉)

Rua Marquês de Ávila e Bolama, Labcom-UBI, Universidade da Beira Interior, 6201-001 Covilhã, Portugal
monica.romaozinho@ubi.pt

Abstract. This reflection provides further insight into the interaction between Jewellery Design and Art, through the exploration of an applied research methodology that did not exclude intuition, experimentation around materials and techniques, and the enhance of emotion and memory. Jewellery and Art gradually approached over the 20th century, a more affirmative aspect since the 60s, with the pioneering use of industrial materials, often intersected in assemblages and the adoption of the multiple, in denial of the single product. New relationships between body and object as well as new meanings were explored, stimulated by even futuristic jewellery concepts that provided a new experience to the user and freed themselves from the exclusive use of precious metals and gems. This involvement has contributed to the affirmation of jewel, one of the least changeable products that communicate something about us and the way we see ourselves in the world, more actively or more implicitly. The literature review and instruments such as exploratory sketches and prototypes play a key role in our applied research, focused on strategies that provide new experiences to the jewelry owner, minimize the environmental impact through options such as upcycling, and affirm a line of research focused on body-object relationship and memory.

Keywords: Assemblage · Jewellery Design · Emotion

1 Art and Jewelley Design: Dialogues and Intersections

Jewellery Design has always benefited from the ruptures and experimentations operated in the field of Plastic Arts, as a possible vehicle of innovative languages and nonconformist attitudes translated into objects that carry meanings and emotions, unique or produced in limited series, thus possibly less volatile and disposable. This is a fundamental issue that increasingly incorporates Design projects since the product is no longer valued exclusively for its functionalities but, also, for the experience it can provide to its user. In the specific scope of this article, we will analyze in greater depth the impact of some principles and compositional techniques on Jewellery Design projects, with particular emphasis on assemblage. The poetry resides in so many everyday objects that can gain, through upcycling, a second life, assuming new functions and meanings. We can contribute this way for a longer useful life cycle of these products. The jewel, as a piece, is situated in an absolutely hybrid territory, working as an extension of our body, a communication of our personality, and a way of looking at the world and things, it has a purpose which is not limited to a utilitarian function. It stimulates our imagination as creators, and implies an aesthetic sensibility, although it moves away from greater constraints unlike what happens in industrial design and mass production, for example: "The utility of the object together with the purposeful intent behind its conception is what distinguishes design from art. Good design is neither art nor barren instrumental device, neither wholly artistic nor wholly utilitarian, but an inseparable union of the two" [1]. In fact, Industrial Design itself faces new challenges in the present, resulting from an awareness of the environmental and social consequences of mass production not directed according to the needs and designs of the community, gradually assuming a character of surgical action on the territory with a focus on a more empathetic relationship between subject and object.

2 Memory and Feelings: Poetic Objects

The creative process exposed throughout this article is born, at first, from a methodical act of researching, searching, inventorying, and, at a second moment of reinventing from objects that transmit individual and collective memories, in a process of true archaeology of objects apparently useless but containers of memories, and emotions as well. Moreira da Silva actually notes that our concern should be related not only to the formal solution but also to the meanings: "Besides formal and functional components, the designer must also pay attention to the meanings that a given object can assume. (...) Objects play other roles in the lives of their users, in addition to the mechanical functions for which they were objectively designed, consequently, they are capable of evoking emotions" [2].

To understand the power of memory, it is enough to recall the scene from the film *Le Fabuleux destin d'Amélie Poulain* (2001), in which Dominique Bretodeau hears the phone ring in a public booth, discovering inside a small container of his childhood toys and amulets [3]. For brief moments, Bretodeau relives a series of past events, positive and also negative experiences, all of which are part of his individual memory and shape who we are today. In Bretodeau's case, that flashback would be a turning point and an awareness of the speed of time, showing that memory is a process of reconstructing perceived images. As António Damásio once said, the memory of a scene that marked us at the visual or auditory level, for example, is achieved through the conversion of explicit images into a "neural code" that later, by acting in the opposite direction, allows a more or less complete reconstruction in the process of recalling the image [4]. The same codes represent, in a non-explicit way, the content of the images and their sequences, being stored in both cerebral hemispheres, in the associative cortices in the occipital, temporal, parietal, and frontal regions [5]. That same author concluded that during the process of recall (or so-called retro activation), we reconstruct a more or less faithful approximation of the original image, using inverse neural pathways, which operate essentially from the regions that contain the memories stored in codes and produce effects in the regions where explicit images are created [6]. And there are infinite circumstances, present or recovered by memory, capable of provoking feelings, the so-called provoked feelings in contrast to those that correspond to the experience of certain aspects of the state of life of an organism and which have the function of reporting the state of life [7]. Damásio demonstrates that feelings are not an independent fabrication of the brain and result from a collaboration between body and brain, which interact thanks to chemical molecules and nerve tracts, thus extolling the value of homeostasis. It reinforces that this unique relationship between the "body" and the "brain", usually ignored, ensures that feelings disturb what, in their absence, would be an indifferent mental flow [7]. Later in his book, he systematizes the immediate causes of the feelings associated with the most striking images of our mind: the current of life processes in our organism, experienced as spontaneous or homeostatic feelings; emotive responses triggered by the processing of numerous sensory stimuli, such as tastes and smells, and tactile, auditory and visual stimuli, emotive responses resulting from drives (such as hunger or thirst), motivations (such as desire), or emotions, such as joy, sadness, fear, fury, envy, jealousy, contempt, compassion or admiration. The last two causes generate provoked feelings (or emotive responses), the first being referred to as the spontaneous variety that derives from homeostatic mechanisms [8]. In terms of feeling, the same author evidences that conventionally, the human cultural enterprise is explicable in terms of the human intellect, a brilliant finish of the organisms generated by non-thinking genetic programs throughout evolution, as if the creative intelligence had materialized without a powerful impulse, moving forward without a background motive other than pure reason and as if creativity were not an integral part of the complex edifice of emotions [9]. Other authors, specifically linked to the scientific area of Design also claim the importance of feeling, and intuition in creative processes. Stuart Walker highlights the danger of overvaluing rationalization and emphasizes the way our rational side is valued at the expense of our creative and imaginative side that contributes to the return of poetry to design: "The suppression of this "other half" has led to a materially abundant but spiritually impoverished world. It is argued here that greater acknowledgment of this "other half" in industrial design can lead to products that are expressive of a more balanced understanding of human needs. Such a shift would not only contribute to a culturally richer material environment, it also would allow us to more effectively address the principles of sustainability" [10].

Jewelry, namely family jewels (an opening medal with a photo inside, a bracelet with the date of baptism, among others), symbolic objects of greater or lesser value, which we carry close to our body, has always been transmitted from generation to generation, providing affective experiences. We emphasize Chapman's words in his work: "Objects are emotional. Whether a wedding ring, an heirloom dining table, or a tatty old teddy bear, objects can bring us a sense of comfort and connection to our past" [11]. However, a product designed by a designer should not result from a deliberate act of "stirring" memories, it should function as a question that can trigger different responses from the public. Jewellery provokes emotions and memories, and in the same way, it is born of emotions and memories because they play a fundamental role in the creative process. Other "unsentimental" objects were present in the past at the service of everyday life during a more finite time horizon. They contain memories and experiences, they were important while fulfilling their function, and they continued to be part of our life although they were kept in a drawer. The project presented throughout this reflection illustrates the attribution of a second life to a set of objects, through upcycling and the application of principles that come from the Plastic Arts and Industrial Design fields. As with Bretodeau's objects which are not precious in their matter, we discover qualities in objects that mean little or nothing to other people. We are interested in these *objets trouvés*, a concept introduced by Marcel Duchamp, but the series that we present later results from an emotional choice of matter and not from indifference or "total aesthetic absenteeism" [12]. It is the process of interpretation that determines the power of your message: "Thus it is our interpretation of the object, and not the object itself, that dictates the character of affective response. (...) Furthermore, this process is highly subjective, shaped by each user's bank of prior experiences and memories" [13]. We avoid explicit messages, well-defined speeches, and staged justifications. The work must provoke questions, and interrogations, as Lisa Walker observes: "I like my work to be open to some interpretation. I don't want to hand the viewer everything. Mysterious isn't really the right word – it's more a question, or a feeling or a presence. I'd like the viewer to experience all of this" [14].

3 The Materials of Everyday Life Gathered in an Assemblage

As it was mentioned in the preamble, the Plastic Arts have always been pioneers in the sense of operating ruptures and the exploration of new formulas that contaminated the universe of jewelry. This reflection focuses on assemblage as a process of exploration of new concepts, not a technique regulated by rules: "Assemblage is art that is made by assembling disparate elements – often everyday objects – scavenged by the artist or bought specially" [15]. This approach was triggered by artists such as Picasso, his work Still Life (1914) being an example of this turning point, or Vladimir Tatline who in 1915 had participated in the exhibition Tramwy V (in the so-called Petrograd) with his counter-reliefs that resulted from his struggle with concepts such as the representation of real space. The origins of what would become the constructivist movement lay precisely in Tatline's research through which he intended to replace the pictorial representation of an object (considered bourgeois after the revolution) by a volumetric complex resulting from the intersection of protruding and angular planes. Tatline symbolically abolished the frame and background with the conviction that these divorced the artistic work from the real world and it is visible the influence that synthetic cubism would exert on these works even because the same artist had been in Picasso's studio in 1913 [16]. As Tafuri refers, for all the avant-garde movements the law of assemblage was fundamental: "And since the assembled objects belonged to the real world, the picture became a neutral field on which to project the experience of the shock sufferes in the city. The problem now was that of teaching that one is not to 'suffer' that shock, but to absorb it as an inevitable condition of existence" [17]. However, Tatline disgusted the two-dimensional versions of analytic and synthetic cubism. If constructivism was as realistic as synthetic cubism, it is because it integrated the materials that constitute it, natural or industrial, validating them as integral elements of a plastic synthesis. Not only did the interpenetration of plans challenge traditional concepts of sculptural mass, but a new construction technique replaced traditional carving and modeling techniques: "Tatlin's respect for materials and his synthesis of their forms was related to his belief in obeying natural laws, in creating forms of universal significance which would therefore appeal to a mass audience" [18].

Next to this line of research was Kurt Schwitters, applying the learning of Cubism and Futurism to his assemblages, an example of which is the work *Revolving* (1919) which goes beyond the two-dimensionality of collage becoming an abstract relief made of real materials [19]. The rejection of Schwitters by the Berlin Dadaists contributed to his involvement on a more personal path personified by his art MERZ, first presented at the Sturm Gallery in 1919 [20]. In the 60s, the connection of contemporary jewellery to movements such as Art Povera and Pop Art contributed to a democratization of jewelry making it more accessible and we witnessed the exploration of new forms, a return of the so-called poor materials, in which some interventions sometimes referred to objects of adornment of prehistory [21]. This artistic heritage is present in the creative process of artists involved in the field of contemporary jewellery, they are also "constructors" of three-dimensional and semantic structures, as happens in the work of Ramón Puig-Cuyás who seeks to achieve a share of order in the civilizational chaos: "What drives me is the need to establish a small measure of order in our chaotic universe, to reach both perfection and clarity in the structure of a piece and also the depth of its mystery, like the music of Bach" [22]. The process of making is always implicit in the assembly and fitting solutions, the barrier between designer and craftsman blurs. This approach often results in the production of unique pieces with their own individuality and brings the owner of the jewelry closer to the designer. As Filomena Silvano observes, new consumption practices are now associated with the emergence of new value regimes, evident in new market logics, such as the sale of used clothes and the production of objects on a small scale and using sustainable techniques, focused on the valorization of objects, knowledge of materials, ways of making and perceptions of the lifetime of broader things [23]. The same author then adds that the empathetic relationship with objects and their materiality also stems from access, even if external, to the processes of making and its complexities [24].

In the *(Des)compasso* project, the chosen objects arouse specific feelings, especially associated with a phase of our still embryonic journey when we learned geometric drawing in class. At that time, geometry laws did not arouse in us great emotion instead of irregular and less disciplined vocabulary of forms. Over time we learned to appreciate their purity and visual strength and to realize that they could integrate infinite compositions and true constellations. In this act of assemblage, the first rehearsals start from chaos, a metaphor of reality, and a certain incoherence from which we are challenged and moved by the desire to find order through economics and synthesis. In this search for a gradual reduction of uncertainties and ambiguities throughout the process, exploratory sketches and intercalary prototypes were our anchors (Fig. 1).

Fig. 1. (Des)compasso #2. 2023. Sketchbook.

The project is born out of an attempt to deviate from standardization. We continue to be moved through life. A compass determines our mechanized routines, with no room for deviations or lack of rigor in the traces of our existence. *(Des)compasso* is an invitation to nonconformism, to deviation metaphorically marked by the deconstruction of an instrument that is part of our experiences, our learning, and our projects at a time when drawing emerged from our hands and our thinking without filters. It is an object that invokes positive and negative memories. The compass as an instrument suitable for technical drawing has always stimulated an interventional role and intuitive manipulation on the part of its user. On the other hand, this project, particularly the brooch concept, allowed us to explore the possibility of movement in the piece because we can rotate some components and opt for diverse positions in the case of the brooch (Fig. 2).

The necklace marks the beginning of the *(Des)compasso* series, being composed of elements of industrial manufacture that articulate with each other, a process of easy understanding, as well as the adaptation to different bodies and the length, can be adjusted: a jewel to assemble and awake memories (Fig. 3). Easily dismantled, it allowed to replace its components without the need for any specific instruction. The object is subject to the same premises of Industrial Design, in which the components articulate with each other through screws, wheels, and knots, among other components that integrate our archive built over time, an inventory of inanimate objects, materials, parts, a true metaphor of the Bretodeau box. We would never classify these objects as useless, they just waited for their application. Some are more perfect than others, some new, some old, and many exhibit the marks of oxidation and wear. All fulfill a function as a part of a whole marked by the abolition of ornament. We find beauty in these imperfections that translated into color or texture. The experimentation around imperfection was equally evident in the field of Plastic Arts, if we think of the work of Antoni Tápies or Alberto Burri, exponents of Informal Art that valued not only the pictorial matter but the medium itself [25].

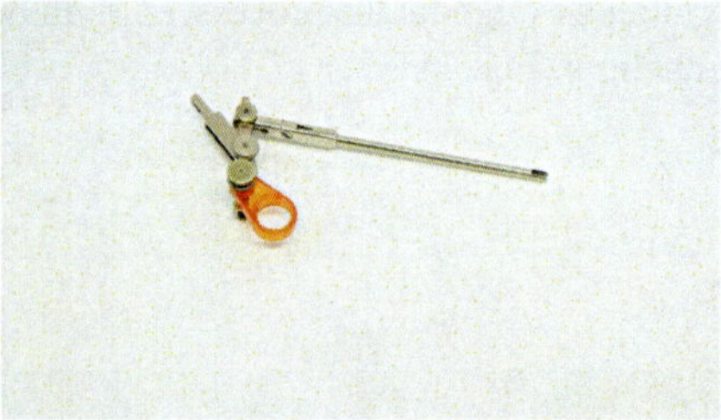

Fig. 2. *(Des)compasso* #2. 2023. Brooch. Steel components extracted from compasses, auxiliary part for Rotring pen nozzle, additional elements.

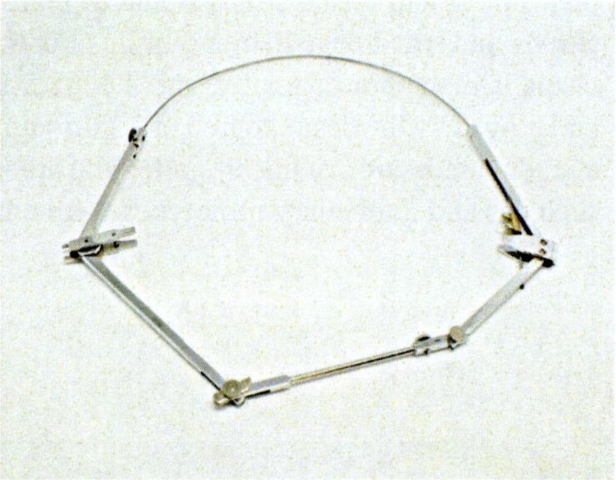

Fig. 3. (Des)compasso #1. 2022. Necklace. Steel components extracted from compasses, additional elements, steel cable.

Because objects and materials communicate: "Yet, beyond technical capacities, materials possess a deeper and more subjective dimension. All materials have meaning. Or, more specifically, all materials interact with their context of use and, through this interaction, elicit meaningful associations within the mind of their user" [26].

In this series, the objects fulfill the purpose of the ambition of an impact on society through the recognition of components and materials that integrated a moment of their experience (school background in this case), by the fact that each piece is unique, more humanized and has its own identity by incorporating components discarded by society sometimes belonging to distinct disciplinary universes: "Jewellery belongs with all the other areas that make up an expression of personal identity: fashion/clothing, makeup, hair, your home, your car, your friends, where you holiday, your job and so on" [27]. Chinese ink technical drawing pens, compasses, rulers, and various stencils (of curves, letters, and furniture) have integrated our past experiences. We are also interested in this exploration of assemblage applied on the mimetic scale of an earring, dissociated from a pair logic (Fig. 4). Emotion and intuition are the motor in this more organic building process: "Increasingly, the Design project is concerned with providing each individual with solutions suited to their needs and desires, and sometimes characteristics, a more humanistic design, but at the same time contemplating emotion and intuition" [28]. We played with the wire giving lightness to its structure, with the straight lines that it raises horizontally, then vertically, and finally diagonally, causing tensions between linear elements. These micro-objects achieve dynamic qualities, moved by the action of our body, yet disciplined when it comes to balance. The work of the American artist Alexander Calder is an example of the infinite potential of a simple matter like a wire. Calder embarked on the exploration of kinetic art present in the concept of surrealist design mobiles, applied to different scales including jewellery, emphasizing the structure of these objects. The same artist dedicated himself for more than five years to the project of a miniature circus, the so-called Cirque Calder, a true laboratory for the exploration of new formulas and techniques, a microcosm that contained human and animal characters executed in awire, complemented by canvas, wood, cork, labels, fragments of old metal and rubber pieces [29]. These constructions had an impact both on his first wired jewelry, often on tests for large-scale sculptural pieces, and in his first tests of kinetic structures.

In the 60s, there were increasing points of contact between artistic experimentation and traditional jewellery that allowed a new relationship between the subject and the jewellery, sometimes more uncomfortable, most often facilitating interaction with a "new" world. According to Cappellieri, the experimentations of Surrealism, Kinetic Art, Informal and Pop Art, as well as the CoBrA Group, have evident repercussions in the jewellery sector, and many of its main exponents have designed jewellery: "Moreover, while it had mainly been artists – from Calder to Picasso – who approached jewellery in the 1940s, now it was jewelers who were approaching art, changing both its plastic and its semantic features" [30]. The construction of more malleable wire structures takes us back to childhood in which we quickly and organically constructed figures of great abstraction that could be adapted fluidly to different positions and movements, stimulating, on the other hand, the study of gravity, the function of weight and counterweight, the laws of asymmetric balance and movement.

Fig. 4. (Des)compasso #3-4-5-6. 2023. Earrings. Steel/acrylic components extracted from compasses, wire.

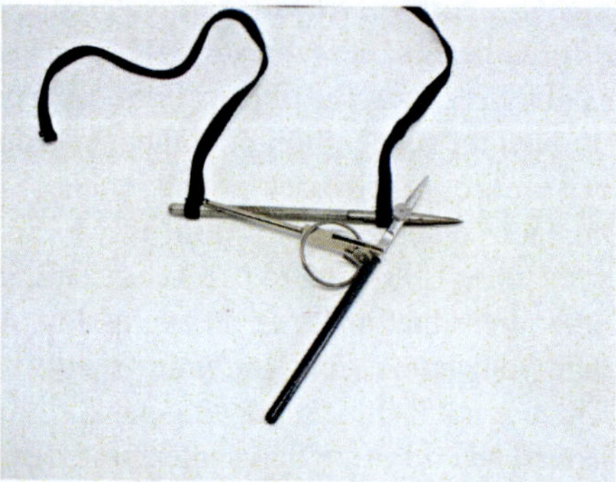

Fig. 5. (Des)compass #7. 2022. Necklace. Compass components, strip-lines, ring, velvet ribbon.

However, if the series *(Des)compasso* is born of positive memories, it also refers us to negative experiences if we think of the permanent frustration we felt when we used the line strip in geometric drawing while still learning in high school and when the drawings quickly became a smeared shape (Fig. 5). The same frustration felt later whenever a chinese ink technical drawing pen clogged on the eve of an architectural project delivery at the college. The negative experience can also be part of the design as a creative process, contributing to the enrichment and intensity of its semantic value and for the sensory and more humanized experience of the user. An error in the execution phase is considered, at first glance, as a defect, an argument strong enough for them to be discarded immediately in a series production line. An error that does not represent a danger to the appropriation by the user may correspond to a new path in the experience provided by the object, similar to what happens in the artistic field. The materials of real life thus form an assemblage full of meanings. The world is not perfect and the oxidizing components of old bars mirror this state of affairs that we cannot place on the plane of black and white, certainty and uncertainty, but rather of intermediate and unpredictable tones: "Often times it is the idiosyncratic nature of error that gives character to products, lifting them from anonymous mass-produced homogeneity to something greater. We see this in things such as air bubbles in handblown glass or knots in an oak table. Many designers work at the threshold of error, to celebrate such partially controlled processes, working at the limits of a machine's ability and forcing it to fail, thus producting extraordinary results" [31].

4 Conclusions

Firstly, we conclude that Design in the field of Jewellery continues to benefit from its historical relationship with the Plastic Arts, being stimulated by its ruptures, nonconformisms, experimentalisms, non-submission to processes of massification and acceptance of the imperfect and deviation from normative standards. We sought to explore creative processes that (re)value the role of intuition, of feeling, applying concepts derived from Art such as assemblage, in the development of limited series of products or unique and identity pieces, such as the (Des)compasso series. This close relationship intersects with the lessons we draw from Industrial Design and the urgency to adopt progressively less disposable and repairable solutions, capable of triggering emotional bonds. Flusser also highlights the consequences of a growing awareness of the effects of consumption and waste. According to him, maybe the awareness of the ephemeral side of everything that is created (and as such of immaterial projects) can contribute for a future scenario in which things are projected with more responsibility, giving way to a culture in which the objects of use are less and less obstacles and more and more vehicles of interpersonal relationship [32]. We recognize the value of other typologies of industrial objects, external to the universe of jewellery that can affirm the function of memory in the jewellery project, not necessarily from the perspective of the creation of sentimental jewellery mainly linked to family pieces. In this creative process that is based on a cross-play of memories, we privilege the synthesis and economy of contents, in an attitude of affirmation of the jewel as an idiosyncratic object that mirrors something about its creator, that stimulates us to discover the invisible and the unspeakable, that provokes questions and invites us to write a new narrative.

References

1. Walker, S.: How the other half lives: product design, sustainability, and the human spirit. Des. Issues **16**(1), 54 (2000). https://doi.org/10.1162/074793600300159673. Accessed 27 Mar 2023
2. Almendra, R., Silva, F.-M.: Emotion: a vital component in design decision making. In: Raposo, D., Martins, N., Brandão, D. (eds.) Human Dynamics and Design for the Development of Contemporary Societies. AHFE (2022) International Conference, vol. 25, p. 172. AHFE Open Access, USA (2022). https://doi.org/10.54941/ahfe1001386. Accessed 27 march 2023
3. Le Fabuleux destin d'Amélie Poulain, 2001. [film] Directed by Jean-Pierre Jeunet. France: UGC
4. Damásio, A.: A Estranha Ordem das Coisas: A vida, os sentimentos e as culturas humanas. 12nd ed. Temas e Debates, Lisboa, p. 137 (2017)
5. Damásio, A.: A Estranha Ordem das Coisas: A vida, os sentimentos e as culturas humanas, 12nd ed. Temas e Debates, Lisboa, p. 138 (2017)
6. Damásio, A.: A Estranha Ordem das Coisas: A vida, os sentimentos e as culturas humanas. 12nd ed. Temas e Debates, Lisboa, p. 150 (2017)
7. Damásio, A.: A Estranha Ordem das Coisas: A vida, os sentimentos e as culturas humanas. 12nd ed. Temas e Debates, Lisboa, p. 27 (2017)
8. Damásio, A.: A Estranha Ordem das Coisas: A vida, os sentimentos e as culturas humanas. 12nd ed. Temas e Debates, Lisboa, p. 145 (2017)

9. Damásio, A.: A Estranha Ordem das Coisas: A vida, os sentimentos e as culturas humanas. 12nd ed. Temas e Debates, Lisboa, p. 30 (2017)
10. Walker, S.: How the other half lives: product design, sustainability, and the human spirit. Des. Issues **16**(1), 52 (2000). https://doi.org/10.1162/074793600300159673. Accessed 27 Mar 2023
11. Chapman, J.: Meaningful Stuff: Design That Lasts, p. 81. The MIT Press, Cambridge (2021)
12. Moure, G.: Marcel Duchamp, p. 17. Ediciones polígrafa, S.A., Madrid (1988)
13. Walker, S.: How the other half lives: product design, sustainability, and the human spirit. Des. Issues **16**(1), 73 (2000). https://doi.org/10.1162/074793600300159673. Accessed 27 Mar 2023
14. Rhodes, K., Themelios, N.: Another Unreliable Guidebook to Jewellery by Lisa Walker, p. 391. RMTI Design Hub Gallery, RMTI University, and Perimeter Editions, Melbourne (2021)
15. Art terms: https://www.tate.org.uk/art/art-terms. Accessed 27 Mard 2023
16. Moszynska, A.: Abstract Art, 2nd edn., pp. 72–73. Thames & Hudson, London (2004)
17. Tafuri, M.: Design and Capitalist Development, p. 86. The Mit Press, Cambridge (1976)
18. Moszynska, A.: Abstract Art, 2nd edn., p. 73. Thames & Hudson, London (2004)
19. Moszynska, A.: Abstract Art, 2nd edn., p. 68. Thames & Hudson, London (2004)
20. Furor Dada: Colecção Ernst Schwitters, Lisbon. [exhibition], p. 68. 28 Jan–4 Apr (1999)
21. Filipe, C.: Joalharia Contemporânea em Portugal: Das vanguardas de 1960 ao início do século XXI, p. 78. MUDE-Museu do Design e da Moda, Lisboa (2019)
22. Pastor, C. (ed.): Maestros de la joyería contemporánea, p. 154. Promopress, Barcelona (2014)
23. Silvano, F.: Antropologia da moda, p. 170. Documenta, Lisboa (2021)
24. Silvano, F.: Antropologia da moda, p. 171. Documenta, Lisboa (2021)
25. Ferrari, S.: Guia de História da Arte Contemporânea: Pintura, escultura, Arquitectura, Os Grandes Movimentos, p. 94. Translated by M.J.Figueiredo. Editorial Presença, Lisboa (2001)
26. Chapman, J.: Meaningful Stuff: Design That Lasts, p. 66. The MIT Press, Cambridge (2021)
27. Rhodes, K., Themelios, N.: Another Unreliable Guidebook to Jewellery by Lisa Walker, p. 271. RMTI Design Hub Gallery, RMTI University, and Perimeter Editions, Melbourne (2021)
28. Almendra, R., Silva, F.-M.: Emotion: a vital component in design decision making. In: Raposo, D., Martins, N., Brandão, D. (eds.) Human Dynamics and Design for the Development of Contemporary Societies. AHFE (2022) International Conference, vol. 25, p. 173. AHFE Open Access, USA (2022). https://doi.org/10.54941/ahfe1001386. Accessed 27 March 2023
29. Lemaire, G.-G.: Alexander Calder. Translated by R. Ibero. Ediciones Polígrafa. Barcelona (1998)
30. Cappellieri, A.: Twentieth-century Jewellery: From Art Nouveau to Contemporary Design in Europe and the United States, p. 45. Skira Editore S.p.A, Milano (2010)
31. Walker, S.: How the other half lives: product design, sustainability, and the human spirit. Des. Issues **16**(1), 91–92 (2000). https://doi.org/10.1162/074793600300159673. Accessed 27 March 2023
32. Flusser, V.: Uma filosofia do design: A forma das coisas, p. 61. Translated by S. Escobar. Relógio d'Água Editores, Lisboa (2010)

Emotional Design in Fashion: Memories and Experience in a Redesign Project

Liliane Gonzaga Sommermeyer[1](✉), Maria Cecília Loschiavo dos Santos[1], and Joana Cunha[2]

[1] Faculdade de Arquitetura e Urbanismo, Universidade de São Paulo, São Paulo, SP 05508-220, Brasil
{lilianegonzaga,closchia}@usp.br
[2] Centro de Ciência e Tecnologia Têxtil, Universidade do Minho, 4800-058 Guimarães, Portugal
jcunha@det.uminho.pt

Abstract. The fashion industry annually generates tons of textile waste and is considered one of the most polluting industries in the world. This unsustainable scenario is largely due to fast fashion, which encourages consumers to get rid of their clothes, even if they have emotional ties with them. However, as people experience remarkable experiences with their pieces, this distancing becomes more complex. Thus, the objective of this study was to report and analyze how affective ties can influence a creative redesign process in a fashion workshop, understanding the emotions of participants in the process of transforming their own disused pieces, through upcycling. This investigation was carried out through an action-research in a 4-day workshop applied in a fashion studio at a university with a total of 8 participants. Through three different tools, we identified the emotions of the participants regarding their disused clothes prior, during and after its redesign process. The workshop results indicate that clothing is considered very important for most participants. Regarding the emotions experienced during the redesign process, the most mentioned emotions were: Excited, Stimulated and Insecure. Results also show that although most participants try to get rid of disused pieces, they do so by trying to stay close to them, revealing a strong emotional bond with clothes.

Keywords: Problem solution · upcycling · contextual inquiry · fashion and sustainability

1 Introduction

Human relationships, as well as their interactions with products, are made up of emotions. These emotions can be positive or negative and experienced at different levels of intensity, processing and psychological states [1–4]. So, when we use a product and have an emotional experience, it is common for an affective bond to be established between the experienced situation and the product. With the fashion product, it is not different [5]. However, nowadays, it is more and more challenging to keep a piece of cloth in good condition for a long period of time, since the majority of clothing production comes from the fast fashion industry, based on products with a short life cycle and bad quality.

When we talk about product development, design thinking is currently one of the most used approaches in the field of design, as it is human-centred and seeks to solve problems through empathy, ideation, prototyping and testing. It is often used in the areas of design, fashion and technology to create innovative products, services or processes that meet people's needs in a creative and effective way [6, 7]. Upcycling, on the other hand, is the process of taking waste or products that are no longer useful and turning them into something of greater value [8]. This creative technique has been widely used by sustainable fashion brands, seeking to minimize the large volume of waste materials and products from the industry [9].

By understanding their emotions, needs and desires, designers can identify opportunities to create beautiful, sustainable and, above all, affective clothing and accessories. In this sense, design thinking and upcycling can be used to develop innovative ways to create sustainable fashion products with a longer life cycle.

The research intends to validate the hypothesis that, even if a designer has an affective bond with a garment, they manage to transform it. Thus, this paper aims to report and analyse how affective ties can influence a creative redesign process in a fashion workshop, held at the University of Minho with volunteer students from the second year of the Fashion Design course, for 4 days. To carry out the activities in the workshop, the students selected personal garments, with which they maintained a strong affective relationship. These pieces were in disuse that were not discarded because of the affective bonds.

The resignification of clothes can happen from the moment that designers want to resignify emotions experienced with fashion items or when they realize that the fashion piece has lost its function (whether due to the deterioration of the material or to the simple fact that it no longer fits their current personal style) and that there are techniques, such as upcycling, that can be applied to modifying clothes. Finally, resignifying clothes is paramount because it is imperative to minimize the consequences of the high generation of waste resulting from the consumption and production of the fast fashion industry.

This study follows an action research approach, based on Tripp [10] and Stringer [11]. It also uses the work of Krueger [12] as a methodological reference for setting up the focus groups. In addition, we also rely on the studies by Osgood, Suci and Tannenbaum [13] and Likert [14], to collect and analyse the emotional aspects of the participants. To solve the proposed problem: how affective bonds can influence a creative redesign process in a fashion workshop, we highlight as fundamental the studies on design and emotion by Desmet and Hekkert [1], Norman [2], Damásio [3] and Burcikova [4], seeking to establish a direct relationship with the theories and show how they can be related in research in the field of fashion. For the redesign of clothes, we rely on creative fashion design practices with a focus on upcycling [8] and design thinking [6, 7].

2 Emotional Design

According to Desmet and Hekkert [1], the concept of affect encompasses a broad range of psychological states, including emotions, feelings, moods, and passions. Norman [2], believes that emotional design comprises three levels of processing, namely visceral, behavioural, and reflective. The visceral level refers to the initial emotional response that a person feels, while the behavioural level involves the experience of the emotion. Finally, the reflective level brings awareness to the emotion experienced. Damásio [3] divides emotions into three categories, which are background emotions, primary emotions, and social or secondary emotions.

Desmet and Hekkert [1] proposed a theory that suggests that product design can be thought of as a stimulus that triggers emotional responses in the user and that certain design features can influence the specific emotions that are elicited. They propose four different design dimensions that can influence emotional responses: sensory, semantic, pragmatic and behavioural. The authors suggest that emotional responses to a product can be evaluated along two dimensions: valence and arousal. Valence refers to whether the emotion is positive or negative, while arousal refers to the intensity of the emotional response.

As mentioned by Desmet and Hekkert [1], emotional responses to a product can be influenced by various design features, and designers can use this knowledge to intentionally elicit specific emotional responses in users. For example, a product might be designed to evoke a sense of calm or nuisance through its visual and tactile qualities, or it might be designed to evoke a sense of luxury or exclusivity through its symbolic meaning. Overall, the theory of design and emotion provides a framework for understanding the complex interplay between product design and emotional responses. By considering the various design dimensions that can influence emotional responses, designers can create products that elicit specific emotions in users and enhance their overall experience with the product.

About Norman's theory [2], we can say that the visceral level refers to the initial, instinctive emotional response that a user feels when they first encounter a product. This response is often based on sensory factors such as colour, shape, and texture, and can be either positive or negative. The behavioural level involves the user's experience of the product, and the emotions they feel while using it. This level is more complex and can be influenced by a wide range of factors such as usability, functionality, and user satisfaction. The emotional response at this level can have a significant impact on the user's overall perception of the product. At last, the reflective level involves the user's awareness and reflection on their emotional response to the product. This level can be influenced by a wide range of factors, including personal experiences, cultural values, and social norms. Norman [2] suggests that designers should aim to create products that elicit positive emotional responses at all three levels of processing. By doing so, designers can create products that not only meet the functional needs of users but also create a deeper emotional connection with them. This emotional connection can lead to increased user satisfaction, loyalty, and brand recognition.

For Damásio [3], background emotions are the constant, underlying emotional states that are present throughout our lives. These emotions include feelings of pleasure, pain, and general mood states, such as happiness or sadness. Primary emotions are the more intense, immediate emotional responses that we experience in response to specific events or situations. These emotions include fear, anger, joy, and sadness, and are often accompanied by physiological changes such as changes in heart rate or breathing patterns. In his last category, Damásio [3] states that social or secondary emotions are the emotions that arise from our interactions with other people. These emotions include empathy, guilt, shame, and pride, and are often influenced by cultural and social norms.

According to Damásio [3], emotions play a critical role in decision-making. When faced with a decision, emotions provide us with important information about the possible outcomes of our choices. This emotional information is then combined with more rational cognitive information to guide our decision-making processes. Damásio [3] demonstrates that emotions are not simply a by-product of our thoughts and behaviours, but rather an integral part of our decision-making processes. This theory has important implications for fields such as psychology, neuroscience but also for the field of design, where a better understanding of the role of emotions can help improve our understanding of human behaviour and decision-making, and improve the skills of fashion designers when they plan a product.

In addition to the aforementioned authors, we also have the work of Burcikova [4], whose research focus is on emotionally durable design in fashion. The author challenges the concept of disposability in fashion and emphasizes the importance of creating clothes that can be continuously used and valued over time. According to Burcikova [4] following Chapman's work on *Emotionally durable design: Objects, experience and empathy*, emotionally durable design is a concept that emphasizes the emotional connection that people develop with products and how this connection can influence their behaviours towards these products. This approach encourages designers to create products that can be enjoyed and used for a long time, rather than quickly discarded.

Burcikova [4], in line with Chapman's ideas, says that in the context of fashion design, emotionally durable design can be applied by creating garments that are not only aesthetically pleasing but also have a personal significance or emotional value to the wearer. This can be achieved by using high-quality materials, creating garments that fit well, and incorporating design elements that have a personal meaning to the wearer. Burcikova [4] highlights the potential for creating more sustainable and socially responsible fashion practices. By creating clothing that is valued and cared for over time, the need for constant consumption and disposal of clothing can be reduced, leading to a more sustainable and responsible fashion industry. This approach challenges the idea that fashion is disposable and instead emphasizes the creation of long-term value through the continuous use of familiar clothing items.

2.1 Emotional Measurement Tools

Osgood, Suci and Tannenbaum [13] developed instruments capable of analysing aspects that are part of a bipolar continuum, discriminating perspectives that refer to the direction and intensity of the individual in relation to the evaluated object or experience. The authors developed the semantic differential (SD) as a tool for measuring people's attitudes towards various incentives including emotions. The SD consists of a set of bipolar adjectives that describe the meaning of a particular concept or stimulus. Participants rate the stimulus on a seven-point scale between the bipolar adjectives, such as "happy-sad," "active-passive," or "pleasant-unpleasant." The ratings are then analysed to determine the perceived emotional valence (positive or negative) and arousal level (high or low) of the stimulus. Table 1 shows the 22 positive and negative emotional valence proposed by Desmet and Hekkert [1], based on the theory developed by Osgood, Suci and Tannenbaum [13].

Table 1. Emotional valence (adapted from Ortíz Nicolás, Aurisicchio & Desmet [15])

Positive Emotions	Negative Emotions
Admiration	Contempt
Anticipation	Displeasure
Amusement	Frustration
Confidence	Shame
Fascination	Fear
Relaxed	Hostility
Inspiration	Sadness
Joy	Boredom
Pride	Disgust
Satisfaction	Disappointment
Surprise	Dissatisfaction

The semantic differential [13] has been widely used in research on emotion, attitude, and perception, and it remains a popular tool for measuring emotional responses to stimuli. However, it is important to note that the SD is not without its limitations, and researchers should consider alternative measures and methods to complement their findings.

The Likert Scale [14] is a commonly used method of measuring attitudes, opinions, and perceptions in social science research. The scale typically consists of a series of statements or items that respondent's rate on a 5- or 7-point scale ranging from strongly agree to strongly disagree. In terms of measuring emotions, the Likert Scale has been adapted to ask respondents to rate their emotional experiences on a similar 5- or 7-point scale. For example, respondents might be asked to rate how strongly they feel happy, sad, angry, or anxious. One advantage of using the Likert Scale (1932) for emotion measurement is that it allows for a standardized and quantitative assessment of emotional experiences. This can be useful in research settings where researchers want to compare emotional experiences across individuals or groups. Additionally, the Likert Scale can be administered quickly and easily to large groups of people, making it a practical tool for researchers. "Likert scale (measures human attitude) are the examples of such scales in Psychometrics used widely in the social science & educational research" [16].

However, some emotions may be difficult to capture using the Likert Scale, as they may be more complex or nuanced than can be captured by a simple rating scale. In general, the Likert Scale can be a useful tool for measuring emotions, but it should be used in conjunction with other methods to provide a more complete understanding of emotional experiences.

3 Design Thinking and Upcycling

Design Thinking is primarily an internal problem-solving tool that has positive external consequences, resulting in a greater sense of coherence for consumers within the brand they choose to purchase from [6, 7]. It is particularly beneficial for forward-thinking companies that aim to gain a competitive edge. The approach is characterized by three key stages: Inspiration, which involves grasping the context and identifying the issue; Ideation, which entails discovering the solution or concept; and Implementation, which involves creating the structure that will manifest this concept.

Upcycling is a sustainable tool used to transform waste or products, which are no longer useful and would be discarded, into a new product [8]. In the context of fashion design, upcycling involves taking disused or discarded clothing or fabrics and turning them into new, on-trend pieces. Combining design thinking and upcycling in fashion design can lead to sustainable innovations and creative solutions. By reimagining and repurposing materials, designers can create unique and one-of-a-kind fashion items that appeal to environmentally conscious consumers. This approach can also reduce waste and promote a more circular fashion economy.

To apply design thinking and upcycling in fashion design, designers should conduct research to understand the needs and desires of their target audience. Designers can then envision and prototype potential upcycling concepts, exploring different ways to transform disused clothing into new fashion items. Combining design thinking and upcycling in fashion design can lead to innovative and sustainable products that meet consumers' needs in creative and effective ways. In addition, by using instruments capable of identifying and measuring emotions, it is possible to understand how people relate to clothes loaded with emotional ties and how they feel in the process of transforming these pieces. We will see confirmation of this in the results of the participants' work.

4 Methodology

The research was developed through an action-research methodological approach, which is a type of methodology where participants get involved in a participatory way [10, 11], applied in a fashion redesign workshop. To carry out this research, we also used the work of Krueger [12] as a methodological reference. From this methodological crossing, it was possible to elaborate the format of the workshop, which was coordinated by three teachers and a garment manufacturing technician, structure the activities and establish the priorities and tools available to collect data from the participants. In addition, we defined a schedule for carrying out each stage of the workshop, which took place over four days at the University of Minho.

We recruited a group of 8 volunteer students that signed up for the workshop using a google form. According to Krueger [12], focus groups usually involve a moderator who leads a discussion with a small group of participants (usually 6 to 10 people) and encourages them to share their thoughts, opinions and experiences. Krueger [12] emphasizes the importance of carefully planning the focus group interview, starting with identifying the research question and objectives.

The tools used for data collection consisted of two word clouds (generated by Mentimeter, an interactive technological tool) and two questionnaires (available through google forms), applied on the first and last day of the workshop. Also, at the end of the event we collected the testimonies of the participants through a video recording, to understand what emotions were present during the transformation of the clothes. Furthermore, we also registered in notes, our impressions and observations during the activities.

The word cloud initially generated by Mentimeter, served as a warm-up exercise to understand the first emotions of the participants in relation to the selected pieces, and collect information visually. During the focus group interview, Krueger [12] emphasizes the importance of creating a relaxed, non-threatening environment to encourage open and honest discussion. He recommends starting with warm-up questions to help participants get comfortable and gradually moving on to more complex or sensitive topics. At the end of the workshop, we asked the students to answer the Mentimeter again and we generated a new word cloud. With this, we were able to compare the results and perceive the different perceptions and sensations between the beginning and the end of the activity.

By generating polar semantic differential questions with answers such as "good or bad", "yes or no", with scales from 1 to 5 with opposite adjectives at their ends (Likert scale), it was possible to measure and classify the experience lived by the participants in the process of transforming the garments. These types of questions, with simple and objective language, combined with open text questions, allowed participants to deepen the subjects explored in the workshop. Both the Likert Scale [14] and the tools proposed by Osgood, Suci and Tannenbaum [13] were useful to understand the emotions of workshop participants. However, they alone would be insufficient to understand the complexity of these emotions. Therefore, we also used the students' self-reports, to understand which emotions were present in the process of redesigning the clothes. With the reports, we were able to visualize the students' emotions more deeply and expand our perception of the emotional experiences of a group in relation to their clothes.

In addition to data collection, we also had a remote presentation on sustainability and upcycling in fashion, lasting 50 min and carried out through the Zoom platform. The content presented was intended to clarify possible doubts about the upcycling technique and bring examples of fashion brands that develop products based on this technique. Based on the principles of design thinking and upcycling, research should begin by understanding the target audience needs and desires. In the case of the workshop, the target audience was the participants themselves. In order for them to understand their current needs, they needed to reflect on why those pieces of clothing were in disuse. After that, we started a brainstorming session to come up with ideas for the garments and to explore creative possibilities (draft ideas, creative tests, studies and design proposals). Tables 2 and 3 summarize all the activities carried out on the four days of the workshop.

On the first day, we had a dynamic consisting of signing the informed consent, data collection, presentation session and practical activities carried out by the participants. During the 2nd and 3rd days they were able to experiment with different materials, techniques and designs to find the best solutions and to assemble the final piece. In this phase, the work dynamics of each participant and the group interactions were observed. These observations were registered, through notes and collection of images, throughout the sessions.

Table 2. Activities developed on day one in the workshop "Fashion, emotion and circularity"

	Activity	Goal	Tools	Time
Day one	Sign Informed consent	Description and authorization to participate in scientific research	Pencil	5 min
	Reply Mentimeter	Warm-up exercise	Mentimeter	5 min
	Reply Google form	Understand the quantity, frequency of purchases and type of clothing; durability and disposal; importance of clothing; interest in sustainable fashion; feelings about the pieces before transformation (emotions and memories) and physical characteristics of the pieces	Google form	30 min
	See an exhibition session on sustainability and upcycling in fashion	Clarify possible doubts about the upcycling technique and bring examples of fashion brands that develop products based on this technique	Zoom	50 min
	Brainstorming + draft ideas + creative tests + studies + design proposals	Experience ideas and creative possibilities in selected garments	Mannequin, paper, pencil and pins	2 and a half hours

The questionnaires used were made available to participants on Google Forms, based on the semantic differential tool and the Likert Scale, in order to understand how emotional responses to different stimuli occurred. The questions were elaborated in open and closed questions. This set of questions was chosen to obtain qualitative and quantitative data. In Table 4 we summarize all the information collected in the questionnaires, seeking to detail the discussed topics explored, as well as the scope and typology of the questions. With this survey of information from the participants, it was possible to better

Table 3. Activities developed on day two, three and four in the workshop

	Activity	Goal	Tools	Time
Day two and three	Working on the clothes	Make the new outfit	Mannequin, scissors, pins, sewing machine, trimmings	8 h
Day four	Finalize and present the new outfit	Show how the clothes turned out and present the worked ideas	Mannequin, scissors, pins, sewing machine, trimmings	3 h
	Reply Mentimeter	Understand what were the final emotions of the participants in relation to the selected pieces, and collect information visually	Mentimeter	5 min
	Reply Google Form	Understand how the participants felt during and after transforming the clothes, if there was a creative block, how was the redesign process; what were the advantages and difficulties, how was the final result and physical characteristics of the new pieces	Google form	30 min
	Give testimonials	Capture, clarify and retain all the emotions involved in the redesign process	Video and audio recorder	5 min

understand their emotions and memories in relation to the pieces of clothing. In addition, we could also notice the different types of clothes consumed, the physical characteristics of the clothes and, finally, the redesign process as a whole.

Finally, the testimonies were collected on the last day of the workshop and were recorded on video, in order to retain the maximum amount of information from the students to be analysed after the event. These statements were analysed in order to find patterns of behaviour and/or emotions that we were unable to perceive via Mentimeter or Google forms. After the workshop was held, we analysed the data by transcribing

Table 4. Information collected in the questionaries.

Discussion topics	Scope of questions	Typology of questions
Fashion, consumption and disposal	Quantity of piece of each participant	Closed
	Purchase frequency	Closed
	Most purchased type of clothing	Closed
	How to get rid of clothes	Closed
	Importance	Closed
	Interest in sustainable fashion	Closed
Emotions and Memories	Related to parts selected for redesign and the final piece of cloth produced (Positive or negative)	Likert scale + Open question
Physical characteristics of parts selected for redesign	Shape, material and texture related to selected parts	Likert scale + Semantic differential (SD)
Redesign process: *creative blocks*	Timing and proposed solutions	Closed + Open
Redesign process: *facilitators or obstacles*	Creative point of view	Likert scale
	Emotional point of view	Likert scale
Final results	From an expectation point of view	Likert scale
	From a usability point of view	Closed
Plans to wear the new outfit	Participant's intentions to wear the new outfit	Closed

the discussion, identifying key themes and patterns, and comparing and contrasting responses from different participants, as Krueger (2002) recommends. At the end of the activity, we seek to obtain feedback from the participants to understand the strengths and weaknesses of the workshop.

5 Results, Analysis, and Discussion

First, we analysed the participants' responses to the questionnaires and the word cloud, crossing the information obtained by these two tools, both at the beginning and at the end of the workshop, as well as between these two moments. Then we watched the recordings of each participant and transcribed the session directly into an Excel spreadsheet. We coded the data using categories such as: pain points, needs, etc.

In addition, we registered expressions, quotes, and behaviours of the participants from the moment they started the workshop. We coded the data and proceeded to the analysis, cross-referencing it with the theoretical references from Desmet and Hekkert (2002), Norman (2004) Damásio (2004) and Burcikova (2019) emotional design principles.

5.1 Participants' Attitudes Towards Clothing Consumption, Disposal and Sustainability Techniques

In terms of consumption, we found that most participants (6 of them) buy clothes whenever they feel like it and only two buy them once a month. This indicates a high level of consumption, as the fast fashion market offers new collections every week and is always encouraging consumers to purchase new products. The results of the workshop also indicate that clothing is considered very important for most participants. The first questionnaire pointed out that half of the participants donate the pieces to family members. These results show that, although most participants try to get rid of disused pieces, they do so by trying to stay close to them, revealing a strong emotional bond with the clothes.

Participant 1 said that what they found most important in the workshop is that sustainability is much more than what they learn, that generally they have a very small view of what sustainability actually is. This participant also stated that they really enjoyed learning about upcycling and that they intended to explore this technique further in the future.

Participant 2 pointed out that they learned from the workshop, that in addition to realizing that upcycling is something very important for fashion, they also learned about some sewing techniques, how to make frills and hems, which was something they had no experience with and that for the first time used a sewing machine. This participant also stated that what they found most important to have learned in the workshop was that the issue of sustainability is very urgent and to learn about sewing techniques.

The main learning pointed out by participant 3 was related to handling the sewing machine, "because despite seeing my mother use it, I had never tried it". They also pointed out that they had some difficulties in pattern making and how to assemble the piece, but that they found it very good to have learned how to do it.

Other participants pointed out that time was a problem to carry out the activity and that they would need more time to finish the pieces. They also pointed out that they miss a greater stimulus in the course to carry out creative activities, outside the conventional classes and have more craft practices. We believe that because of this, all the participants indicated at the end of the workshop that they would participate in an experience like that again.

5.2 Attitudes Towards the Redesign Process: Blockage, Solutions and Interactions

Only one participant found it difficult to redesign the pieces from a creative point of view. Half of the participants felt some kind of blockage during the creative process. Blockage occurred early on for everyone who had these blockages, and for some this blockage occurred again later in the process. During the workshops, we also identified interactions between students leading to a collaborative attitude in their creative processes.

The participants' answers to the final form showed that, although they had emotional bonds with the pieces, this was not a creative blocking factor and that, as we thought, transforming the pieces was easy, or easier than they expected, both from the creative and emotional point of view. However, we cannot say that these results are unanimous and represent the emotions of all volunteers, since 2 of them did not attend the last meeting and did not respond to the second form.

The results of the workshop confirm the theory of Desmet and Hekkert (2002), that a product can trigger emotional responses in the user. However, we realized that these emotions did not prove to be limiting from a creative point of view, even if they generated some blockages (overcome during the redesign process). We also realized the importance of using complementary tools to survey emotions, such as recording testimonials on video, that enable us to understand more clearly what happened with each participant.

5.3 Redesign Experience, Emotions and Importance for the Future

From the intersection of research on emotional design, we noticed that the emotional experiences of the students during the workshop were more connected to the reflexive level proposed by Norman (2004), of conscience for the experienced emotion. This is because the pieces evidenced memories and emotions lived, therefore conscious. We also realized that the emotional experiences in the workshop were connected to the social or secondary emotions proposed by Damásio (2004), since from this interaction, it generated emotions such as pride, joy and empathy.

While working on the clothes selected for transformation, the participants were asked to reflect on their emotional attachments. Burcikova's (2019) work was instrumental in understanding the connection between the participants and the clothes they selected, as it addresses the concept of emotionally durable design and fashion. We realized that through the sustainable and responsible fashion practices used in the workshop, participants were encouraged to create products personally meaningful.

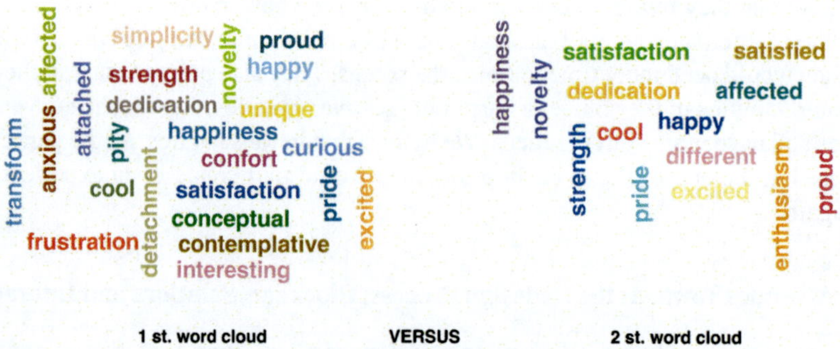

Fig. 1. Word clouds generated by Mentimeter

A significant difference was noted in terms of the initial and final emotions of the participants, as we can see in Fig. 1. Analysing the first word cloud, it was noted contrasting emotions ranging from negative feelings such as: Anxious, Frustration and Attached to positive feelings such as: Interesting, Curious and Happiness. In the second word cloud, it can be seen that the students' emotions are already fully positive with words like: Enthusiasm, Satisfied, Strength, Happy, Excited and Pride.

Words such as: Excitement, Strength, Affected, which emerged from the first word clouds, suggest an intensity in the emotions experienced by the participants, which also corroborates the theory of Desmet and Hekkert (2002). When asked about the experience in redesigning clothes, participant 1 said: "It was really good! I was really surprised because I didn't think it was going to be that usable". This participant initially proposed to make two pieces, but ended up making three, a top, a jacket and pants. Participant 1 also said that they learned that they could do much more than what they thought. According to that participant: "My head is much more than I think it is". The main feeling detached by this participant at the end of the workshop was pride.

About the redesign experience, participant 2 said: "I think it was interesting, because I felt sorry for having these pieces at home and not using them, and I liked some of them, so I think it was good to give them a new life". About the main feeling, this participant also said it was good, a bit of pride for having managed to transform the pieces and giving a new opportunity to pieces that were in disuse. When asked about the experience in redesigning clothes, participant 3 affirmed: "It was quite satisfactory, because the piece turned out as I imagined it, even if it is not completely finished now, but I can already visualize it and imagine that it is something I would use. It was very similar to my drawing". About the main sensation, participant 3 said that he liked it, because the piece turned out as he had imagined and that he managed to create not just 1, but 3 pieces that he will now use.

Other participants also pointed out that the experience "was good", "I wasn't expecting the result I got", that the most challenging was "giving a new life to the piece I didn't like" and "I was surprised with my result, because I didn't think I was going to make it." In addition, they also highlighted that they were happy because "we can give new forms to what we don't use", and also pointed out that the feeling when finishing the piece was "satisfaction", "good" and "pride", "I liked it a lot", " it was very intuitive", "super-happy".

The result indicates that, at the end of the workshop, the feelings and emotions were positive, such as: Interesting, Curious and Hopeful. Regarding the emotions experienced during the redesign process, some emotions mentioned were: Excited, Stimulated and Insecure. We believe that both valence and arousal, two dimensions of emotional design proposed by Desmet and Hekkert (2002) and also by Osgood, Suci and Tannenbaum (1957), may have influenced the participants' positive (those who completed the pieces) and negative (those who dropped out of the workshop) responses.

6 Conclusion

Redesigning clothes is not just about aesthetics or functionality. It can also involve emotional issues that designers need to consider. In particular, emotions can play a significant role in how people perceive and wear clothes and how they relate to themselves and others through their appearance. In projects based on design thinking, it is possible to approach problems from the perspective of the other and thus better understand their feelings. In addition, from redesign, fashion designers can create clothes with an emotional purpose and, through creative tools such as upcycling, promote consumer connection with their personal stories and values. Besides developing skills in sustainable techniques, students were also able to develop critical thinking, teamwork and promote reflection on the fashion industry, emotional design and sustainability. And also, improve the ability to solve problems arising from the fashion industry and develop upcycling skills and other creative solutions.

By accessing the affective memories related to the use of the selected pieces, the participants allowed themselves to relive experiences and emotions. They were also able to perceive that, just as we can reframe objects, it is also possible to give a new look to something already experienced. This process led them to reflect on the clothes and, later, reframe emotions and transform the pieces, giving the products a new life. Therefore, the study confirmed our hypothesis that it is possible to transform clothes, even if the person has emotional bonds with the pieces. Also, when we transform something, we have the opportunity to learn new things.

By adopting this approach of reframing products through reframing emotions, fashion designers can contribute to a more sustainable fashion industry, reducing waste and promoting a circular economy. Through this process, it is possible to explore and express emotions in a tangible way, creating meaningful and unique pieces that reflect the identity of the consumer. Furthermore, fashion design can be used as a tool for emotional expression and, through creativity, transform not only wardrobes, but also people's lives. The memories and experiences lived in the workshop were positive. This leads us to believe that proposing practical experiences, such as these, are important for fashion design students to broaden their vision of sustainability and also realize that by transforming a clothing product they also transform themselves.

Acknowledgements. I would like to thank CAPES for the financial support for carrying out this research. This work is also financed by Project UID/CTM/00264/2021 of 2C2T – Centro de Ciência e Tecnologia Têxtil, funded by National Funds through FCT/MCTES.

References

1. Desmet, P.; Hekkert, P.: The Basis of Product Emotions (2002). https://doi.org/10.1201/9780203302279.CH4
2. Norman, D.A.: Emotional Design: Why We Love (or Hate) Everyday Things. Basic Books, New York (2004)
3. Damásio, A.R.: Emotions and feelings: a neurobiological perspective. In: Manstead, A.S.R., Frijda, N., Fischer, A. (eds.) Feelings and Emotions: The Amsterdam Symposium, pp. 49–57. Cambridge University Press (2004). https://doi.org/10.1017/CBO9780511806582.004
4. Burcikova, M.: Mundane Fashion: Women, Clothes and Emotion all Durability. PhD thesis, School of Art, Design and Architecture, University of Huddersfield (2019)
5. Cunha, J., Providência, B.: Percursos do Design Emocional, Edição 2C2T/Lab2PT – Universidade do Minho, ISBN 978-989-54168-2-0 (2020). http://hdl.handle.net/1822/65288
6. Buchanan, R.: Wicked Problems in Design Thinking, Design Issues, vol. 8, No. 2, pp. 5–21. The MIT Press (1992). https://web.mit.edu/jrankin/www/engin_as_lib_art/Design_thinking.pdf

7. Brown, T., Kātz, B.: Change by Design: How Design Thinking Transforms Organizations and Inspires Innovation. Harper Business, New York (2009)
8. McDonough, W., Braungart, M.: The Upcycle: Beyond Sustainability-Designing for Abundance. North Point Press, New York (2013)
9. Roberts, G.H.: Upcycling for sustainability: Q&A with Yolande Klaassen, founder and CEO, Revive Clothing Lab. Int. J. Sustain. Fashion Textiles **1**(2), 295–303 (2022)
10. Tripp, D.: Pesquisa-ação: uma introdução metodológica. Educ. Pesqui. **31**(3), 443–466 (2005)
11. Stringer, E.T.: Action Research: A Handbook for Practitioners. Sage (1996)
12. Krueger, R.A.: Designing and Conducting Focus Group Interviews. University of Minnesota (2002)
13. Osgood, C.E., Suci, G.J., Tannenbaum, P.H.: The Measurement of Meaning. University of Illinois Press, Urbana (1957)
14. Likert, R.: A technique for the measurement of attitudes. Archiv. Psychol. **22**(140), 1–55 (1932)
15. Ortíz Nicolás, J.C., Aurisicchio, M., Desmet, P.M.: How users experience great products. In: IASDR 2013: Proceedings of the 5th International Congress of International Association of Societies of Design Research" Consilience and Innovation in Design", Tokyo, Japan, 26–30 Aug 2013. International Association of Societies of Design Research (2013)
16. Joshi, A., Kale, S., Chandel, S., Pal, D.K.: Likert scale: explored and explained. Curr. J. Appl. Sci. Technol. **7**(4), 396–403 (2015). https://doi.org/10.9734/BJAST/2015/14975

Black Box: Facilitating Tool for Testing, Using Neuroimaging to Support Research in Design and Emotion

Ana Luísa Sabim[1], Bernardo Providência[1(✉)], and Wellington de Medeiros[2]

[1] Landscape, Heritage and Territory Laboratory, University of Minho, Guimarães, Portugal
providencia@eaad.uminho.pt

[2] Programa de Pós-Graduação em Design, Universidade Federal de Campina Grande, Campina Grande, Brasil
wellington.medeiros@ufcg.edu.br

Abstract. This paper presents a tool developed for testing user interaction with 3D printed abstract shapes and collecting data from brain processing.

Human perceptual mechanisms allow users to establish an interaction relationship with any artifact.

This occurs because part of our perceptual system seeks information, while the other part seeks meanings, in this sense this study resorts to neuroimaging as a potential mapping against the understanding of the meanings attributed by our brain to the forms and objects with which we interact. In this context, the designer, as a designer and researcher, approaches emotion, aiming to build holistic products where the interaction with the user responds not only to the pragmatic issues of functionality and usability, but also to the constitution of meanings. The exercise of emotion evaluation through neuroimaging almost always resorts to the use of 2D images, limiting the use of other stimuli, which can be tested on 3D objects from a trilogy of interactions – visual, tactile and acoustic – with the user.

In this experiment, the Wireless EEG was used, adapting the existing evaluation protocols, considering specific requirements for haptic evaluation. For this purpose, an artifact (Black Box, controlled by light periods) was designed to trigger "trigger" moments during the use of the objects, synchronizing the different evaluation mechanisms. This box also allowed minimizing dispersing stimuli by canceling distracting elements in the focus of the evaluation object.

Finally, this article aims to identify the main advantages and disadvantages between the evaluation protocols recurrently used and the evaluation proposal adapted with the introduction of a Black Box and the use of wireless EEG, in the evaluation of tangible and intangible three-dimensional products.

Keywords: Design · Electroencephalography (EEG) · Emotions · 3D Objects · Context · Environment

1 Introduction

Brain response refers to the physiological and psychological changes that occur in response to stimuli, such as thoughts, emotions, and external events. These changes can include changes in neurological activity, hormone levels, and behavior [1].

Surveying brain response using neuroimaging technologies/tools is done by capturing electrical, magnetic, or hemodynamic signals generated by neuronal activity in the brain [2].

Regarding the possibility of measuring and subsequently understanding the brain activity caused by various stimuli, there are already some techniques used such as electroencephalography (EEG), functional magnetic resonance imaging (fMRI), positron emission tomography (PET), transcranial magnetic stimulation (TMS), magnetoencephalography (MEG) and near infrared spectroscopy (NIRS). These technologies allow us to map brain activity in real time and identify the areas of the brain that are active in response to certain stimuli or tasks. This information can be used to better understand brain function in different contexts, from processing sensory information to making complex decisions [3].

It is worth noting that different techniques may be better suited for different types of research questions and experimental designs. When it came to this research paper, the technology that proved most convenient was EEG. The main reasons why this technology was selected over any other were:

- The high temporal resolution, i.e. its ability to record changes in brain activity in real time. This allows for accurate analysis of brain activity patterns that occur during the perception of 3D objects, including the detection of fast and subtle brain responses x [4].
- The fact that this is a non-invasive and relatively affordable technique compared to other neuroimaging techniques, such as functional magnetic resonance imaging (fMRI) or positron emission tomography (PET). This makes EEG a viable option for studies involving the analysis of a large number of participants [4].
- This technology also carried the advantage of being within easy reach of the team that developed this project, unlike others already listed. However, it is worth clarifying that, if there was an option, it was still selected thanks to the two aforementioned points.

This records the electrical potentials generated by nerve cells in the cortex region [5]. The recording is made through an electroencephalograph, a device that has a set of electrodes (Fig. 1) placed on the scalp/scalp, in order to fully measure the brain

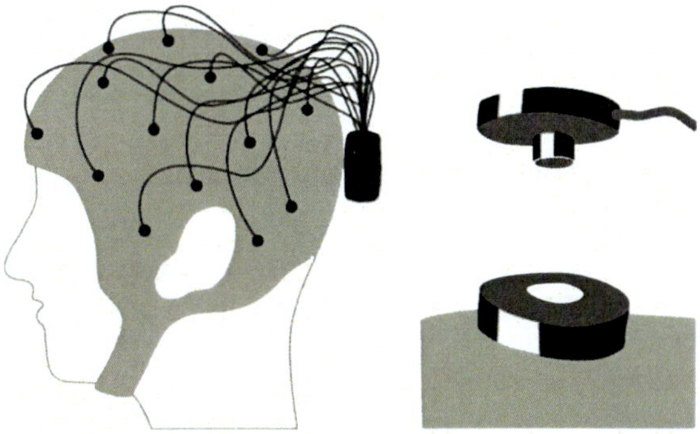

Fig. 1. Representation of an electroencephalograph and its electrodes.

information produced. Thus, the electrical activity in the scalp is detected, amplified and recorded numerically or in graphs (Fig. 2). This record allows the observation of brain electrical activation in vivo, i.e., in real time, of the acquisition period for each electrode.

In order to enable the understanding and analysis of the data provided by this technology, it is necessary to understand the types of waves produced, as well as their meaning. These are the "Beta" type, observed when the patient is awake and alert, or concentrated, the "Alpha" type that indicates a state of relaxation and decreased anxiety, which are normally produced in a state of wakefulness, the "Theta" type, commonly recorded when the individual produces reduced brain activity, almost asleep; and the "Delta" type, normally associated with the state of deep sleep. This type of recording is one of the most effective non-invasive methods for studying brain functionality and allows the understanding of normal [6, 7] and pathological [8, 9] physiological phenomena.

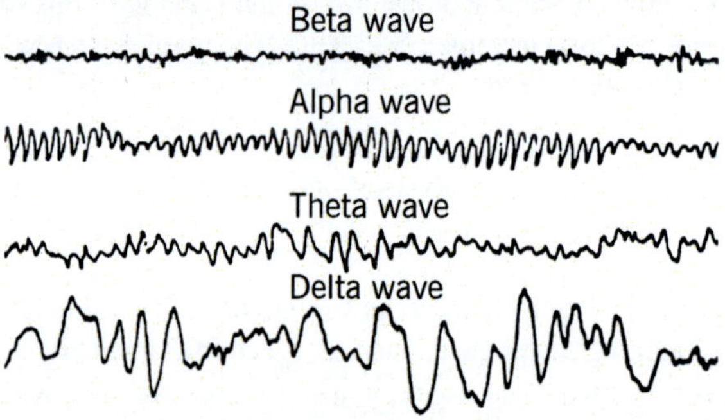

Fig. 2. Representation of the different waves captured by an EEG.

In order to enable the understanding and analysis of the data provided by this technology, it is necessary to understand the types of waves produced, as well as their meaning. These are the "Beta" type, observed when the patient is awake and alert, or concentrated, the "Alpha" type that indicates a state of relaxation and decreased anxiety, which are normally produced in a state of wakefulness, the "Theta" type, commonly recorded when the individual produces reduced brain activity, almost asleep; and the "Delta" type, normally associated with the state of deep sleep. This type of recording is one of the most effective non-invasive methods for studying brain functionality and allows the understanding of normal [6, 7] and pathological [8, 9] physiological phenomena.

2 Evaluation of Emotional Response, in Interaction with the Object: Self-report vs Brain Mapping

The assessment of emotional response is often performed using two main methodologies: self-report and brain mapping. Self-report involves the participant reporting their own subjective experiences, while brain mapping measures brain activity in response to an emotional stimulus.

2.1 Self-report – Context

Beyond self-reports the brain may contain more reliable information as was proven in the brain mapping studies of "Oliver Sacks" [11] where the mixing of two musical compositions during brain mapping created some confusion to himself, counteracting the According to Darren Bridger [10], the fact that our "non-conscious" mind constantly processes what it sees, shaping many of our reactions outside the realm of consciousness, means that clarity of the information processed in the brain. The vast majority of the time, we are not aware of the mechanisms in our brain that cause a preference for one particular design over another.

In this regard, [12] have proven that even when subjects are aware of a particular stimulus and response, they may not know how the former influenced the latter. There are, however, situations in which subjects are not even aware that they have been influenced by a stimulus or that they have produced a response. Nisbett and Wilson conclude with three controversial assertions:

- subjects may not be aware of an important stimulus influencing a response;
- subjects may not be aware of the occurrence of a response;
- even when they are aware of the occurrence of a stimulus and a response, they may not correctly relate the two elements present.

The resulting consequence, in terms of the value of introspection, is quite clear: "introspective access, if it exists, is not sufficient to produce adequate reports about critical stimuli in response to questions asked within minutes or seconds after the stimulus has been processed and the resulting response produced." [12] From this perspective, if introspective access is insufficient to produce any effect on questions associated with processes that have just occurred, then it will be totally incapable of producing adequate reports about processes that have occurred a long time ago, or in response to questions associated with patterns of behavior. [12].

2.2 Brain Mapping – Context

Brain mapping is a technique that allows visualizing brain activity in response to certain stimuli or tasks. To do this, the preferred neuroimaging tool is chosen, installed on the participant's head, and the participant is asked to perform specific tasks while images of the brain are acquired in real time. From these images, it is possible to generate maps of the brain activity (Fig. 3), identifying the areas that are active in response to the tasks [13].

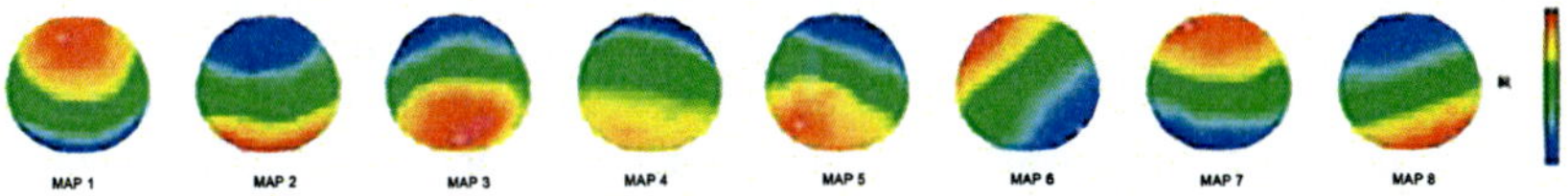

Fig. 3. Example of a brain mapping resulting from EEG testing.

Thanks to the increasingly evident technological advance, as well as the growing appreciation of the potential of the results from brain mapping, neuroimaging technologies have also evolved to more capable and versatile equipment. In this sense, Wireless electroencephalograms have appeared on the market that allow a connection to the data acquisition device without cables, i.e., it collects data through the helmet without the need for a physical connection to the computer. This advance has allowed new paradigms regarding the testing models, no longer being strictly necessary and limiting the use of traditional laboratory space, with magnetic field isolation and space conditioning. The EEG can now be used in completely different contexts, allowing for the testing and collection of data within practical situations (e.g., daily activities of the individuals tested) and analysis of three-dimensional objects and/or spaces.

With the advance that disaggregated the electroencephalogram from the fixed space, new challenges emerged correlating the problematic inherent to the necessary conditions for a "clean" testing and without excessive external interferences such as visual noise. Despite the advantages brought by the wireless component, EEG continues to be a sensitive technology with particular needs in terms of surroundings.

It was thanks to the awareness of these needs, as well as the interest in evaluating 3D objects in an informal environment, that triggered the need to create an artifact/container, easy to build, low cost and easy to transport, and focused on improving external conditions in a testing situation using EEG.

2.3 Self Report vs Brain Mapping

One of the main advantages of self-report is that it provides detailed information about the participant's subjective experience, including their emotions, feelings, and thoughts in relation to the emotional stimulus. However, self-report may be susceptible to biases, such as the participant's tendency to respond to the socially desirable or difficulty expressing complex emotions in words. [14].

On the other hand, brain mapping can provide objective information about brain activity associated with emotional response, without relying on the subjective account of the participant. This technique can be useful for identifying the areas of the brain that are active in response to a specific emotional stimulus and for better understanding the neurological mechanisms involved in emotional response. However, brain mapping also has its limitations, such as the lack of specificity for particular emotions, the need for high-cost, specialized equipment, and appropriate testing environments [15].

A comparison study between these two techniques was conducted by R.Brouwer, Hogervorst, Van Erp, & Heffelaar in 2014. The researchers concluded that self-report and brain mapping provide complementary information about emotional response, with self-report as more sensitive input for complex emotions and brain mapping more sensitive to more basic emotions. In addition, they found that combining the two techniques can provide a more complete understanding of emotional response [16].

Another study comparing the two techniques was conducted by S. L. Raio and colleagues in 2013 [17] They found that the emotional response measured by self-report and brain mapping can be different in situations where the emotion felt by the participant differs from the emotion expressed by the emotional stimulus.

Overall, these studies suggest that using both techniques can provide a more complete understanding of emotional response in different contexts.

3 Methodologies

3.1 Traditional Assessment Method – Standardized Optimal Conditions for Testing with EEG (Electroencephalogram)

Several researchers who have addressed the topic of electroencephalography or have used it as a tool have described processes and testing requirements in agreement with each other in terms of "standardized norms". These standards consist of a list of requirements for space and environmental conditions that are considered essential for successful EEG testing.

Several studies discuss the importance of standardizing the EEG laboratory environment to ensure reliable and reproducible results. They mention that the EEG laboratory should be designed to minimize external electromagnetic interference, have an adequate air conditioning system and sound insulation. In addition, recording should be performed in a separate room with adequate constant light, regulated temperature, and normal humidity conditions. Intermittent disturbing events should also be avoided [18–20].

"The identification and elimination of artifacts is crucial for proper quantitative analysis of EEG recordings. With optimal recording conditions, artifacts are only incidental and appear infrequently." [20].

3.2 Black Box – Tool

The "Black Box", as the name implies, is a box or container specifically designed for an experiment conducted in a laboratory at the University of Minho. Its configuration consists of a roofmate structure and painted in black to minimize light and parasitic noise from focusing on an illuminated object at the time of testing. The design of the artifact not only excludes external light, but also provides a directed attention thanks to the focus of an installed led lamp (Fig. 4), controlled by an Arduino (Fig. 5) that allows to trigger moments of light inside. In this way, the container provides an atmosphere of greater concentration and narrowing of the participant's attention, because it isolates the field of vision, allowing the only possible space to observe is the one where the object to be analyzed is located.

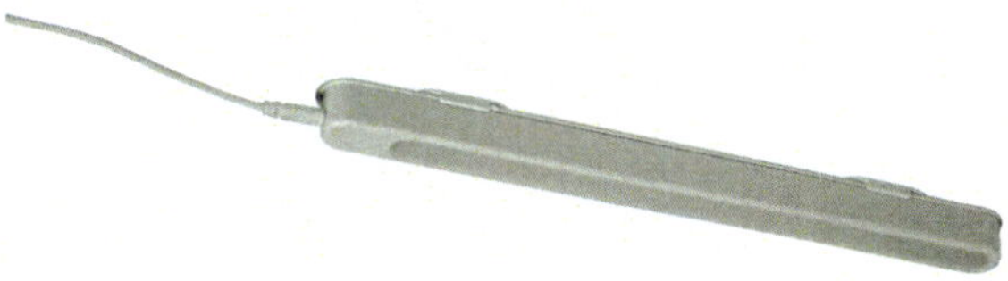

Fig. 4. LED lamp installed on the ceiling of the Black Box.

Fig. 5. Arduino used for programming the led lamp.

This innovative proposal by Black Box is based on the transposition of the "chiaroscuro" (light/shadow) technique applied by Rembrandt in the painting "Dr. Tulp's Anatomy Lesson" in 1632, and, inherited from Caravaggio, conferring and enhancing the objects, the people involved in the painting and the resulting shadows, "at the limit of the absolute become pure obscurity".

Besides the added value from the point of view of light direction and consequent attention of the participant, this container is dismountable and was thought to allow and facilitate the logistics of the exercise of analysis of 3D objects mimicking the identity of a theater stage, through the black draperies that allow to hide or divide parts of the container (Figs. 6, 7 and 8). In this way, the participant only has visual access to the front part (front half), with the back half reserved to house the objects to be analyzed. On the opposite side from the participant and outside the "Black Box" there is also a technician who, through two black sleeves, accesses the inside of the box and manipulates the objects to be exposed during the evaluation process.

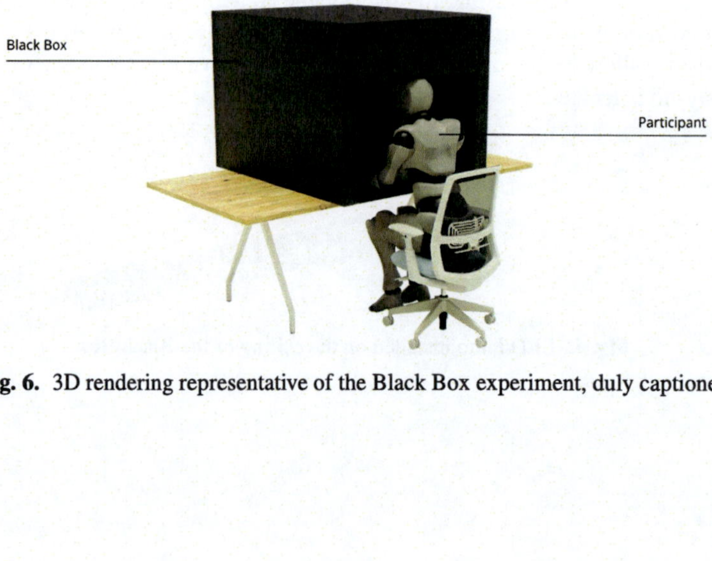

Fig. 6. 3D rendering representative of the Black Box experiment, duly captioned.

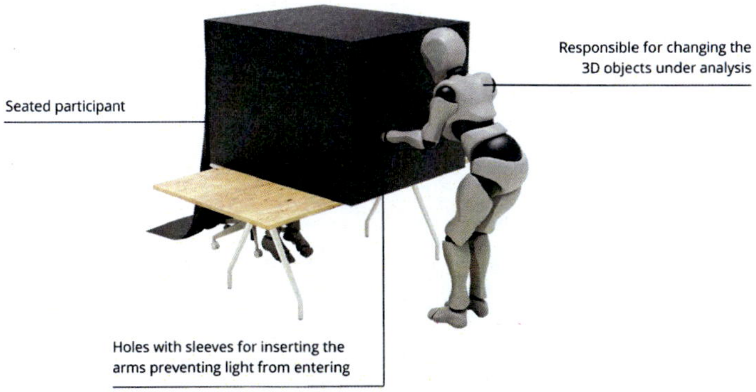

Fig. 7. 3D rendering of the Black Box experiment, properly captioned.

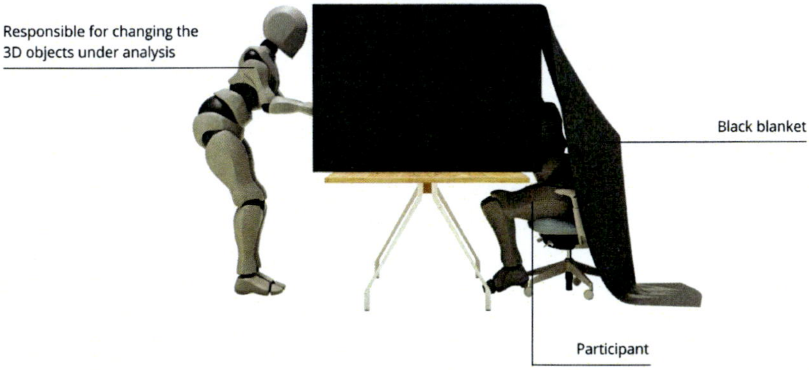

Fig. 8. 3D rendering of the Black Box experiment, properly captioned with participant.

The container (Fig. 9) in question was designed, adjusted and then built with the measures that made it possible to meet all the requirements for this type of testing, namely with three-dimensional objects in a non-laboratory space. The measures are indicated in the technical drawing (Fig. 10):

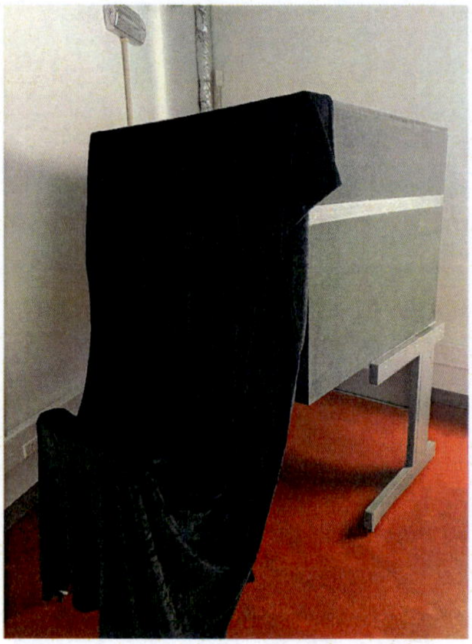

Fig. 9. Result of the black container built by the research designers.

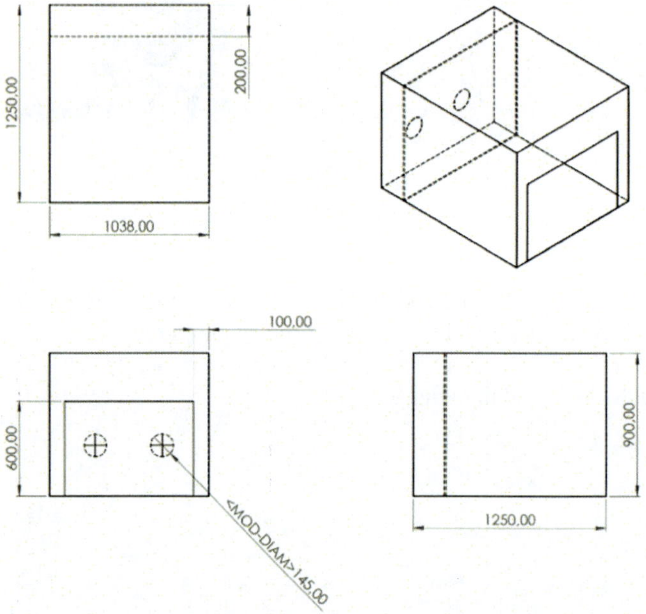

Fig. 10. Technical drawing of the black container – Black Box.

4 Testing

4.1 Study Case

It was with the use of self-report and semi-structured interviews, submitted to the participants, that several tests were run on this container, making it possible to adjust its components and to ascertain the contribution that this tool means to the type of testing in question.

5 Data Evaluation and Analysis

During the evaluation periods, participants mentioned their success in creating a completely dark space, which during the "baseline" period (period that allows for reduced brain activity between the evaluation of two stimuli) eliminating visually distracting factors and, directing during the analysis period, attention to the object. The spotlight was also approved for this purpose.

> "The light is strong and as soon as she turned on, I immediately looked at the object in front of me. From the first object, I waited anxiously for it to turn on and illuminate the next object that is in the center."

During the first tests some interference at the sound level was noticed, which raised the need to add (wax) earplugs during the test. The arm movement was praised as being easy to perform, as well as the effectiveness in changing the object from x in x time. This change is performed by the element on the opposite side of the container, and through two holes drilled, silently switches products when the light goes out. This replacement of objects, initially time-consuming and somewhat noisy, has been improved thanks to the addition of a cavity, refining the positioning in view of the participant. This cavity prevents objects of little stability (e.g. regular sphere) from moving out of the central location during the process.

Finally, this container underwent consecutive practical/actual tests with different participants, who were simultaneously answering questions about the functionality of each part, until all requirements were met, namely the acquisition of brain waves.

5.1 Possible Objects to Valuate

In order to determine the objects to be used for testing, abstract objects were designed for 3D printing. The design started from the assumption that abstract shapes would be free of immediate associations with known objects. Thus, canonical and non-canonical geometric, organic and hybrid forms were designed.

This container was designed to evaluate a collection of 3D objects, printed in PLA (Fig. 11), with the characteristics needed for the research under development.

It was of interest to survey the user-object interaction at the level of touch and vision, simultaneously. Considering that the test included more than one object, it was necessary to think the container in a modular way, assigning different tasks to both parts. In the evaluation compartment, the participant can observe and manipulate the object for 15 s with the help of the lighting (led lamp) installed on the module's ceiling. In the second compartment, there is a smaller space for the orderly "storage" of the objects, which pass through to the analysis module when stipulated. This second module also had two perforations in order to allow the insertion of the arms that will manually exchange the objects in an orderly manner (Fig. 12).

The dimensions of the figures were stipulated so as not to make handling difficult, taking into account the space for movement of the upper limbs, nor to make visual analysis difficult because of the small size.

Fig. 11. Three-dimensional objects used in testing with the Black Box.

Fig. 12. Three-dimensional objects used for testing in the Black Box using electroencephalography.

6 Relationship of the "Black Box" to the Senses

This project was motivated by the intention to create a more suitable environment for testing neuroimaging equipment as a tool to obtain data about people's responses to both tactile and visual interaction with shapes that are usually primary constituents of design products.

In a laboratory evaluation context, in people's perception of many different kinds of things, the evaluation object is usually presented on a computer screen that controls exposure times in synchrony with other equipment or according to the evaluation protocol. In this case, given the characteristics of magnetic field isolation or equipment interconnection, the participant's mobility is reduced and the evaluation environment is similar to that of a sound booth with perforated walls, which may influence the interaction of the objects.

Given the technological evolution evident to date, it is now possible to use Wireless EEG equipment that allows not only mobility during the evaluation, but also its use in spaces not confined to the laboratory. For this project, it was necessary to create a space (container) that allowed the isolation of parasitic noises, as well as the free manipulation of three-dimensional objects, expanding the possibilities of user-product interaction during the evaluation period, using the senses of touch and vision simultaneously.

Both senses operate in the establishment of the relationship with space and volume, as well as in the construction of the texture and meaning itself – experiences considered essential to the testing of three-dimensional objects using electroencephalography.

Quoting the famous author Fernando Rodrigues, "it is in touch that the product takes shape". This is also the greatest sense of our body [21] and, for this very reason, an element of interest in the study of the interaction and emotional relationship user-artifact.

According to Ackerman (1997) [22], there are several receptors in this sensory system that allow us, as a whole, to apprehend very specific information about shapes, textures, touch, among others. In the cerebral cortex, the parietal lobe is primarily responsible for receiving and processing sensory inputs such as touch. Cutaneous sensitivity is fundamental for a functional human life at an emotional and social level [23] and, therefore, so important for what we aim to introduce in this work.

F. Rodrigues also tells us in his book "Neuromarketing" that vision is the sense with the greatest capacity to collect information, having the ability to collect 25x more information than the others.

Vision plays an important role in the tactile and spatial perception of objects. This same statement can be proven through the analysis performed in the study "The influence of vision on the haptic perception of spatial layout" where participants performed haptic perception tasks in a dark and a light environment, while monitoring brain activities related to haptic perception. The results showed that this was influenced by the presence of light, suggesting that vision plays an important role in tactile and spatial perception. The authors conclude that vision and touch work together to create a complete representation of the physical environment.

Another example that proves not only the importance of vision and touch alone, but also the significant partnership between these two senses, is the study "Multisensory integration in the perception of texture: neuromagnetic evidence". This sought to investigate multisensory integration in the perception of texture, using brain imaging techniques to measure neurological activity in response to the presentation of simultaneous tactile, visual and tactile-visual stimuli. The results showed that the simultaneous presentation of these stimuli produced significantly greater brain activity compared to the isolated presentation of each. The authors conclude that multisensory integration between vision and touch is fundamental to tactile perception and that joint analysis of these two senses can provide a more complete representation of physical objects.

7 Discussion

Taking into account the lack of solutions that would respond to the absence of a properly equipped laboratory environment, appropriate for testing using electroencephalography, this project resulted in a structure of easy construction, use, transport and reduced cost.

Since it is a container that can be "thrown away" and is small in size, compared to any space previously built for this purpose, it becomes possible to invest in this type of area in a more practical way, and the ownership of a laboratory infrastructure is no longer an obstacle.

This tool also brings a component of innovation associated with the fact that it was previously designed for testing with three-dimensional objects. This testing is, so far, a segment of little investment by researchers and consequent research projects using EEG (or similar technologies). Admittedly, until now this market has focused on the analysis of computerized (2D) images using monitors and, for this reason, no solution similar to the one proposed here has yet been thought of.

Although the vast majority of the requirements for the moments of analysis with EEG, performed so far, remained relevant to the environment needed when testing 3D objects, there were issues associated with touch and vision that needed solutions adapted to provide the most favorable conditions for a correct participant-object interaction. In particular, the need to create a light focus that was directed at the 3D object and, in addition to creating better analysis conditions, simultaneously allowed the participant a "baseline" period between objects. This was possible thanks to the Arduino's programming that allows the lamp to be turned on and off in a regular, pre-established time sequence. Moreover, this space was designed so that the participant could move his upper limbs, necessary for the tactile analysis of the object, in a free and comfortable way.

It is also worth mentioning the process of creating the container that required a specific method for its development, being an example of the kind of attitude that researchers in the design field need to adopt in order to execute unprecedented approaches to specific investigations. In this case, the team's repertoire of designers was crucial to the creation of the artifact, as it replicated the design process common in the field, involving material research, sketches, prototyping, and testing. Although it is not uncommon to develop devices for test execution in research in other areas, it is possible to verify the distinctiveness of the artifact developed as a consequence of production by research designers.

8 Conclusion

The proposal of a tool capable of creating a favorable environment for the process of testing three-dimensional objects using electroencephalography intends to positively change the practical panorama for the type of research reported here. The practicality and reduced cost, besides the creation of new requirements thinking about the analytical relationship with the 3D object, contributes to increase the possibility of this type of testing and consequent investment in the area.

More than contributing with new scientific material for this research sector, the result of this experience provided the possibility for research designers to explore tools that are usually out of their reach, for being confined to laboratory environments.

Although the vast majority of tests and projects related to electroencephalography and similar technologies make use of computer equipment (2D figures/images) to test the user's brain, this container still challenges product designers and neurodesign enthusiasts to put touch and interaction with the physical product in the foreground.

8.1 Future Perspectives

In short, the Black Box played the role of facilitator of data collection and analysis of physical characteristics of 3D objects, with the marked added value of allowing and accentuating the exploration of sensory issues that add value to the testing process.

One of the ongoing projects using the Black Box and electroencephalography deals with the analysis of the participant-object relationship and subsequent comparison with another testing console conducted in a controlled laboratory environment, with 2D figures on computer monitors replacing sensory contact.

This breakthrough appears to hold strong promise for progression in what is a better understanding of the user-product relationship, and consequent improvement in the designer's ability to project. EEG testing is now more accessible outside the laboratory environment, opening up a variety of possibilities for exploring new research projects.

It also provides the appropriate environment and encourages greater investment in the specific exploration of three-dimensional objects/products.

References

1. Acharya, J.N., Hani, A., Cheek, J., Thirumala, P., Tsuchida, T.N.: American clinical neurophysiology society guideline 2: guidelines for standard electrode position nomenclature. J. Clin. Neurophys. **33**(4), 308–311 (2016). https://doi.org/10.1097/WNP.0000000000000316
2. Prakriti Gaba, B.S.P., De Simone, C.V.: Novel techniques in epilepsy management: venous pacing and capture of electrical activity in the primate cortex. J. Neurol. Neurophys. **7**(3), 373 (2016). https://doi.org/10.4172/2155-9562.1000373
3. Thatcher, R.W.: Neuroimaging: introduction and overview. In: Aminoff, M.J., Boller, F. (eds.) Handbook of Clinical Neurology: Neurology of Cognitive and Behavioral Disorders, pp. 3–18. Elsevier (2015)
4. Luck, S.J.: An Introduction to the Event-Related Potential Technique. The MIT Press, Cambridge, MA (2014)
5. Schomer, D.L., da Silva, F.L.: Niedermeyer's Electroencephalography: Basic Principles, Clinical Applications, and Related fields. Wolters Kluwer Health/Lippincott Williams & Wilkins, Philadelphia (2010)
6. Campbell, B.A.: EEG recording and operation: EEG fundamentals. J. Neurother. **13**(2), 125–175 (2009). https://doi.org/10.1080/10874200902831560
7. Pereda, E., et al.: Non-linear behaviour of human EEG: fractal exponent versus correlation dimension in awake and sleep stages. Neurosci. Lett. **250**(2), 91–94 (1998). https://doi.org/10.1016/s0304-3940(98)00435-2
8. Engel, J.: Mesial temporal lobe epilepsy: what have we learned? Neuroscientist **7**(4), 340–352 (2001). https://doi.org/10.1177/107385840100700410
9. Yuvaraj, R., Murugesan, S., Madhivanan, K.: EEG based automatic sleep stage classification using energy efficient features. J. Med. Syst. **38**(10), 1–9 (2014)
10. Bridger, D.: Neuromarketing. Autêntica Business, São Paulo (2018)
11. NOVA | NOVA Short | Inside Oliver Sacks's Brain: Youtube. Nova PBS Official. https://www.youtube.com/watch?v=wc6m0Uyis-8 (2010)
12. Nisbett, R.E., Wilson, T.D.: Telling more than we can know: verbal reports on mental processes. Psychol. Rev. **84**(3), 231–259 (1977). https://doi.org/10.1037/0033295x.84.3.231
13. Poldrack, R.A., Farah, M.J.: Progress and challenges in probing the human brain. Nature **526**(7573), 371–379 (2015). https://doi.org/10.1038/nature15692
14. Brück, C., et al.: Emotional voices: the tone of (true) feelings. In: Armony, J., Vuilleumier, P. (eds.) The Cambridge Handbook of Human Affective Neuroscience, pp. 265–285. Cambridge University Press (2013). https://doi.org/10.1017/CBO9780511843716.015
15. Johnstone, T., et al.: Failure to regulate: counterproductive recruitment of topdown prefrontal-subcortical circuitry in major depression. J. Neurosci. **27**(33), 8877–8884 (2007). https://doi.org/10.1523/jneurosci.2063-07.2007
16. Hogervorst, M.A., Brouwer, A.-M., van Erp, J.B.: Combining and comparing EEG, peripheral physiology and eye-related measures for the assessment of Mental Workload. Front. Neurosci. **8**, 322 (2014). https://doi.org/10.3389/fnins.2014.00322

17. Raio, C.M., et al.: Cognitive emotion regulation fails the stress test. Proc. Natl. Acad. Sci. **110**(37), 15139–15144 (2013). https://doi.org/10.1073/pnas.1305706110
18. Jobert, M., et al.: Guidelines for the recording and evaluation of Pharmaco-EEG data in man: the international Pharmaco-EEG society (IPEG). Neuropsychobiology **66**(4), 201–220 (2012). https://doi.org/10.1159/000343478
19. ACNS: Guideline 8: guidelines for recording clinical EEG on Digital Media. J. Clin. Neurophysiol. **23**(2), 122–124 (2006). https://doi.org/10.1097/00004691200604000-00009
20. Seeck, M., et al.: The standardized EEG electrode array of the IFCN. Clin. Neurophysiol. **128**(10), 2070–2077 (2017). https://doi.org/10.1016/j.clinph.2017.06.254
21. Taylor, J.H.: The sense of touch. Am. J. Psychol. **32**(1), 24–67 (1921). https://doi.org/10.2307/1413702
22. Ackerman, J.M.: A tactile working memory system in humans. Brain Res. **769**(2), 398–405 (1997). https://doi.org/10.1016/s0006-8993(97)00714-5
23. Rodrigues, F.: Influência do Neuromarketing nos Processos de Tomada de Decisão. Psicosoma (2011)

Tactile and Visiotactile Quality Perceived in Textile Materials of Nightdress from the Confections Pole of the Agreste Region of Pernambuco/Brazil

Aline Paiva Rodrigues da Silva and Germannya D'Garcia Silva(✉)

Federal University of Pernambuco, Recife, Brazil
{aline.paiva,germannya.asilva}@ufpe.br

Abstract. This article presents the results of a preliminary study that evaluated the perceived tactile and visiotactile quality of textile materials applied to sleeveless nightgowns at the garment production center in Pernambuco (PE) and also, how users perceived the quality of sleep wearing products with two different types of fabric: natural and synthetic fiber. Recent research highlights the interference of sleepwear on sleep quality and, when it comes to sleepwear, the choice of materials is a key step in the fashion design. The proposed investigation method, adapted from Nogueira (2011) was divided into three stages: 1. Data collection; 2. Definition and Application of the tactile and visiotactile protocol and 3. Subjective evaluation of fabric interference in sleep quality. The dialogues established between the tangible and intangible aspects of the dressed body brought an understanding of the perceptions stimulated by the materials, such as: thermal comfort, roughness, softness and feelings of pleasure and relaxation. As a result, it was found that there is a preference for the polyamide + spandex composition, due to the pleasure felt when touching the fabric reported by the users. The one to the detriment of cotton; even this one possessing technical attributes that benefit sleep.

Keywords: Textile materials · sleepwear · sensory · sleep

1 Introduction

The pandemic caused by COVID-19 has taken everyone around the world by surprise. The high transmission capacity of the new coronavirus has made social isolation the most recommended combative measure by health authorities. This public health recommendation has caused thousands of people to have their work routines and activities impacted, needing to adapt to remote work/home office. According to the Brazilian Institute of Geography and Statistics/IBGE – PNADCOVID-19, in May 2020 Brazil had 8.7 million workers in remote activity. This abrupt change in the lifestyle of Brazilians, associated with the entire emotional process of adaptation, from fear, anxiety and stress, made the sleep issue come to the fore (Jornal da USP, 2020).

Sleeping well is one of the crucial factors in keeping the immune system stable. However, in a study of 432 people in Austria, Germany and Switzerland, which aimed to analyze the effects of the 'lockdown' on sleep and rest relationships, reported that there was a worsening in the sleep of participants in the first period of the pandemic (Blume et al., 2020).

This new lifestyle caused changes in consumption behavior and suggests that the fact that people are staying at home longer has increased the demand in the sleepwear segment. In May 2020, there was a considerable increase in searches for the term "pajamas", by 142% on Google Trends, (a tool that maps consumption trends) compared to the same month in 2019.

Studies by Shin et al. (2016) and Chow et al. (2019) prove that there is a direct link between textile materials applied in pajamas and sleep quality, which interferes in several aspects with the quality of life of individuals who use them. In cold environments, the textile material should insulate the body and improve the wearer's sense of comfort. In hot environments, the textile material must have the ability to intervene in the exchange of heat and humidity between the body and the environment; promoting relaxation at bedtime.

Hot environments have a negative impact on the relaxation process for sleep to begin, and on thermoregulation during sleep, making the individual more agitated and unable to have a restorative sleep. Thus, the technical and sensory information of fabrics for sleepwear must be in the design phase of the project. This must happen so that all materials chosen for the product are consistent with the objective of enabling good usability, improving sensation and thermal comfort.

It is up to the clothing designer to consider the beginning of the first stage of sleep, when the body needs to be at a high level of relaxation, that is, there must be the absence of any environmental and physical discomfort so that the individual goes through this process in a fluid way (Udale, 2015; Martins; Almeida 2010).

The authors of this research believe that the assumptions of clothing ergonomics associated with the appropriate choice of textile materials can have a positive impact on the quality of sleep of users. This article presents the results of a tactile and visiotactile test with users from the municipality of Caruaru, the largest city in the wild region of Pernambuco, to evaluate: 1. The perceived tactile and visiotactile quality of the fabric applied in sleeveless nightdresses; 2. To what extent did the natural cotton fiber and polyamide + spandex synthetic fiber nightdresses interfere with the sleep quality of the users participating in the research and 3. What is the preference in the choice of users?

2 Theoretical Reference

2.1 The Sleepwear Segment of the Confections Pole of the Agreste Region of Pernambuco/Brazil

According to data from the State Government and the Management Center for the Textile and Apparel Chain in Pernambuco – NTCPE, in 2017, the state of Pernambuco hosted 3.3% of the companies in the Brazilian textile chain, with emphasis on the clothing sector, with 4, 2% of the national total. The clothing segment represented 94.0% of the business in the apparel sector. In the municipalities of Caruaru, Santa Cruz do Capibaribe, Recife and Toritama, there were 71.3% of the apparel manufacturing units and 71.1% of the jobs in the State. However, the sector had a low rate of skilled people, being only 2% (IEMI, 2021).

According to the IBGE's Central Register of Companies/CEMPRE, in 2018 there were 2,291 clothing and apparel production units in the wild region of Pernambuco; 235 of which were intended for the production of underwear, and most of them in the sleepwear segment. The main companies in the segment, with a significant monthly production and monthly export of millions of pieces to the whole country, are concentrated in two municipalities: Santa Cruz do Capibaribe (153) and Caruaru (36). In this work, the city of Caruaru was the space chosen for being one of the main clothing commercialization centers in the state of Pernambuco.

Despite this economic potential, the sector still has problems in product quality and specification, as companies do not have size standards for clothing designs, nor are they concerned with usability issues of textile materials used in products.

2.2 Comfort and Pleasantness of Clothing

Dressing is a complex sensory experience that permeates almost all the senses. The sight can arouse desire; the sense of smell can activate the sensation of freshness through the aroma of clean clothes and the touch through textures. Which can provide the comfort of touch when the fabric comes in contact with the skin. Therefore, the feeling of comfort in use and the pleasantness of this clothing by users should be investigated by the designers of these products (Mariano, 2016).

Pleasantness, a term derived from the ergonomics of the product, concerns the user's interaction with the product, and should bring emotional benefits, be pleasant to the consumer and promote comfort and aesthetic-symbolic pleasure (Iida, 2016).

On the other hand, comfort in clothing is defined by three aspects: Physical, Physiological and Psychological. The physical aspect is related to the sensations caused by the contact of the skin with the fabric, and the modeling applied to the body and its movements. The physiological is linked to how clothing interferes with the physiological characteristics of the body, mainly with the thermoregulatory factor. And, finally, the psychological that portrays aspects related to aesthetics, social and cultural environment (Martins, 2005).

The technical aspects of the textile material are associated with physical and physiological comfort such as grip, elasticity, malleability and weight (Saltzman, 2004 apud Martins, 2005). The aesthetic aspects are related to the combination of shapes, colors, textures and finishes. The symbolic aspects permeate the identification of the product with certain ethnicities, classes, groups, social values, status or regions, existing associations through psychological aspects. These associations are important for the construction of the users' perceptions, and, in this way, it allows the subjective evaluation of the product's quality (Jordan, 2002).

This interaction with the body is detected and processed by the central nervous system, which transmits information about these stimuli to the cerebral cortex for several sensory receptors. Despite being classified in different receptor systems, they work together, transforming interactions into sensations. It is through the ability to feel that the individual understands the world in which he is inserted, receiving stimuli through smells, flavors, touch, sight and sounds. When a sensation is accompanied by an interpretation, which leads the individual to access memories and build opinions through their experiences, what is called perception is given rise (Belmonte; Ceveró, 2005).

Perception is related to influences built during the individual's experience, such as culture, religion, references found in various aspects involving life. This world interpretation is particular and complex and cannot be generalized. Even if two individuals receive the same stimuli, in the same environment, have the same age group, they capture, organize and interpret this repertoire by an individual perceptive process, relating their expectations, needs and values (Nogueira, 2011). This complexity generates the understanding that in order to understand how fashion is perceived, it is necessary to understand how the body physically and physiologically reacts in contact with the product being researched.

2.3 Perception of the Clothed Body and the Influence on Sleep

The textile product, when in direct contact with the skin, is able to stimulate the groups of sensory touch receptors, changing the user's perception according to the characteristics present in the fabric. The different possibilities of sensations resulting from this skin-textile interaction are caused by the different receptors present in the skin. These receptors have a direct connection with the brain, through nerve fibers, which can be activated depending on how this interaction is stimulated in the skin by the textile material (Dias, 2009).

The feeling of roughness or discomfort, such as feeling the skin "prick", are quickly detected because they cause irritation to the skin's sensory receptors. The perception of moisture, in this tissue-skin relationship, increases friction and causes more receptors on touch; which can cause materials that were previously considered comfortable to no longer be comfortable, depending on the time of use (Nogueira, 2011).

Consumers, at the time of purchase, are guided by several perceptive experiences with the product. Therefore, despite the choice of material being important during the development of the project, other aspects that configure the product need to be considered.

In 1999, Hes developed a system of evaluation of textile products, more specifically flat fabrics, focusing on thermophysiological functions. This author evaluated the classification of flat textile products in four categories: 1. Subjective evaluation of textile comfort and the disadvantages that this implies; 2. Objectively assessing sensory comfort, considering the thermal touch; 3. Definition of thermal comfort and the physiology of the human body; and 4. "Equation of the human body's thermal balance and mechanisms of its cooling in high heat conditions" (Hes, 1999 apud Martins, 2005).

At the end of his study, the author found (Hes, 1999 apud Martins, 2005, pg. 67):

> [...] that the "thermal absorption" does not depend exclusively on the structure and composition of the fabrics, but on the textile materials and finishes used, which sometimes cause greater changes in the "thermal sensation" of contact provided to the users.

This claim was challenged in the study developed by Chow et al. (2019), whose participants classified the materials studied (wool, cotton and polyester) as soft, rough or wet. Even though the study found polyester to be the material that most negatively interfered with sleep, participants indicated, in subjective ratings, that it was slightly softer and smoother than the other two. Fabrics with low hygroscopic properties (absorptive capacity) may increase friction and discomfort with use. These are caused by the lower heat exchange capacity between the skin and the external environment, allowing for less thermal comfort. In the case of polyester, this capacity is less than one percent (1%), while cotton fibers absorb about twenty-four percent (24%) of moisture (Nogueira, 2011; Chow et al., 2019).

3 Methodological Trajectory

According to Nogueira (2011), the sensory analysis or examination of organoleptic characteristics is characterized as an examination of the characteristics of a product by the sense organs. It is an important tool for evaluating consumer preferences. It studies in an orderly and structured way the properties of a product, using the human being as an instrument of measurement through the use of the senses (sight, touch, hearing, smell and taste). This analysis profile is widely used by the cosmetics and food industry, as a tool to evaluate the attributes obtained by the senses of taste and smell (Carvalho, 2015).

According to Carvalho (2015), for the sense of touch, relating the texture of materials, the analysis methods are the same, and are characterized as: Descriptive – Evaluates the attribute's intensity. Evaluators must be trained to qualify and identify the attribute through a scale; Discriminatory – Performs the evaluation through sensory differences between two or more products, such as a sample and a standard; Affective – Evaluates the acceptance and preference of evaluators with respect to an attribute. Applying sensory analysis techniques in a structured way, it is possible to transform subjective data into objective, evaluating sensory properties in a more tangible way (Nogueira, 2011).

In the thesis developed by Nogueira (2011), at the University of Minho (UMinho), the author used scale, comparison and statistical evaluation methods, and sought to understand the behavior of Portuguese and French consumers, based on tactile, visual and visiotactile perception. For this research, the proposed research method is divided into 3 steps: 1. Data collection; 2. Definition and application of the tactile and visiotactile protocol; and 3. Subjective evaluation of tissue interference with sleep quality. The steps used by this method are presented in Table 1.

Table 1. Adaptation of the Nogueira method (2011).

Research Method	
Step 1 – Data collection	• Identify the main sales areas of fashion products in Caruaru – PE; • Carry out an on-site survey of stores of all types of clothing; • Select, visit and buy nightdresses at selected sleepwear stores; • Carry out a survey of the materials present in the sleeveless nightdress sold in the city
Step 2 – Tactile and Visiotactile Evaluation Protocol	• Define the samples present in the sensory panel, based on materials found in sleeveless nightdress sold in the city of Caruaru; • Define the semantic differential attributes present in the tactile and visiotactile evaluation; • Structure the evaluation following the security protocols established to combat COVID-19; • Select participants to take part in the experiments; • Apply experiments
Step 3 – Protocol for evaluating tissue interference in the sleep quality of the participants	• Based on the two samples with opposing ratings, buy a sleeveless nightdress of each material; • Assemble the protocol and define the steps for the experiment; • Select participants interested in testing the products; • Apply the experiment protocol with selected participants; • Follow the participants through self-report for a week

4 Field Research

4.1 Step 1. Data Collection

During the period from January 9 to 18, 2021, six shopping centers in the city of Caruaru were visited: Feira de Caruaru, Fábrica da Moda, Polo Comercial, Centro da Cidade, Caruaru Shopping and Shopping Difusora. In each of these centers there is an intense movement of people and different audiences, both tourists and residents of the city. During the visits to the shopping centers, 25 stores selling sleepwear were selected. In this research the main focus is the textile material (fabric), not the modeling.

4.2 Step 2. Tactile and Visiotactile Evaluation of Fabrics used in Nightdresses

Participant Selection

The group of volunteers was composed of ten women aged between 20 and 53 years old.

All practitioners of physical activity, with a balanced diet, socially active, non-smokers, and who exercised some remunerated economic activity. These criteria were chosen based on the understanding that the factors listed above, when not present, negatively influence sleep quality, generating potential interference in the research result. In the initial planning, the participants would be selected in usual environments of physical activity practice (gyms and/or pilates studios). However, in view of the scenario of social isolation and for the safety of all involved, a structured questionnaire was applied to select the participants using the Google Forms tool.

Selection of Fabrics Samples

A total of ten fabric variations were found in the 25 stores in the city of Caruaru. A selection of six samples was chosen because it represents 90.6% of the variations in compositions found: Cotton, Polyester, Polyester + Spandex, Polyester + Viscose, Viscose + Spandex and Polyamide + Spandex. Another criterion for the number of six samples was the need for time control for the test, approximately thirty minutes per person. Nogueira (2011) reports that after this time, there is a loss of tactile sensitivity, which may interfere with the results.

Creation of the Sensory Panel: Evaluation Scale and Fabrics Attributes

In this research, tactile and visiotactile attributes were used, prioritizing the nine attributes that appear in both. In order to be evaluated consistently, samples were handled according to each attribute. Figure 1.

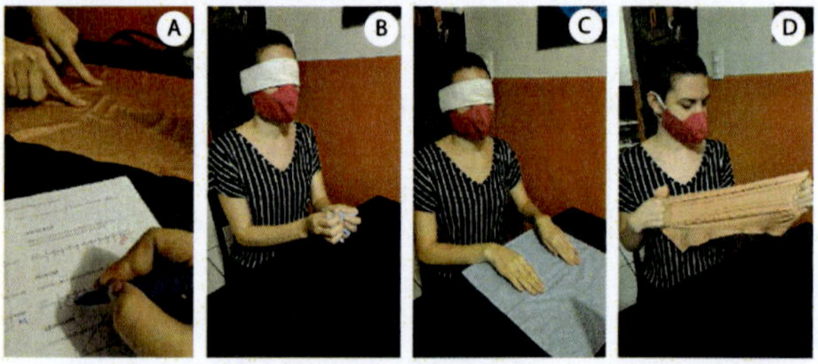

Fig. 1. Sequence of moments of the tactile and visiotactile evaluations. A-Sticky/ B-Fresh-Hot/ C-Rough-Soft/ D-Elasticity. Source: Authors

The order of the attributes was as follows: **Fresh-Hot** – The participant holds the fabric with both hands for 10s, then analyzes the heat transfer. **Slippery** – The participant manipulates the fabric, passing it from one hand to the other, analyzing the ease of the hand in sliding on the fabric. **Thin-Thick** – The participant's thumb is placed on one side of the fabric and the index finger on the other, feeling the thickness. Light-Heavy – The fabric sample is held in both hands, lifted and the weight of the fabric is felt. **Elasticity** – Force is applied horizontally, vertically, and transversely across the sample. **Sticky** – The fingers are slightly lifted, after 5s, and checked if the material remained stuck to them. **Smooth-Rough** – With the material resting on a surface, the participant is asked to slide her hands over the sample. **Rough-Soft** – Participant is asked to slide her hands across the sample in all directions. **Crumpled** – The fabric is compressed with both hands, placed on the support surface and analyzed for the amount of creases.

Field Research and Safety Procedures to Combat COVID-19

Due to the limitations faced by the pandemic scenario, the safety protocols to deal with COVID-19 were prioritized and the evaluations were applied in different places than previously planned. In order to respect the rigor of the evaluations and to reduce possible alterations in the results due to the interference of the location and/or environmental conditions, some measures were adopted. Only the researcher and the evaluator were present at the indicated place, both wearing a mask and respecting the social distance of two meters. The evaluator only handled the samples after proper hand and support surfaces had been sanitized.

After this step, the thermal sensation was observed. The environment was configured so that it would be similar across all applications. The experiments were carried out at similar times and the fan was kept on. Verbal confirmation of the evaluator's thermal comfort was also requested, so that she would not feel excessively cold or heat. In all experiments, the support surface of the samples had a polished finish, usually made of glass, so that there was no change in texture. As explained in Fig. 2, the same sequence of actions defined in the protocol was applied to the ten participants.

Tactile and visiotactile evaluation protocol

Step 1. Before starting the experiment, the participant washes her hands with neutral pH soap, to neutralize the pH of her skin.
Step 2. The participant is blindfolded.
Step 3. One tissue sample is delivered at a time. The participant is guided by the evaluator while manipulating the tissue sample according to the attributes of the evaluation and classifying them on a scale from 0 to 10.
Step 4. The participant withdraws the blindfold.
Step 5. Tissue samples are delivered, one at a time, in the same order delivered in step 3 (the order of samples varies among participants, but the same assessment is repeated in steps 3 and 5). The participant is guided by the evaluator, while manipulating the tissue sample according to the evaluation attributes and rating them on a scale from 0 to 10.

Fig. 2. Tactile and Visiotactile Evaluation Protocol. Source: Authors

After the evaluator was seated and blindfolded, the tactile evaluation was started. The evaluators were asked to interact with the material according to the step shown in Fig. 2. The meetings lasted approximately 30 min for both stages to be carried out. These were structured in terms of the number of samples, aspects and the researcher's conduct, so that there was no extrapolation of time. During the evaluations, the samples were delivered to the evaluators in different orders, with the aim of reducing the possible influence of this sequence on the results.

4.3 Step 3 – Perception of Fabrics in the Clothed Body and its Interference with Sleep

To carry out the third stage, only two participants from the group of sensory evaluators volunteered. The fabric perception process took place while the participants wore and slept in their nightdresses for a week. At the beginning of each day following the use of one of the nightdresses, each of them reported the experience of the previous night.

According to the results of tactile and visiotactile evaluations, two textile materials were chosen. Each of the nightdresses was exclusively composed of one of these materials and was given to the participants; as shown in Fig. 3. Since the participants were not carrying out the experiment in a controlled laboratory, some guidelines were given to reduce sleep disruption variables.

Both participants had their own bedroom, but the environments were different: one slept using a fan and the other air-conditioning. External stimuli similar to those prior to the experiment were maintained and, therefore, there was no change throughout the experiment.

Fig. 3. Polyamide + spandex nightdress (left) and Cotton nightdress (right). Source: Authors

The questioning applied to the participants on the following day varied between the questions: Did you wake up during the night? Did you feel hot? Did you sweat more? Did you feel more restless while sleeping? What do you think of the nightdress texture? Did the nightdress texture help you get into the relaxation process? Did you feel a difference in your mood during the day? Have you become sleepier or more willing?

5 Results

5.1 Data Collection

The fabric samples found in the 25 sleepwear stores in the city of Caruaru showed different compositions. We found 56 pieces with materials that were mostly repeated, resulting in nine variations of textile compositions. The mixed synthetic and mixed synthetic + artificial fibers composed six of the nine identified fabrics, corresponding to 74.9% of the textile materials found. Among the other fabrics, it is worth highlighting the presence of 100% ordinary cotton in 10.7% of the nightdresses. Cotton nightdresses with other processing profiles were not found on the market.

All samples were categorized as: Natural Fibers (vegetables and animals); Blended Fibers (synthetic, synthetic + natural and synthetic + artificial) and Non-Natural Fibers (artificial and synthetic). It was not possible to find sleepwears with animal fibers (silk, for example), nor pieces with 100% artificial fibers (like viscose) in the stores visited. Figure 4 shows the distribution of samples by category.

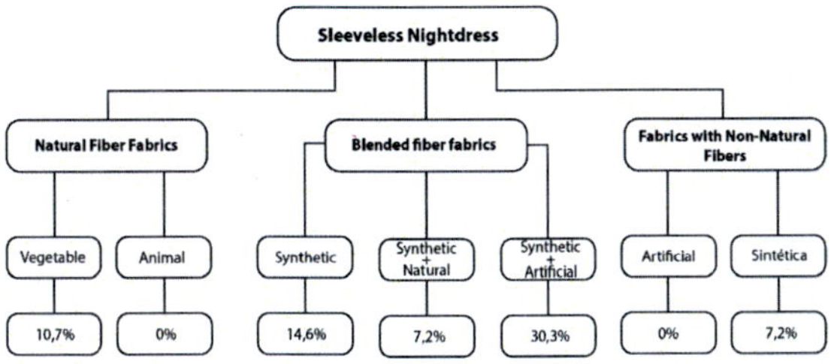

Fig. 4. Flowchart of material categories with percentages. Source: Authors

When choosing nightdresses, aspects related to freshness, elasticity and texture are mainly taken into account. And it is practically unanimous that salespeople recommend cold knit nightdresses (polyamide + spandex). The argument used by them in the statement is that these nightdresses are perceived as cooler and more comfortable.

5.2 Tactile and Visiotactile Evaluation of Fabrics Used in Nightdresses

In the first moment, the participants received the six samples of fabrics blindfolded and performed the test with only the touch. Before the start of the tests, the participants were instructed on how they should handle the fabric. They felt the fabric and then answered, on a scale from 0 to 10, how they felt about the aspect mentioned by the researcher. In the second moment, they removed the blindfolds and the same protocol was applied. Each sample had nine features. To reach the result of the analysis, a weighted average calculation was carried out, which made it possible to build the graphs shown in Fig. 5. In them, it is possible to identify how the samples were evaluated and identify the preponderant characteristics.

The graphics were created based on the materials and characteristics evaluated. Bipolar attributes have the mentioned term first at 0 and secondly at 10 (example: Fresh (0)-Hot (10)). In isolated attributes: Little is equivalent to 0 and Much is equivalent to 10. From the results it is possible to observe the differences in the tactile and visiotactile evaluations. The samples chosen to exemplify this process, visibly show the changes in perceptions with tactile stimuli and uniting vision with tactile.

In sample 4 – Polyester: there was a reduction in the perceived quality of 3 attributes in the visiotactile evaluation (Fresh – Hot; Thin – Thick and Elasticity). The 3 attributes that remained stable were (Slippery; Light – Heavy and Rough – Soft) and the 4 attributes that increased were (Sticky; Smooth – Rough; Rough – Soft and Crumpled). In sample 5 – Cotton: most attributes remained stable in both evaluations.

In sample 6 – Polyamide + Spandex: the material showed extreme evaluation in some attributes in both evaluations. 6 aspects were maintained (Fresh – Hot; Slippery;

Thin – Thick; Light – Heavy; Elasticity and Smooth – Rough) and the 3 attributes that increased in the visiotactile were (Sticky; Rough – Soft and Crumpled).

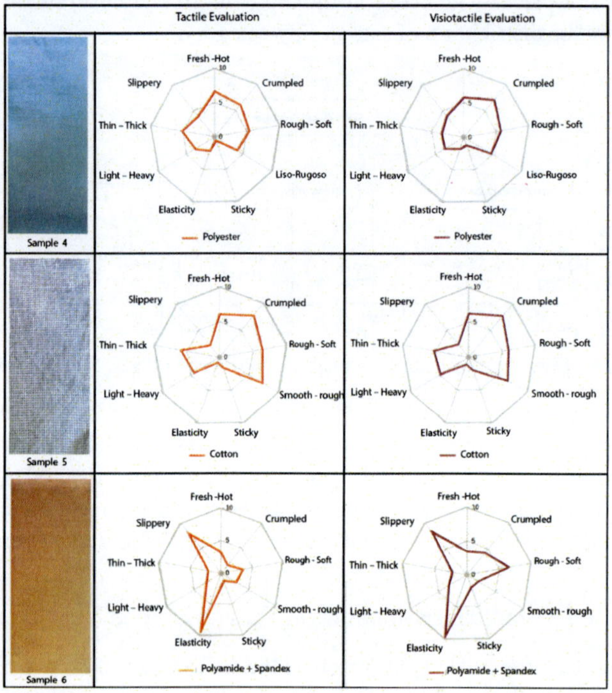

Fig. 5. Compilation of tactile and visiotactile test results. Source: Authors

5.3 Perception of Fabrics in the Clothed Body and Their Interference with Sleep

The experiment was carried out by the participants for four consecutive days, in which they slept two nights in a row with the cotton nightdress and two nights with the polyamide + spandex nightdress. The day after wearing each nightdress, participants sent a self-report of the experience. The experiences with each material are explained below.

Cotton Nightdress

On the first day, participants showed similar perceptions about the nightgown used that night. Both entered the relaxation process quickly and did not take long to fall asleep. Despite reporting the material as hot and having felt more heat, one stated that she sweated more and the other that she did not feel sweating. Both perceived the nightdress as soft and comfortable, did not feel restless during sleep and were sleepier the next day. On the second day, the evaluations were maintained. Both reported the nightdress as warm but comfortable and soft. Participant 1 felt sleepier during the day, while participant 2 felt more willing. They did not wake up during the night and felt more relaxed during pre-sleep. Participant 1 stated that she found it difficult to get out of bed when she woke up because she was very thermally comfortable.

In the tactile and visiotactile evaluations, cotton was also evaluated as a warm and soft material, maintaining this same perception in the dressed nightdress. Despite this, it was observed that the participants did not experience difficulty sleeping, which possibly indicates a perceptual alteration of the material. Cotton is more hygroscopic and performs a better exchange of heat between the environment and the user's body, resulting in better thermal comfort.

Polyamide + Spandex Nightdress

When sleeping in the second nightdress, participants provided several concordant reports. They stated that they perceived the nightdress as soft, fresh, comfortable and pleasant to the touch. That they didn't feel heat or sweat; and that the nightdress's texture helped with the relaxation process, allowing them to quickly enter the first stage of sleep. On the first day, both reported that they had woken up during the night. Participant 1 felt more agitated during sleep, and after waking up took a while to return to sleep. Participant 2, on the other hand, did not report agitation during sleep and found it easy to go back to sleep. Both felt less willing the next day. On the second day, participants offered similar reports. However, even feeling more agitated during sleep and waking up at night, both reported greater disposition the next day, unlike the first day. Sensations experienced, both said they preferred the sensation of the polyamide + spandex fabric, popularly known as "cold mesh". According to the participants, this choice is mainly related to the texture because it is very soft, fresh and pleasant to feel in the body and in the touch.

6 Conclusions

Conducting the study relating the use of nightdresses to the sleep quality of users, through self-report, was important to understand if the theorized subject expresses the reality of users in Caruaru. The results of this study corroborate those carried out by Chin and Show in Sydney, even though experiments were carried out with participants from different cultures and experiences. The selection of textile materials in the nightgowns used to sleep brought to light how much talking about sleep is a public health issue. In addition to reinforcing the understanding that the choice of materials applied in projects cannot be guided only by the aesthetic result, but also that they must be guided by how users perceive and communicate with clothing.

The comfort of textile materials considers different properties named mechanical, thermal and surface. However, in contact with the body, the perceptive aspects must be taken into account, in addition to its psychological, cultural and other social factors that form it. All these properties must be seriously considered, observed and analyzed. Through this research we were able to prove that the most appreciated feature by consumers, when buying textile products, is the touch of fabrics. It was also verified that there is a preference, among local market consumers, for nightdresses popularly called "cold knit" (polyamide + spandex). Also, that cotton is classified as a second option.

Nogueira (2011) also discusses why there is this preference for polyamide + spandex over cotton, even though cotton has more characteristics that benefit sleep. The author states that the textile contact with the skin surface stimulates several sensory receptors, mainly touch, which is easily stimulated by the roughness of the product, producing less friction with materials with smoother surfaces.

The textile material placed on the skin surface causes a momentary sensation of heat or cold, the thermal characteristic of the material being also determined by the apparent difference between the material temperature and the skin temperature. This means that, in general, materials with a rough surface are perceived as hotter than those with a smooth surface, even if they have the same composition (Nogueira, 2011). In the tactile and visiotactile evaluations, cotton was indicated as a rough and warm material and polyamide + spandex as smooth and fresh.

This study did not intend to find an absolute truth about the interaction of textile materials applied in nightdresses and the sensorial perceptions of Caruaru users. The aim of this study is to point out how these experiences can be rich and of great value to the Design area. Many possibilities of similar analyzes can be elaborated, exploring other materials, different styles of clothing and different product configurations. In addition to emphasizing the importance of carrying out more studies and projects developed with the aim of being user-centered in a more complete experience. This makes it possible for the industry not only to expand its broad domain, but also to more efficiently adjust its development and production strategies.

References

Belmonte, C., Cerveró, F.: Sistema sensorial: (sensibilidad somática y visceral). In: Tresguerres, J., et al. (ed.) Fisiología humana, 3rd edn. pp. 73–103. Mc Graw Hill-Interamericana, Madrid (España) (2005)

Blume, C., Schmidt, M.H., Cajochen, C.: Effects of the COVID-19 lockdown on human sleep and rest-activity rhythms. Curr. Biol. **30**(14), R795–R797 (2020). https://doi.org/10.1016/j.cub.2020.06.021

Carvalho, F.: Análise sensorial de têxteis: classificação do tecido através do tato. In: 11° Colóquio de Moda – 8ª Edição Internacional. Curitiba. https://docplayer.com.br/52743600-Analise-sensorial-de-texteis-classificacao-de-tecidos-atraves-do-tato.html (2015). Last accessed 10 Feb 2021

Chow, C., et al.: The impact of sleepwear fiber type on sleep quality under warm ambient conditions. Nat. Sci. Sleep **11**, 167–178 (2019)

Dias, M.: 'Percepção dos materiais pelos usuários: modelo de avaliação Permatus'. PhD thesis, Federal University of Santa Catarina, Florianópolis (2009)

Jornal da USP. Distúrbios do sono podem ser agravados pelos efeitos da pandemia. https://jornal.usp.br/atualidades/disturbios-do-sono-podem-ser-agravados-pelos-efeitos-da-pandemia/ (2020). Last accessed 4 Apr 2021

Jordan, P.W.: Designing Pleasurable Products: An Introduction to the New Human Factors, 1st edn. Taylor & Francis, London (2002)

ﬂnstituto Brasileiro de Geografia e Estatística. Estatísticas do Cadastro Central de Empresas – CEMPRE. Homepage. https://sidra.ibge.gov.br/tabela/6450 (2018). Last accessed 5 Dec 2021

ﬂnstituto Brasileiro de Geografia e Estatística. Trabalho – PNAD COVID19 | IBGE. https://covid19.ibge.gov.br/pnad-covid/trabalho.php (2020). Last accessed 5 Dec 2021

IEMI. Estudo da Competitividade dos Setores Têxtil e Confeccionista no Estado de Pernambuco. 20117. https://docero.com.br/doc/80vs801. Last accessed 5 Dec 2021

Iida, I., Guimarães, L.: Ergonomia: projeto e produção, 3rd edn. Blucher, São Paulo (2016)

Mariano, R.: A percepção sensorial do corpo vestido: uma análise têxtil sob o ponto de vista feminino. Universidade de São Paulo, Mestrado (2016)

Martins, S.: Diversidade e versatilidade para o universo dormir: uma proposta sustentável para o segmento sleepwear. In: 14º Colóquio de Moda. 3ª Edição Internacional. Londrina. 2010. http://www.coloquiomoda.com.br/anais/Coloquio%20de%20Moda%20-%202010/71807_Diversidade_e_versatilidade_para_o_universo_dormir.pdf. Last accessed 16 Oct 2019

Martins, S.: O conforto no vestuário: uma interpretação da ergonomia: metodologia de avaliação de usabilidade e conforto no vestuário. PhD thesis, Federal University of Santa Catarina, Florianópolis (2005)

Nogueira, C.: Análise sensorial de produtos têxteis. PhD thesis, Minho University, Guimarães (2011)

Sher, L.: COVID-19, anxiety, sleep disturbances and suicide. Sleep Med. **70**, 124 (2020). https://doi.org/10.1016/j.sleep.04.019. Last accessed 15 Oct 2020

Shin, M., et al.: The effects of fabric for sleepwear and bedding on sleep at ambient temperatures of 17 °C and 22 °C. Nat. Sci. Sleep **8**, 121–131 (2016)

Udale, J.: Tecidos e moda: explorando a integração entre design têxtil e o design de moda, 2nd edn., p. 200. Porto Alegre, Bookman (2015)

Zhai, Y., Du, X.: Addressing collegiate mental health amid COVID-19 pandemic. Psychiatry Res. **288**, 113003 (2020). www.sciencedirect.com/science/article/pii/S0165178120308398, Last accessed 20 Feb 2021

Author Index

J. Cunha et al. (Eds.): CIMODE 2023, *Advances in Fashion and Design Research II*, pp. 553–554, 2024.
https://doi.org/10.1007/978-3-031-43937-7